BASIC PLASMA PROCESSES ON THE SUN

BASIC PLASMA PROCESSES ON THE SUN

Not yet enough!

It is not sufficient to prove a
case, we must also tempt or
raise men to it:
Hence the wise man must learn to convey his wisdom;
and often in such a manner
that it may sound like foolishness!

Nietzsche

INTERNATIONAL ASTRONOMICAL UNION

UNION ASTRONOMIQUE INTERNATIONALE

BASIC PLASMA PROCESSES ON THE SUN

PROCEEDINGS OF THE 142TH SYMPOSIUM OF THE INTERNATIONAL ASTRONOMICAL UNION HELD IN BANGALORE, INDIA, DECEMBER 1–5, 1989

EDITED BY

E. R. PRIEST

Mathematical Institute, University of St. Andrews, Scotland

and

V. KRISHAN

Indian Institute of Astrophysics, Bangalore, India

KLUWER ACADEMIC PUBLISHERS

DORDRECHT / BOSTON / LONDON

Library of Congress Cataloging-in-Publication Data

International Astronomical Union. Symposium (142nd : 1989 :
Bangalore, India)
Basic plasma processes in the sun : proceedings of the 142nd
Symposium of the International Astronomical Union, held in
Bangalore, India, December 1-5, 1989 / edited by E.R. Priest,
V. Krishan.
p. cm.
Includes index.
ISBN 0-7923-0879-4 (HB).
1. Solar wind--Congresses. 2. Space plasmas--Congresses.
3. Astrophysics--Congresses. I. Priest, E. R. (Eric Ronald), 1943-
. II. Krishan, V. (Vinod) III. Title.
QB529.I57 1989
523.7--dc20 90-41849

ISBN 0-7923-0879-4

Published on behalf of
the International Astronomical Union
by
Kluwer Academic Publishers, P.O. Box 17, 3300 AA Dordrecht, The Netherlands.

Kluwer Academic Publishers incorporates
the publishing programmes of
D. Reidel, Martinus Nijhoff, Dr W. Junk and MTP Press.

Sold and distributed in the U.S.A. and Canada
by Kluwer Academic Publishers,
101 Philip Drive, Norwell, MA 02061, U.S.A.

In all other countries, sold and distributed
by Kluwer Academic Publishers Group,
P.O. Box 322, 3300 AH Dordrecht, The Netherlands.

Printed on acid-free paper

Printed in the Netherlands

TABLE OF CONTENTS

FOREWORD

Much of the excitement in modern Solar Physics has come from the realisation that the Sun is a plasma and that this plasma is interacting with the magnetic field in a wide variety of subtle ways. As well as being of great interest in their own right the observed plasma phenomena on the Sun are of much wider importance, since they reveal to us details of basic phenomena that are expected to be occurring throughout the universe. It was with this in mind that 173 solar physicists from 17 countries gathered together in Bangalore with an air of anticipation.

We were not disappointed as we received the warmest of welcomes from our graceful and charming host,Vinod Krishan. She and her colleagues worked tirelessly to make our stay a most memorable one and to ensure that the meeting ran with calm and efficiency. In addition to being stimulated by an excellent series of talks on the up-to-the-minute advances in our subject, it was a pleasure to make new friendships from so many countries and to learn, in particular, of the Solar Physics being done in India which has a great tradition and is of a high standard. Furthermore, we enjoyed hearing about Indian culture and appreciating its beauty, especially on our day's tour into the countryside to visit some Hindu and Jain temples.

The meeting began with a "Good Morning" welcome from Vinod Krishan and an interesting address by Professor Bhattacharyya describing Bangalore and its scientific heritage, both of which have been reprinted here. Each session included review talks and contributed talks with plenty of time for lively discussion periods. All these talks have been refereed and I am most grateful to authors and referees for enabling me to send the camera-ready manuscripts off to the publisher only four months after the meeting in order to ensure a speedy publication. Many of the poster papers too have been included here.

The Scientific Organising Committee consisted of V Krishan (chair), S M Chitre, M Kuperus, A Mangeney, E R Priest, R Rosner, Ch V Sastry, L Vlahos, V V Zheleznyakov, and the members of the Local Organising Committee were J C Bhattacharyya (chair), M H Gokhale, R C Kapoor, P K Kaw, V Krishan, G Krishna, R Nityananda, A R Choudhuri, Ch V Sastry, K R Sivaraman, R K Varma and were well served by the secretaries, S Rajasekaran and M Ramani. We are most gateful to our sponsors, namely IAU Commissions 10, 12, 40, 49, Indian National Science Academy (New Delhi), Department of Science and Technology (New Delhi), Indian Institute of Astrophysics (Bangalore), Raman Research Institute (Bangalore), Physical Research Laboratory (Ahmedabad), Inter-University Centre for Astronomy and Astrophysics (Pune), Indian Space Research Organisation (Bangalore), National Aeronautical Laboratory (Bangalore) and the Institute for Plasma Research (Gandhinagar), Council for Scientific and Industrial Research (New Delhi) and Department of Atomic Energy (Bombay). Also I am extremely thankful to Shiela Wilson and Pat Dunne for secretarial help undertaken with their usual patience and efficiency.

St Andrews, March 1990 Eric Priest

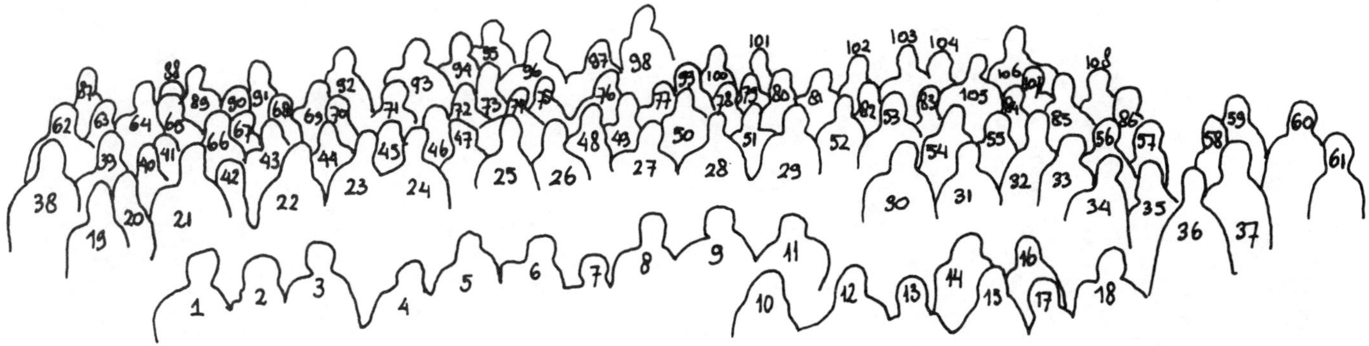

1. Li Jianke 2. Guha Thakurtha M 3. E Ebenezer 4. M H Gokhale 5. A Taktakishvili 6. A Paranjpye 7. D F Smith
8. P K Manoharan 9. A V Stepanova 10. V Krishan 11. V V Zaitsev 12. M Ryutova 13. F Chiuderi 14. M Dryer
15. Xu Min Jian 16. E R Priest 17. Zhou Aihua 18. S K Saha 19. Y Leblanc 20. C Oberoi 21. Chou Chih-Kang
22. V K Verma 23. A K Gupta 24. Hari Om Vats 25. C Debi Prasad 26. K B Ramesh 27. C Sivaram 28. R V Bhonsle
29. Ch V Sastry 30. B Buti 31. Ranu Kundu 32. S Hasan 33. L Mestel 34. S K Bose 35. E I Mogilevsky 36.
37. J Javaraiah 38. 39. K R Bondal 40. G C Joshi 41. E Ya Zlotnik 42. Anita Joshi 43. H M Hiremath
44. L N Kotcherlakota 45. 46. K Sasidharan 47. 48. B R Madhava Rao 49. N D N Prasad
50. 51. G B Gelfreikh 52. K R Lang 53. J V Hollweg 54. T G Forbes 55. M R Kundu 56. Hong Sik Yun
57. B N Dwivedi 58. F Verheest 59. R R Rausaria 60. J Kuijpers 61. C T Vanajakshi 62. B Roberts 63. Udit Narain
64. J N Desi 65. V D Kuznetsov 66. W Kalkofen 67. 68. Nagendra Kumar 69. J Heyvaerts 70. Zdenka Smith
71. W Unno 72. P Venkatakrishnan 73. R T Gangadhara 74. C Chowdappa 75. R P Lin 76. P Hoyng 77. A R Choudhuri
78. R Kariyappa 79. R Vasundhara 80. K Sundara Raman 81. V G Kurt 82. B Lokanadham 83. R K Varma 84. K Sinha
85. 86. R Steinitz 87. R C Kapoor 88. 89. V P Gaur 90. 91. P K Raju
92. R Pallavicini 93. Franz Kneer 94. R K Shevgaonkar 95. M N Vahia 96. N O Weiss 97. D Montgomery 98. S P Bagare
99. Sydney D'Silva 100. M Anamma 101. M N Leony 102. 103. D Mohan Rao 104. K E Rangarajan
105. P S M Aleem 106. V B Semikoz 107. B S Nagabhushana 108. R S Narayanan.

I. INTRODUCTION

GOOD MORNING

V Krishan
Indian Institute of Astrophysics
Bangalore-560034, India.

Good Morning. I am Vinod Krishan and this is the IAU Symposium on Basic Plasma Processes on the Sun. Make sure you are at the right conference! During the last year, I have exchanged several messages with most of you. In the beginning, there was chaos, as we received queries about this symposium and that symposium. We first proposed a symposium with the title "From Solar to Astrophysical Plasmas" inspired partly by the availability of a tremendous amount of high-quality observations and theory for the Sun and because most of the basic plasma physical processes (like magnetic reconnection, plasma physical acceleration, radiation, generation of magnetic fields and so on) which are being studied in depth on the Sun also occur in the rest of the Universe. The rest of the Universe, however does not reveal itself the way the Sun does. Studies of the solar interior have brought out the role of rotation and the magnetic field. One also is aware how the rotation and magnetic field could turn a core collapse into an outgoing wave; hence, according to Fowler and Hoyle, our best chances of testing pulsar and supernova ideas may be in the solar context. We would like to know the signatures of particle acceleration processes such as might occur at the supernovae shock waves, turbulence generated in supernova ejecta or of the accelaration by strong pulsar electric fields. In the near future, we will not be able to make in situ observations of, for example, a supernova shock wave. However, despite the limited range of parameters, the study of acceleration processes in the heliosphere may provide a good foundation. Thus the Sun is not only our Provider, it is also our Teacher.

However, the IAU Executive Committee did not agree with our proposal. It suggested that we might narrow down our horizons to Basic Plasma Processes on the Sun, but we have still sneaked in references to the rest of the universe. In fact, for example, Prof. Zheleznyakov has not changed the title of his talk, which is "From Solar Plasmas to Plasmas at Neutron Stars". Perhaps he should hold a symposium on "From Solar to Astrophysical Plasmas" in future.

Anyway, we have plenty to discuss in concentrating on the Sun. We have ventured to study the Sun from inside out. The interior of the Sun, which as William Press stated, is more distant than the most distant galaxies, being separated from the surface by an optical depth of 10^8. The medium in between is a highly turbulent one, with complex interactions of velocity and magnetic fields. In fact, the whole Sun Quiz can be neatly divided into three rounds consisting of: (i) nuclear interactions; (ii) interactions of velocity and magnetic fields; and (iii) interaction of charges with electromagnetic fields. But the boundaries of their play are not so well defined. This is why we are here.

We believe that we know enough about these processes to transfer our knowhow to a study of other objects, while the ensuing discussions will reveal how much more needs to be discovered. This dichotomy of thought, has seeped into the scientific programme. We have here, in the participants, experts from a variety of fields, from superconductivity to supernovae; for electrons when in Rome behave exactly as at home; for inhomogeneity induces impulse.

E. R. Priest and V. Krishan (eds.), Basic Plasma Processes on the Sun, 3–4.

The Scientific Programme has been prepared by the Scientific Organising Committee with lot of trading in cash and kind. My anxiety must have reached its peak in August because that's when I started thinking about the conference all day long, except while sleeping because then I dreamt about it. In one dream, I found that none of the first-day speakers had arrived. So, I thought that at least I could begin with my talk, but at that very moment amnesia struck me. I could not recall a word about velocity and magnetic fields. In this state of acute agony, I woke up.

I am extremely happy to see all of you here. We have a packed programme and it may even be overbooked. Let us work for the next few days without a care in the world and in the midle we shall go for a picnic. But first let us meet our captain, the host of this Conference, our Director, Professor J C Bhattacharyya.

WELCOME ADDRESS

J C Bhattacharyya
Indian Institute of Astrophysics
Bangalore-560034, India

Participants of the IAU Symposium # 142 on 'Basic Plasma Processes in the Sun', Ladies and Gentlemen.

Most hearty welcome to the beautiful city of Bangalore, which is aptly described as the present-day scientific capital of India. Although it is beginning to suffer from the inevitable effects of a world-wide population explosion, Bangalore still retains some of its old charms, which had all along attracted the cream of intelligentzia from all over the country. They came in search of a serene atmosphere, where undisturbed minds and unfettered thoughts could flourish; they have contributed to some of the gems of human creativity from this part of the world.

The present city of Bangalore has grown out of a settlement of a chieftain owing allegiance to the mighty Vijayanagar rulers in the sixteenth centuary A.D. Four hundred years ago, the founder, Kempe Gowda, had dreamt of a giant metropolis; he had laid four corner stones of the future city, which can be seen even today. The city has now spread far beyond the founder's dreams, but the original charm still lingers around. The rulers of the state of Mysore, the great challengers to British expansionism in India, were Hyder Ali and his son, Tippu Sultan. They had poured out their hearts in beautifying the city; relics of their devotion are evident among the gardens and old edifices scattered all over. The British had found it an ideal place for locating their army base, which still persists as a sprawling cantonment, occupying a large area at the heart of the present metropolis. Sir Winston Churchill, the saviour of Britain in Second World War, had spent some of his younger years here, as a British army officer.

The salubrious climate and the serene atmosphere had attracted other interests. At the beginning of the present century, Sir Jamshedji Tata, the pioneer industrialist of the country, decided to establish an institution for higher learning. It was against the advice of many of his friends, who advocated that the money be spent in feeding the hungry and clothing the poor. But Sir Jamshedji had insisted on creating an institution where the best talents in the country would find a heaven for blossoming. The Indian Institute of Science, located in the north-western corner of the city has kept his vision alive.

Bangalore has welcomed several other great minds of India. It is here that the Nobel Laureate poet Rabindranath Tagore has composed some of his soul-stirring poems and songs which still ring in the hearts of millions all over India. One of his famous novels, "Sesher Kavita", meaning "The concluding poem" was largely composed here in a house named "Balabruiee", very close to this place. The novel has an unusual theme; its main character Amit Ray is an unconventional and unorthodox philosopher and a poet. Through his words, Tagore has given a beautiful definition of the quality known as culture. Amit says: "The stone which makes up a diamond is learning; but the sparkle which emanates from it is culture".

The high tide of scientific culture flowed into Bangalore with Sir C V Raman, when he moved over from Calcutta in the thirties. He was associated first with the Indian Institute of Science and later established his own laboratory. Besides building a very active school of researchers, he established a Science Academy, with activities covering

E. R. Priest and V. Krishan (eds.), Basic Plasma Processes on the Sun, 5–7.

the important aspect of scientific publications. His research institute and the Academy, both are forging ahead; the latter having completed its golden jubilee five years ago.

It will not be possible to give you a fair account of the scientific activities in Bangalore within the limited time which has been allotted to me. But I can tell you that as far as the subject of Astronomy and Astrophysics is concerned this is the Mecca of India. As many as five of the research and educational institutions in the city have large aspects of this subject in their curriculum activities. The international journal of Astrophysics and Astronomy is published by the Academy of Sciences here; even the editorial work of the Bulletin of the Astronomical Society of India, the other astronomical journal in India, is also done here. A majority of the community of astronomers in present-day India is attached to these five institutions in the city. From very low-frequency radio waves to high-energy gamma rays, all bands of electromagnetic spectra are covered by them.

One of these five institutions in Bangalore is the Indian Institute of Astrophysics, which is playing host to the present symposium. The Institute shifted its centre of activities to Bangalore from Kodaikanal only fifteen years ago, but it has a chequered history of more than two hundred years. In the penultimate decade of the eighteenth century, a small observatory had started in Madras and continued there till the end of the next century. The names of Goldingham, Taylor and Pogson are associated with it, who had carried out astronomical activities over those lean years in India. Among notable achievements are the Madras Star Catalogue of 1845 and discoveries of some asteroids and variable stars. The observatory was shifted to Kodaikanal along with the dawn of the present century, where work on solar physics, a subject more directly relevant to the present symposium, was started. It was there, in Jan 1909, that John Evershed made systematic studies of sunspot spectra and discovered the radial motion of solar matter; it was perhaps the first observation of solar plasma movements.

I shall not burden you with the details of solar work which has been continued at Kodaikanal till today. It will perhaps suffice to say that five years ago a small group of solar physicists from USA was surprised to find in the archives of the Kodaikanal Observatory an unbroken series of photoheliograms and spectroheliograms covering almost eight sunspot cycles. The discovery has led to a collaborative program between the Hale Observatories and Kodaikanal, aiming at solving the riddle of solar rotation.

One more international collaborative investigation is also being pursued between the Pulkovo Observatory in USSR and Kodaikanal, where the large-scale distribution of solar magnetic fields over the past eight sunspot cycles is being deciphered from the old H-alpha spectroheliograms.

I must point out that even before Kodaikanal started functioning, several important discoveries in solar physics were achieved on Indian soil. On 18th August 1868, the central track of a total solar eclipse crossed over the present state of Andhra Pradesh. On that occasion two European teams, a British team led by Col. Tennant and a French one led by the famous astronomer Janssen, had their camps in the vicinity of Guntur town. Both the teams detected a bright yellow line in the flash spectra, which did not match any line known from laboratory experiments. Based on these findings, Sir Norman Lockyear predicted the existence of a new element, Helium. That was twenty seven years before the element was isolated in laboratories by Ramsay. The event is often described as the birth of solar astrophysics.

Another total solar eclipse three years later yielded a proof of existence of the F-corona; the discoverer was Janssen who had once again come to India for the event. Still a third eclipse in January 1898 resulted in the discovery of the ultra-voilet extension of chromospheric spectra by a British team; young Evershed was in the group; his experience had paved the way for his return to Kodaikanal eight years later.

The 1898 eclipse is very important to India from a different point of view too. It was at this eclipse that Kavasji Naegamvala, an Indian scientist, independently organized and led a succcessful team of observers. He recorded excellent photographs and spectrograms of the event. Although no outstanding results came out of his effort, he managed to break a psychological barrier, which had discouraged native scientists in attempting experiments in the contempory frontiers of astronomy. His success enthused many young hearts, who started cherishing dreams of becoming astrophysicists. The culmination came twenty years later, when a young lecturer in Physics, in the University of Calcutta, solved the persistent puzzle of the anomalous strengths of ionized calcium lines in chromospheric spectra. The young scientist was Megh Nad Saha, who went on to establish stellar spectroscopy on a solid foundation.

I feel I have given you enough details about the venue and the local interest in the subject of the symposium; I wish to add just one more point of the special significance of the Sun in the Indian mind. As in many other civilizations the world-over, the Sun is worshipped as an all-powerful god in this country too. The traditional hymn in praise of the Sun, is Gayatri Mantra, which generations of Hindus have been reciting daily over the past few millenia. The key words of that mantra are:

"Tat sabitur varenyam, bhargo devasya
dheemahi, dheeyo yo nah prachodayat."

Let us meditate on that heavenly splendour, the glorious Sun, so that he may inspire our wisdom. The hymn is mainly an acknowledgement of the vital role played by this heavenly body in our very existence, but the underlying message of searching our minds for the ultimate truth is abundantly clear. It is the same idea which has brought us together, from all parts of the globe, to understand this enigmatic process, which is our Sun. The details of questions have kept on changing, but the query remains the same comprising the eternal musings of the human mind.

With these words, I welcome you all once more to the Symposium.

THE PLASMA UNIVERSE

Carl-Gunne Fälthammar
Dept of Plasma Physics
Royal Institute of Technology
S-100 44 Stockholm, Sweden

ABSTRACT

The term "Plasma Universe", coined by Hannes Alfvén, emphasizes the fact that plasma phenomena discovered in the laboratory and in accessible regions of space, must be important also in the rest of the universe, which consists almost entirely of matter in the plasma state.

Relevant aspects of this concept will be discussed. They include the response of the plasma to electric currents, the support of magnetic-field aligned electric fields, violation of the frozen-field condition, rapid release of magnetically stored energy, acceleration of charged particles, chemical separation, filamentary and cellular structures, and critical velocity interaction.

1. Introduction

The term Plasma Universe was coined by Hannes Alfvén (1986) to be the symbol of a change of paradigm (Alfvén 1983), resulting from progress in plasma physics, and especially from new empirical knowledge gained by in situ measurements in space plasmas.

It is a well known fact that almost all the matter that constitutes our universe is in the plasma state. Virtually all astrophysical phenomena take place in a plasma environment, and many of them are of an essentially plasma physical nature.

Exciting discoveries are being made by means of modern remote sensing techniques using formerly inaccessible parts of the electromagnetic spectrum. The phenomena behind them involve matter in various extreme plasma states. Translation of such observations into understanding of the physics of the phenomena involved has to rely on whatever knowledge we have of how matter behaves in the plasma state.

In building the necessary foundation of knowledge, the empirical knowledge so far gained - and that which is still to be gathered - in laboratories and in the Earth's own magnetosphere plays a crucial role.

E. R. Priest and V. Krishan (eds.), Basic Plasma Processes on the Sun, 9–20.

To fully recognize the role of such empirical knowledge of plasmas, it is useful to briefly look back at the evolution of plasma physics.

2. Classical versus modern plasma physics

The evolution of plasma physics falls into two distinct epochs, which we may characterize by the terms classical and modern plasma physics. Classical plasma physics was based on a very limited empirical basis derived essentially from experiments with cool, weakly ionized plasmas. On this slim empirical basis was built an comprehensive theoretical superstructure. This classical plasma theory was assumed to have general validity, i. e. to be applicable to parameter ranges that had not yet been attained experimetally.

From the classical plasma theory, applied to extremely hot plasmas, such as are needed for producing thermonuclear fusion energy, it was in the 1950's generally concluded that the problem of magnetic confinement of hot plasma was rather simple.

As soon as the thermonuclear effort made it possible to conduct experiments in the new parameter range of very hot plasmas, the limited validity of the classical plasma theory became evident. A number of "anomalous" phenomena, not foreseen in the framework of classical theory, were found. The resulting "thermonuclear crisis" led to the start of a new era in thermonuclear research. The lesson learned was that even self-consistent, sophisticated and elegant theories can be completely misleading unless based on sufficient empirical knowledge. Therefore, the new era in in thermonuclear research is characterized by a close interplay between experimental and theoretical efforts.

Thus, by making available a new parameter range for empirical study, the thermonuclear effort became one of the foundations of modern plasma physics.

The other foundation was the space effort, which made it possible to make in situ measurements, and even active experiments, in the space plasma. This led to an even more drastic widening of the parameter ranges in which empirical knowledge can be gained.

Like in the case of fusion research, there are lessons to be learned from the way in which magnetospheric research developed. In this case too, some generally accepted theoretical deductions turned out to be misleading. In retrospect we know that the reason was that relevant plasma physical processes were in some cases known but disregarded, in other cases not yet discovered.

3. The metamorphosis of the magnetosphere

The magnetosphere as we now know it (Fig. 1) is very different from the simple concepts of it that prevailed before the space age.

The pre-space age concepts of our neighbourhood in the cosmos were based on generally accepted classical theories, which were

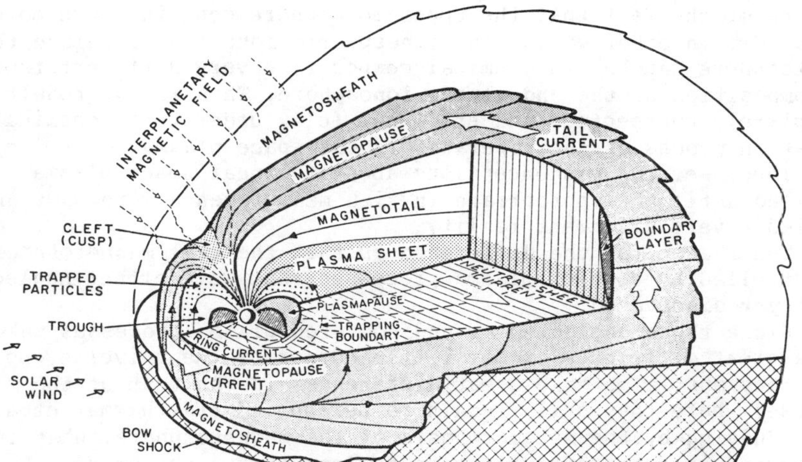

Fig. 1. The Earth's magnetosphere as we now know it. This dynamic and richly structured cosmical plasma system provides a unique opportunity of laying a safer empirical foundation for understanding more distant cosmical plasmas.

believed to be generally valid. On this basis it was concluded that the electric field would be a secondary parameter of little relevance. As a consequence, electric fields, and in particular magnetic-field aligned electric fields, which we now know to be of crucial importance, were long disregarded.

Much of this delay could have been avoided, if certain results already known from laboratory experiments had been applied to the space plasma. For example, Hannes Alfvén proposed already in 1958 that electric double layers, a phenomenon that had been observed in the laboratory, could exist above the ionosphere and cause energization of auroral primary electrons (Alfvén 1958). This suggestion, now known to be valid, was almost universally refuted because it was incompatible with generally accepted theories.

One more example may be quoted, since it has obvious repercussions in astrophysics. Quite late in the space age, it was discovered that, over the auroral oval, there are veritable fountains of upflowing ionospheric ions. Indeed, the outflux is so great that at certain times large parts of the magnetosphere is dominated by ionospheric ions, i. e. by matter terrestrial - not solar - origin. Furthermore, the most prevalent ion is usually O+, in spite of the fact that the upper ionosphere contains much more He+ and H+. In other words, the ionospheric ions that populate the magnetosphere can have a chemical composition very different from the composition of the underlying ionosphere. This is the result of a completely unexpected, and still not fully understood, chemical separation processes that operates in the space plasma.

Thus, serious misconceptions about the near space plasma prevailed until the appropriate in situ measurements were made and revealed a very different reality.

The reason for the difference is that the real magnetosphere is controlled by complex plasma processes that were either neglected or not yet discovered.

Since those neglected or undiscovered plasma processes make such a great difference in our little corner of the universe, no doubt they must also make great differences in the rest of the universe, where they still remain to be applied (Fälthammar et al. 1978). Judging by the metamorphosis of the magnetosphere, what is to be expected in astrophysics may be no less than a change of paradigm, as predicted by Alfvén.

In the rest of this paper we shall discuss some of the plasma phenomena that are now known to play a role in our own corner of the Plasma Universe.

4. Electric currents in cosmical plasma

Except in very limited regions, for example parts of planetary environments, the magnetic field is not curl-free. This means that in almost all cosmical plasmas the existence of the magnetic field depends on currents that flow in the plasma itself. This means that the ability of the plasma to carry electric current is essential, and we now know that this ability can be much more limited than classical theory would predict. It is therefore important to explicitly account for the currents (Alfvén 1977).

5. Current limitation and magnetic-field aligned electric fields

Anomalous resistivity

The first widely recognized mechanism for radically changing the current-conducting capability of a plasma was anomalous resistivity. Current-driven instabilities were assumed to drive microinstabilties that developed to a state of microturbulence with electric fields large enough to reduce the mobility of the conduction electrons by several powers of ten.

This phenomenon was invoked in explaining a number of phenomena, such as acceleration of auroral particles and dissipation in singular regions where MHD breaks down. However, in the light of direct measurements it is doubtful whether anomalous resistivity, plays any significant role in the magnetosphere (Coroniti 1985).

To determine whether or not anomalous resistivity is important in astrophysical plasmas, we need to understand much better than now the nonlinear development of the instabilities concerned. Laboratory experiments and observations in the space plasma combined with numerical simulations may provide the required insights.

Electric double layers

From laboratory experiments it is known that current-driven plasma instabilities may lead to a completely different end state than anomalous resistivity , possibly with anomalous resistivity as a transient intermediate state (Bohm and Torvén 1987). This alternative end state is a state with one or more double layers (for a recent review, see Raadu 1989).

Electric double layers in non-magnetized plasma have been known from the time of Langmuir in the beginning of this century. However, their existence in magnetized plasmas was proved experimentally as late as 1978 (Coakley et al. 1978; Torvén and Anderson 1979). Unlike anomalous resistivity, which, if it exists, allows the plasma to support a distributed potential drop, the electric double layer is a higly localized potential structure with a thickness of only tenths of Debye lengths. A classical case of such a structure observed in the laboratory is shown in Fig. 2.

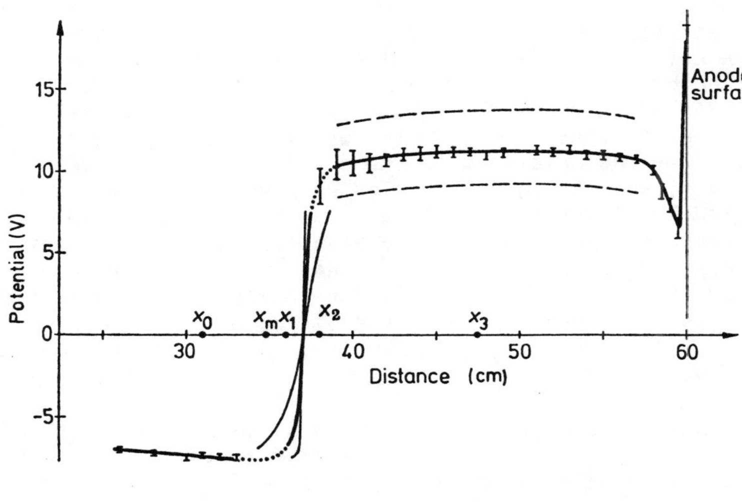

Fig. 2. Example of a strong electric double layer observed in the laboratory (Torvén and Lindberg 1980). Formation of electric double layers can drastically change the properties of a plasma in terms of its ability to (1) carry electric current, (2) accelerate charged particles and (3) release magnetically stored energy.

In the laboratory, the consequences of double layer formation depend on the curciut of which it is an element. In a low inductance circuit, the current is disrupted. In a high inductance circuit, large overvoltages are created, and the continued flow of current through these voltages rapidly drains the magnetic energy in the circuit (Torvén et al. 1985). Situations corresponding to both these cases are conceivable in cosmic plasmas.

There also exist electric double layers without any net current. The simplest example is the wall sheath of a plasma column.

Based on an overwhelming body of evidence (for a review see for example Fälthammar 1983) there is now an almost complete consensus among space physicists that electric double layers play an important role in auroral particle acceleration.

Electric double layers have also been invoked to explain phenomena in various astrophysical plasmas. A few examples may be mentioned.

The first suggestion that strong double layers may be responsible for the rapid release of magnetically stored energy that takes place in solar flares was made by Jacobsen and Carlqvist (1964). The idea was further developed by Alfvén and Carlqvist (1967), Carlqvist (1969, 1986) and Hasan and ter Haar (1978).

In the context of X-ray pulsars it has been suggested that an electric double layer is established in the accreting matter (Williams et al. 1986). In this case it is a double layer with no net current.

The possibility of electric double layers being responsible for the huge energy release in so-called double radio sources has been suggested by Alfvén (1978). In a detailed study of this idea Borovsky (1986) suggests that both electric double layers and anomalous resistivity exist, in a kind of symbiotic relation, in the extragalactic jets associated with the double radio sources.

Acceleration of charged particles to cosmic ray energies has been suggested by Alfvén (1978). In an analysis of relativistic double layers, Carlqvist (1986) has estimated that double layers may allow acceleration of up to 10^{14} eV per unit charge.

A thorough treatise on electric double layers and their role in astrophysics has recently been given by Raadu (1989).

The magnetic mirror effect

In a collisionless plasma there is a third mechanism by which electric currents can be limited and magnetic-field aligned electric fields supported, namely the magnetic mirror effect. (For a systematic discussion of these mechanisms, see Fälthammar, 1978.) For a current flowing out of a magnetic mirror, the most important carriers of the current, namely the electrons, are impeded by the magnetic mirror force, and only those in the loss cone can contribute to the current. Under certain assumptions, one of which is continual replenishment of the loss cone, there exists a simple relation (Knight 1973; Fridman and Lemaire 1980) between the current density in the mirror neck and the total applied voltage. One feature of the relation is that there exists a saturation current density that cannot be exceeded however large voltage is applied.

In some cases, observed with the Viking satellite this saturation seems to have been reached (Brüning et al. 1989).

In the case of the Earths auroral oval the relation has a linear part which corresponds to a constant conductance of typically 3 ($\mu A/m^2$)/kV (Fälthammar 1977), in good agreement with observations.

If the scattering in the source plasma is insufficient to keep the loss cone filled, the Knight relation becomes invalid, and the current can choke to arbitrarily small values.

It is well established that the magnetic mirror effect plays an important role in the magnetosphere. Since inhomogeneous magnetic fields are common in astrophysical plasmas, this effect may well be important there, too.

Violation of the frozen-in field condition

The existence of magnetic-field aligned electric fields means that the frozen-in magnetic-field condition can be violated (see e.g. Alfvén and Fälthammar 1963, Ch. 5.4.1). Such violation has been well established in the magntosphere, and may be important in other parts of the universe, wherever one or more of the three above-mentioned mechanisms operate.

One example where unfreezing of magnetic field lines is invoked is in Alfvéns theory of the saturnian rings. In this theory, unfreezing is a requisite for the so-called partial corotation (Alfvén 1981, Ch. III.3).

6. Filamentary structure

From the laboratory it is well known that current-carrying plasma has a tendency to break up into filaments.

In the magnetosphere the most conspicuous example of filamentary structure is that of the aurora. Unfortunately, direct measurements of electric currents in the space plasma are not yet possible, and the only information available is that deduced from single-satellite measurements of the magnetic field. Therefore the exact relation between auroral currents and auroral filamentation remains to be determined.

A theoretical relation applicable to the filamentation problem, in the presence of magnetic, gravitation and pressure forces, has been derived by Carlqvist (1988). It includes the Bennett relation and Jeans' criterion as special cases.

7. Cellular structure

One of the surprising discoveries in the space plasma is the prevalence of sharply defined boundaries between plasma regions, often regions with different properties such as density, temperature or composition. Such boundaries are often marked by electric sheet currents. Examples are the magnetopause, the geomagnetic neutral sheet and the solar wind "sector boundaries".

It has been pointed out by Alfvén (1981, Ch II.10) that the cosmical plasma thus has a tendency toward dividing itself up into "compartments". As this tendency has been found to be present throughout the regions of space that happen to be within the reach of spacecraft, it would be surprising indeed, if it did not also exist throughout the cosmical plasma. If it does, one important consequence is that a matter-antimatter symmetric universe may be possible, because matter and antimatter may occupy different "compartments", separated by "Leidenfrost layers" (Alfvén 1981, Ch IV 9.6).

8. Chemical separation

One of the great surprises in magnetospheric research was the discovery of large outflows of oxygen ions from the ionosphere and of the fact that large regions of the magnetosphere are sometimes dominated by oxygen plasma that has originated in the Earth's own atmosphere (see e.g. Chappel et al., 1987 and references therein).

One reason why auroral currents cause an exchange of matter is that they can drive instabilities, which cause transverse acceleration of ions. As a consequence of their increased orbital magnetic moment these ions are expelled into the magnetosphere by the magnetic mirror force (see e.g. Kaufmann, 1984 and references therein).

This energization and subsequent expulsion of ions is highly selective. Therefore, the ions that reach the magnetosphere have a chemical composition very different from the one prevailing where they came from. This constitutes a very efficient chemical separation mechanism, which was entirely unknown and unexpected until it was empirically discovered in the magnetosphere. This has great potential implications on the interpretation of abundance ratios in astrophysical plasmas.

9. The Critical Velocity phenomenon in plasma neutral-gas interaction

Alfvén's (1942) theory of the origin of planets and satellites (Alfvén and Arrhenius 1976) was based on the hypothesis that a plasma and a neutral gas in relative motion experience a strong interaction (and a violent ionization of the neutral gas), if the relative velocity exceeds a certain critical value, even if binary collisions are entirely insufficient for any appreciable momentum exchange. The value of this critical velocity was assumed to be the one where the kinetic energy of the plasma ions, in the rest frame of the neutral gas, equals the ionization energy of the latter. This hypothesis had no support whatsoever in the physics known at time it was invoked. Nevertheless, its validity was discovered in a laboratory experiment by Fahleson (1961) and later confirmed in many other experiments.

Thanks to dedicated laboratory experiments (for reviews see Sherman 1977 and Raadu 1978), the phenomenon is largely understood but important unsolved problems remain (Brenning and Axnäs, 1988). The phenomenon has also been found to operate in the space plasma.A recent review of space experiments on the phenomenon has been given by Torbert (1988).

The Critical Velocity phenomenon has been invoked in a number of cosmical applications, such as the formation of an ionosphere at the Jovian satellite Io (Cloutier et al., 1978), the interaction of the solar with gas clouds (Lindeman, 1974; Gold and Soter, 1976), with comets (Haerendel, 1986; Galeev et al., 1986), with planetary atmospheres (Luhmann 1988) and with the interstellar medium (Petelski et al., 1980). Thus, the phenomenon may have important astrophysical implications, but these cannot be evaluated in detail until a full understanding of the phenomenon has been achieved.

10. Concluding remarks

Discoveries made by in situ observations of the near Earth space plasma has necessitated dramatic revisions of our concept of how cosmic plasmas behave. The reason for the unexpected behaviour is plasma processes that were previosly either neglected or not yet known. As our universe, like the Earth's space environment, consists almost entirely of plasma, drastic revisions of astrophysics, too, may be needed before we have a true concept of our Plasma Universe.

17. References

Alfvén, H., On the Theory of Magnetic Storms and Aurorae, Tellus, 10, 104, 1958.

Alfvén, H., Electric Currents in Cosmic Plasmas, Rev. Geophys. Space Phys., 15, 271, 1977.

Alfvén, H., Double Radio Sources and the New Approach to Cosmical Plasma Physics, Astrophys. Space Sci., 54, 279, 1978.

Alfvén, H., Cosmic Plasma, D. Reidel Publ. Co., Dordrecht, Holland, 1981.

Alfvén, H., Paradigm Transition in Cosmic Plasma, Geophys. Res. Lett., 10, 487, 1983.

Alfvén, H. , The Plasma Universe, Phys. Today, 39, 22, 1986.

Alfvén, H. and Arrhenius, G., Evolution of the Solar System, NASA SP-345, Washington, D.C. 1976.

Alfvén, H. and Fälthammar, C.-G., Cosmical Electrodynamics, Fundamental Principles, Oxford, 1963.

H. Alfvén and P. Carlqvist, Currents in the Solar Atmosphere and a Theory of Solar Flares, Solar Physics, 1, 220, 1967.

Bohm, M. and Torvén, S., Extended Potential Drops Preceding Double Layer Formation in a Triple Plasma Device, XVIII Int. Conf. on Phenomena in Ionized Gases, Swansea, Contributed Papers 2, 318,, 1987.

Borovsky, J.E., Parallel Electric Fields in Extragalactic Jets: Double Layers and Anomalous Resistivity in Symbiotic Relationships, Astrophys. J., 306, 451, 1986.

Brenning, N., and Danielsson, L., Experiment on the interaction between a Plasma and a Neutral Gas II, Phys. Fluids, 18, 661, 1975.

Brenning, N. and Axnäs, I., Critical Ionization Velocity Interaction: Some Unsolved Problems, Astrophys. Space Sci., 144, 15, 1988.

Brüning, K., Block, L.P., Marklund, G.T., Eliasson, L., Pottelette, R., Murphree, J.S., Potemra, T.A. and Perraut, S., Viking Observations Above a Post Noon Aurora, accepted for publication in J. Geophys. Res., 1989.

Carlqvist, P., Current Limitation and Solar Flares, Solar Physics, 7, 377, 1969.

Carlqvist, P., On the Acceleration of Energetic Cosmic Particles by Electrostatic Double Layers, IEEE Transactions on Plasma Science, PS-14, 794, 1986.

Chappell, C.R., Moore, T.E., and Waite, J.G., Jr., The Ionosphere as a Fully Adequate Source of Plasma for the Earth's Magnetosphere, J. Geophys. Res., 92, 5896, 1987.

Cloutier, P.A., Daniell, R.E., Dessler, A.J., and Hill, T.W., A Cometary Ionosphere Model for Io, Astrophys. Space Sci., 55, 93, 1978.

Coakley, P., Hershkowitz, N., Hubbard, R. and Joyce, G., Experimental Observations of Strong Double Layers, Phys. Rev. Lett., 40, 230, 1978.

Coroniti, F.V., Space Plasma Turbulent Dissipation: Reality or Myth?, Space Sci. Rev., 42, 399, 1985.

Fahleson, U.V., Experiments with Plasma Moving Through Neutral Gas, Phys. Fluids, 4. 123, 1961.

Fridman, M. and Lemaire, J., Relationship Between Auroral Electrons Fluxes and Field-Aligned Electric Potential Difference, J. Geophys. Res., 85, 664, 1980.

Fälthammar, C.-G., Problems Related to Macroscopic Electric Fields in the Magnetosphere, Rev. Geophys. Space Phys., 15, 457, 1977.

Fälthammar, C.-G., Generation Mechanisms for Magnetic-Field-Aligned Electric Fields in the Magnetosphere, J. Geomagn. Geoelectr., 30, 419, 1978.

Fälthammar, C.-G., Magnetic-Field-Aligned Electric Fields, ESA Journal, 7, 385, 1983.

Fälthammar, C.-G., Akasofu, S.-I., and Alfvén, H., The Significance of Magnetospheric Research for Progress in Astrophysics, Nature, 275, 185, 1978.

Galeev, A.A., Gringauz, K.I., Klimov, S.I., Remizov, A.P., Sagdeev, R.Z., Savin, S.P., Sokolov, A.Yu., Verigin, M.I., and Szego, K., Critical Ionization Velocity Effects in the Inner Coma of Comet Halley: Measurements by Vega-2, Geophys. Res. Letters, 13, 845, 1986.

Gold, T. and Soter, S., Cometary Impact and the Magnetization of the Moon, Planet. Space Sci., 24, 45, 1976.

Haerendel, G., Plasma Flow and Critical Velocity Ionization in Cometary Comae, Geophys. Res. Letters, 13, 255, 1986.

Hasan, S., S., and ter Haar, D., The Alfvén-Carlqvist Double-Layer Theory of Solar Flares, Astrophys. Space Sci., 56, 89, 1978.

Jacobsen, C., and Carlqvist, P., Solar Flares Caused by Circuit Interruptions, Icarus, 3, 270, 1964.

Kaufmann, R.L., What Auroral Electron and Ion Beams Tell Us About Magnetosphere-Ionosphere Coupling, Space Sci. Rev., 37, 313, 1984.

Knight,S., Parallel Electric Fields, Planet. Space Sci., 21, 741, 1973.

Lindeman, R.A., Vondrak, R.R., Freeman, J.W., and Snyder, C.W., The Interaction Between an Impact-produced Neutral Gas Cloud and the Solar Wind at the Lunar Surface, J. Geophys. Res., 79, 2287, 1974.

Luhmann, J., An Assessment of the Conditions for Critical Velocity Ionization at the Weekly Magnetized Planets, Paper XIII.1.6 at the XXVIIth COSPAR Meeting, Helsinki, 1988.

Petelski, E.F., Fahr, H.J, Ripken, H.W., Brenning, N., and Axnäs, I., Enhanced Interaction of the Solar Wind and the Interstellar Neutral Gas by Virtue of a critical Velocity Effect, Astronomy and Astrophysics, 87, 20, 1980.

Raadu, M., The Role of Electrostatic Instabilities in the Critical Ionization Velocity Mechanism, Astrophys. Space Sci., 55, 125, 1978.

Raadu, M., A., The Physics of Double Layers and Their Role in Astrophysics, Physics Reports, 178, 25-97, 1989.

Sherman, J.C., Review of the Critical Velocity of Gas-Plasma Interaction II: Theory, Astrophys. Space Sci., 24, 487, 1973.

Torbert, R., Review of Ionospheric CIV Experiments, Paper XIII.2.1 at the XXVIIth COSPAR Meeting, Helsinki, 1988.

Torvén, S. and Andersson, D., Observations of Electric Double Layers in a Magnetized Plasma Column, J. Phys. D: Appl. Phys., 12, 717, 1979.

Torvén, S. and Lindberg, L., Properties of a Fluctuating Double Layer in a Magnetized Plasma Column, J. Phys. D: Appl. Phys., 13, 2285, 1980.

Torvén, S., Lindberg, L., and Carpenter, R.T., Spontaneous Transfer of Magnetically Stored Energy to Kinetic Energy by Electric Double Layers, Plasma Phys. and Controlled Fusion, 27, 143, 1985.

Williams, A., C., Weisskopf, M., C., Elsner, R., F., Darbro, W., and Sutherland, P., G., Accretion onto Neutron Stars with the Presence of a Double Layer, Astrophys. J., 305, 759, 1986.

II. THE SOLAR INTERIOR

INTERIOR STRUCTURE OF THE SUN.

Jørgen Christensen-Dalsgaard
Astronomisk Institut, Aarhus Universitet
DK-8000 Aarhus C
Denmark

ABSTRACT. Observations of solar oscillations have provided us with detailed information about the solar interior. Here I consider three examples of results obtained in such helioseismic investigations: i) the effect of the equation of state on the comparison between observed and theoretical frequencies; ii) a determination of the depth of the solar convection zone; and iii) indications of deviations from standard models of the structure of the solar core.

1. Introduction.

Although the theory of stellar evolution is often regarded as a well-established part of astrophysics, it is important to realize that stars are extremely complex, compared with standard calculations of stellar models. Hence the possibility that important aspects of stellar structure or evolution might have been neglected in the calculations should be kept in mind. To test this, detailed relevant observations are required.

Observations of solar oscillations have given us a large amount of very precise data on the properties of the solar interior; this has opened up the possibility of testing computations of stellar evolution in great detail for a single star. Recent compilations list over 2000 observed frequencies, with estimated errors that are in some cases less than 0.01 per cent. This must be compared with the other relevant data that are available: the solar mass, radius and luminosity, which are known with comparable precision, and the neutrino flux, which, as is well known, is subject to considerable observational and theoretical uncertainties. Furthermore, the physical nature of the oscillations is in general well understood. The observed modes correspond to standing acoustic waves, or *p modes*. Given a solar model it is relatively straightforward to compute its oscillation frequencies. Thus, the frequencies provide a clean diagnostics of conditions inside the Sun.

On the basis of these data we may study the basic processes that determine the structure of the solar interior. In this way we can test the assumptions, of perhaps questionable validity, on which computations of stellar evolution are based. Also, such calculations require information, which is

E. R. Priest and V. Krishan (eds.), Basic Plasma Processes on the Sun, 23–34.

often uncertain, about the properties of matter under the conditions in stellar interiors. Since the frequencies are sensitive at a significant level to even quite subtle details of the equation of state or the opacity, it is possible to use the observations to study properties of plasmas under conditions so extreme that they cannot be reproduced in the laboratory.

2. Properties of the solar interior.

It is useful to review very briefly normal calculations of solar models, and their possible shortcomings (see also Bahcall & Ulrich 1988; Turck-Chièze *et al.* 1988; Bahcall 1989; Turck-Chièze 1990). It is assumed that the model is in hydrostatic and thermal equilibrium. Evolution is controlled by the gradual fusion of hydrogen into helium; it is assumed that there is no mixing in the solar interior, so that the composition in any given mass-shell is determined solely by the local nuclear burning. With these assumptions the structure is largely determined by the *microphysics* of the solar interior, *i.e.*

- the equation of state
- the opacity
- the nuclear energy generation rates.

In addition, the computation requires that the solar mass is known, as well as the initial chemical composition, which is assumed to be uniform. The goal is to compute a model at the age of the present Sun, which is also assumed to be known, with the observed radius and surface luminosity.

In practice, the initial helium abundance Y_0 cannot be determined independently and must be regarded as a free parameter of the calculation, as must the "mixing-length" parameter α which measures the efficiency of convective energy transport near the solar surface. Y_0 and α are adjusted until the model of the present Sun has the correct radius and luminosity. In this way one obtains what is sometimes called a "standard solar model". It is evidently dependent on the uncertainties in the assumed microphysics, but is otherwise well-defined.

The computation of standard solar models ignores, or grossly simplifies, a number of processes that might be labelled the *macrophysics* of the Sun. These include

- energy transport
- dynamics of convection
- convective overshoot
- molecular diffusion
- core mixing
- magnetic fields.

Energy transport by convection is treated in a rather crude way, with furthermore depends on the *a priori* unknown parameter α. Near the surface convection is probably sufficiently vigorous to have dynamic effects on the average hydrostatic equilibrium, yet such effects are often ignored. At the lower boundary of the convection zone motion is normally supposed to stop at the point where convective instability ceases; hence convective overshoot, which must surely be present, is neglected. Molecular diffusion might affect the composition profile in the convectively stable region, yet with a few exceptions has been ignored. Instabilities in the deep interior could lead to material mixing, affecting the composition profile and hence solar evolution. Finally, one probably cannot totally exclude the possibility of a magnetic field in the solar interior of sufficient magnitude to have an effect on the overall structure of the Sun. - It should be noted, however, that except for the composition profile in the radiative interior of the model, much of the uncertain macrophysics is concentrated very near the surface.

3. Properties of solar oscillations.

The observed solar oscillations[1] have cyclic frequencies ν between about 1500 and 5000 μHz; their behaviour on the solar surface ranges from spherically symmetric oscillations to oscillations whose wavelengths are at the limit of observational resolution. Recent reviews of the theory and observations of solar oscillations have been given by Libbrecht (1988a), Vorontsov & Zharkov (1989) and Christensen-Dalsgaard & Berthomieu (1990).

The oscillations correspond to standing waves, or *normal modes*, of the Sun. A mode of oscillation is characterized by three wave numbers: the *radial order* n which, roughly, gives the number of zeros in the eigenfunction in the radial direction; the *degree* ℓ; and the *azimuthal order* m, ranging between $-\ell$ and ℓ, which measures the number of zeros in longitude. The degree is related to the horizontal wavenumber k_h and wavelength λ of the mode at radius r by

$$k_h = \frac{2\pi}{\lambda} = \frac{L}{r}, \tag{1}$$

where $L = \sqrt{\ell(\ell+1)}$.

Apart from damping or excitation, the time dependence of a single mode is harmonic, as $\cos(\omega t)$. The *angular frequency* ω is related to the cyclic frequency ν through $\omega = 2\pi\nu$. In general $\omega = \omega_{n\ell m}$ depends on all three wave numbers. However if rotation or other departures from spherical symmetry are ignored, $\omega_{n\ell m}$ does not depend on m. I shall adopt this approximation here.

In calculations of solar oscillation frequencies it is common to ignore a number of complicating features that are so far badly understood, such as·

- nonadiabaticity
- excitation, more generally
- dynamical effects of convection
- detailed atmospheric behaviour
- magnetic fields.

More detailed calculations show that these approximations may change the frequencies by several μHz. Thus they have a substantial effect on comparisons between observed and computed frequencies. On the other hand, the complications are essentially all located near the solar surface. Thus they add to the uncertainty of the surface region already found in the calculation

---

[1]with the exception of oscillations at long periods whose identity is still in doubt

of the model but do not directly affect the properties of the oscillations in the deeper solar interior.

With this simplification the computation of the oscillation frequencies is a straightforward numerical problem. However, to understand the results it is useful to consider the asymptotic theory of the oscillations. The observed p modes can be approximated locally by plane sound waves, with the dispersion relation $k^2 \equiv k_r^2 + k_h^2 = \omega^2/c^2$. Here k_r and k_h are the radial and horizontal components of the wave vector, and c is the adiabatic sound speed. Using equation (1), we obtain $k_r^2 = \omega^2/c^2 - L^2/r^2$. This shows that there is a *turning point* where $k_r = 0$ and the wave propagates horizontally, the location $r = r_t$ of which is determined by

$$\frac{c(r_t)}{r_t} = \frac{\omega}{L}. \tag{2}$$

It corresponds to a point of total internal reflection; for $r < r_t$, $k_r^2 < 0$, and the mode decays exponentially. At the surface the wave is reflected by the steep density gradient. A mode of oscillation is a standing wave, formed as an interference pattern between waves travelling between the internal and the surface reflection. It is trapped between the surface and r_t, and hence its frequency depends largely on conditions in this region.

Since c decreases with increasing radius, the turning point radius increases with increasing ℓ. Modes at highest observed values of ℓ are confined to the outermost fraction of a percent of the solar radius, whereas the lowest-degree modes penetrate essentially to the centre. Hence, by considering different modes of oscillation one is effectively sampling different parts of the Sun. In fact, it has been possible to invert the frequencies, to infer the sound speed as a function of position in most of the Sun (*e.g.* Christensen-Dalsgaard *et al.* 1985; Brodsky & Vorontsov 1987).

In analyzing frequency differences between observations and theory, one must correct for the fact that with increasing r_t the modes extend over a smaller fraction of the solar mass, and hence their frequencies are easier to perturb. This effect may be eliminated by considering scaled frequency differences $Q_{n\ell}\delta\omega_{n\ell}$, where $Q_{n\ell}$ is a measure of the inertia in the mode, relative to the inertia of an $\ell = 0$ mode of the same frequency (*e.g.* Christensen-Dalsgaard 1988a; Christensen-Dalsgaard & Berthomieu 1990). For modifications to the model which are confined close to the surface, the resulting $Q_{n\ell}\delta\omega_{n\ell}$ is mainly a function of frequency. The condition for this to be true is that the extent of the region over which the modification is significant is much smaller than the depth of penetration of the modes considered. It follows that if $Q_{n\ell}\delta\omega_{n\ell}$ *does* depend on ℓ for a set of modes, the change in the model extends at least to the lower turning point of those modes. Also it may be shown that low-frequency modes are insensitive to modifications that are confined to the superficial layers of the model (Libbrecht 1988b; Christensen-Dalsgaard 1988b).

These properties of the oscillations are particularly important in the light of the errors near the surface of the model. These errors may be expected to lead to scaled frequency errors that are essentially independent of ℓ and small at low frequency. Frequency errors that do not have these properties therefore indicate errors in the bulk of the model.

4. Results.

Here I can only give a brief summary of some of the results on solar internal structure that have been obtained recently from helioseismic investigations. In addition to the references given, it might be noted that some of these issues were discussed in more detail by Christensen-Dalsgaard (1988a) and Christensen-Dalsgaard (1989).

4.1 Effects of the equation of state.

It is of obvious interest to compare observed frequencies of solar oscillations with frequencies of representative models. To illustrate also the effect of using different formulations of the equation of state, I consider two such models. One model uses the simple EFF (Eggleton, Faulkner & Flannery 1973) equation of state, the other the far more elaborate MHD equation of state (Hummer & Mihalas 1988; Mihalas, Däppen & Hummer 1988; Däppen *et al.* 1988). In both cases opacities from the Los Alamos Opacity Library (Huebner *et al.* 1977) were used. The computational procedures, and the detailed effects of the equation of state on the model and frequencies, were discussed by Christensen-Dalsgaard, Däppen & Lebreton (1988).

Figure 1 shows differences, scaled to remove the ℓ-dependence of the mode inertia, between selected observed frequencies from Duvall *et al.* (1988) and Libbrecht & Kaufman (1988), and computed frequencies for these two models. The most obvious feature is the reduction in the ℓ-dependence of the differences when going from the EFF to the MHD equation of state, particularly for $\ell \gtrsim 50$. This strongly indicates that the error in the interior of the model, particularly in the convection zone where modes of degree higher than about 50 are trapped, has been reduced by using the MHD equation of state (Christensen-Dalsgaard, Däppen & Lebreton 1988). For modes with $\ell < 40$, which penetrate beneath the convection zone, there remains a substantial ℓ-dependence, indicating that there are significant errors in the radiative interior of the model. In fact, as discussed in the following section, the convection zone appears to be too shallow in this model.

Apart from this difficulty, there is a striking agreement between the observed frequencies and those of the MHD model. For this model the errors for modes of degree exceeding 100 are comparable with the estimated observational errors. Some care is required, however, when interpreting this result, given the known inadequacies in the model and frequency computations. It is likely that the agreement in Figure 1b is fortuitous, resulting from a partial cancellation of several sources of error. However, such errors would mainly affect the frequency-dependence of the frequency differences. Given the relatively small variation of the differences with degree it is likely that there are no gross errors in the bulk of the model. Indications of problems in the core are discussed in section 4.3 below.

4.2 The depth of the solar convection zone.

The transition from adiabatic stratification of the temperature in the convection zone to subadiabatic stratification in the radiative region below takes place quite rapidly. There is a similar transition in the gradient of sound

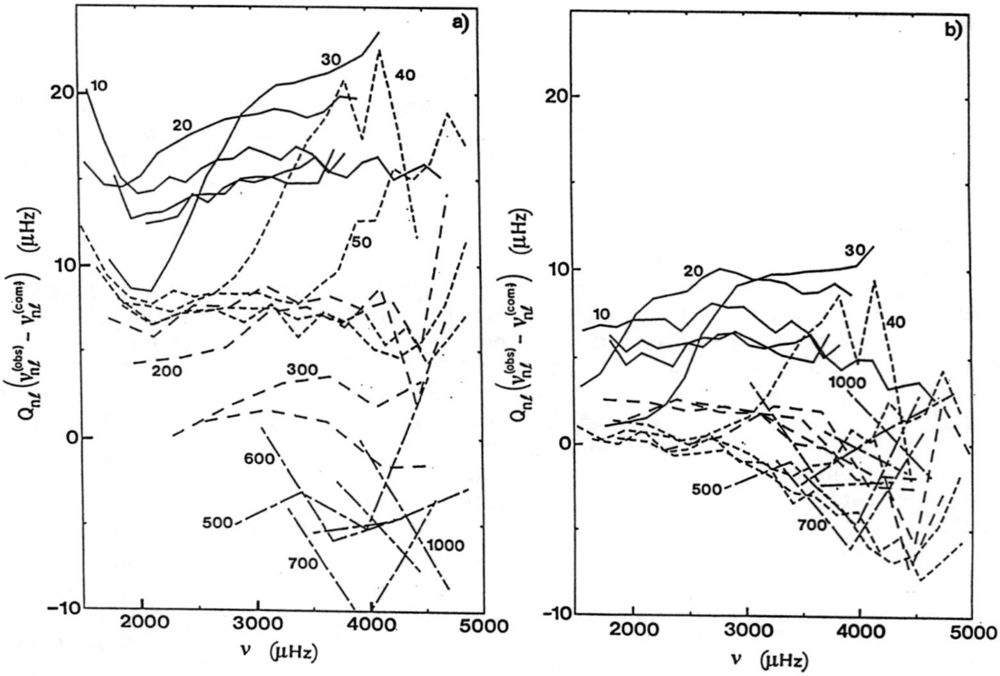

Figure 1. Scaled differences between observed and computed frequencies for selected values of ℓ. Panels a) and b) are based on models computed with the EFF and the MHD equations of state, respectively. Points corresponding to a given value of ℓ have been connected, according to the following convention: ℓ = 0, 5, 10, 20, 30 (————————); ℓ = 40, 50, 70, 100 (- - - - - - - - - -); ℓ = 150, 200, 300, 400 (— — — — — — — — —); and ℓ = 500, 600, 700, 800, 900, 1000 (—— - —— - —— -).

speed, which can be located in the sound speed determined from inversion of the oscillation frequencies. It was noted by Gough (1984, 1986) that this transition is particularly evident in the quantity

$$W_0 \equiv \frac{r^2}{GM}\frac{dc^2}{dr}, \tag{3}$$

where G is the gravitational constant and M is the mass of the Sun. In the adiabatically stratified part of the convection zone, outside ionization zones of abundant elements, $W_0 \approx -2/3$, and W_0 increases rapidly beneath the convection zone. This behaviour is clearly visible in W_0 as inferred from the inversion. Christensen-Dalsgaard, Gough & Thompson (1990) made careful tests, based on artificial data, of this method for determining the depth d_b of the convection zone and applied to the observed solar frequencies. The result was that in the Sun d_b = 0.287 ± 0.003 R, where R is the radius of the Sun. The inversion also provided a measure of the sound speed at the bottom of the convection zone. From this, the temperature T_b at this point can be determined, provided that the chemical composition is known. If, as suggested by calibration of solar models, the hydrogen abundance X is in the range 0.7 - 0.74, T_b = 2.13 - 2.28×10^6 K.

The depth of the convection zone in the MHD model considered in the previous section was 0.267 R, *i.e.* significantly lower than the solar value. This is consistent with the pattern of frequency differences shown in Figure 1b. An explanation could be that the opacities are too low in the region around the base of the convection zone; if instead the Cox & Tabor (1976) tables are used, d_b is 0.285 R, very close to the inferred value for the Sun. It should be noted, however, that what is measured is the extent of the adiabatically stratified region. This probably includes most of the region of convective overshoot beneath the convectively unstable layer. In fact, the increase in d_b required to bring the model illustrated in Figure 1b into agreement with the solar d_b is not inconsistent with the, admittedly uncertain, estimates of the extent of the convective overshoot (*e.g.* Schmitt, Rosner & Bohn 1984; Roxburgh 1985).

4.3 Conditions in the solar core.

Only those p modes that have the lowest degree penetrate to the energy-generating core of the Sun. Hence this region, which is of course of special interest for understanding solar evolution and finding a solution to the neutrino problem, is particularly difficult to probe with five-minute oscillation data. Nevertheless some information has recently emerged, although the consequences for our understanding of the deep solar interior are still unclear.

Figure 2 shows results obtained by Gough & Kosovichev (1988) from inversions for the sound speed c and the density ρ in the solar core, in terms of relative deviations from a standard solar model. Rather similar results for the sound speed have been obtained with different inversion methods (Christensen-Dalsgaard, Gough & Thompson 1988; Vorontsov 1988; Dziembowski, Pamyatnykh & Sienkiewicz 1990). In these inversions there is a tendency for a subtantial positive sound speed difference very near the centre; this, however, could well be an artifact of the analysis. Thompson (1990) carried out a similar analysis for the density difference, with very similar results. He pointed out that considerable care is required when interpreting the results: the inferred $\delta\rho/\rho$ is in fact an average with fairly complex weight functions, extending over a substantial range in r. Thus, for example, it is likely that the actual density difference at the centre is greater, perhaps by a factor of two, than the 10 per cent difference shown in Figure 2b.

It is so far not clear how this deviation from the standard model has come about. Gough & Kosovichev (1988) determined the difference in hydrogen abundance X by imposing the equations of stellar structure, assuming the opacity and energy generation rate to be given. The resulting δX, which was similar in shape to $\delta c/c$ as shown in Figure 2a, corresponds approximately to what would result from a small amount of partial mixing of the

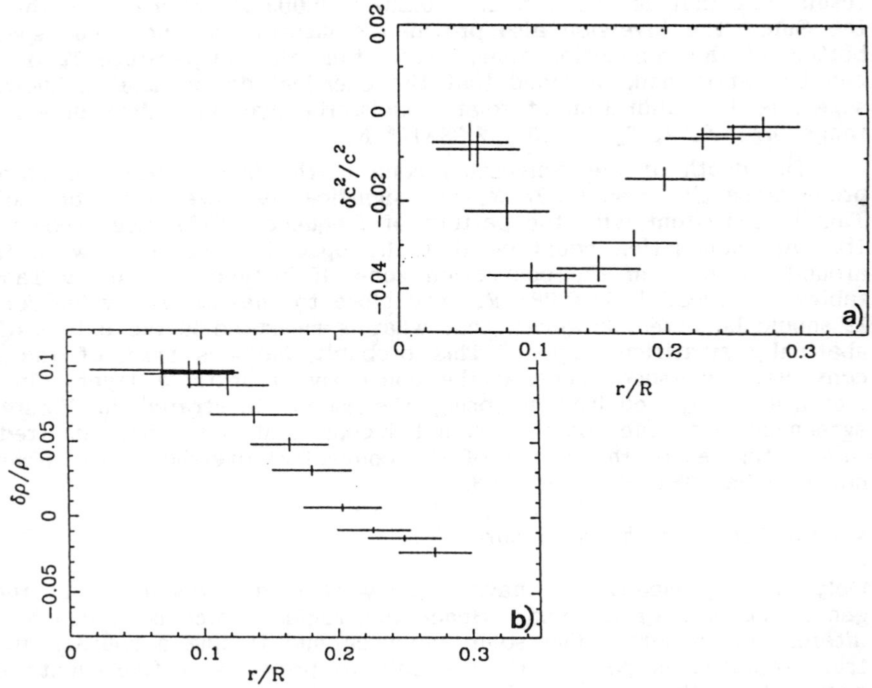

Figure 2. Relative difference in sound speed c (panel a) and density ρ (panel b) between the Sun and a standard solar model (Model 1 of Christensen-Dalsgaard 1982), obtained from inversion of frequencies of low-degree solar p modes. These results are averages over the actual differences, the horizontal size of the crosses giving a measure of the extent of the average; the vertical size gives estimates of the standard errors of the results. (From Gough & Kosovichev 1988).

solar core. From this Gough & Kosovichev (1990) inferred the temperature difference between the Sun and the model and hence were able to estimate the expected neutrino flux; the result was somewhat lower than for a standard model, although still in excess of the observed value. A possible difficulty with the interpretation in terms of partial mixing is that this, at least in some cases, leads to a *decrease* in the central density (*e.g.* Christensen-Dalsgaard 1986), in contrast to the results shown in Figure 2b.

Dziembowski, Pamyatnykh & Sienkiewicz (1990) used a slightly different inversion for the sound speed, with a positive sound-speed difference at the centre, to determine the variation of pressure $p(r)$ and $\rho(r)$. Given $p(r)$ and $\rho(r)$, using a parameterized expression for the hydrogen abundance $X(r)$, and imposing the constraint that the total luminosity should have the observed value, they determined the structure of the core such that the neutrino flux was minimized. They found that for realistic parameters this minimum flux was considerably *higher* than for the standard model. The reasons for the discrepancy with Gough & Kosovichev are so far unclear, but it is not unlikely that the differences in the assumed $c(r)$ near the centre play a substantial role.

5. Discussion.

The comparison of observed and theoretical frequencies shows that it is possible, amongst standard solar models, to match the observations to within less than about 0.3 per cent. Thus it appears that there are no gross errors in normal calculations of solar evolution. This is no ground for complacency, however. The remaining errors in the computed frequencies are systematic, and are furthermore considerably higher than estimated intrinsic errors in either the observations or the computations. Thus there remain significant problems in the model. The accuracy of the observations is such that even fairly subtle changes in the equation of state can have significant effects on the frequencies. Similarly the computed frequencies are sensitive to the assumed opacities. Thus it is possible to test descriptions of the properties of plasmas on the basis of the observations. Also, it is of some interest that it has been possible to determine the extent of the convection zone, which is of importance for studies of convection zone dynamics, on the basis of the observations.

The solar core poses particular difficulties. There seems to be substantial evidence that the central density is somewhat higher than in standard models, and that there is a region of lower sound speed in the outer parts of the core. However the interpretation of these results, in terms of their origin and their effect on the neutrino flux, is currently rather unclear. It is likely that improved observations of low-degree p modes will help; but we also need a better understanding of the relations between various changes in the core of the Sun.

Acknowledgements: I am very grateful to M. J. Thompson for useful discussions.

References.

Bahcall, J. N., 1989. *Inside the Sun, Proc. IAU Colloquium No. 121*, eds Berthomieu G. & Cribier M., in the press.

Bahcall, J. N. & Ulrich, R. K., 1988. *Rev. Mod. Phys.*, **60**, 297.

Brodsky, M. A. & Vorontsov, S. V., 1987. *Pis'ma Astron. Zh.*, **13**, 438 (English translation: *Sov. Astron. Lett.*, **13**, 179).

Christensen-Dalsgaard, J., 1982. *Mon. Not. R. astr. Soc.*, **199**, 735.

Christensen-Dalsgaard, J., 1986. *Seismology of the Sun and the distant Stars*, p. 23, ed. Gough, D. O., Dordrecht, D. Reidel Publ. Co..

Christensen-Dalsgaard, J., 1988a. *Seismology of the Sun & Sun-like Stars*, p. 431, eds Domingo, V. & Rolfe, E. J., ESA SP-286.

Christensen-Dalsgaard, J., 1988b. *Multimode Stellar Pulsations* p. 153, eds Kovacs, G., Szabados, L. & Szeidl, B.; Kultura Press and Konkoly Observatory, Budapest.

Christensen-Dalsgaard, J., 1989. *Inside the Sun, Proc. IAU Colloquium No.*

121, eds Berthomieu G. & Cribier M., in the press.
Christensen-Dalsgaard, J. & Berthomieu, B., 1990. In *Solar interior and atmosphere*, eds Cox, A. N., Livingston, W. C. & Matthews, M., Space Science Series, University of Arizona Press, submitted.
Christensen-Dalsgaard, J., Duvall, T. L., Gough, D. O., Harvey, J. W. & Rhodes, E. J., 1985. *Nature*, **315**, 378.
Christensen-Dalsgaard, J., Däppen, W. & Lebreton, Y., 1988. *Nature*, **336**, 634.
Christensen-Dalsgaard, J., Gough, D. O. & Thompson, M. J., 1988. *Seismology of the Sun & Sun-like Stars*, p. 493, eds V. Domingo & E. J. Rolfe, ESA SP-286, Noordwijk, Holland.
Christensen-Dalsgaard, J., Gough, D. O. & Thompson, M. J., 1990. *Mon. Not. R. astr. Soc.*, submitted.
Cox, A. N. & Tabor, J. E., 1976. *Astrophys. J. Suppl.*, **31**, 271.
Duvall, T. L., Harvey, J. W., Libbrecht, K. G., Popp, B. D. & Pomerantz, M. A., 1988. *Astrophys. J.*, **324**, 1158.
Dziembowski, W. A., Pamyatnykh, A. A. & Sienkiewicz, R., 1990. *Mon. Not. R. astr. Soc.*, submitted.
Däppen, W., Mihalas, D., Hummer, D. G. & Mihalas, B. W., 1988. *Astrophys. J.*, **332**, 261.
Eggleton, P. P., Faulkner, J. & Flannery, B. P., 1973. *Astron. Astrophys.*, **23**, 325.
Gough, D. O., 1984. *Mem. Soc. Astron. Ital.*, **55**, 13.
Gough, D. O., 1986. *Seismology of the Sun and the distant Stars*, p. 125, ed. Gough, D. O., Reidel, Dordrecht.
Gough, D. O. & Kosovichev, A. G., 1988. *Seismology of the Sun & Sun-like Stars*, p. 195. eds Domingo, V. & Rolfe, E. J., ESA SP-286.
Gough, D. O. & Kosovichev, A. G., 1990. *Solar Phys.* (special issue, IAU Coll. 121), submitted.
Huebner, W. F., Merts, A. L., Magee, N. H. & Argo, M. F., 1977. *Astrophysical Opacity Library*, Los Alamos Scientific Laboratory report LA-6760-M.
Hummer, D. G. & Mihalas, D., 1988. *Astrophys. J.*, **331**, 794.
Libbrecht, K. G., 1988a. *Space Science Rev.*, **47**, 275.
Libbrecht, K. G., 1988b. *Seismology of the Sun & Sun-like Stars*, p. 3, eds Domingo, V. & Rolfe, E. J., ESA SP-286.
Libbrecht, K. G. & Kaufman, J. M., 1988. *Astrophys. J.*, **324**, 1172.
Mihalas, D., Däppen, W. & Hummer, D. G., 1988. *Astrophys. J.*, **331**, 815.
Roxburgh, I. W., 1985. *Solar Phys.*, **100**, 21.
Schmitt, J. H. M. M., Rosner, R. & Bohn, H. U., 1984. *Astrophys. J.*, **282**, 316.
Thompson, M. J., 1990. *Astrophys. J.*, in preparation.
Turck-Chièze, S., 1990. *Solar Phys.* (special issue, IAU Coll. 121), submitted.
Turck-Chièze, S., Cahen, S., Cassé, M. & Doom, C., 1988. *Astrophys. J.*, **335**, 415.
Vorontsov, S. V., 1988. *Seismology of the Sun & Sun-like Stars*, p. 475, eds Domingo, V. & Rolfe, E. J., ESA SP-286.
Vorontsov, S. V. & Zharkov, V. N., 1989. *Sov. Sci. Rev. E. Astrophys. Space Phys.*, **7**, 1.

DISCUSSION

ALURKAR: What's the status of the global oscillation of period 2h 40min ? How effective are they for studies of the solar interior?

CHRISTENSEN-DALSGAARD: Long-period oscillations of the Sun are standing gravity waves, with large amplitudes in the solar core. Thus they are potentially very effective as probes of the core, and definite observation of these modes is highly desirable. Unfortunately, the observational situation is unclear. There have been reports of several oscillations with long periods, in addition to the initially discovered 2h 40min oscillation. However, the identification, or even solar nature, of these oscillations is still questionable. It is likely that only the forthcoming observations from networks or from space will settle the issue - we all hope that the modes are there!

VAN BALLEGOOIJEN: You find agreement between theory and observation regarding the structure near the base of the convection zone. Can you put an upper limit on the extent of convective overshooting?

CHRISTENSEN-DALSGAARD: The inferred depth includes the adiabatic part of the overshoot region. Hence the extent of this region cannot be estimated without reference to solar models, and it is strongly dependent on the opacity, which determines the depth where the transition to convective stability takes place. Estimates of the extent of the transitions to purely radiative stratification probably depends strongly on the details of the transition. I would guess that substantial deviations from the behaviour in normal solar models over more than $0.01R_{\odot}$ would have been detectable. To test this requires careful calculations of frequencies, and subsequent inversion of them, for models with various forms of overshooting.

CHOUDHURI: Is it possible to conclude anything about the strength of overshooting from the sharpness of the kink in the curve for W_0 (equation 3)?

CHRISTENSEN-DALSGAARD: With the present observations, the shape of the kink is consistent with that of a normal solar model, without overshooting. To study details of the overshooting better observations will be required.

VARMA: If one could identify and monitor magnetoacoustic oscillations in the Sun, would it be possible to draw information about the structure of magnetic fields inside the Sun?

CHRISTENSEN-DALSGAARD: There is no evidence for global magnetoacoustic oscillations of the Sun, as a separate class of modes. However, the observed p-modes would be perturbed by the presence of a large-scale magnetic field, and from observations of such frequence shifts information about the field might be obtained.*

---

*Editor: A detailed description of these effects has been given by Roberts and Campbell (Nature 323,603,1986), Campbell and Roberts (Ap J 338,538,1989), Gough and Thompson (MNRAS, in press) and by several other authors at IAU Symp. 123 (Ed. JC-CHRISTENSEN-DALSGAARD and S Frandsen).

MOGILEVSKIJ: What is your opinion about the structure of the plasma and magnetic field in the zone between the core and the base of the convection zone?

CHRISTENSEN-DALSGAARD: A large-scale magnetic field would contribute to the splitting of frequencies according to azimuthal order m, giving rise to an even component in the expansion of splitting as a function of m. There is some observational evidence for such an even component. From this Dziembowski & Goode have inferred the evidence of a strong localized magnetic field at the base of the convection zone, at perhaps 1MG. However, the observations are probably still not decisive.

Fine-structure vertical magnetic flux tubes near the solar surface would scatter the modes and hence contribute to the damping; it might also cause a frequency shift, as shown by Bogdan & Zweibel (1985). This, however, must be considered as a part of the uncertainty introduced by the solar surface region.

ORAEVSKY: In 1991 USSR is launching a satellite called "Coronas" for global observations of the Sun. This includes measurements of global Sun oscillations. The Institute of Terrestial Magnetism, Ionosphere and Radio Wave Propagation of USSR Academy of Sciences welcomes all research Institutes and groups to take part in this mission.

CHRISTENSEN-DALSGAARD: We look forward very much to the results of these observations.

THE ELECTRODYNAMICS OF NEUTRINOS IN DISPERSIVE MEDIA

V.N. Oraevsky, V.B. Semikoz

1. Neutrinos interacting with the vacuum of vector bosons and leptons possess important vacuum electromagnetic characteristics: viz., the anomalous magnetic moment $\Delta\mu_\nu^{vac}$ and the mean-square radius of charge distribution $<r^2>^{1/2}$. As a result, neutrinos, just like neutrons, may interact with an external magnetic field.

The magnitude of any electromagnetic characteristic of a neutrino is proportional to the Fermi weak interaction constant $G_F = 10^{-5}/m_p^2$*), so that in a vacuum electromagnetic interactions determine only radiation corrections (RC) to the principal Born contribution to the cross sections of weak processes.

However, in a dispersive medium (DM), say in a plasma, the electromagnetic interaction of a neutrino with charged particles and the external magnetic field is much more pronounced. Furthermore, in the low-energy range, where the neutrino de Broglie wavelength $\lambda_\nu = p^{-1}$ is greater than or equal to the mean distance between particles, $n_0^{-1/3}$, the contribution of electromagnetic processes with the neutrinos involved may be higher than the contribution due to ordinary weak interactions, i.e. the electromagnetic contribution can no longer be considered a small correction, as in the case of RC in a vacuum.

2. In the case of DM, a neutrino acquires new characteristics; an induced electric charge [1,2] and a magnetic moment [3], depending on the frequency spectrum and on the polarization of the external field, and also on the polarization of electromagnetic radiation of the neutrino itself.

Special emphasis is given to the polarization origin of electric charge, due to an inhomogeneous electron concentration around a neutrino.

The magnitude of the induced electric charge

* Here we use the system of units in which h = c = 1.

E. R. Priest and V. Krishan (eds.), Basic Plasma Processes on the Sun, 35–38.

$$e_\nu^{ind} = \frac{G_F(1+4\xi)e}{4\pi d\sqrt{2}\, r_D^2}$$

per electron charge e_ν^{ind}/e) is equal approximately to 10^{-16} for metal or the Sun, and 10^{-8} for a collapsing star. Here $\xi = \sin^2\theta_w$ is the parameter of Weinberg-Salam's model ($\xi \sim 0.23$ from experiment); $\alpha = e^2 = 137^{-1}$; r_D is the Debye radius of the plasma.

The concept of neutrino "electric charge" is illustrated by considering the electromagnetic part of the total amplitude of elastic neutrino scattering from nuclei in a plasma. It is shown that, if the contribution due to neutral currents is neglected, the situation is completely analogous to Mott's scattering of an electron from a Coulomb centre in the same medium [4].

We obtain the total electromagnetic vortex of a neutrino in an isotropic medium, which determines the corresponding electromagnetic current [5]. The expression for the electromagnetic current is used to derive lower-order multipole moments of a neutrino in an isotropic medium [6]. We consider also a specific neutrino electromagnetic moment (Zeldovich anapole) caused by the non-conservation of parity [7].

3. The many-particle equations of motion of neutrinos and electrons - the relativistic kinetic equations (RKE) - are derived in [8]. The most important feature of electrodynamics of neutrinos in a DM is the presence in the RKE of the contribution due to the self-consistent electromagnetic field, which has never been taken into account in the conventional treatment of relativistic neutrino kinetics. The RKE for neutrinos and electrons, obtained by the Bogolyubov method using Feynman diagrams for the νe-system, contain macroscopic effects of spatial parity violation due to pseudovector currents in the scattering. We derive also the integral of νe collisions under dynamic polarization of the medium. This integral (of the Lenard-Balesku type) is a typical integral occurring in plasma electrodynamics, and is of great importance for applications in the theory of collapsing stars.

4. The applications concerning elementary processes in a DM with neutrinos are considered. The cross-section of elastic neutrino scattering from spinless nuclei under the collapse of a presupernova star is calculated [4]. It is shown that a change in the electromagnetic structure of electron neutrinos in a DM reduces (screens) the cross section of νA scattering by approximately a factor of 10 as compared with the standard Born cross section calculated in the standard model for the collapse conditions. As for ν_μ-neutrino, the cross section of νA-scattering changes slightly in comparison with the standard Born cross section. Therefore, in a dense plasma the diffusion of neutrinos of distinct flavours occurs differently.

5. In the symmetric model of electroweak interactions with right currents we predict a change of neutrino helicity in a plasma (left neutrinos $\leftrightarrows$ right neutrinos) without including the vacuum magnetic moment $\Delta\mu_\nu^{vac}$ of the neutrino [9].

A change of neutrino helicity in a supernova star results in a reduction of the observed left neutrino flux and in a restriction for the mixing angle of a left- and right W-boson ($\sin 2\zeta \leq 0.4E/p_F$) which is the important parameter of a symmetric model. Here E is the neutrino energy, p_F is the Fermi momentum of electrons with density $n_0 = p_F^3/3\pi^2$. In the symmetric mode a neutrino has two helicity states due to the existence of an effective neutrino rest mass [10]:

$$m_\nu^{eff} = \frac{G_F(\varepsilon^2-1)}{\sqrt{2}} m_e \int \frac{d^3p}{\varepsilon_p}\left[f_0^{(e)}(\varepsilon_p) + f_0^{(\tilde{e})}(\varepsilon_p)\right]$$

where $\varepsilon = (1+tg\zeta)/(1-tg\zeta)$ is the parameter of the same model; m_e is the electron rest mass; $f_0^{(e)}(\varepsilon_p)$, $f_0^{(\tilde{e})}(\varepsilon_p)$ are the Fermi distribution functions of electrons and positrons. In a vacuum ($f_0^{(e)} = f_0^{(\tilde{e})} = 0$) or without right currents ($\zeta = 0$, $\varepsilon = 1$) the mass m_ν^{eff} is zero.

6. Thus, in a dispersive medium a neutrino acquires electromagnetic properties which resemble, in many respects, the properties of an electron under the same conditions: the electromagnetic interaction of a neutrino with charged particles can be screened at distances of the order of the Debye radius; in a medium a neutrino can emit and absorb electromagnetic waves (Vavilov-Cerenkov radiation and Landau's damping of plasma waves), etc. For an electron which, besides electric charge, possesses vacuum weak charge (both vector and pseudovector) weak interaction with other electrons and nuclei is only a small correction to the electromagnetic interaction in a DM. On the contrary, for neutrinos (under the same conditions) the electromagnetic interaction with electric charges is stronger than the weak interaction caused by the existence of only a bare weak nuetrino charge in vacuum. This distinguishes the behaviour of low-energy neutrinos in a DM from the corresponding behaviour in a vacuum (or in a dielectric) where the electromagnetic interaction is only a RC as compared with the point Fermi weak interaction.

References

1. V N Oraevsky and V S Semikoz, *Physica* **142A**, 135 (1987).
2. V N Oraevsky and V B Semikoz, Proceedings of the Joint Varenna-Abastumani International School and Workshop on "Plasma Astrophysics" Sukhumi, USSR, 19-28 May 1986, p.321.
3. V B Semikoz, *Sov. J. Nucl. Phys.* **46**, 946 (1987).
4. L B Leinson, V N Oraevsky and V B Semikoz, *Phys. Lett.* **209B**, 80 (1988).
5. V N Oraevsky, A Yu Plakhov, V B Semikoz and Ya A Smorodinsky, *Zh. Eks. Theor. Fiz.* **93**, 1157 (1987) (*Sov. Phys. JETP*, **66**, 890 (1987)) Errata in *Zh. Eks. Theor. Fiz.* **95**, p.39 (1989); ibid. **95**, N6, (1989).
6. V. B Semikoz and Ya A Smorodinsky, *Zh. Eks. Theor. Fiz.* **96**, 35 (1989) (*Sov. Phys. JETP*, **68**, 20 (1989).

7. V B Semikoz and Ya A Smorodinsky, *Pis'ma Zh. Eksp. Theor. Phys.* **48**, 361 (1988) (*Sov. Phys. JETP Lett.* **48**, 399 (1988)).
8. V B Semikoz, *Physica* **142A**, 157 (1987).
9. V B Semikoz, *Pis'ma Zh. Eksp. Theor. Phys.* **49**, 254 (*Sov. Phys. JETP Lett.* **49**, 288 (1989)).
10. V N Oraevsky, V B Semikoz and Ya A Smorodinsky, *Phys. Lett.* **227B**, 255 (1989).

DISCUSSION

SIVARAM: (i) How would the presence of a large magnetic field affect your result for helicity flip?
(ii) Can you comment on the limits of the neutrino magnetic moment in these processes?

SEMIKOZ: (i) If a magnetic field is taken into account, the collective interaction of neutrinos with matter (introduced by us) would be more important than a simple inclusion of the vacuum magnetic moment of neutrinos which interacts with the magnetic field too. The helicity flip occurs both due to the vacuum magnetic moment and due to our effective magnetic moment. However the vacuum helicity flip is suppressed by dense matter (which is the main reason why the convective zone is considered) for this effect!), but our channel of helicity flip increases with electron density.
(ii) The laboratory upper limit (Reines *et al*) is known to be

$$\mu_\nu^{vac} \leq 10^{-10} \text{ Bohr magneton } (\mu_B).$$

The astrophysical restriction is stronger. The cooling times of white dwarfs (about ~ 400 stars) imply

$$\mu_\nu^{vac} \leq 10^{-11}\mu_B \quad \text{(S Blinnikov (USSR), 1988).}$$

Also the restriction of a possible right (sterile) neutrino flux emitted by SN 1987A, which could be converted into left neutrinos at the Earth (after crossing the Galaxy's magnetic field) gives the limit (D Nötzold, 1988)

$$\mu_\nu^{vac} \leq 10^{-12}\mu_B.$$

ORAEVSKY: In a medium an induced charge of the neutrino is created: its value is proportional to the Fermi Constant and the electron density. There is also an induced magnetic moment, its value for the Supernova is ~10^{-10} of a Bohr magneton, which is 9 orders of magnitude higher than the vacuum value. In a medium, due to coherent interaction with particles, an induced neutrino mass is created.

Problems of Solar Convection

Wasaburo Unno
Research Institute for Science and Technology
Kinki University
3-4-1 Kowakae, Higashi-Osaka-city, 577 Osaka
Japan

ABSTRACT. Kinetic energy of convection is transported inwards in the main body of convection zone. The temperature gradient becomes super-radiative at the top of the overshooting zone. These two effects make the solar equilibrium model much less sensitive to the assumed mixing length in Xiong's eddy diffusion theory. Observed brightening of downflow at high level in the surface region over intergranular lanes seems to be consistent with the overshooting model. Momentum transport by convective motion is shown to be crucial in pushing down magnetic flux tubes against buoyancy. Also, subadiabatic layers are possibly formed temporarily in the middle of the convection zone, exciting oscillations and generating chaotic motions.

1. Hydrodynamics of Solar Convection

Solar convection is a three dimensional turbulent motion on which rotation, circulation, and oscillations are superposed and in which magnetic fields are embedded. Equations governing the motion are well known, but the complete treatment of the equations are still very difficult. In numerical studies (e.g. Nordlund, 1978; Glatzmaier, 1987), we have to choose a finite mesh size below which eddy diffusivities should be assumed. Different approximations with proper size of the resolution are appropriate for different problems.

The local mixinglength theory (Vitense, 1953) assumes the representative eddy size to be of the order of the scale height and neglects structures below the mixing length. It is the most convenient theory for constructing equilibrium model. Justification of the theory, however, was given much later by Antia, Chitre, and Pandey (1981) and Narashimha and Antia (1982). They showed that eddies of larger horizontal size essentially occupy deeper levels, if eddy diffusivities are taken into account. However, sensitivity of the equilibrium model on the assumed mixing length makes the local mixinglength theory unsatisfactory. Xiong's (1979) nonlocal mixinglength theory (better be called as the eddy diffusion theory) is advisable for constructing solar convection zone model (Unno, Kondo and Xiong, 1985). Xiong's theory differs from Vitense's theory in that it includes eddy diffusive transport of kinetic and thermal energies and temperature-velocity correlations and in that it can treat

E. R. Priest and V. Krishan (eds.), Basic Plasma Processes on the Sun, 39–44.

time-dependent problems. The model is insensitive to the assumed mixinglength, and it agrees well with observations. The hydrodynamic structure of the models has been studied by Unno and Kondo (1989). There are two reasons for the insensitivity of the model to the assumed mixing length. One is the structure of the overshooting layer in which the upper part shows larger temperature gradient than the radiative one (super-radiative) because the convective energy flux is negative (downward). The other reason is that the kinetic energy transport is directed downward in the main body of convective zone. Both of these effects require larger efficiency of convective transport (large Nusselt number) and therefore larger mixing length. Since the structure of the model saturates to the adiabatic one with the increase of mixing length, the largest possible length of the order of the distance to edge of the convection zone should be chosen. The characteristics of overshooting convection and the role of kinetic energy transport will be discussed in the following sections.

2. Structure and Dynamics of Overshooting Convection

Suemoto, Hiei, and Nakagomi (1987; 1989) observed brightness variations in continuum in the CaII K line wavelength region in its outer wing($|\Delta\lambda| > 3\AA$) and inner wing ($3\AA > |\Delta\lambda| > 0.5\AA$) for granules and intergranule space. They found that excess brightness of granules (hot layer) does not extend beyond $3\AA = |\Delta\lambda|$ or $\tau = 0.3$ and there are hot regions at high levels ($\tau < 0.1$) in intergranular lanes. The convective flux is downward for $\tau < 0.3$ as expected from Xiong's theory. The hydrodynamics of high levels of intergranular lanes seems to resemble that of plate-tectonics in which high density plates sink liberating gravitational energy through stable stratification. A similar situation was discussed by Ribes and Unno (1986) to interpret facular brightness by the downdraft along the magnetic flux tube. There are also some other problems associated with the convective overshooting. The layer is important not only for providing boundary conditions to hydrodynamics of deeper layers but also for exciting large scale motions. Kondo and Unno (1982; 1983) studied the penetrative convection for idealized two layer models composed of convectively unstable and stable layers (represented by subscripts 1 and 2, respectively, in Figure 1) with different constant viscosity and conductivity in each layer. The critical Rayleigh number defined by the unstable layer is greatly reduced from the usual value of the order of 10^3 for rigid and free boundaries of fixed temperature, if the viscosity of stable layer is very small. The penetration depth into weakly stable layer (small β_2) is large for eddies with large horizontal size, reducing viscous dissipation. With significant thermal conductivity, temperature fluctuations are smoothed out in up and downdrafts, giving rise to overstability of long period. No overstability of supergranulation has been reported, perhaps because upper layer is very stable. The physics of penetrative convection may better be applied to the main body of convective zone below the hydrogen ionization zone, as will be discussed later.

Now, coming back to the problem of kinetic energy transport, we should estimate how large the effect should be. With the help of the gradient diffusion approximation, the kinetic energy flux, $F_c^{(K)}$, is estimated to be

$$F_c^{(K)} = \frac{1}{2}\overline{\rho v^2 v_z} = \frac{1}{2}\overline{\rho (v^2)' v_z} = -\frac{\rho \kappa_e}{2}\frac{\partial \overline{v^2}}{\partial z}, \tag{1}$$

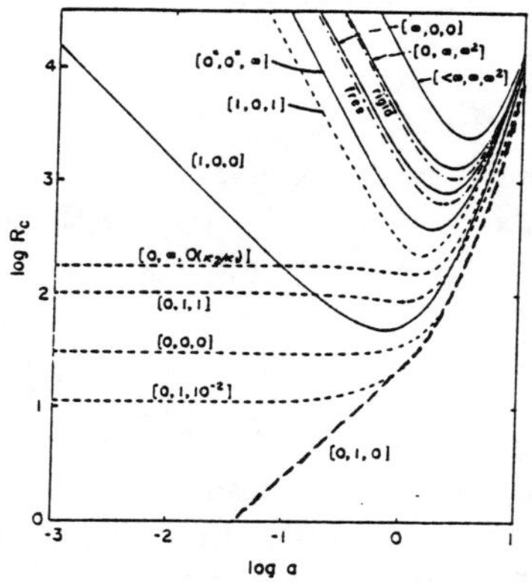

Fig. 1: The critical Rayleigh number R_c, ($R \equiv \alpha\beta_1 g z_1^4/\kappa_1\nu_1$), for various values of $[\nu_2/\nu_1,\ \kappa_2/\kappa_1,\ |\beta_2/\beta_1|]$, as functions of horizontal wavenumber a. Dashed curves represent cases of overstability, where $\mathrm{Re}\lambda = 0$ and $\mathrm{Im}\lambda \neq 0$, and solid curves the case of exchange of stability. Note low R_c values for $\nu_2 = 0$ and $\beta_2 = 0$ (Kondo and Unno, 1983).

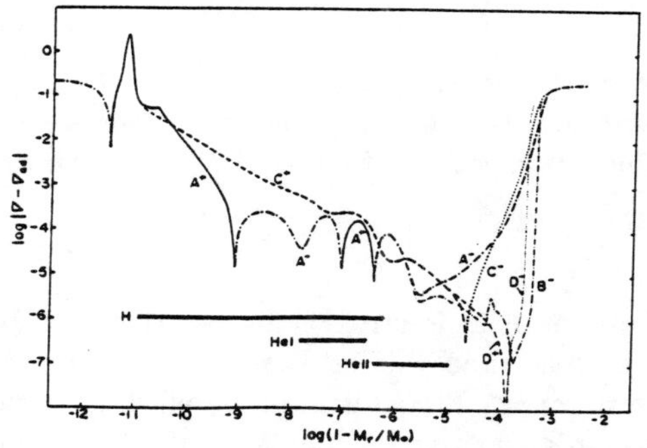

Fig. 2: $(\nabla - \nabla_{ad})$ in the solar convection zone. Case A and B (C and D) are for longer (shorter) mixing length but differ in the lower boundary conditions. A^- etc. show the internal subadiabatic layers (Unno and Kondo, 1989).

where the eddy conductivity is estimated to be

$$\kappa_e = 0.25(\overline{v^2})^{1/2} l_e, \tag{2}$$

(Nakano et. al, 1979), while the thermal energy flux, $F_c^{(th)}$, is given by

$$F_c^{(th)} = \overline{\rho W v_z} = \overline{\rho W' v_z} = \overline{c_p T' v_z} = \alpha^{-1}\gamma(\gamma-1)^{-1}\overline{\rho(v^2)^{3/2}}. \tag{3}$$

We obtain the ratio, $F_c^{(K)}/F_c^{(th)}$, to be (Unno and Kondo, 1989)

$$\frac{F_c^{(K)}}{F_c^{(th)}} = -0.02\alpha^2 \tag{4}$$

where α denotes the mixing length-scale height ratio. Numerical experiments of highly nonlinear compressible convection with a fixed flux boundary condition by Hurlbert et al. (1984), by Yamaguchi (1985), and by Chan and Sofia (1986) give the ratio to be negative and of some 4–8%. The value increases with increasing mixing length (Roxburg, 1987). In Unno and Kondo (1989), the ratio amounts up to some 20% in the lower part of the convection zone. The effective α value adopted may have been a little too large.

The inward flux of the kinetic energy in the result of continuity with faster speed in a narrower downdraft area and slower speed in a wider updraft area. In the gradient diffusion approximation, a faster downdraft is the result of larger velocity fluctuations in the parent region lying one mixing length above. In the same context, therefore, the confinement of magnetic tubes in the convection zone can be discussed by means of net momentum transport of anisotropic turbulence. The average vertical momentum transport per unit horizontal area $\overline{\rho v_z v_z}$ can be estimated in the gradient approximation as

$$\overline{\rho v_z v_z} = -\overline{\rho}\nu_e \frac{\partial(\overline{v^2})^{1/2}}{\partial z} = -\frac{\alpha}{5R_e}\overline{\rho v^2}, \tag{5}$$

where the effective Reynolds number R_e may be taken to be 10. A magnetic flux tube of the field strength B and the thickness d receives magnetic buoyancy balancing the above momentum transport and satisfies the following relation in equilibrium,

$$\frac{B^2}{8\pi} = \frac{1}{5R_e}\frac{\ell_e}{d}\overline{\rho v^2}. \tag{6}$$

If the field is stronger or the thickness is larger, the flux tube is lifted up. A thin weak flux tube will be pushed downward and finally find a rest position, flattened, if $Bd = const.$. Flat tubes will be piled up, forming a thick tube which will then expand or float to upper layers. At a depth of 1.5×10^{10}cm from the surface where $\rho = 7 \times 10^{-2}$gcm^{-3} and $v^2 = 6 \times 10^7$cm^{2}s^{-2}, the equilibrium field of $1.5 \times 10^3 G$ is given by equation (5) for a flux tube of $d = \ell_e$. Thus, the momentum transport by convection is an important factor in the dynamo mechanism.

3. Subadiabatic Layer in Convection Zone

With a large mixing length, subadiabatic layers appears in the middle of convection zone (Figure 2). It may be the result of too large mixing length assumed. However, such a stability reversal is seen in the Boussinesq convection for large Rayleigh numbers (Herring, 1963). Urata (1986) made numerical simulation of three-dimensional time-dependent Boussinesq convection with Rayleigh number of 2×10^4 and the horizontal aspect ratio of 2. Starting with the linear Benard cell pattern, the motion soon changes the mean temperature profile to be negative (subadabatic) in the central portion sandwiched with steep conductive temperature-gradient regions on both upper and lower edges. Then, the convective flux becomes oscillatory with the frequency of gravity wave mode standing in the stable stratification. The oscillation becomes much reduced after some time, Then, chaotic variations start to grow to large amplitudes. The chaotic variation are initiated with the growth of a new mode which shows large vertical vorticity without any initially imposed angular momentum to the system, a mode which was first found by Lopez and Murphy (1986). Coupling of three kinds of motion, Benard convection, gravity wave, and horizontal vorticity seems to be the cause of chaotic variations (Figure 3a and 3b). The subadiabatic layer in the middle of the solar convection zone, if it happens even temporarily, may be responsible for some of the observed oscillations or chaotic variations. Vertical vorticity mode should be important to convert toroidal field into poloidal field. It is interesting to note that vertical vorticity as well as sink and source patterns are visible in the SOUP flow vectors (Simon et al., 1988).

4. Nonlinear Magnetoconvection

Nonlinear magnetoconvection was studied and applied to solar convection problems by Rudraiah, Kumudini, and Unno (1985). It is found that turbulent magnetoconvection

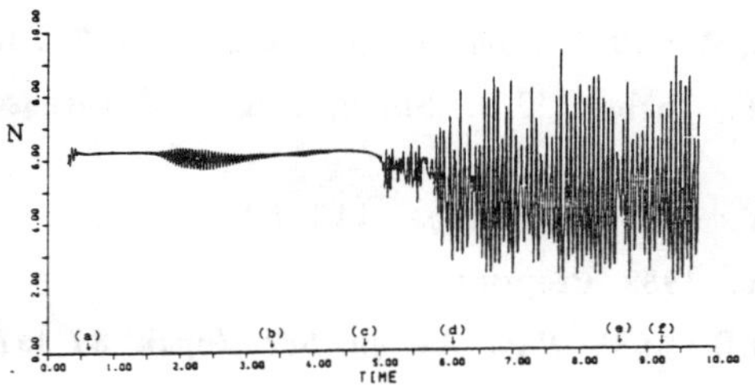

Fig. 3a: Time series of horizontally averaged total heat flux for $Pr = 1$ (Urata, 1987).

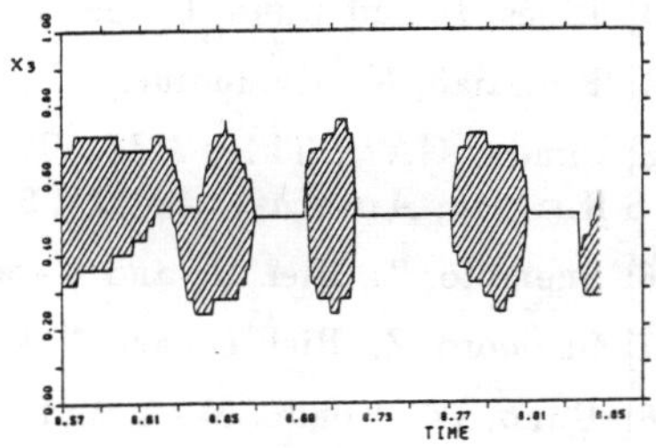

Fig. 3b: Variations of temporary stable (hatched) layer around the phase (e) in Fig. 3a (Urata, 1987).

is characterized by the effective Rayleigh number of about 4 times the critical Rayleigh number and by the effective Nusselt number of about 2. The amplification of the field strength is estimated to be of factor 5 in each cell ($\sim$ one scale height) so that e.g. 50G fields at the bottom of the convection zone could be amplified to a few thousand gauss until the magnetic flux tube reaches the surface layer. Also, darkness of sunspots can be interpreted by the nonlinear magnetoconvection with the effective Nusselt number of about 2. Another important effect of convection is that faster speed of down flow than upflow pushes down a magnetic tube very effectively, as discussed before. A flux tube can stay in the convection zone long enough for dynamo mechanism to work.

References

[1] Antia, H.M, Chitre, S.M. and Pandey, S.K.: 1981, *Solar phys.*, **70**, 67.

[2] Chan, K.L. and Sofia, S., 1986, *Astrophys. J*, **307**, 222.

[3] Glatzmaier, G.A.: 1987, *The Internal Sol. Ang. Vel.*, eds. B. R. Durney and S. Sofia, (D. Reidel Pub. Co.), p. 233.

[4] Hurlbert, N.E., Toomre, J., and Massaguer, J.M., 1984, *Astrophys. J.*, **282**, 557.

[5] Herring, J.R.: 1963, *J. Atmos. Sci.*, **20**, 325.

[6] Kondo, M.-A., and Unno, W.: 1982, *Geophys. Astrophys. Fluid Dynamics*, **22**, 305.

[7] Kondo, M.-A., and Unno, W.: 1983, *Geophys. Astrophys. Fluid Dynamics*, **27**, 229.

[8] Lopez, J.M. and Murphy, J.O.: 1985, Monash University Preprint.

[9] Nakano, T., Fukushima, T., Unno, W. and Kondo, M.: 1979, Publ. Astron. Soc. Japan, **31**, 713.

[10] Narashimha, D., and Antia, H.M.: 1982, *Astrophys. J.*, **262**, 358.

[11] Nordlund, A., 1978, Astron. Papers dedicated to B. Strömgren, eds. A. Reiz and T. Andersen, Copenhagen Univ. Obs., Copenhagen.

[12] Roxburg, I.A.: 1978, *Astron. Astrophys.* **65**, 281

[13] Ribes, E. and Unno, U.: 1976, *Astron. Astrophys.*, **53**, 197.

[14] Rudraiah, N., Kumudini, V., and Unno, W.: 1987, *Publ. Astron. Soc. Japan*, **37**, 183.

[15] Simon, G.W., Title, A.M., Topka, K.P., Tarbell, T.D., Shine, R.A., and Ferguson, S.H.: 1988, *Astrophys. J.*, **327**, 964.

[16] Suemoto, Z., Hiei, E., and Nakagomi, Y.: 1987, *Solar Phys.*, **112**, 59.

[17] Suemoto, Z., Hiei, E., and Nakagomi, Y.: 1989, Preprint.

[18] Unno, W., Kondo, M.-A. and Xiong, D.R.: 1985, *Publ. Astron. Soc. Japan*, **37** 197.

[19] Unno, W., and Kondo, M.-A, 1989, *Publ.Astron. Soc. Jap.*, **41**, 197.

[20] Urata, K., 1987: *Prog. Fluid Mech.*, **2**, 150.

[21] Vitense, E.: 1953, *Z. Astrophys.*, **32**, 135.

[22] Xiong, D.R.: 1979, *Acta Astron. Sinica*, **20**, 238., [= *Chines Astron.*, **4**, 234, 1980]

[23] Yanaguchi, S.: 1985, *Publ. Astron. Soc. Japan*, **37**, 735.

DISCUSSION

WEISS: Recently there have been some impressive numerical simulations of three-dimensional compressible convection (Nordlund & Stein; Brummell, Cattaneo, Hurlburt & Toomre) which also show a downward kinetic energy flux associated with rapidly sinking plumes. Do you think it will be possible to calibrate mixing-length type theories by comparison with such models run on supercomputers?

UNNO: One way is to compare the kinetic energy transport to the total convective flux.

DALSGAARD: The principal quantity, as far as solar structure is concerned, obtained from mixing length theory is the entropy jump across the superadiabatic part of the convection zone. It is interesting that the detailed calculations give roughly the same value as obtained from mixing length theory, once this is calibrated to obtain the correct solar radius.

UNNO: However, use of the solar radius as a calibration is not entirely satisfactory from a theoretical point of view.

MOGILEVSKIJ: What is the difference in convection between active and quiet regions?

UNNO: The active-region magnetic field (except for sunspots) has practically no influence on large eddies like supergranule cells. Granules are smaller in size and elongated vertically by a magnetic field. For the strong field of a sunspot, heat transport is much reduced, if the temperature is fixed at the bottom of a convection cell.

MECHANISMS FOR DYNAMO MODE EXCITATION

P. HOYNG
Laboratory for Space Research
Sorbonnelaan 2, 3584 CA Utrecht
The Netherlands

(extended summary)

Dynamo theory of the solar magnetic field finds itself in a period of reappraisal since about 1980. There are two reasons for this: (1) The required increase of the angular speed Ω with decreasing r is neither found in numerical simulations nor measured by helioseismology. (2) Mean field modeling shows often little regard for inherent physical restrictions and selfconsistency.

In recent years, attention has shifted from the convection zone to the boundary layer between convection zone and radiative interior as a more likely location for the solar dynamo (Rai Choudhuri 1990a,b). The topic of dynamo mode excitation, however, is model-independent and starts from the observation that kinematic $\alpha\Omega$ dynamos are strictly periodic in time. Only the fundamental mode is excited and it has an infinite phase memory, which is not satisfactory.

The reader is referred to the proceedings of the Kiev symposium for the complete text of my review (Hoyng 1990). Here I shall restrict myself to an extended summary, giving only new references that came to my attention during the last half year.

1. Mode Excitation in Mean Field Theory

The transport properties of the mean field $\langle \mathbf{B} \rangle$ are determined by the dynamo equation. Its basic form is:

$$\partial_t \langle \mathbf{B} \rangle = \mathbf{D} \cdot \langle \mathbf{B} \rangle + \mathbf{F} \; ; \qquad \mathbf{D} = \nabla \times (\mathbf{v}_0 \times + \alpha - \beta \nabla \times) \, . \tag{1.1}$$

Here, $\mathbf{v}_0$ is the mean flow, and α and β are the usual dynamo coefficients related to the turbulent velocity $\mathbf{v}$ superposed on $\mathbf{v}_0$. Tradition has it that $\langle \mathbf{B} \rangle$ is a spatial average over an unspecified scale in between the radius R of the dynamo and the eddy size λ_c. The term $\mathbf{F}$ collects all terms due to the fact that, among other things, $\langle f \langle g \rangle\rangle \neq \langle f \rangle \langle g \rangle$. $\mathbf{F}$ is always neglected on the ground that it is formally of

E. R. Priest and V. Krishan (eds.), Basic Plasma Processes on the Sun, 45–49.

order $\lambda_c/R \ll 1$. The growth rate of an eigenmode of eq. (1.1) is determined by the dynamo number ($\Delta\Omega$ is the difference in angular velocity in the dynamo):

$$D = \alpha\Delta\Omega R^3/2\beta^2 \quad ; \qquad D_c \simeq (2\pi)^3 \ . \tag{1.2}$$

(a). Instability ($D \geq D_c$). One or more eigenmodes will grow until nonlinear effects become important. A popular idea has been that these reduce D to D_c, *e.g.* by reducing α. Only the fundamental mode is then excited. The dynamo has a single frequency ω and an infinite quality factor $Q = \omega/\Delta\omega$. The fine tuning of D may not be stable, leading to complicated nonlinear interactions between magnetic and fluid modes. In that case the dynamo has several frequencies, each with a finite Q.
(b). External forcing ($D < D_c$). All eigenmodes are damped so that $<\mathbf{B}> \rightarrow 0$. Subcritical mode excitation may occur if $\mathbf{F}$ in eq. (1.1) is sufficiently large. $\mathbf{F}$ depends on $\mathbf{v}$ and has the character of a random forcing term. This would lead to excitation of many eigenmodes, each with a finite Q.

Another external forcing mechanism could be a relic field in the radiative core, which imposes a nonzero boundary condition on eq. (1.1) at the base of the convection zone. This boundary condition could even be periodic, in which case the phase stability of the dynamo at that frequency may be very high.

2. Observations

Variations in the 22-yr period. From the epochs of sunspot extrema one finds $\delta P_{rms}/P \sim 0.1$, where P = mean half cycle period. It is not known if subsequent period variations δP are independent or whether the cycle is rather a passive and noisy reflection of a high Q oscillator in the solar core. There are marginal indications for a random walk in phase (which would exclude a core oscillation), but the data set (*AD* 1610 - present) is really too small to draw a conclusion. The phase stability of the 22-yr cycle may ultimately be determined from the ^{14}C data and of the ^{10}Be record discovered in Greenlandic ice cores.
Longer periods. Several proxy records indicate irregular modulations in the level of the activity cycle with a time scale between one and a few hundred years. From an analysis of the 88-yr cycle Feynman and Gabriel (1990) have argued that the solar dynamo is operating in a region close to the transition between period doubling and chaos. Attempts have been made to determine attractor dimensions from the sunspot and ^{14}C records but the problem is again the length and quality of the data set. It is recalled that Barnes *et al.* obtained a remarkable similarity with the observed solar cycle, including Maunder-type minima, by simulating yearly sunspot numbers from narrowband Gaussian noise.
Periods shorter than 22 year. Stenflo and coworkers have analysed 25 years of synoptic magnetograph data and discovered that the surface magnetic field exhibits well-defined periods down to 1.5 year. This has been interpreted as excitation of overtones of the dynamo equation. Gokhale and Javaraiah (1990) have extended

the dataset by using magnetic field data which they simulated from the Greenwich sunspot data (but they have not yet analysed the power diagram at even ℓ).

3. Nonlinear mode excitation

I shall restrict myself to nonlinear mean field models. Three types of nonlinearity have been considered since the very beginning: quenching of the α-effect, flux losses through buoyancy, and suppression of the *differential* rotation. Early nonlinear studies established the existence of stable nonlinear dynamo waves, but did not systematically explore the behaviour of the dynamo as a function of position in parameter space. That was only done after about 1980, when mean field modeling fell in love with nonlinear dynamics. Initial studies in this area truncated all spatial dependence by replacing ∇ by ℓ^{-1}. In return for extreme simplification they discovered that the solar dynamo might have chaotic properties. Ruzmaikin has suggested that the 'grand minima' in the solar cycle may correspond to a strange attractor, entirely in line with the fashion of that time. It took many years before more realistic attempts were undertaken in which the spatial structure is taken into account. Three such models have recently been published (Schmitt and Schüssler 1989, Brandenburg *et al.* 1990, and Belvedere *et al.* 1990).

Nonlinear mean field theory is still in a very early stage of development. On the one hand, very interesting dynamical behaviour is discovered, sometimes strikingly similar to what is observed, but on the other hand the solutions are often very sensitive to the adopted functional form of the nonlinearity and/or the values of parameters. To make things worse, the nonlinearities are always introduced phenomenologically. It is therefore not yet possible, for example, to predict the magnitude of $<\mathbf{B}>$. Dynamical selfconsistency is the outstanding problem in nonlinear mean field theory, and in my view the prospects for progress are not very bright.

4. Mode excitation by external forcing

External forcing may be effectuated through boundary conditions or through fluctuations in the turbulent convection. A fossil field in the core would impose a boundary condition $<\mathbf{B}> = \mathbf{B}_{core}$ at the base of the convection zone. The field will be smeared out by the turbulent dynamo, but there remains a net polarity and intensity asymmetry in the activity cycle. Models have been constructed along these lines, and the conclusion is that $\mathbf{B}_{core} \lesssim 1$ G at the bottom of the convection zone. It has also been suggested that the solar core performs a 22-yr torsional oscillation with the poloidal component of the field acting as a spring. The dynamo in the convection zone would then be in a state of forced oscillation. This remains an interesting idea, although a credible model has never been elaborated.

External forcing by fluctuations in the turbulence (the term $\mathbf{F}$ in (1.1)) has received little attention. One possibility is that a *sudden burst* in the cyclonic convection causes a jump in α, and it has been shown that this may induce a reversal of the (mean) geomagnetic field in linear theory. The conventional wisdom of today is that things like reversals in α^2 dynamos and aperiodicity in $\alpha\Omega$ dynamos are caused by nonlinearities. However, even in linear theory the periods of adjacent cycles of an $\alpha\Omega$ dynamo must differ, as the realisation of the turbulence is different. It should be possible to formulate linear mean field theory in the same way as a scalar diffusion process, with a continuous loss of memory.

This is again the effect of the random forcing term $\mathbf{F}$, but it is extremely difficult to evaluate its magnitude. This problem has recently been tackled by using *ensemble* rather than spatial averaging. It has been shown that the large scale field of an $\alpha\Omega$ dynamo behaves as a *quasiperiodic* random function. The period stability of the solar dynamo is currently being evaluated. A rough estimate gives $\delta P_{rms}/P \sim 1$, highlighting the potentially enormous influence of $\mathbf{F}$ in the linear theory.

References

Belvedere, G., Pidatella, R.M. and Proctor, R.M.E.: 1989, *Geophys. Astrophys. Fluid Dynamics*, to appear.

Brandenburg, A., Moss, D. and Tuominen, I.: 1989, *Geophys. Astrophys. Fluid Dynamics*, to appear.

Feynman J. and Gabriel, S.B.: 1990, *Solar Physics*, to appear.

Gokhale, M.H. and Javaraiah, J.: 1990, *M.N.R.A.S.*, to appear.

Hoyng, P.: 1990, in *Solar photosphere: Structure, Convection, and Magnetic fields*, ed. J.O. Stenflo, Kluwer (Dordrecht), p. 359.

Rai Choudhuri, A.: 1990a, these proceedings.

Rai Choudhuri, A.: 1990b, *Astrophys. J.*, to appear.

Schmitt, D. and Schüssler, M.: 1989, *Astron. Astrophys.* *223*, 343.

DISCUSSION

VAN BALLEGOOIJEN: What are the prospects for direct numerical 3D simulations, without invoking the mean-field approximation?

HOYNG: These 3D simulations fail at present to reproduce an acceptable solar cycle. One of the reasons is suspected to be inadequate spatial resolution. The contribution of subgrid processes must always be parametrised (one might say that a kind of mean field approximation is implicit, albeit on a rather small spatial scale). It is my understanding that there is no hope of being able to resolve eddies of the required size in the foreseeable future in solar dynamo models.

WEISS: Modern supercomputers are powerful enough to model selfconsistent nonlinear dynamos in idealized geometries with parametrized diffusion coefficients. However, it seems premature to embark on such a detailed computation until we understand more about the location of the dynamo and the variation of angular velocity near the base of the convection zone.

HOLLWEG: Is the 153-day periodicity of solar flares a proxy indicator of yet another fundamental dynamo period?

HOYNG: It could be.

SILVARAM: Would you clarify the role of ^{10}Be in detecting variations in activity? What range of periods does it cover and how does it compare with ^{14}C?

HOYNG: ^{10}Be and ^{14}C are produced by (solar) cosmic rays interacting with the Earth's atmosphere. ^{10}Be gets attached to aerosols and is removed by precipitation after a typical time of 1 or 2 years (much faster than ^{14}C). Because of its half-life time of 1.5×10^6 year (as compared to 5730 yr for ^{14}C) the ^{10}Be proxy record in ice cores reaches much farther back in time, at least in principle. Variations in ^{14}C and ^{10}Be records over the last 10^4 year show a very encouraging correlation.

AL-KHASHLAN: If the number of relevant dynamical variables is very low what is the behaviour of the dynamo?

HOYNG: Models for the solar dynamo with only a few dynamical variables have been published for example by Ruzmaikin (Comm. Astrophys. 9 (1981)85), Krause and Roberts (Adv. Space Res. 1 (1981)231) and by Weiss, Cattaneo and Jones (Geophys. Astrophys. Fluid Dyn. 30 (1984)305). The nature of dynamo action is field amplification by shear flows (of the right kind), to balance loss processes such as ohmic decay, losses through buoyancy, etc.

LOCATING THE SEAT OF THE SOLAR DYNAMO

Arnab Rai Choudhuri
Department of Physics
Indian Institute of Science
Bangalore 560012. India

ABSTRACT. By analyzing the hypothesis that the solar dynamo operates in a thin layer at the bottom of the convection zone, we conclude that this hypothesis can fit observational facts only if there is turbulence with the length scale of a few hundred kilometers in and around the dynamo region.

1. INTRODUCTION

The traditional view that the solar magnetic fields are generated by an $\alpha\omega$ -dynamo operating in the convection zone has been subject to criticisms in recent years because of the following difficulties: (i) magnetic buoyancy would remove any magnetic flux from the convection zone quickly (Parker, 1975), and (ii) self-consistent dynamical models of convection zone dynamos are unable to reproduce the surface rotation pattern and the butterfly diagram simultaneously (Gilman, 1983). In response to these difficulties, it has been suggested that the dynamo operates in a thin layer at the bottom of the convection zone or underneath it rather than operating in the main bulk of the convection zone. The overshoot region underneath the convection zone is supposed to be a suitable location for the operation of the solar dynamo, since the subadiabatic gradient there can suppress the magnetic buoyancy (Spiegel and Weiss, 1980; van Ballegooijen, 1982). Alternative effects such as thermal shadows (Parker, 1987) and the drag due to the meridional flow (van Ballegooijen and Choudhuri, 1988) have also been proposed in order to suppress magnetic buoyancy in a thin layer at the bottom of the convection zone. The differential rotation and the α-coefficient driving the dynamo action there are also expected to have correct signs, at least in the lower latitudes, so as to make the dynamo waves go in the correct direction (Gilman et al., 1989).

E. R. Priest and V. Krishan (eds.), Basic Plasma Processes on the Sun, 51–55.

Though the hypothesis that the dynamo operates in a thin layer at the bottom of the convection zone may solve old problems, it raises several new questions. Choudhuri and Gilman (1987) found that even if the flux is initially created at low latitudes at the bottom of the convection zone, the Coriolis force makes the flux rise parallel to the rotation axis and emerge at high latitudes poleward of where sunspots are seen. In § 2 we present recent work on the question whether the flux can be made to emerge at sunspot latitudes. Then § 3 discusses whether the dynamo operating in a thin layer can have the correct period and wavelength. We shall see that completely different arguments in § 2 and § 3 point to the existence of turbulence with the same length scale of few hundred kilometers in the dynamo region.

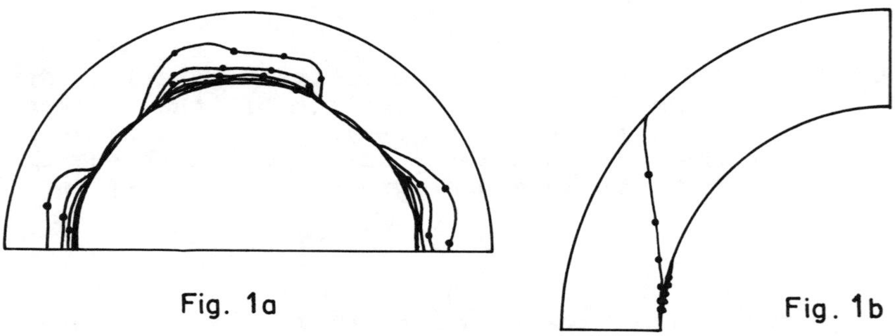

Fig.1. Evolution of a partially anchored flux ring with initial magnetic field 1.7 × 10^4 G. The successive configurations in (1a) or the dots in (1b) are at intervals of 14 days.

2. RISE OF FLUX THROUGH THE CONVECTION ZONE

The calculations of Choudhuri and Gilman (1987) were carried out for flux rings symmetric around the rotation axis. In order to investigate whether the Coriolis force could be suppressed by taking account of departures from axisymmetry, Choudhuri (1989) considered initial states in which the flux rings possessed non-symmetric undulations and had parts embedded in the stable layers underneath the convection zone. Though those parts remained anchored there, the upper parts of the flux rings rising in the forms of loops still moved parallel to the rotation axis. Figure 1 shows the evolution of such a non-axisymmetric flux ring starting from 5° latitude with an initial magnetic field of 1.7×10^4 G. Figure 1a is a polar plot of the successive configurations in the (r, φ) plane, whereas Figure 1b shows the trajectories in the (r, θ) plane of the highest and lowest points of the flux ring with the growing loops. We clearly see that we are still unable to get the flux out at the typical sunspot latitudes.

More recently, however, Choudhuri and D'Silva (1990) studied the interactions of the rising flux with the turbulence in the surrounding convection zone and found that the flux can be brought out radially if there is turbulent exchange of angular momentum between the flux tube and the surrounding fluid sufficiently rapidly, say in 10 hours or so. Such turbulent exchange is possible only (i) if the flux rises in the form of thin tubes of radius 500 km or so, and (ii) if there is enough turbulence at that length scale of a few hundred kilometers. It is then necessary to satisfy these two conditions if we want the flux created at the low latitudes to emerge at low latitudes on the surface.

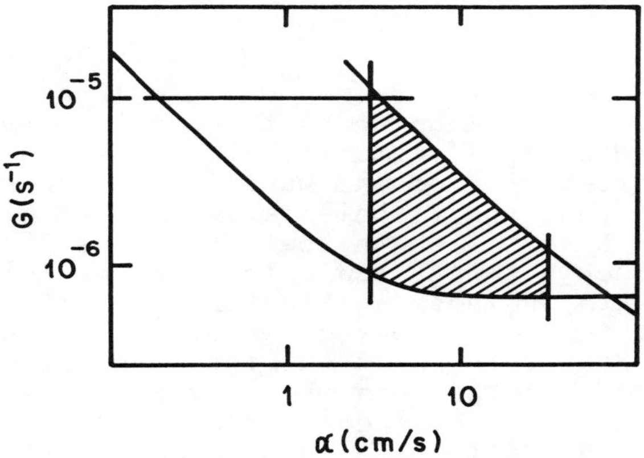

Fig.2. The allowed region (shown by shading) in velocity shear G vs. α-coefficient parameter space.

3. DYNAMO WITH CORRECT PERIODICITY AND WAVELENGTH

Since the dynamo region at the base of the convection zone is supposed to have a thickness of 10^4 km, Choudhuri (1990) has solved the kinematical dynamo equations for a rectangular slab of such thickness. By demanding a period of 22 years and a half-wavelength of 40° in the θ-direction, one can put some constraints on the allowed values of various parameters appearing in the kinematical dynamo equations. Figure 2 shows the allowed regions of the parameter space for the α-coefficient and the velocity shear G. The allowed values of different parameters suggest that the two source terms in the generation of the toroidal field (i.e. the differential rotation and the α-effect terms) have to be of the same order of magnitude, making the solar dynamo a truly $\alpha^2\omega$-dynamo. Since $\alpha\omega$-

dynamos produce very little poloidal field and the poloidal field of the sun is small, it used to be assumed that the solar dynamo is of the $\alpha\omega$-type. However, if the dynamo operates at the base of the convection zone, then even if it produces significant poloidal fields, such fields may not leak to the surface. Hence there is no difficulty in invoking an $\alpha^2\omega$-type solar dynamo, which may produce substantial poloidal fields in its region of operation.

Choudhuri (1990) also found that the α-coefficient and the turbulent diffusion η have to be constrained within about one factor of ten, the median values being $\alpha \sim 10$ cm s^{-1} (see Figure 2) and $\eta \sim 10^{10}$ cm^2 s^{-1}. On the basis of mixing length arguments, such values imply reasonable turbulent velocities of about 30 m s^{-1}, but rather small turbulent length scales of the order of 300 km.

4. CONCLUSION

We thus see that from two completely different considerations - by demanding that the flux created at low latitudes emerge at low latitudes, and by demanding that the dynamo have correct period and wavelength - we come to the same conclusion that the turbulence in and above the dynamo region should have a length scale of a few hundreds of kilometers. Such a length scale is much smaller than all the scale heights and goes against the current theoretical prejudice that the solar convection should mainly involve giant cells. We merely point out that the hypothesis that the solar dynamo operates in a thin layer at the base of the convection zone can be made consistent with observational facts only if such small-scale turbulence exists.

REFERENCES

Choudhuri, A.R.: 1989, Solar Phys. (in press).
Choudhuri, A.R.: 1990, Astrophys. J. (in press).
Choudhuri, A.R., and D'Silva, S.: 1990, Astron. Astophys. (submitted).
Choudhuri, A.R., and Gilman, P.A.: 1987, Astrophys. J. 316, 788.
Gilman, P.A.: 1983, Astrophys. J. Suppl. 53, 243.
Gilman, P.A., Morrow, C.A., and DeLuca, E.E.: 1989, Astrophys. J. 338, 965.
Parker, E.N.: 1975, Astrophys. J. 198, 205.
Parker, E.N.: 1987, Astrophys. J. 312, 868.
Spiegel, E.A., and Weiss, N.O.: 1980, Nature 287, 616.
van Ballegooijen, A.A.: 1982, Astron. Astrophys. 113, 99.
van Ballegooijen, A.A., and Choudhuri, A.R.: 1988, Astrophys. J. 333, 965.

DISCUSSION

PRIEST: Taking a field greater than 10^5G at the base of the convection zone would seem a good solution to your problem of making flux tubes rise through the sunspot zones rather than moving to the north of them. Why should not partial evacuation make the field strength larger than the equipartition value? After all, twenty years ago we did not dream that photospheric field strengths outside sunspots could be as large as 1kG due to such an effect. Also B Roberts and others feel that helioseismology observations may imply such larger fields at the base of the convection zone.

CHOUDHURI: It is true that a field of value 10^5G would have sufficiently strong magnetic buoyancy to make it come out radially and would solve the problem of making the flux emerge at sunspot latitudes. I, however, personally feel somewhat uncomfortable with the idea of such a large field at the base of the convection zone. Firstly, such a field would completely inhibit convection and the dynamo process. Secondly, bipolar regions are believed to be caused by parts of flux tubes coming through the surface whereas other parts remain anchored deep down. A field of 10^5G is too strong even to be anchored in the overshoot region (see Choudhuri [1989]). Still I agree that we know too little about the conditions at the base of the convection zone and perhaps we should keep our minds open on this issue.

GOKHALE: The emergence of field portions from the base of the convection zone to the surface creates a radial flux across the convection zone. What happens to it as time proceeds?

CHOUDHURI: This question has been discussed at length by Parker (1984,*Ap.J.* **281**,839; 1987 *Ap.J.* **312**,868). Though some flux in the neighbouring active regions may be able to escape due to reconnection, Parker believes that the major part of the flux is again pulled below the surface. Hence the repeated emergence of flux in the same active site may involve considerable reprocessing. When Parker considered the question whether a thin layer at the base of convection zone can store enough flux to explain all the flux that we see on the surface, he concluded that a very thin layer would not be sufficient for such storage even if we allow considerable reprocessing. This is another difficulty with the thin-layer dynamo idea which gets more aggravated if the flux is not reprocessed efficiently.

HELIOSEISMOLOGICAL DETERMINATION OF STRATIFICATION AND DYNAMIC PROCESSES IN THE SOLAR CORE

A.G. KOSOVICHEV
Crimean Astrophysical Observatory
334413, Nauchny, Crimea, USSR

Helioseismology provides a powerful tool for investigating stratification and dynamics of plasma in the Sun's interior. The observed frequencies of the 5-min oscillations depend on density, sound speed, chemical abundances and angular velocity distributions inside the Sun. These parameters are estimated using inversion techniques.

The inversion of the frequencies presented by Duvall *et al.* (1988) gives some evidence for material redistribution in the energy-generating core as compared with a standard solar model (Gough & Kosovichev, 1989). The helioseismological inversion results lead to conclusions that the evolutionary age of the Sun is perhaps 10 per cent greater than it is generally accepted and the neutrino flux from ^8B decay may be about 60 per cent of the standard model value as a result of mixing. Outside the solar core the sound speed is about 1 per cent greater than it is in the standard solar model. The result is evidence that the energy transport is less effective in the layers between 0.4 and 0.7 R (Christensen-Dalsgaard, 1988).

The rotational frequency splitting data of Duvall and Harvey (1984) and Brown and Morrow (1987) have been analyzed to determine the angular velocity in the Sun as a parametric function of radius and latitude using the asymptotic inversion technique (Kosovichev, 1988). The results show that the equatorial rotation rate slowly decreases below the photosphere. The main changes in rotation take place near the bottom of the convective zone where the solar dynamo might operate and in the solar core. The latitudinal differential rotation is close to the surface rotation in the convective zone for $r/R > 0.7$ and decreases rapidly in a region below the convective zone for $0.4 < r/R < 0.7$. The decrease of the latitudinal differential rotation might be connected with a Kelvin-Helmholtz instability (Dziembowski & Kosovichev, 1987).

REFERENCES

Brown, T.M. & Morrow, C.A. 1987. Astrophys. J. Lett., **314**, L21.
Christensen-Dalsgaard, J. 1988, *Seismology of the Sun & sun-like stars*, ESA Publ. SP-286, p. 431.
Duvall Jr, T.L. & Harvey, J.W. 1984. Nature, **310**, 19.
Duvall Jr, T.L. *et al.* 1988. Astrophys. J., **324**, 1158.
Gough, D.O. & Kosovichev, A.G. 1989, *Inside the Sun.* (in press).
Dziembowski, W. & Kosovichev, A.G. 1987. Acta Astron., **37**, 341.
Kosovichev, A.G. 1988. *Sov.* Astron. Lett., **14**, 145.

E. R. Priest and V. Krishan (eds.), Basic Plasma Processes on the Sun, 56.

INTERNAL LARGE-SCALE TOROIDAL MAGNETIC FIELD OF THE SUN

V.N.KRIVODUBSKIJ
Kiev Shevchenko University
Astronomical Observatory
Observatornaya Street, 3
Kiev - 53, 252053
Ukraine, U.S.S.R.

The evolution of the large-scale toroidal magnetic field B_φ of the Sun generating in the radiative zone by acts of differential rotation on the relict radial magnetic field B_r is investigated regarding the magnetic buoyancy and ohmic dissipation.

The magnetic buoyancy plays main role in the constraint of the magnetic induction amplitude of the toroidal magnetic field generated on the up-to-date stage of the evolution of the Sun.

The condition of the stationarity $\partial B_\varphi / \partial t = 0$ yields the maximum estimation of the magnetic induction of toroidal field in the range 10^7 - 2×10^8 G when the value of the radial gradient of the angular velocity near the core of the Sun is about 5×10^{-16} rad/s·sm (which is infered from helioseismology investigations) and when the relict radial field induction is about 1×10^{-3} - 5×10^4 G. The characteristic time scales of reaching established stationary regime for these fields change in the range from 50 millions years to 16 years.

Toroidal magnetic fields due to energetic constraints must be concentrated in relative thin magnetic flux tubes with diameters 16,000 - 800 km and with magnetic fluxes 2×10^{25} - 1×10^{24} Mx.

E. R. Priest and V. Krishan (eds.), Basic Plasma Processes on the Sun, 57.

MAGNETIC BUOYANCY WITH VISCOSITY AND OHMIC DISSIPATION AND FLUX TUBE FORMATION

V.D.KUZNETSOV
IZMIRAN, Troitsk, Moscow Region 142092, USSR

In the framework of the magnetohydrodynamic equations with dissipative terms in the form of turbulent viscosity (ν_t) the linear stage of the instability of a subphotospheric field with respect to magnetic buoyancy is considered. For an exponential in z, isothermal plane-parallel atmosphere with a constant Alfvenic velocity

$$c_s^2 = kT/m = \text{const}, \quad \alpha = B_0^2/(8\pi\rho_0) = v_A^2/2c_s^2 = \text{const}$$

the perturbations of the form

$$f(z)\exp[i\omega t + ik_x x + ik_y y]$$

are described by differential equation with constant coefficients. The qualitative dependence of the growth rate of instability on transverse wave number is determined (see Figures 1 and 2) and characteristic scales of the magnetic tubes are evaluated at the linear stage of instability (Kuznetsov, 1987, 1988).

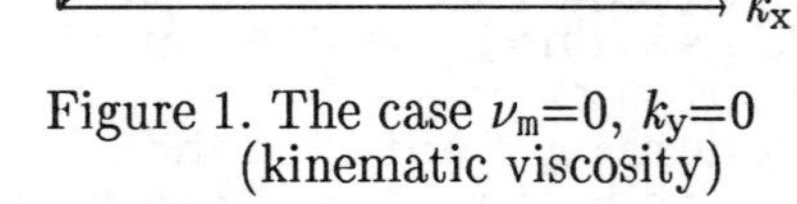

Figure 1. The case ν_m=0, k_y=0 (kinematic viscosity)

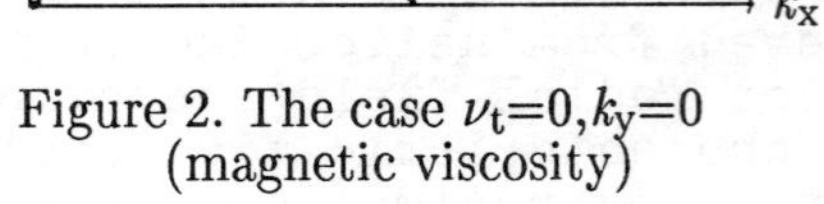

Figure 2. The case ν_t=0,k_y=0 (magnetic viscosity)

The perpendicular wavelength is given by

$$\lambda_\perp \approx 2\pi(\nu_t c_s/g)^{1/2} = 1.4\cdot 10^8 \text{ cm},$$

E. R. Priest and V. Krishan (eds.), Basic Plasma Processes on the Sun, 58–59.

where

$$\nu_t = 6 \cdot 10^{12}\ \mathrm{cm^2 s^{-1}}, \qquad H = c_s^2/g = 1.4 \cdot 10^8\ \mathrm{cm}.$$

The parallel wavelength lies in the range (Parker, 1979)

$$2 \cdot 10^9\ \mathrm{cm} \approx 14.5H \approx 8\pi H/\sqrt{3} < \lambda_{\parallel} < 4\pi H/\sqrt{\sqrt{2}-1} \approx 19.5H \approx 2.7 \cdot 10^9\ \mathrm{cm}$$

For the above parameters characteristic of the top of the solar convective zone, the longitudinal and transverse scales of the emerging flux tubes are in good agreement with the sizes of the magnetic structures observed as arches and loops. The fragmentation of the subphotospheric magnetic field into flux tubes must be most effective in the hydrogen ionization zone where the value of the adiabatic index of the gas γ is minimum and the elasticity of the gas cannot stabilize the instability. Taking at the depth $h = 3 \cdot 10^8$ cm under the photosphere the parameters $p_0 = 10^8$ dyn/cm^2, $T = 2.5 \cdot 10^4$ K, $\gamma = 1.09$ we find that for the instability of the longitudinal perturbations (Parker 1979)

$$\gamma < \Gamma = (1+\alpha)^2/(1+3\alpha/2) \approx 1+\alpha/2$$

the value of the magnetic field

$$B_0 \geq [16\pi p_0(\gamma-1)]^{1/2} \simeq 2 \cdot 10^4\ \mathrm{G}$$

is needed, which is quite acceptable.

References

Kuznetsov, V.D. (1987) 'A typical sizes of the flux tubes in the unstable atmosphere with turbulent viscosity and ohmic dissipation', *Magnitnaya Gidrodinamika* 2, 13–18 (in Russian).

Kuznetsov, V.D. (1988) 'Magnetic field fragmentation into flux tubes in sub photospheric layers of the Sun', in: E.I.Mogilevskiy (ed.), *Fizika Solnechnoy Aktivnosti*, Nauka, Moscow, pp. 30–35 (in Russian).

Parker, E.N. (1979) '*Cosmical Magnetic Fields*', Clarendon Press, Oxford.

EFFECT OF TURBULENCE ON EMERGING MAGNETIC FLUX TUBES IN THE CONVECTION ZONE

Sydney D'Silva and Arnab Rai Choudhuri
Dept. of Physics
Indian Institute of Science
Bangalore, India.

Abstract: Working under the hypothesis that magnetic flux in the sun is generated at the bottom of the convection zone, Choudhuri and Gilman (1987; Astrophys. J. 316, 788) found that a magnetic flux tube symmetric around the rotation axis, when released at the bottom of the convection zone, gets deflected by the Coriolis force and tends to move parallel to the rotation axis as it rises in the convection zone. As a result, all the flux emerges at rather high latitudes and the flux observed at the typical sunspot latitudes remains unexplained. Choudhuri(1989; Solar Physics, in press) finds that non-axisymmetric perturbations too cannot subdue the Coriolis force. In this paper, we no longer treat the convection zone to be passive as in the previous papers, but we consider the role of turbulence in the convection zone in inhibiting the Coriolis force. The interaction of the flux tubes with the turbulence is treated in a phenomenological way as follows: (1) Large scale turbulence on the scale of giant cells can physically drag the tubes outwards, thus pulling the flux towards lower latitudes by dominating over the Coriolis force. (2) Small scale turbulence of the size of the tubes can exchange angular momentum with the tube, thus suppressing the growth of the Coriolis force and making the tubes emerge at lower latitudes. Numerical simulations show that the giant cells can drag the tubes and make them emerge at lower latitudes only if the velocities within the giant cells are unrealistically large or if the radii of the flux tubes are as small as 10 km. However, small scale turbulence can successfully suppress the growth of the Coriolis force if the tubes have radii smaller than about 300 km which may not be unreasonable. Such flux tubes can then emerge at low latitudes where sunspots are seen.

Results: (a) The drag on the flux ring due to the updraught of the large scale turbulence (eg: giant cells) is modelled as the drag on a cylinder moving perpendicular to its axis.

E. R. Priest and V. Krishan (eds.), Basic Plasma Processes on the Sun, 60–61.

The flux ring released at the bottom of the convection zone moves radially out only if the velocity in the updraught is above a critical value u_c, so that the outward drag dominates over the Coriolis force. Fig. 1 shows that u_c increases with increasing flux tube cross-sectional radius σ_0. u_c is independent of B_0 except at unreasonably large magnetic fields of a 100 kG. It is clear that for the typical giant cell velocities of a 100 ms^{-1}, flux tubes cannot be dragged out radially.

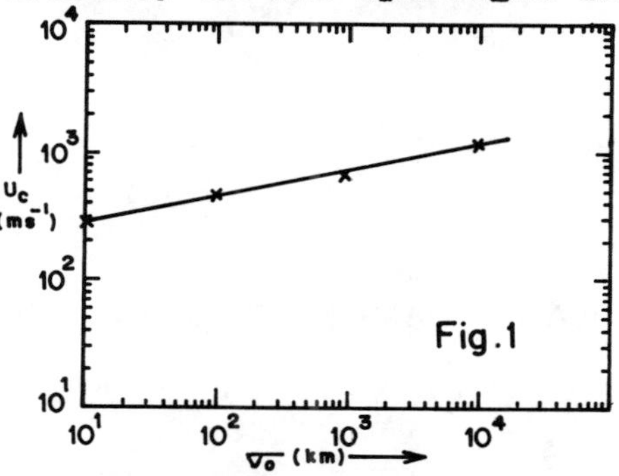

Fig. 1

(b) The efficiency of the small scale turbulence in suppressing the growth of the Coriolis force is measured by the angular momentum exchange time τ. Smaller the value of τ, greater is the efficiency of the process.

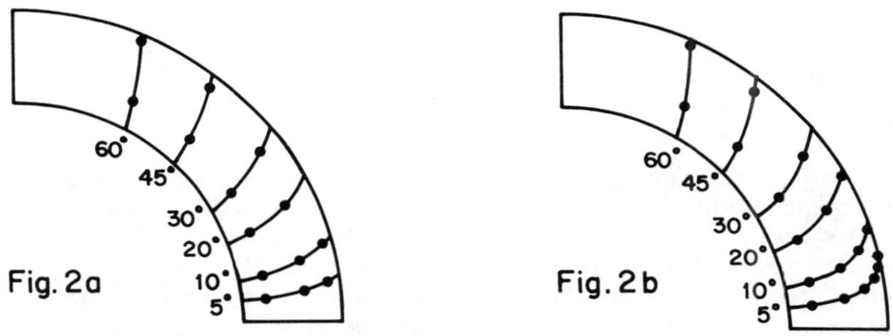

Fig. 2a Fig. 2b

Fig 2a shows the trajectories of flux rings when released at the bottom of the convection zone at different latitudes. There is no transverse drag but τ = 13 hrs. The growth of the Coriolis force is successfully suppressed. Fig 2b shows how the mechanism fails when τ is increased to 17 hrs. B_0 = 17 kG in both cases. The critical τ (i.e. τ_c), above which this mechanism fails, (1) increases with B_0 (43 hrs for B_0= 54 kG) and (2) can be taken as σ_0^2 divided by the turbulent viscosity. This dependence of τ_c on σ_0 puts an upper cutoff on the radii of the flux tubes that this mechanism can transfer out radially. It turns out that this mechanism fails for tubes of radii larger than a few hundred kms.

Conclusions: If the dynamo operates at the base of the convection zone and if we want the flux initially created at low latitudes to emerge at low latitudes, then

(1) The flux tube sizes should be smaller than a few hundred kms (~ 300 km)

(2) There should be small scale turbulence at the same scale (~ 300 km).

PLASMA DAMPING OF GRAVITATIONAL WAVES

C.Sivaram
Indian Institute of Astrophysics
Bangalore 560034, India

Abstract

The possibility of the damping of plane gravitational waves while propagating in a plasma medium is considered. The gravitational plasma frequency, $\omega_G^2 \approx G n_o m_n$, is for a neutron star medium $\sim 10^3$ Hz, which is the same as the frequency of the gravitational waves emitted by a collapsing star. So resonant damping of such waves within a collapsing star is probable. Estimates are made for the damping length for dense and dilute plasmas (also in the presence of magnetic fields). Analogies with Landau damping are made. Applications to other astrophysical situations are outlined.

The propagation of electromagnetic waves $\omega \approx \omega_P$ in a plasma is drastically affected, leading to a shielding effect. Relative velocity between particles can fluctuate when a plane gravitational wave passes. For dense neutron star matter, the gravitational plasma frequency is $\omega_G^2 \approx G n m_n \approx 0.2 \cdot 10^4$ Hz and as a collapsing star emits gravitational waves with $\omega \approx 0.1 \cdot 10^4$ Hz, this would drastically affect the propagation of such waves through the dense medium. The damping time is found to be proportional to the square of the wave frequency and varies inversely as the cube of the particle velocity. Perturbations in the charged particle trajectories due to the gravitational wave also generate electric currents in the plasma. In the presence of a magnetic field, as the gravitational waves are coupled to the electromagnetic waves, they can be cyclotron damped. The damping time is proportional to: $\omega_B / n T^{1/2} m_n^{3/2}$; and for conditions in a neutron star is $\sim 10^2$ sec. So these effects could be important for consideration of gravitational waves generated by a collapsing star.

E. R. Priest and V. Krishan (eds.), Basic Plasma Processes on the Sun, 62.

PROPAGATION AND OSCILLATION OF NEUTRINOS WITH MAGNETIC MOMENT INSIDE THE SUN.

C.Sivaram
Indian Institute of Astrophysics
Bangalore 560034

Abstract: The VVO mechanism suggests that the existence of a neutrino magnetic moment would lead to variations of the observed solar neutrino flux with the magnetic activity of the Sun. The MSW mechansm suggests neutrino oscillations in the dense plasma solar interior. Assuming a non-vanishing neutrino magnetic moment and a magnetic field gradient increasing towards the solar core as required in some dynamo models, the combined effect of the oscillation and the flipping on neutrino propagation in the solar interior is studied. The effect of charge screening in scattering cross-section is also considered.

The required value of $\sim 10^{-10}\mu_B$ for the VVO mechanism is just below the experimental bound from neutrino scattering and constraints from SN 1987 A already appear to rule out such a large value[1]. The MSW mechanism involves a density dependent oscillationlength. However if in addition to an electron density varying with depth we have a magnetic field gradient increasing towards the solar core then the combined effect of the varying density and magnetic field on the neutrino propagation has to be studied. A large magnetic field ($\sim 10^5$ G) if present in the solar core would cause the neutrinos to flip helicity over a distance $\sim 3 \times 10^5$ m., so that one would detect only half of the emerging flux. For combined presence of varying electron density and Field, the oscillation length is:

$$losc = 2\pi/[(\mu_{12} B_{\perp})^2 + [(2m_2^{\ 2}-m_1^{\ 2})/2E - 2^{1/2} G_F \, NY_e F]^2]^{1/2}$$

The quantities N and B are defined in the radial sense in the above equations:

$$\langle N \rangle = (1/R) \int_0^R dr \; n(r), \quad B = (1/R) \int_0^R dr \; B(r),$$

where gradients are taken into account. Estimates give $B \simeq 10^8 (\Delta m/\Delta\mu) G$, Δm is in ev and $\Delta\mu$ in units of μ_B ev/G. Again charge screening takes place in the solar plasma where the virtual photon acquires a mass $m_\gamma = [(4\pi\alpha/\langle kT \rangle)\Sigma ni]$, ni is species present. The neutrino mfp

E. R. Priest and V. Krishan (eds.), Basic Plasma Processes on the Sun, 63–64.

is $\langle \ell \rangle = [\sum_i \sigma_i (m_i)]^{-1}$ where σ_i is spin flip C.S. Result is $\langle \ell \rangle \gg l_{osc}$. Spin flip probability by scattering on nuclei is only ~ 0.1. These effects are, however, very important for supernova neutrinos[1].

References:

1. Sivaram,C. 1989. 'Proc. of Third ESO-CERN Symposium" - n.Caffo et al ed. Kluwer Acad. p.471.

2. Voloshin, M. et al 1986. Sov. J. Nucl. Phys. 44, 440.

III. STELLAR PLASMAS

MAGNETIC BRAKING

L MESTEL
Astronomy Centre, University of Sussex
Falmer, Brighton BN1 9QH, England

ABSTRACT. The theory of angular momentum transport by magneto-thermo-centrifugal winds is reviewed, with special reference to recent observations of rapidly-rotating young solar-type stars.

1. INTRODUCTION

Although this is a Symposium on *Basic Plasma Processes on the Sun*, I need make no apology for extending my mandate to include other solar-type stars. The growing wealth of observations in the optical, radio and X-ray bands has firmly established the "Solar-Stellar Connection", in particular demonstrating the close correlation between the strength of solar-type magnetic activity and the star's rotation α. Single stars in young clusters show both stronger chromospheric activity and more rapid rotation (Noyes *et al.* 1984), supporting strongly the theoretical picture of a two-way interaction: magnetic coupling transfers angular momentum from the star, and the strength of the dynamo-built field declines with α. Further evidence comes from close binary systems, where the near synchronization (tidal or magnetic) of spin and orbital motion keeps the rotation rate high, so that older stars such as the RS CVn stars - some evolved off the main sequence - nevertheless remain magnetically active. Detailed comparison of braking models with observation may shed light on some of the most difficult problems in the area, such as the dependence on stellar rotation of the strength (and even the structure) of the dynamo-built fields, and the efficiency of coronal heating.

2. BRAKING BY WINDS: MATHEMATICAL THEORY

The basic idea of magnetic braking is now very familiar (Schatzman 1962; Mestel 1967a,b; 1968; Weber & Davis 1967; Goldreich & Julian 1970, and many other papers since). Gas outflowing as a wind from a rotating star but not subject to any torques conserves its angular momentum and so develops a rotational shear. In the simplest model the magnetic field permeating the gas is taken as symmetric about the rotation axis. Because of approximate field-freezing into the highly conducting gas, the shear acting on the meridional ("poloidal") component $\underset{\sim}{B}_p$ generates an azimuthal ("toroidal") component $\underset{\sim}{B}_t$, and the twisted field exerts torques which try to establish corotation of the gas with the star. The process is thus a competition between the speed v_p of the

E. R. Priest and V. Krishan (eds.), Basic Plasma Processes on the Sun, 67–76.

outflow that generates the shear and the Alfvén speed $v_{Alf} = B_p/(4\pi\rho)^{\frac{1}{2}}$ at which the field lines try to straighten. Intuitively one therefore expects that the gas will be kept in near corotation with the star as long $v_p << v_{Alf}$, so that a modest rate of mass loss can yield a disproportionately large rate of loss of angular momentum.

The essence of the mathematical theory describing the quasi-steady braking of an axisymmetric star is as follows.

The *steady-state magneto-kinematic equation* $\nabla \times (\underset{\sim}{v} \times \underset{\sim}{B}) = 0$ yields

$$\underset{\sim}{v} = \kappa\underset{\sim}{B} + \tilde{\omega}\alpha\underset{\sim}{t}, \tag{1}$$

equivalent to

$$\underset{\sim}{v}_p = \kappa\underset{\sim}{B}_p, \tag{2}$$

$$\Omega \equiv v_t/\tilde{\omega} = \kappa B_t/\tilde{\omega} + \alpha, \tag{3}$$

where $\underset{\sim}{t}$ is the unit toroidal vector, $\tilde{\omega}$ the axial distance, κ a scalar function, and α a constant on individual field lines. Equation (2) identifies poloidal field lines and the wind streamlines; equation (3) generalizes Ferraro's law of isorotation.

The *continuity equation* $\nabla.(\rho\underset{\sim}{v}) = 0$ constrains κ:

$$\rho\kappa \equiv \rho v_p/B_p = \eta = \text{constant on poloidal field-streamlines.} \tag{4}$$

The *toroidal component of the non-viscous equation of motion*

$$\rho\underset{\sim}{v}.\nabla(\Omega\tilde{\omega}^2) = \tilde{\omega}\ [(\nabla \times \underset{\sim}{B}_t) \times \underset{\sim}{B}_p/4\pi].\underset{\sim}{t} \tag{5}$$

combines with (2) and (4) to yield

$$-\tilde{\omega}B_t/4\pi + \eta\Omega\tilde{\omega}^2 = -\ \beta/4\pi \tag{6}$$

where β is again constant on field-streamlines. Multiplication by the strength $d\Phi = B_p dA$ of an infinitesimal poloidal flux tube of local area dA converts (6) into

$$(-\tilde{\omega}B_t B_p/4\pi)dA + (\rho v_p\Omega\tilde{\omega}^2)dA = (-\beta/4\pi)d\Phi, \tag{7}$$

interpretable as showing that the transport across dA of angular momentum by the gas outflowing along $\underset{\sim}{B}_p$, plus the similar transport by the moment of the Maxwell stresses (Lüst & Schlüter 1955), yields the total constant transport $-\beta/4\pi$ per unit poloidal magnetic flux.

Equations (3), (4) and (6) combine into

$$\Omega(1 - 4\pi\eta^2/\rho) = (\alpha + \eta\beta/\rho\tilde{\omega}^2), \tag{8}$$

where

$$4\pi\eta^2/\rho = 4\pi\rho v_p{}^2/B_p{}^2 = v_p{}^2/v_{Alf}{}^2. \tag{9}$$

Near the star the wind speed $v_p << v_{Alf}$, but with ρ decreasing outwards $4\pi\eta^2/\rho$ increases, becoming unity at the Alfvénic point P_A where

$$v_p = v_{Alf} = B_p/(4\pi\rho)^{\frac{1}{2}} \quad \text{and} \quad \rho = \rho_A = 4\pi\eta^2. \tag{10}$$

Thus by (8), Ω (and likewise B_t) will become singular at P_A unless

$$\alpha = -\eta\beta/\rho_A\tilde{\omega}_A{}^2 = (-\beta/4\pi)/\eta\tilde{\omega}_A{}^2 \tag{11}$$

by (10), or

$$(-\beta/4\pi)d\Phi = (\rho v_p/B_p)\alpha\tilde{\omega}_A{}^2(B_p dA) = (\rho v_p dA)\alpha\tilde{\omega}_A{}^2: \tag{12}$$

the angular momentum transported per second along the flux tube is equivalent to that carried by the steady matter flux ($\rho v_p dA$) if it were kept rotating with angular velocity α out to the Alfvénic point - "effective corotation" out to P_A. At any point the actual values of Ω and B_t are given by

$$\frac{\Omega}{\alpha} = \frac{(1 - \rho_A\tilde{\omega}_A{}^2/\rho\tilde{\omega}^2)}{(1 - \rho_A/\rho)} = \frac{[1 - (v_p/v_A)(B_A\tilde{\omega}_A{}^2/B_p\tilde{\omega}^2)]}{(1 - \rho_A/\rho)} \tag{13}$$

on use of (4) (with $(B_p)_A, (v_p)_A$ shortened to B_A, v_A), and

$$\tilde{\omega}B_t = -4\pi\eta\alpha\tilde{\omega}_A{}^2(1 - \tilde{\omega}^2/\tilde{\omega}_A{}^2)/(1 - \rho_A/\rho). \tag{14}$$

$B_p\tilde{\omega}^2$ is approximately constant along a field-streamline, while ρ_A/ρ, v_p/v_A both decrease strongly towards the coronal base; hence close to the star $\Omega \simeq \alpha$, and by (14) the magnetic transport term in (6) dominates by the factor $\simeq (\tilde{\omega}_A/\tilde{\omega})^2$. Sufficiently far beyond P_A, $\Omega\tilde{\omega}^2 \simeq \alpha\tilde{\omega}_A{}^2$, and the material transport term in (6) ultimately dominates. Over a wide domain encompassing P_A both terms are comparable, and even well within P_A, Ω already lags markedly behind α. It must be emphasized that this does not contradict the basic result (12): "effective corotation" is an exact prescription for computing the angular momentum *transport* along a field-streamline, carried *jointly* by the magnetic and material stresses (Mestel 1967a,b; 1968).

For simplicity we assume the wind zone to be isothermal with sound speed a_w. The wind speed then satisfies the generalized *Bernoulli integral*

$$H \equiv \tfrac{1}{2}(v_p{}^2 + \Omega^2\tilde{\omega}^2) + a_w{}^2\log\rho - GM/r - \alpha\Omega\tilde{\omega}^2 = E$$

$$= \text{constant on field-streamlines.} \tag{15}$$

The new terms are the local rotational energy $\Omega^2\tilde{\omega}^2/2$ per gram and the term $\alpha\Omega\tilde{\omega}^2$ coming from $(\underset{\sim}{j} \times \underset{\sim}{B}/c).\underset{\sim}{v} = \tilde{\omega}\alpha(\underset{\sim}{j} \times \underset{\sim}{B}/c).\underset{\sim}{t}$, the rate of working per gram of the same magnetic torque that gives the outflowing gas angular momentum. If the structure of the field $\underset{\sim}{B}_p$ is supposed known, then substitution of equations (4) and (13) converts equation (15) into a relation $H(\rho,r) = E$ between ρ and the convenient monotonic coordinate r. Once the choice (11) has ensured that Ω and B_t are non-singular, then all non-singular solutions for ρ pass automatically through the Alfvénic point P_A. The critical points of the curve $H = E$ are at the intersections (ρ_s,r_s), (ρ_f,r_f) of $\partial H/\partial r = 0$ and $\partial H/\partial\rho = 0$, corresponding respectively to the slow and fast magnetosonic wave speeds v_s and v_f. Smoothness of flow through these points yields the two conditions

$$H(\rho_s, r_s) = H(\rho_f, r_f) = E \tag{16}$$

which suffice to fix the solution along each field-streamline in terms of the non-dimensional parameters

$$\ell_w = GM/Ra_w{}^2, \quad k = \alpha^2R^3/GM, \quad \zeta_w = B_0{}^2/8\pi(\rho_0)_w a_w{}^2, \tag{17}$$

where B_0 and $(\rho_0)_w$ are standard values, eg at the point where the field-line considered emerges from the coronal base (Weber & Davis 1967; Goldreich & Julian 1970; Sakurai 1985).

In a slow rotator such as the Sun, the terms in Ω in (15) are small compared with the thermal pressure term, and so merely perturb slightly Parker's (1963) thermally-driven wind: the gas undergoes strong acceleration through the slow point where the speed is essentially the sound speed, followed by a much slower subsequent acceleration. By contrast, in a moderately rapid or very rapid rotator, the terms in Ω take over before the Alfvénic point is reached, yielding a net extra driving term $\alpha^2\tilde{\omega}^2(\Omega/\alpha)(1 - \Omega/2\alpha)$. Well within P_A where $\Omega \simeq \alpha$ this term $\simeq \frac{1}{2}\alpha^2\tilde{\omega}^2$, and even at P_A it is typically of the same order. This "centrifugal wind" effect must be included in estimates of the dependence of the rate of braking on the star's rotation.

From (4) and (10), at a general point

$$\tfrac{1}{2}\rho v_p{}^2/(B_p{}^2/8\pi) = 4\pi(\rho v_p/B_p)^2/\rho = \rho_A/\rho = (v/v_A)(B_A/B_p). \qquad (18)$$

The Alfvénic surface S_A (defined by all the points P_A) thus separates the domains nearer the star where the magnetic energy dominates from the more distant regions where the kinetic energy dominates. This suggests that near the star the wind will not distort the field but will rather be channelled parallel to $\underset{\sim}{B}_p$, whereas well beyond S_A the field will be passive, being pulled out to follow the wind. The detailed structure of the magnetic field should be determined by the remaining component of the equation of motion:

$$B_p{}^2(\nabla \times \underset{\sim}{B}_p) - [(\nabla \times \underset{\sim}{B}_t)\cdot\underset{\sim}{B}_p]\underset{\sim}{B}_t = -4\pi\rho\underset{\sim}{B}_p \times \underset{\sim}{F}_p, \qquad (19)$$

this being the balance of the trans-$\underset{\sim}{B}_p$ component of the total magnetic force density and the corresponding component of the density $\rho\underset{\sim}{F}_p$ of non-magnetic (including inertial) forces. Well within S_A the term in $\underset{\sim}{B}_t$ causes a deviation from the vacuum condition $\nabla \times \underset{\sim}{B}_p = 0$ of order $(B_t/B_p)^2 \simeq (\tilde{\omega}_A B_A/\tilde{\omega} B_p)^2(\alpha\tilde{\omega}_A/v_A)^2 << 1$ if $\tilde{\omega} << \tilde{\omega}_A$; and likewise the effect of the thermal, inertial and gravitational terms in $\underset{\sim}{F}_p$ is small for $\tilde{\omega} << \tilde{\omega}_A$. Thus in the simplest model, with $\underset{\sim}{B}_p$ having a dipolar angular distribution over the stellar surface, the field near the star is well approximated by $\underset{\sim}{B}_p = (B_0R^3/r^3)(\cos\theta, \frac{1}{2}\sin\theta, 0)$ in spherical polars. Curl-free dipolar field lines leaving the star at colatitudes $\theta_s \simeq \pi/2$ will reach the equator at $r = R/\sin^2\theta_s << r_A$, so that the approximation $\nabla \times \underset{\sim}{B}_p = 0$ is still valid, whereas those leaving near to the poles will extend to regions close enough to S_A for the forces driving the wind to be able to distort the field into a more rearly radial structure. We are thus led to a multi-component corona, comprising a "dead zone" with hot gas trapped by closed field lines, and a "wind zone" with outflowing, somewhat cooler gas (Mestel 1968; Pneuman & Kopp 1971). This picture is vindicated qualitatively by X-ray observations of bright domains contiguous with the coronal holes associated with the wind.

3. BRAKING LAWS

The direct loss of angular momentum J from the star comes from the outflowing gas in the wind zone. With the radial structure adopted for the field at the Alfvénic surface S_A, approximated as a sphere, the "effective corotation" result (12) yields

$$-dJ/dt \simeq (8\pi/3)(\rho_A v_A r_A{}^2)\alpha r_A{}^2 \qquad (20)$$

where α is now identified with the assumed uniform rotation of the star. Along each field-streamline

$$4\pi\rho_0 v_0/B_0 = 4\pi\rho_A v_A/B_A = B_A/v_A \tag{21}$$

from (4) and (10), so that

$$-dJ/dt \simeq (2/3)(B_A r_A{}^2)^2\alpha/v_A. \tag{22}$$

The simplest field models (Weber & Davis 1967; Sakurai 1985) replace Fig. 1 by a field that is radial all the way back to the star, so that $B_A = B_0(R/r_A)^2$ and (22) becomes

$$-dJ/dt \simeq (2/3)(B_0R^2)^2\alpha/v_A. \tag{23}$$

In a slow rotator the mainly thermal driving yields $v_A \simeq (2 - 3)a_w$, virtually independent of α. From (23), $-\dot{J}$ is independent of ρ_0: with the radial field structure, a higher ρ_0 at fixed a_w yields a greater mass loss rate, but the associated reduction in r_A compensates exactly. If further one assumes a linear relation $B_0 \propto \alpha$ for the dynamo-built field, then with $J = kMR^2\alpha$, (23) yields $d\alpha/dt \propto -\alpha^3$, which integrates to $\alpha \propto t^{-\frac{1}{2}}$, identical with the Skumanich law (1972), inferred from comparison of the solar rotation with observations of late-type stars in the younger Pleiades and Hyades clusters.

The field structure of Fig. 1 ensures that in fact only part of the flux emanating from the star is coupled with the wind, so that $B_A r_A{}^2 < B_0R^2$ and by (22) the braking rate is reduced. Also, in a rapid rotator the centrifugal driving term in (15) forces the gas to reach the local Alfvénic speed at a smaller value of r_A, again reducing the braking; this is shown by v_A (a monotonically decreasing function of r) appearing in the denominator in (22). The model of Mestel and Spruit (1987) takes some account of these effects. The approximately dipolar field has $B \simeq B_0(R/r)^3$ out to the radius $\bar{r} < r_A$, and $B \simeq B_0(R/\bar{r}^3)(\bar{r}/r)^2$ beyond $\bar{r}$. (Since along dipolar field lines $\tilde{\omega}^2 \propto r^3$, $B\tilde{\omega}^2 \simeq$ constant along a field line, as quoted above). Thus (22) becomes

$$-dJ/dt \simeq (2/3)\alpha(B_0R^2)^2/(\bar{r}/R)^2 v_A \tag{24}$$

$$= (2/3)(B_0{}^2\alpha^3{}_\odot)R^4(\alpha/\alpha_\odot)^3/(\bar{r}/R)^2 v_A \tag{25}$$

if again $B_0 \propto \alpha$. The details of the wind speed, and in particular its value v_A at P_A, depend on the value of $a_w{}^2/\alpha^2\tilde{\omega}_A{}^2$. In the limit with this parameter zero, $\Omega/\alpha = (1 - \surd(19/27))$ at P_A, yielding the value $4\alpha^2\tilde{\omega}_A{}^2/27$ for the centrifugal driving term in (15). With a_w finite, even in a rapid rotator there will always be a cone about the axis in which thermal driving dominates. A tolerable approximation for v_A is given by

$$v_A/a_w = \{[v_{th}(r_A)/a_w]^2 + \alpha^2\tilde{\omega}_A{}^2/3a_w{}^2\}^{\frac{1}{2}} \tag{26}$$

where $v_{th}(r_A)$ (given by (15) with the terms in Ω dropped) is $(2 - 3)a_w$ typically. Substitution of (26) into (24) yields an equation for r_A and so also for v_A/a_w, which now depends not only on a_w and on α but also on ρ_0. -As a convenient alternative to (24) or (25), we can write

$$-dJ/dt \simeq [(8\pi/3)\alpha_\odot(\rho_0)_\odot R^4(v_0/a_w)(a_w)_\odot]K(\alpha/\alpha_\odot), \tag{27}$$

where the bracket is the angular momentum that would be carried off from the Sun by the solar wind if there were no solar magnetic field, and the function $K(\alpha/\alpha_\odot)$ takes care of the different effects studied above. In Mestel & Spruit (1987) the coronal sound speed a_w was assumed independent of α, but the relation $\rho_0 \propto B_0 \propto \alpha$

was adopted as the simplest way of accounting for the optically thin X-ray emission rates $L_X \propto \rho_0{}^2$ which are observed to be consistent with $L_X \propto \alpha^2$ (Pallavicini *et al.* 1981; Micela *et al.* 1984); whence $K = (R/\bar{r})(r_A/R)^2(\alpha/\alpha_\odot)^2$.

The remaining unknown is the extent $\bar{r}$ of the dead zone. On the equator beyond $\bar{r}$ the oppositely directed field lines are held apart by a thermal pressure equal to the magnetic pressure $[B_r(r,\pi/2)]^2/8\pi$. At $\bar{r}$ these field lines start to diverge from the equator to form the separatrix S between the wind and dead zones, with the analogous boundary condition

$$p + B^2/8\pi = \text{continuous across S.} \tag{28}$$

The difference in p across S becomes especially marked when the rotation is high. Just outside S, in the wind zone, we have seen that the centrifugal force of near corotation assists in the outward acceleration of the gas, so reducing ρ and p below that in a thermally driven wind. By contrast, in the dead zone the same centrifugal force must be included in the condition of hydrostatic balance that holds along each field line. Again we take the dead zone isothermal, with the sound speed a_d. At the point (r,θ) along the field line leaving the star's surface at (R,θ_0)

$$\frac{\rho}{\rho_o} = \exp\left[-\left[\frac{GM}{Ra_d{}^2}\right](1-\frac{R}{r}) + \tfrac{1}{2}\left[\frac{\alpha^2R^2}{a_d{}^2}\right]\left[\frac{r^2\sin^2\theta}{R^2} - \sin^2\theta_o\right]\right]. \tag{29}$$

Beyond the point where the components along $\underset{\sim}{B}$ of gravity and centrifugal force balance, ρ and p exponentiate outwards, requiring by (28) a corresponding increase in the relative jump in B. To make a first estimate of $\bar{r}$ the field lines are given the vacuum dipolar form $\sin^2\theta/r$ = constant, and condition (28) is applied just at the point $(\bar{r},\pi/2)$ in the form $(B^2/8\pi)_w = (\rho a^2)_d$ with ρ given by (29). As for the wind zone, the relations $B_0 \propto \alpha$, $(\rho_0)_d \propto B_0 \propto \alpha$ are adopted. As expected, the growth in B_0 with α yields an increasing dead zone, but later the growth in $\bar{r}$ is halted and reversed by the strong centrifugal effect on the dead zone pressure.

The angular momentum loss, fixed by the function K in (27) can be parametrized by $\dot{\alpha} \propto -\alpha^p$, with the index p increasing from $\simeq 2.3$ to $\simeq 2.9$ as $\alpha/\alpha_\odot$ increases from unity to the maximum possible value $\simeq 80$. Over the same range K increases by 8×10^4, as compared with the value 5×10^5 predicted by the Skumanich law p = 3. If instead of increasing like B_0, ρ_0 is assumed independent of α, equation (27) yields smaller values for p, reducing the maximum braking rate by another factor $\simeq 6$. The reason is partly an increase in $\bar{r}$ but primarily the increasing importance of the centrifugal driving term, which prevents the increase in r_A with α from being large enough to compensate for the constancy of ρ_0 (*cf.* the remark following equation (23)).

4. COMPARISON WITH OBSERVATION

Too much attention should not in fact be paid to the precise values of the index p. As pointed out by Vilhu & Moss (1986), allowing a_w rather than ρ_0 to vary suitably with α can yield again the Skumanich index p = 3. Because of the many uncertainties – eg in dynamo theory and in the theory of coronal heating – one is well advised to be guided by observation. Support for the dead zone model comes from observations of non-thermal radio emission from stellar coronae (*cf.* Mutel in Havnes *et al.* 1988). Further support comes from combined optical and X-ray studies of the rapidly rotating (P = .5d) G8–KO dwarf star AB Doradus (Collier Cameron *et al.* 1988). The EXOSAT data show that the part of the spectrum 1–10 keV is well fitted by thermal bremsstrahlung from gas at $T \simeq 1.7 \times 10^7$K, with an emission measure that yields an emitting volume greater than 10 times the volume of the star.

This high energy emission is not subject to eclipse, consistent with its origin in a dead zone extending out to 2–3 R (rather than in the cooler, more compact regions associated with star–spots and like them subject to eclipse, which are responsible for the .1–1 keV part of the spectrum). Observations in Hα (Collier Cameron & Robinson 1989a, b) show transient absorption features, which are consistent with clouds of HI transiting the stellar disk and scattering chromospheric Hα photons. The crucial observation is that the line–of–sight velocity component of the absorbing material always matches that of the obscured stellar limb as the cloud enters and emerges from transit, showing clearly that it is in *near corotation* rather than in a Keplerian orbit. From the rate at which they cross the profile the clouds appear to form at 3–4R, outside the Keplerian corotation radius $r_k = (GM/\alpha^2)^{1/3} \simeq 2.6R$.

It is gratifying to a theorist to have further confirmation of the picture of magnetic control of rotation. There is also evidence of slow outward motion of the Hα clouds, suggesting that the "dead" zone in fact makes some contribution to mass and angular momentum loss: this is discussed further below.

The work of Section 3 assumes the simple dynamo law $B_0 \propto \alpha$, which is in fact responsible for most of the α–dependence of the Skumanich law and its modifications. In a recent survey, Stauffer and Hartmann (1987) (see also Gray 1982) have produced evidence for a rapid early spin–down on the main sequence, with a decline from $35\alpha_\odot$ to $5\alpha_\odot$ occurring in a few x 10^7yr, and at a rate weakly dependent on α, with subsequent braking given by the Skumanich law $\dot{\alpha} \propto -\alpha^3$. These results *prima facie* set problems for standard wind theory. One can get a weak dependence on α by supposing that the dynamo dipole component saturates at $\alpha = \tilde{\alpha}$; an essentially thermal wind would then yield $\dot{\alpha} \propto -\alpha^p$ with $p \leqq 1$ for $\alpha > \tilde{\alpha}$, while when α is large enough for centrifugal driving to dominate, the very weak dependence $p = 1/3$ results. But all the deviations from the conditions leading to the Skumanich law – the dead zone, centrifugal driving and now dynamo saturation – also reduce $\dot{\alpha}$, and it is not easy to produce an early braking time of a few x 10^7yr. For example, if the present–day Sun were losing half its angular momentum in 3 x 10^9yr, and if dynamo saturation occurs at $\tilde{\alpha} = 20\alpha_\odot$, then spin–down of the whole Sun from $70\alpha_\odot$ to $20\alpha_\odot$ would require 7.5 x 10^7yr; while if $\tilde{\alpha} = 10\alpha_\odot$, spin–down to $10\alpha_\odot$ requires 4.8 x 10^8yr. The problem is made still harder by revised estimates (Pizzo *et al.* 1983) that put the solar braking time at $\simeq$ 2 x 10^{10}yr (implying $r_A \simeq 12R$ and a surface field strength $\simeq$ 1.4 Gauss in the wind zone). Stauffer and Hartmann suggest that their observations imply an initial spin–down of just the convective envelope, which has only $\simeq$ 1/20 of the total moment of inertia of the Sun. The consequent constraints on core–envelope magnetic coupling necessary for such a relative shear to persist for a few x 10^7yr are severe though perhaps not completely implausible (*cf.* Mestel & Weiss 1987, where the same problem is discussed but with a "fossil" differential rotation supposed maintained for 2 x 10^9yr). Even so, one is prompted to look for a more efficient braking process, operative at high rotations.

5. EXPLORATIONS

There are many possible variations on the simple wind models discussed above. The high coronal temperature inferred for AB Doradus could imply also a coronal base density for rapid rotators higher than the $\rho_0 \propto B_0$ relation adopted by Mestel & Spruit. As noted above, for a given surface field strength an increase in ρ_0 has primarily the effect of moving the Alfvénic surface inwards without changing much the rate of braking. If the dipolar component of the dynamo–built field saturates at $\alpha = \tilde{\alpha}$, but quadrupolar and higher components increase, one can imagine models in which the "dead" zones extend further than the Alfvénic surface in the wind zone, i.e. $\bar{r} > r_A$.

For a given total magnetic flux F_A crossing S_A, the rate of braking given by expression (22) increases as v_A decreases. This result must be treated with caution: a hypothetical cool corona ($T < 10^5$K) with the same base density as in the Sun will not expand under its thermal pressure, for the density exponentiates down to such low values that the pressure far from the star would be balanced by that of the interstellar medium (Parker 1963). However, once a hot dense wind that has passed through the slow magnetosonic point has ceased to receive heat from the star, it will tend to cool adiabatically so that its Mach number increases still further. If the gas is then shocked, the compressed subsonically moving gas behind the shock cools radiatively and so becomes still denser. The shock may then define a new "coronal base" from which the gas expands into a new standard wind flow (Li Jianke & Mestel, in preparation), but again in a rapid rotator centrifugal driving will force v_A up, and there is no obvious reason why the net braking rate should increase.

Probably the most hopeful prospect for enhanced braking is by a development of the "dead zone" picture. As already noted, beyond $r \simeq r_K$ equation (29) predicts that ρ increases outwards along a field line. Cool HI condensations as observed in AB Doradus may form by radiative recombination and the associated energy loss; a reduction in a_d would lead to further compression, to be made up by further outflow of heated gas from the star into the lower corona. The centrifugal pull $\rho\Omega^2 r$ on a high density condensation in turn causes further distortion of the closed field loops, leading ultimately to field line reconnection and detachment – an effective "macro-resistivity", allowing flow across the field, driven by centrifugal force against the restraining Lorentz force. Although such a trans-field flow cannot be described by the perfect conductivity theory of Section 1, nevertheless one expects again that the slower the magnetically-controlled outflow across $\underset{\sim}{B}$, the greater the distance to which the field can enforce near corotation. And if also $\tilde{\bar{r}} > r_A$, this transport of angular momentum by HI clouds forming in and breaking out of the "dead zone" will dominate over the normal braking from the wind zone.

Further calculations (Barker & Mestel, in preparation) should clarify the details of the expected field structure and the formation of condensations. The very high temperature 1.7×10^7K inferred from the X-ray emission for AB Doradus would severely limit the density increase predicted by (29), but the above thermal instability could set in if the coronal dead zone is inhomogeneous, with parts of it having a temperature closer to the solar value of $\simeq 2 \times 10^6$K. The likely sensitivity of the results to the assumed magnetic field structure and to the coronal base density may yield clues as to dynamo action and to coronal heat input at high rotations.

REFERENCES

Collier Cameron, A., Bedford, D.K., Rucinski, S.M., Vilhu, O. & White, N.E., 1988. *Mon. Not. R. astr. Soc.*, **231**, 131.

Collier Cameron, A. & Robinson, R.D., 1989a,b. *Mon. Not. R. astr. Soc.*, **236**, 57 and **238**, 657.

Goldreich, P. & Julian, W.H., 1970. *Astrophys. J.*, **160**, 971.

Gray, D.F., 1982. *Astrophys. J.*, **261**, 259.

Havnes, O. *et al.* (eds.), 1988. *Activity in Cool Star Envelopes*, Kluwer Academic Publishers, Dordrecht.

Lüst, R. & Schlüter, A., 1955. *Zeits. f. Astrophys.*, **38**, 190.

Mestel, L., 1967a. *Mem. Soc. R. Sci. Liège*, **5**, 15, 351.

Mestel, L., 1967b. In: *Plasma Astrophysics*, ed. Sturrock, P.A., Academic Press, London.

Mestel, L., 1968. *Mon. Not. R. astr. Soc.*, **138**, 359.

Mestel, L. & Spruit, H.C., 1987. *Mon. Not. R. astr. Soc.*, **226**, 57.

Mestel, L. & Weiss, N.O., 1987. *Mon. Not. R. astr. Soc.*, **226**, 123.

Micela, G., Sciortino, S. & Serio, S., 1984. In: *X-Ray Astronomy - 84*, eds. Oda, M. & Giacconi, R.,

Institute of Space and Astronomical Sciences, Tokyo.
Noyes, R.W., Hartmann, L., Baliunas, S.L., Duncan D.K. & Vaughan, A.H., 1984. *Astrophys. J.*, **279**, 763.
Pallavicini, R., Golub, L., Rosner, R., Vaiana, G.S., Ayres, T. & Linsky, J.L., 1984. *Astrophys. J.*, **248**, 279.
Parker, E.N., 1963. *Interplanetary Dynamical Processes*, Interscience, New York.
Pizzo, V., Schwenn, R., Marsch, E., Rosenbauer, H., Mühlhäuser, K.-H. & Neubauer, F.M., 1983. *Astrophys. J.*, **271**,335.
Pneuman, G.W. & Kopp, R.A., 1971. *Sol. Phys.*, **18**, 258.
Sakurai, T., 1985. *Astr. Astrophys.*, **152**, 121.
Schatzman, E., 1962. *Ann. Astrophys.*, **25**, 18.
Skumanich, A., 1972. *Astrophys. J.*, **171**, 565.
Stauffer, J.R. & Hartmann, L.W., 1987. *Astrophys. J.*, **318**, 337.
Vilhu, O. & Moss, D.L., 1986. *Astr. J.*, **92**, 1178.
Weber, E.J. & Davis, Jr., L., 1967. *Astrophys.J.*, **148**, 217.

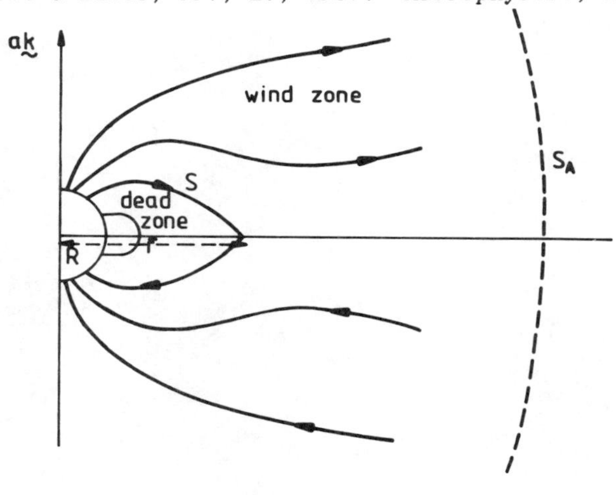

Figure 1. Schematic magnetic field model

DISCUSSION

WEISS: Some models of young stars on the main sequence assume that only the envelope is spun down, so that strong differential rotation persists for 10^8 years or longer. Do you think that this is likely?

MESTEL: One reason why I am trying to increase the efficiency of braking in the earliest phases (*i.e.*when the star is rapidly rotating) is because I am unhappy about the interpretation of observations that requires spin-down of just the envelope. You recall that in our paper we noted what severe constraints are imposed on internal magnetic fields coupling core and envelope, if such a strong shear is supposed to persist for so long.

CHITRE: The helioseismological data seem to suggest that the differential rotation observed at the solar surface is impressed even in the interior. Is this indicating the influence of the magnetic field linking the core with the envelope on the rotation profile?

MESTEL: My impression is that the interpretation of the helioseismological data is still ambiguous. Maybe Dr Christensen-Dalsgaard will comment.

DALSGAARD: I agree with your impression that the question of the core rotation has not been settled. I think we can definitely say that the region outside $0.2R_{\odot}$ cannot be rotating at substantially more than the surface rate. Some observations indicate splitting for $\ell = 1$ modes corresponding to almost twice the surface angular velocity. If true, these observations would indicate a very small core rotating quite rapidly.

MESTEL: That would still restrict severely the strength of the field penetrating into this very small core.

ENERGY RELEASE IN STELLAR FLARES

R. PALLAVICINI
Osservatorio Astrofisico di Arcetri
Florence, Italy

ABSTRACT. Flare-like events similar to those observed on the Sun occur on many different types of stars, particularly on late K and M dwarfs. Although the physical mechanisms responsible for these events remain largely unknown, it is likely that the flare energy derives from dissipation of magnetic fields as is the case for solar flares. We review the basic observational facts that suggest an analogy between solar and stellar flares and we discuss how the different physical conditions occurring on stars may affect the application of current solar-type models to the stellar case. We show that, in spite of a qualitative agreement found between model predictions and observations, there is still no convincing evidence that stellar flares are simply scaled-up versions of solar flares. Major advances in the observations of stellar flares are required before this fundamental question can be safely addressed.

1. Introduction

Over the past several years it has become increasingly evident that solar-type phenomena occur in a variety of different astrophysical contexts. In particular, stars of spectral type later than F show evidence for chromospheres, coronae, winds, magnetic fields and surface inhomogeneities such as spots, plages and flares. Since these stars are rotating magnetized gaseous bodies with a subphotospheric convective zone, it is generally accepted that some sort of dynamo action is at work, as believed to occur for the Sun. The dynamo-generated magnetic fields emerging at the stellar surface are thought to be responsible for the observed activity phenomena.

This paper will focus only on one particular form of activity, i. e. stellar flares, and will be further restricted to only one class of stars, the so-called UV Ceti-type flare stars. These are late K and M dwarfs in the solar neighbourhood that are known to be characterized by intense flaring activity at optical, XUV and radio wavelengths. The reason for selecting UV Ceti-type stars while discussing stellar flares in the context of a solar meeting is mainly dictated by the fact that flares from UV Ceti-type stars present the greatest analogy to solar flares, in contrast to the case of other stellar sources (e.g. RS CVn and Algol-type binaries and pre-main sequence objects) for which there are probably more differences than similarities with respect to the solar case (see, e.g., Mullan 1985, Pallavicini and Tagliaferri 1989, Byrne 1989).

There are, in fact, many analogies between solar flares and flares from UV Ceti-type stars, as has been most recently discussed at the Conference held in 1988 at Stanford University (IAU Coll. No. 104, see proceedings in Solar Physics Vol. 121, 1989 and in Haisch and Rodonò 1989). There are also significant differences, however, between solar and stellar flares. Hence the fundamental question arises: are these differences due to the different physical environments in which the same basic physical mechanisms occur, or rather are the environments so different that even the basic flare mechanisms may be different in the solar and stellar case?

1.1 Flare energetics

Before trying to answer the above question, one has to make a fundamental assumption, i. e. that the energy released in stellar flares derives from dissipation of magnetic fields as believed to occur for solar

E. R. Priest and V. Krishan (eds.), Basic Plasma Processes on the Sun, 77–92.

flares. There is no compelling evidence in the stellar case to believe that this is actually true (and, in fact, a variety of exotic, basically non-solar models were proposed in the past, cf. Gurzadyan 1980). However, in the absence of any other viable alternative, we shall assume that this is the case in spite of the fact that the observational evidence for this remains quite meager.

Under the assumption above, a substantial difference appears to exist between the solar and stellar case when considering the energetics of the flare phenomenon. The total radiative energy released in the largest solar flares is on the order of $\approx 10^{32}$ erg, while the total radiative energy released by large flares on M dwarf stars may be as high as $\approx 10^{35}$ - 10^{36} erg (cf. Gershberg 1989; note that these energies are lower limits, since the contribution of mass motions to the energy budget of stellar flares is unknown). If this energy has to come from dissipation of magnetic fields, the energy ΔE released by the flare must be related to the magnetic field strength B and to its variation ΔB by the simple relationship:

$$\Delta E = \frac{B\,(\Delta B)}{4\,\pi}\,V \qquad (1)$$

where V is the flare volume.

For the Sun, $\Delta E \approx 10^{32}$ erg and the typical volume for a large 2-ribbon flare is $V \approx 10^{29}$ cm^3 (see, e.g., Svestka 1976). Hence $B(\Delta B) \approx 10^4$. If $B \approx 300$ Gauss (a reasonable value for solar flares), a variation $\Delta B \approx 30$ Gauss is sufficient to account for the flare energy release. This is not so in the stellar case, for which $\Delta E \geq 10^{35}$ erg. In fact, if the volume were the same as in the solar case, $B(\Delta B)$ would be $\approx 10^{7}$, and in this case one should dissipate entirely a magnetic field of ≈ 3000 Gauss in order to do the job. More likely, if we want to maintain the same ratio 1:10 that we have assumed above for the solar case, one has to postulate magnetic fields as high as $\approx 10^4$ Gauss in M dwarfs and variations $\Delta B \approx 10^3$ Gauss. At present we cannot exclude that magnetic fields as high as 10^4 Gauss may exist on M dwarfs (cf. Mullan 1984) but, if so, this certainly is a rather non-solar situation.

Alternatively, the flare volume may be much larger than in the solar case. For instance, if the volume were one order of magnitude larger, say $\approx 10^{30}$ cm^3, $B(\Delta B) \approx 10^6$ and one needs to dissipate ≈ 300 Gauss in a field of ≈ 3000 Gauss. Note that a flare volume $V \approx 10^{30}$ cm^3 or larger implies that the flare linear size is comparable to the stellar radius in an M dwarf (since in these stars $R_* \leq 0.5\ R_o$); this is again a quite non-solar situation. In either case, it is hard to escape the conclusion that larger volumes or higher magnetic fields or both are required by stellar flares.

1.2 Flare diagnostics

Another important point to be kept in mind is that the primary energy release process is difficult to test observationally even in the case of the Sun (not to speak of stars). The available observational tests, therefore, refer almost exclusively to secondary processes and energy transport mechanisms. The relevant question is what kind of flare diagnostics are accessible in the case of stellar flares.

In the solar case, there is a wealth of observational diagnostics of flares, ranging from γ-rays to radio wavelengths and involving a variety of different spatial, spectral and temporal resolutions (see reviews in Sturrock 1980 and Kundu et al. 1989). This is not the case for stellar flares. There is one observational datum (optical continuum emission) that is probably better known for stellar than for solar flares (only recently systematic high-sensitivity observations of solar white-light flares have started to become available; cf. Neiding 1989). Other diagnostics (optical line emission, UV line and continuum emission, soft X-rays, radio emission) are also sufficiently well known for stellar flares. However, there is a large number of other diagnostics (e.g. hard X-rays, γ-rays, particle emission, mass ejections) that remain completely unaccessible in the stellar case, in addition to the fact that no spatial information is of course available. Our knowledge of magnetic fields and mass motions remain also extremely limited, if at all available. These fundamental observational limitations need to be taken into account when comparing solar and stellar flares and when trying to determine how successful solar models are in explaining stellar flares.

2. Observations

It would not be possible in the limits of a short review paper to summarize all observational data that have been collected in the past several years on stellar flares. Only a few observational results will be presented here with emphasis on the constraints that these data can put on stellar flare models. For a more comprehensive discussion of stellar flare observations and a complete list of references the reader is referred to a number of excellent conference proceedings such as Byrne and Rodonò (1983), Gondhalekar (1986), Havnes et al. (1988), Haisch and Rodonò (1989), Mirzoyan (1989).

In order to illustrate the main points with some specific examples, Figs. 1 and 2 show results of coordinated multi-wavelengths observations of stellar flares as published by Rodonò (1989) and Kahler et al. (1982), respectively. These are perhaps the best cases of stellar flares observed so far simultaneously at various frequencies. Fig.1 refers to a flare observed on 28 March 1984 from AD Leo: complete coverage was obtained at optical wavelengths (both photometry and spectroscopy) as well as at microwaves (2 cm and 6 cm). Infrared observations were also obtained in the K-band (2.2 μ), while UV observations were limited to the Mg II lines. Unfortunately, no X-ray observations could be obtained for this event. Fig. 2 shows a flare observed from YZ CMi on 25 Oct 1979. In addition to observations in optical continuum and Balmer lines, soft X-rays were observed, thus providing information also on the coronal portion of the flare. In the same event (Fig. 2) a decimetric burst was detected which was delayed by 17 minutes with respect to the optical continuum emission. Other relevant flare features which appear in Figs. 1 and 2 will be discussed later on.

2.1 Optical continuum emission

In contrast to the case of solar flares, stellar flares have been most extensively studied in broad-band optical continuum. Since stellar flares are a blue phenomenon while UV Ceti-type stars are intrinsically red, the flares are best seen in the U and B bands where the flare amplitude may be as high as five to six magnitudes (Byrne 1983, Pettersen 1988, 1989, Gershberg 1989, Shakhovskaya 1989). Statistical studies over a large number of flares (Lacy et al. 1976, Gershberg and Shakhovskaya 1983, Pettersen et al. 1984) indicate that the energy emitted in the U band is only slightly larger than in the B band ($E_U \approx 1.2\ E_B$) while the total energy released at optical wavelengths $E_{opt} \approx 3\text{-}5\ E_U$. The total bolometric luminosity of a stellar flare is estimated to be $E_{bol} \approx 6\ E_{opt} \approx 20\text{-}30\ E_U$. On the basis of these empirical relationships it is estimated that the total radiative energy released by flares on UV Ceti-type stars ranges from less than $\approx 10^{29}$ erg to more than $\approx 10^{35}$ erg. The intrinsically brighter stars flare more frequently than the intrinsically fainter stars and produce more energetic flares. While the latter result may be partially due to a selection effect, the higher flare frequency of intrinsically brighter stars suggests a dependence on the stellar radius, as might occur for instance if larger areas of the star are covered by active flaring regions. Although periodicities have occasionally been reported, there is no convincing evidence that stellar flares repeat regularly in time. Surprisingly enough, no periodicities related to the stellar rotation and/or to stellar cycles have yet been found. The available evidence suggests instead that stellar flares occur randomly in time.

The light curves of optical flares (see, e. g., Fig. 1a) present a large variety of different shapes with typically impulsive, often structured, profiles. The observed rise times are in the range ≈ 1-100 sec and the observed decay times are on the order of ≈ 10-1000 sec. The large flares tend to have longer time scales than small flares, as also observed for solar 2-ribbon flares. Moreover, flares observed in the optical continuum are typically more impulsive and of shorter duration than flares observed in optical Balmer lines or in soft X-rays. This is similar to what observed in solar flares, where white-light emission has a much shorter lifetime (and is localized in much smaller areas) than Hα emission. In solar flares there is a very good temporal correlation between white-light flares and hard X-ray emission. If the same were true for stellar flares, we could use the optical continuum emission as a proxy for hard X-rays (which are otherwise unobservable at the present sensitivity levels). Although there is no observational support for this claim, the solar analogy suggests that optical continuum emission could indeed be used to derive information on the primary energy release and particle acceleration in stellar flares.

An important observational finding recently reported by Rodonò (1989) is also illustrated in Fig. 1a. A "negative" flare is observed at IR wavelengths at the time of the optical continuum flare. There is so far no simple explanation for this effect, which is probably related to a change in opacity as a consequence of the

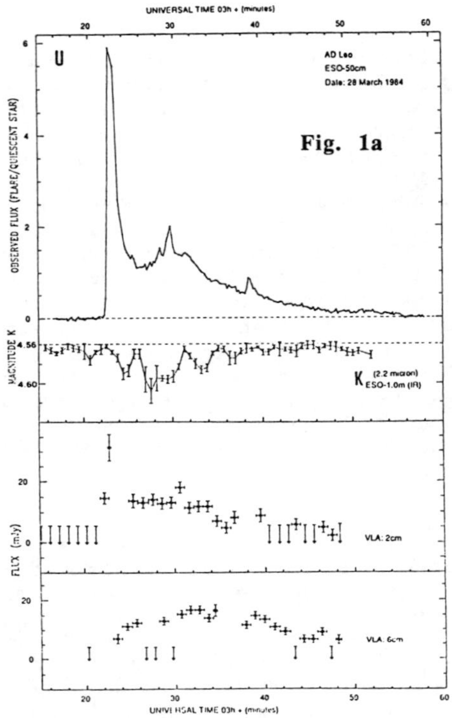

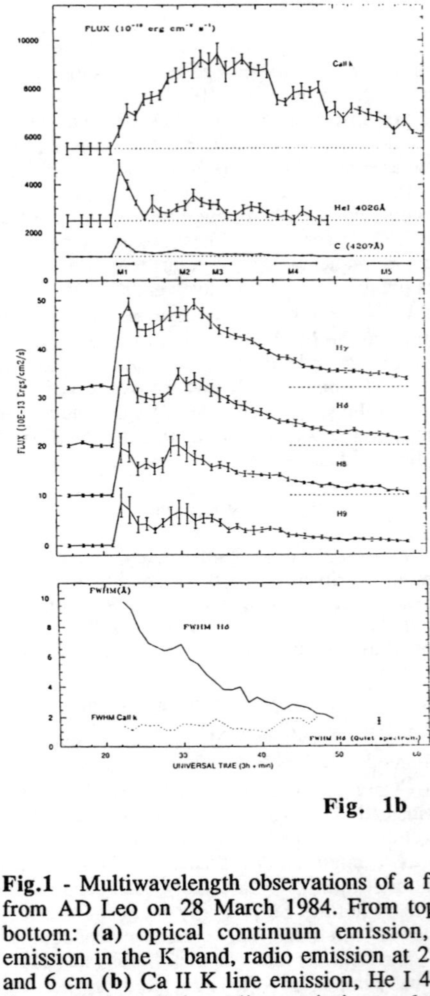

Fig. 1b

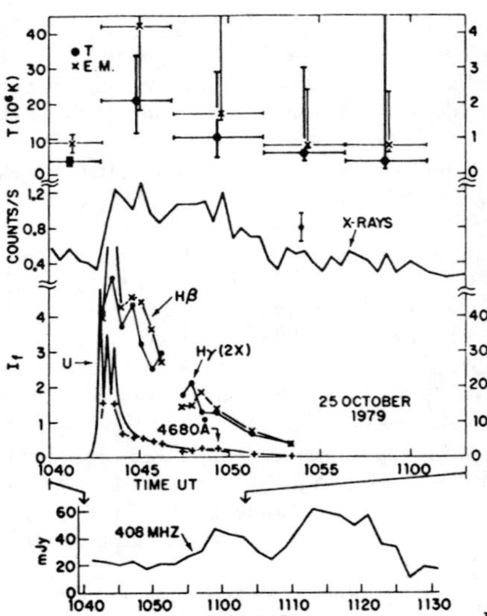

Fig. 2

Fig.1 - Multiwavelength observations of a flare from AD Leo on 28 March 1984. From top to bottom: **(a)** optical continuum emission, IR emission in the K band, radio emission at 2 cm and 6 cm **(b)** Ca II K line emission, He I 4026 line emission, Balmer line emission and line widths as a function of time. The time scale is in minutes (from Rodonò et al. 1989)

Fig. 2 - Multiwavelength observations of a flare from YZ CMi on 25 October 1979. From top to bottom: coronal temperature and emission measure, soft X-ray emission, Balmer line and continuum emission, radio emission at 408 MHz (from Kahler et al. 1982).

different conditions occurring in a flaring atmosphere. It is not known whether similar phenomena occur in solar flares. Curiously enough, there are virtually no observations of solar flares at IR wavelengths. A search for solar counterparts of these "negative" IR flares could provide important insights into the physical conditions of both solar and stellar flares.

2.2 *Optical/UV line emission*

Optical line emission (Balmer series, Ca II H and K lines, He I, etc.) appear to be enhanced during stellar flares (Giampapa 1983, Worden 1983, Foing 1989). Typically, line emission peaks later than optical continuum emission, has a more gradual profile and a much longer duration (cf. Figs. 1 and 2). While an impulsive component is often present in the Balmer lines (in addition to the late gradual phase, cf. Fig. 1b), the Ca II lines show a more smooth behaviour and are further delayed with respect to the Balmer and continuum emission. The energy released in the Balmer lines is only a small fraction of the total energy released by the optical flare. Typically $E_{Balmer} \approx 3\ H\alpha \approx 0.1\ E_{opt}$. The electron densities estimated from the Balmer lines are in the range $\approx 10^{13}$ - 10^{14} cm^{-3}, not much larger than the typical densities derived for solar flares.

As shown by Fig. 1b, the Balmer lines are strongly broadened during the early-flare phases and the broadening decreases with time during the flare evolution. The Ca II lines are not broadened and this suggests that the broadening may be largely due to the Stark effect to which the Ca II lines are much less sensitive than the Balmer lines (Byrne 1989, Robinson 1989). Red asymmetries have also been observed during the early flare phases (Byrne 1989, Foing 1989). As will be discussed later, Stark-broadened profiles and red-asymmetries are important constraints on flare models.

UV lines (Lyα, Mg II , C II, C IV, N V, Si II, Si IV, He II) are also enhanced during flares indicating an increase of pressure in the stellar chromosphere and transition region at the time of flares. Unfortunately, UV observations of stellar flares with IUE have been severely hampered by the lack of sufficient sensitivity for time-resolved observations. In most cases, the typical exposure times needed to observe the star in the UV were comparable to the flare lifetime, with the consequence that only a coarse comparison could be made between flaring and quiescent conditions. In addition to UV lines, a strong UV continuum over the spectral band 1150 - 1950 Å is also observed during stellar flares (Bromage et al. 1986, Phillips 1989), similarly to what observed in stellar flares, but several orders of magnitude stronger.

Orders of magnitude estimates (Pettersen 1988, Byrne 1989) indicate that roughly equal amounts of energy are released in optical emission, in transition region UV emission and in soft X-rays. However, it is important to keep in mind that these estimates are largely based on observations of individual events and that simultaneous observations were available in only a few cases and only for a very limited number of spectral bands. Large variations are expected to occur from flare to flare. Considerable caution, therefore, should be exerted in using flux ratios at different frequencies for constraining flare models, at least until a much larger number of simultaneous multi-wavelength observations will become available.

2.3 *X-ray emission*

Flares from UV Ceti-type stars have been observed at soft (≤ 10 KeV) X-ray energies by both the *Einstein* and EXOSAT Observatories. On the contrary, no hard X-ray observations of stellar flares have yet been possible, and there is no great hope that the situation will change in the near future. In solar flares the ratio of hard to soft X-rays is typically $\approx 10^{-5}$. If the same ratio applies to stellar flares, one would need detector areas as large as a soccer field. The hope is that the flux of hard X-rays from stellar flares may be larger (and the spectrum harder) than in solar flares. Even so, detection of hard X-rays from stellar flares would require major technological advances which at present are beyond the most optimistic expectations.

Fig. 2 shows an example of a soft X-ray flare observed with the *Einstein* Observatory. The flare coronal temperature reached a peak value of $\approx 20 \times 10^6$ K and the emission measure was a few times 10^{51} cm^{-3}. The *Einstein* observations of stellar flares were reviewed by Haisch (1983). With the launch of the EXOSAT Observatory our knowledge of stellar X-ray flares has increased enormously. The EXOSAT observations have been discussed most extensively by Pallavicini et al. (1989). There is a large variety of total energies and time scales for stellar X-ray flares. The time scales range from 10^2 to 10^4 sec and are similar to the typical time scales for solar X-ray flares. Also the coronal temperatures are similar to solar values and typically in the range 20 - 40 x 10^6 K. On the contrary, the emission measures and the total energies are much larger than for solar flares. The total X-ray energies range from 10^{30} to 10^{34} erg and the emission measures are in the range from 10^{51} to 10^{53} cm^{-3}. There is no direct measurement of flare coronal densities (as could be derived from ratios of density sensitive lines). Indirect estimates, based on radiative cooling times, suggest electron densities in the range 10^{11} - 10^{13} cm^{-3}.

The EXOSAT observations have suggested the existence of two different classes of stellar flares which might be the analog of solar compact and 2-ribbon flares (Pallavicini et al. 1989). Although this classification is based only on morphological evidence (the time scales and total energies being different in the two cases), it is tempting to speculate that these two classes may also correspond to different physical mechanisms, as believed to occur for solar flares (Pallavicini et al. 1977, Poletto 1989). We will come back later to the modelling of these two classes of stellar flares.

In analogy to what observed at optical wavelengths, there is a tendency of the largest X-ray flares to occur only on the most active stars. Observational selection effects may prevent small flares to be detected in intrinsically brighter stars. However, it is also true that in the EXOSAT sample no large flare was observed from intrinsically fainter stars, although one of them (UV Ceti) was observed for longer time than any other flare star. Note that the most active flare stars (in terms of their quiescent X-ray emission) are also the optically brightest since the quiescent X-ray luminosity appears to be proportional to the bolometric luminosity (Agrawal et al. 1986, Pallavicini et al. 1989). The average flare frequency distribution for all flare stars detected by EXOSAT is similar to the frequency distribution for optical flares, i. e. it follows a power-law $N(>E) \propto E^{-\alpha}$ with $\alpha = 0.7 \pm 0.1$ (in the optical case, α varies from 0.4 to 1.2 depending on the star while in the solar case $\alpha \approx 0.8$).

2.4 Radio emission

Radio flares from UV Ceti-type stars have been observed from centimetric to metric wavelengths (Gibson 1983, Dulk 1985, Kundu et al. 1987, 1988, Kuipers 1989, Lang 1989). Although energetically unimportant (the ratio of microwave to X-ray flux being typically on the order of 10^{-7} - 10^{-8}), microwave observations of stellar flares possess great diagnostic capabilities for the primary energy release process. Unfortunately, the interpretation of radio data is not straightforward and is usually heavily model dependent. Fig. 1a shows an example of a solar flare observed at 2 cm and 6 cm with the VLA. In this particular case, a good correspondence appears to be present between the optical continuum emission and microwave emission, especially at 2 cm. At first sight this correspondence might remind us of the good correlation known to exist for solar flares between microwave and hard X-ray bursts (if we take the stellar optical continuum emission as a proxy for hard X-ray emission, see above). However, it is important to stress that the correspondence shown by Fig. 1a is not typical of what observed in stellar flares. In most other cases there is a very poor correlation between radio and optical (as well as X-ray) flares (Nelson et al. 1986, Kundu et al. 1988).

Observations from the VLA have in fact provided ample evidence that fundamental differences exist between solar and stellar radio flares. Microwave flares from UV Ceti-type stars are often narrow-band ($\Delta\nu/\nu \leq 0.1$), have high brightness temperatures ($T_b \approx 10^{12}$ - 10^{15} K) and a high degree of circular polarization (up to 100%). These characteristics can only be explained by a coherent emission mechanism which remains largely unknown (electron cyclotron maser or some form of plasma radiation are the most frequently cited candidates, cf. Kuipers 1989). On the contrary, microwave flares from the Sun are most often produced by an incoherent emission mechanism which has been identified as non-thermal gyrosynchrotron emission. The poor correlation found in most cases between stellar microwave flares and optical/X-ray flares is consistent with the coherent nature of most stellar radio flares (since optical and X-ray emission are certainly incoherent).

We also recall that stellar flares at metric wavelengths are often delayed with repect to optical flares (an example at decimetric wavelengths is shown in Fig. 2). This is similar of what observed in solar flares and suggests the presence of disturbances moving throught the coronae of UV Ceti-type stars.

3. Models

There is no comprehensive theory capable of explaining the ensemble of the observational data mentioned above. The attempts made so far have been aimed at explaining some particular aspects of the flare problem, e.g. the optical continuum emission or the evolution of the high-temperature coronal region or the mechanisms responsible for the observed radio emission. The large majority of these attempts (with the only possible exception of stellar radio models) have been developed in the framework of the solar analogy by assuming that the same basic physical processes are operating in both cases, although on a vastly different energy scale and under different physical conditions. It may be questioned whether this approach is reasonable in view of the many differences that indubitably exist between solar and stellar flares (e.g. the energies involved are orders of magnitude larger, the relevant length scales and magnetic field strengths are also probably larger, and the radio emission mechanisms are certainly different).

There are several reasons why the solar analogy should be pursued further in studying stellar flares. First, all attempts that were made in the past to explain UV Ceti-type flares with non-solar models (see, e.g., review by Gurzadyan 1980) have encountered severe difficulties and had to be abandoned as soon as more stringent observational constraints became available. Secondly, in spite of the differences, the number of analogies that appear to be present between solar and stellar flares is so high (see section 2 above) that an investigation of solar-type models under stellar conditions is certainly warranted. This is not to say that solar-type models will be able to explain all phenomena observed in stellar flares, but at least it should be possible to determine what can be explained (if anything) by solar-type models and what cannot. Even in the case that solar models should prove totally inadequate, the investigation of them under stellar conditions is important since it will allow us to gain a much better understanding of the basic mechanisms of solar activity.

3.1 Solar flare loop models

Before considering the application of solar-type models to stellar conditions it may be useful to briefly recall the main features of current solar flare models. We will start with loop models, i.e. models which assume that the flare energy is released within one or more loop-like structures which remain magnetically confined throughout the flare (we will consider later the case when the magnetic configuration is assumed to be disrupted at the flare onset; cf. section 3.7 below).

Two very popular flare loop models are the so-called *thick-target* and *thermal* models (see, e.g., Dennis and Schwartz 1989, Emslie 1989). Both assume that energy is injected impulsively in the coronal portion of a magnetic arch whose footpoints are anchored in dense chromospheric and photospheric layers. In the thick-target model the primary energy release produces a large number of accelerated particles (either electrons or protons). In the most commonly accepted version of this model, the particles are mainly low-energy electrons with cut-off energy $E_c \approx 20$ KeV. These electrons are channeled by the magnetic field towards the dense loop footpoints and produce hard X-rays by non-thermal bremsstrahlung (while the tail of the electron distribution produces microwave emission by the gyrosynchrotron process). Most of the electron energy, however, is deposited at the loop footpoints by collisional losses, thus producing the observed optical and UV line and continuum emission, as well as the white-light flare. Collisional heating at the loop footpoints, in turn, produces evaporation of high-temperature plasma which fills the structure and is ultimately responsible for the observed (thermal) soft X-ray emission.

In the thermal model, the energy released produces mainly heating in the coronal portion of a loop. Energy is transferred downward by a conduction front and the lower atmospheric layers (that are responsible for the observed UV and optical emission) are heated by thermal conduction rather than by non-thermal particles. In this model hard X-rays are produced thermally in the high-temperature region created by the initial energy deposition. Also in this case, however, the large heat flux deposited at the loop footpoints produces evaporation of chromospheric material and hence the filling of the structure with high-temperature soft X-ray emitting plasma.

Although we have sketched these two models as completely separate one from the other, there is ample evidence from solar observations that both thermal and non-thermal processes occur in the same events. What is not known is the relative importance of thermal vs. non-thermal processes in the overal flare energy budget. Similarly, if stellar flares are analogous to solar flares, we may expect that the same basic ingredients (i.e. electrons or proton beams, thermal conduction, evaporation of dense chromospheric material etc.) should also be present in the stellar case. The relevant question is thus how these various energy transport mechanisms are operating in the different physical conditions of M dwarfs. We recall that the masses, radii and effective temperatures of these stars are substantially smaller than the corresponding values for the Sun ($M_* \approx 0.1 - 0.5\ M_o$, $R_* \approx 0.2 - 0.6\ R_o$, $T_{eff} \approx 2800 - 3500$ K), the photospheric densities are at least a factor 10 higher and the surface gravities are a factor 2 to 4 higher.

3.2 Heating by low-energy electrons: gasdynamic effects

Katsova and Livshifts were probably the first to call attention to the importance of gasdynamic effects in explaining the observed properties of stellar flares. In a series of papers that span almost a decade (e.g. Katsova et al. 1981, Livshits et al. 1981, Katsova and Livshits 1988, Katsova et al. 1989) they have investigated a model that is basically an estension to stellar conditions of the gasdynamic model originally developed for solar conditions by Kostyuk and Pikel'ner (1974). They assume that energy is deposited in the upper chromosphere by a flux of non-thermal electrons with a power-law energy distribution $N(E) \propto E^{-3}$ with cut-off energy $E_c \approx 20$ KeV. The spectrum and cut-off energy of these electrons are similar to those derived for solar flares from observations of hard X-ray bursts (Dennis and Schwartz 1989).

The high pressure region produced by the heating generates two perturbations, one upward (which drives chromospheric evaporation and leads to the formation of the soft X-ray emitting region) and one downward towards deeper chromospheric layers. The flare evolution is determined by the gasdynamic response of the atmosphere to the heating perturbation and can be described by solving the full set of time-dependent mass, energy and momentum equations for a prescribed geometry and assigned boundary conditions. Since the authors were mainly concerned with the optical properties of the flare, they treated in detail only the perturbation propagating downward through the optically-thick chromospheric layers (for a more detailed treatment of the gasdynamics in the optically-thin upper portion of the flaring structure see section 3.5 below).

From the upper chromosphere, where the non-thermal electrons are collisionally stopped, energy is transported downward by a conduction front and by a shock wave which precedes the thermal wave. A narrow ($\Delta L \approx 10$ Km) high-density ($n_H \approx 10^{15}$ cm^{-3}) region at $T \approx 10^4$ K forms behind the shock front and the emission from this region is capable of explaining the observed properties of the optical flare. An interesting prediction of the model is that the Balmer lines should be Stark-broadened (because of the high density of the emitting region) and should also be red-shifted, in qualitative agreement with stellar flare observations. This two-step process, however, is not very efficient and a large energy flux ($F_o \approx 10^{12}$ erg cm^{-2} s^{-1}) must be deposited in the upper chromosphere to produce a perturbation sufficiently strong to give the optical continuum flare. Moreover, the assumed geometry is somewhat idealized (a plane-parallel open atmosphere) and the flare evolution has been followed only for the first few tens of seconds (the heating is assumed to operate for 10 sec). In spite of these drawbacks, the model is intesting in so far it tries to explain the flare by consistently solving the full set of hydrodynamic equations including (at least partially) radiative transfer effects.

3.3. Heating by thermal conduction

An essential condition for the Livshits and Katsova model to work is that low-energy electrons (with E_c as low as ≈ 20 KeV) can penetrate in a stellar atmosphere as deep as the upper chromosphere. Mullan (1989) has argued that this is not possible in stellar conditions since flaring loops in M dwarfs are expected to be longer and denser than for solar flares. The penetration depth of an electron of energy E_{KeV} is given by

$$\xi_p(E) \approx 7 \times 10^{19}\ (E_{KeV}/20)\ \ \text{cm}^{-2} \tag{2}$$

while the expected column density of a flaring loop is $\xi(\text{loop}) = \langle n\rangle L \approx 10^{21} - 10^{23}$ cm^{-2} $\gg \xi_p(20$ KeV). Hence, low energy electrons will be effectively stopped in the corona and energy will be transported downward by thermal conduction (Mullan 1976, 1977, Kodaira 1977). Note that in this model the observed ratio of X-ray to optical flare luminosity L_x/L_{opt} is a measure of the relative importance of conduction and radiative losses since $L_X/L_{opt} \approx \tau_c/\tau_r$. On average, this ratio is on the order of 1 in stellar flares (cf. Byrne 1989) but substantial variations can occur from flare to flare and even in the course of the same event.

It seems unlikely that a purely conductive model can explain the impulsive continuum emission. Temporal variations as short as 0.1 sec are observed in the light curve of optical flares (cf. Gershberg 1989), while the shortest time variations expected from a conductive model should be of the order of L/c_s, where L is the loop semilength and c_s is the sound speed. Since c_s is of the order of 500 - 1000 Km s^{-1} in typical flare conditions, one expects variations on time scales longer than several tens to a few hundred seconds if the optical flare is heated by thermal conduction.

There are other ways by which one could discriminate between thermal and non-thermal models. As shown by Cram and Woods (1982) and Canfield et al. (1984), heating of the chromosphere by non-thermal electrons produces Stark-broadened Balmer lines with central reversals. On the contrary, heating by conduction would produce narrow Balmer lines with weak or no central reversal. Solar observations (cf. Canfield et al. 1984) favor a non-thermal model during the impulsive phase of flares and the observed broadening of Balmer lines in the initial phase of stellar flares also argue against a purely thermal model. However, thermal conduction may become an important energy transfer mechanism during the later gradual phase of solar and stellar flares.

3.4 Penetration of high-energy particles to deep layers

The argument presented by Mullan (1989) against heating by non-thermal electrons does not hold if the accelerated particles in stellar flares have energies much larger than those typically encountered in solar flare conditions. From an analysis of optical continuum emission of stellar flares, Grinin and Sobolev (1977) concluded the the typical hydrogen densities in optical stellar flares should be as high as $n_H \approx 10^{15}$ - 10^{17} cm^{-3} and the optical thickness in the continuum » 1, in contrast to the solar case where τ_{cont} is always ≤ 1. From this they argue that continuum emission in stellar flares may originate from direct heating of deep chromospheric and photospheric layers by non-thermal particles with energy larger than in solar flares (Grinin and Sobolev 1989). Two possibilities have been considered: either electrons with energy higher than $\approx$ 100 KeV or protons with energy higher than $\approx$ 5 MeV (protons have also been proposed on different grounds by van den Oord 1988 and Simnett 1989).

Calculations carried out by Grinin and Sobolev (1989) show that protons with a power-law energy spectrum $N_p(E) \propto E^{-3}$ at energies $E \geq 5$ MeV could reproduce the optical continuum emission of stellar flares provided the energy flux carried by protons is $F_p \approx 10^{11}$ erg cm^{-2} s^{-1}. This is about one order of magnitude less that what required by the two-step model of Katsova and Livshifts (see section 3.2 above). Notice that in this model the heating of deep atmospheric layers is produced by both the protons and the optical and UV photons created as the beam penetrates the dense atmospheric layers. The heating due to photons actually penetrates deeper and has a larger amplitude than direct particle heating.

It is not possible at present to discriminate between proton heating and a model in which the same energy input is provided by electrons with energy $E_e = E_p\,(m_e/m_p)^{1/2} \approx E_p/50$. Hard X-ray and γ-ray observations of stellar flares could in principle discriminate between these two possibilities, in addition to providing information on the energy spectrum of accelerated particles. Unfortunately, the feasibility of such observations is far beyond the present capabilities.

3.5 Formation of the high-temperature region

The models discussed above were all concerned with the optical flare and hence with the lower atmospheric levels. The advent of X-ray observations from space has raised interest in models wich deal with the upper atmospheric levels and with the formation of the high-temperature soft X-ray emitting region. The physics of this region cannot be decoupled from the processes which occur at the lower levels. In practice, however, the difficulty of treating the full set of gasdynamic equations in the presence of radiative transfer effects has lead to a virtual decoupling of the lower and upper sections of the same flaring structure. Thus, while Katsova and Livshits (1988) were mainly concerned with the physical processes in the optically-thick chromospheric levels, another class of models have been developed to treat the upper sections of a flaring loop structure. These models assume that the plasma is optically-thin (as is true in the transition region and corona) and treat the chromophere only as a boundary layer and a mass reservoir.

Extensive work has been done in the solar case to model the hydrodynamics of the coronal portion of flaring loops both for the case of thermal heating and for heating by non-thermal particles (Nagai 1980, Pallavicini et al. 1983, Cheng et al. 1983, Nagai and Emslie 1984, MacNeice et al. 1984, Fisher et al. 1985, Peres et al. 1987, Peres 1989). An essential feature of these models is the prediction that a soft X-ray emitting region should form as a consequence of chromospheric evaporation when the chromosphere receives more energy than can be radiated away. The predictions of the models have been shown to be in reasonable agreement with the overal evolution of the high temperature flare plasma, in spite of the simplified treatment of many important physical processes (see discussion in Emslie 1989). It is interesting to see whether the same type of models can also reproduce soft X-ray observations of stellar flares. Unfortunately, very little work has yet been made in this area.

Reale et al. (1988) have applied their solar-type hydrodynamic code to a moderate size flare observed from Prox Cen. Formal agreement between model predictions and the time evolution of the soft X-ray flux is obtained for a loop structure with semilength $L=7 \times 10^9$ cm heated for $\approx$ 700 sec (i. e. only during the rise phase of the flare) with a constant heating flux of $F_H = 1.25 \times 10^{10}$ erg cm^{-2} s^{-1}. From this they conclude that the loop structure involved in the flare was much larger with respect to the stellar radius than in the solar case. Although attractive, this model is not free from difficulties. First, the large size of the flaring loop implies that the structure was almost empty in the preflare state, since the derived semilength is about 4 times larger than the hydrostatic scale height at typical preflare temperatures. Moreover, only the early decay phase of the flare was reproduced and additional heating in the late-decay would be required to reproduce the observed long decay of the flare. Finally, and more importantly, the model may not be unique. For instance, a longer lasting heating in a shorter loop will probably reproduce the data at least as well.

The above difficulties show that more numerical simulations need to be carried out for a variety of different stellar conditions. Work is in progress (e.g. Cheng and Pallavicini 1989) to build a grid of stellar hydrodynamic models in order to explore the effects of different physical parameters on the dynamic evolution of stellar flares. Unfortunately, this type of numerical calculations are very demanding in terms of computer resources and progress in this area is occurring at a very slow rate. Furthermore, the larger energy input that is necessary to deposit in the stellar case affects severely the lower sections of the structure down to photospheric levels. It is questionable whether in this case neglecting radiative transfer effects and a proper treatment of the lower atmosphere is an acceptable approach.

Recently, Fisher and Hawley (1988) have developed a simplified algorithm to evaluate in an approximate way the time evolution of coronal temperature, pressure and emission measure without recourse to time consuming numerical calculations. They have applied the method to a large flare on AD Leo predicting the time evolution of coronal temperature and emission measure. While the derived peak values ($T \approx 25 \times 10^6$ K, $EM \approx 10^{52}$ cm^{-3}) are in reasonable agreement with typical values for stellar flares (see section 2.3 above), there were no soft X-ray observations available for this specific event. Therefore, the ability of the model to reproduce actual stellar flare observations remains to be determined.

3.6 Heating by soft X-ray irradiation.

In principle, one could test the predictions of Fisher and Hawley (1988) for the flare on AD Leo by using optical data (that were available for that particular event). In order to do that, one needs to relate the observed optical emission to the parameters of the high-temperature region by assuming a specific mechanism for transferring energy from the upper to the lower levels. This has been done by Hawley (1989) who has assumed that chromospheric heating during the flare results from "backwarming" by soft X-ray and UV radiation (see also early suggestions by Somov 1975 and Cram 1982).

In brief, the reasoning of Hawley (1989) goes as follows. Balmer line emission depends on the temperature structure in the chromosphere. If the chromosphere is heated by soft X-ray irradiation, the chromospheric temperature structure and the Balmer line emission can be parameterized as a function of the coronal temperature at the apex of the flaring loop. From the observed time evolution of the Balmer lines one can predict therefore the time evolution of the coronal apex temperature. For self-consistency, the predicted temperature evolution should be equal to that predicted by gas-dynamic coronal models such as those discussed in the previous section. Preliminary results published by Hawley (1989) show only a limited agreement between the two predicted temperature evolutions. However, the model is interesting both because it attempts to solve the whole flare problem in a self-consistent way and because it calls attention to soft X-ray irradiation as a mechanism for heating the optical region. It is interesting to observe that a good correlation has recently been found between Hγ and soft X-ray emission in stellar flares (Butler et al. 1989). Whether this correlation is due to soft X-ray irradiation of the lower atmospheric levels remains, however, unclear. Note that the mechanism suggested by Hawley predicts that the flare seen in the Balmer lines should cover an area much larger than that of the optical continuum flare, if the latter is produced -as usually assumed- by non-thermal particles. This is reminiscent of what typically observed in solar flares.

3.7 Two-ribbon flares

In the models above we have assumed that the flare occurs in a magnetically confined loop-like structure that remains virtually unchanged troughout the flare evolution. On the Sun, this is believed to occur only for compact short-lived events. There is, however, another class of solar flares, the so-called long-duration 2-ribbon flares, in which the magnetic configuration is disrupted at the flare onset and energy is gradually released as the magnetic field lines relax back to a closed configuration. This type of events were first modelled by Kopp and Pneuman (1976; see also Cargill and Priest 1983). Kopp and Poletto (1984) have shown that the model is indeed capable of reproducing the main features of solar 2-ribbon flares.

Soft X-ray observations of stellar flares with the EXOSAT Observatory have shown the existence of intense long-duration stellar flares which are strongly reminiscent of solar two-ribbon flares (Pallavicini et al. 1989). The long-decay of these flares indicates that energy must be continuosly provided troughout a large part of the flare decay. The interesting question is whether a reconnection model such as that proposed by Kopp and Pneuman could also explain this type of stellar events.

Poletto et al. (1988) and Poletto (1989) have addressed this question by applying the reconnection model to a few long-duration stellar flares as well as to solar-flares observed by full-disk X-ray instruments. In the latter case, the flares are observed as if they were stellar flares, but the parameters predicted by the model can be directly compared with those observed with high-resolution instruments. In all cases, the reconnection model appears to be capable of reproducing the time evolution and intensity of the flare radiative power, but is unable to uniquely determine the best fit parameters of the model and, hence, the interesting physical properties of the flaring region.

The intrinsic limitations of the reconnection model when applied to stellar conditions are readily apparent. There are three free parameters in the model: the size of the flare region, the strength of the magnetic field at photospheric level and the speed at which the system relaxes from an open to the closed configuration. In the solar case the latter quantity is measured by the velocity V_o at which the reconnection point moves upwards as new closed loops form (see details in Poletto 1989). The time profile of the energy released by reconnection is determined almost uniquely by V_o. The other two parameters play only the role of a scaling factor, larger energy releases being obtained by either larger regions or higher magnetic field strengths. Therefore, in the absence of spatial information or other diagnostics (such as, for instance, density measurements), it is impossible to determine uniquely the spatial scale and magnetic field strength in stellar flares. Moreover, the agreement that can be obtained between model predictions and observations cannot be taken as a proof that long-duration stellar flares are indeed produced by a reconnection process as that envisaged by Kopp and Pneuman (1976). As the matter of fact, the same stellar flare (on Prox Cen) has been modelled as a compact flare by Reale et al. (1988) and as a 2-ribbon flare by Poletto et al. (1988).

4. Concluding remarks

From what we have discussed above, it should appear clear that the fundamental question whether stellar flares can be explained by scaled-up versions of solar-type models cannot be answered as yet. This is not only because the extensions of solar models to stellar conditions is still in a rudimentary stage. More importantly, in the stellar case there appears to be a lack a sufficient observational constraints to tests models and to discriminate between various alternatives.

In the solar case, there is a wealth of observational data and a number of sophysticated models to interprete these data. Although we are still far from a complete flare theory, there is some realistic hope that substantial progress can be reached on a relatively short time scale by an appropriate combination of additional observations and improved theoretical models. This is not the case for stellar flares. In this case, the paucity of the observational data is such that finding formal agreement between model predictions and observations may even result relatively simple. However, one has to worry about unicity and physical consistency of the proposed models. We have given several examples of this situation in the sections above.

In our opinion, substantial progress in the study of stellar flares can only be reached by major advances on the observational side, and particularly by the access to new diagnostic capabilities (e.g. hard X-rays, γ-rays) as well as by vastly improved sensitivity and spectral resolution at optical, UV and soft X-ray wavelengths. Unfortunately, this is not a programme that can expected will be completed in the next few years.

References

Agrawal, P.C., Rao, A.R. and Sreekantan, B.V.: 1986, *Montly Not. Roy. Astron. Soc.* **219**, 225.

Byrne, P.B.: 1989, *Solar Phys.* **121**, 61.

Byrne, P.B.: 1983, in *Activity in Red-dwarf Stars* (P.B. Byrne and M. Rodonò eds.), p. 157.

Byrne, P.B. and Rodonò, M. (eds.): 1983, *Activity in Red-dwarf Stars*, Reidel Publ. Co., Dordrecht.

Bromage, G.E., Phillips, K.J.H., Dufton, P.L. and Kingston, A.E.: 1986, *Montly Not. Roy Astron. Soc.* **220**, 1021.

Butler, C.J., Rodonò, M. and Foing, B.H.: 1989, in *Solar and Stellar Flares - Poster Papers* (B.M. Haisch and M. Rodonò eds.), p. 21.

Canfield, R.C., Gunkler, T.A. and Ricchiazzi, P.J.: 1984, *Astrophys. J.* **282**, 296

Cargill, P.J. and Priest, E.R.: 1983, *Astrophys. J.* **266**, 383.

Cheng, C.C., Oran, E.S., Doschek, G.A., Boris, J.P. and Mariska, J.T.: 1983, *Astrophys. J.* **265**, 1090.

Cheng, C.C. and Pallavicini, R.: 1989, in preparation.

Cram, L.E.: 1982, *Astrophys. J.* **253**, 768.

Cram, L. and Woods, T.: 1982, *Astrophys. J.* **257**, 269.

Dennis, B.R. and Schwartz, R.A.: 1989, *Solar Phys.* **121**, 75.

Dulk, G.A.: 1985, *Ann. Rev. Astron. Astrophys.* **23**, 169.

Emslie, G.A.: 1989, *Solar Phys.* **121**, 105.

Fisher, G.H., Canfield, R.C. and McClymont, A.N.: 1985, *Astrophys. J.* **289**, 414.

Fisher, G.H. and Hawley, S.L.: 1988, submitted to *Astrophys. J.*

Foing, B.H.: 1989, *Solar Phys.* **121**, 117.

Gershberg, R.E.: 1989, *Memorie Soc. Astron. It.* **60**, 263.

Gershberg, R.E. and Shakhovskaya, N.I.: 1983, *Astrophys. Space Sci.* **95**, 235.

Giampapa, M.S.: 1983, in *Activity in Red-dwarf Stars* (P.B. Byrne and M. Rodonò eds.), p. 223.

Gibson, D.M.: 1983, in *Activity in Red-dwarf Stars* (P.B. Byrne and M. Rodonò eds.), p. 273.

Gondhalekar P.M. (ed.): 1986, *Flares: Solar and Stellar*, Rutherford-Appleton Laboratories, Abingdon.

Grinin, V.P. and Sobolev, V.V.: 1977, *Astrofizika* **13**, 587.

Grinin, V.P. and Sobolev, V.V.: 1989, in *Solar and Stellar Flares - Poster Papers* (B.M. Haisch and M. Rodonò eds.), p. 297.

Gurzadyan, G.A.: 1980, *Flare Stars*, Pergamon Press, Oxford.

Haisch, B.M.: 1983, in *Activity in Red-dwarf Stars* (P.B. Byrne and M. Rodonò eds.), p. 255.

Haisch, B.M. and Rodonò, M. (eds.): 1989, *Solar and Stellar Flares - Poster Papers*, Catania Astrophys. Obs. Special Publication [see also *Solar Physics* **121** (1989)].

Havnes, O., Pettersen, B.R., Schmitt, J.H.M.M. and Solheim, J.E. (eds.): 1988, *Activity in Cool Star Envelopes*, Kluwer Acad. Publ.: Dordrecht.

Hawley, S.L.: 1989, in *Solar and Stellar Flares - Poster Papers* (B.M. Haisch and M. Rodonò eds.), p. 49.

Kahler, S. and other 30 authors: 1982, *Astrophys. J.* **252**, 239.

Katsova, M.M., Kosovichev, A.G. and Livshits, M.A., 1981, *Astrofizika* **17**, 285.

Katsova, M.M. and Livshits, M.A.: 1988, in *Activity in Cool Star Envelopes* (O. Havnes et al. eds.), p. 143.

Katsova, M.M., Livshits, M. A, Butler, C.J. and Doyle, J.G.: 1989, in *Solar and Stellar Flares - Poster Papers* (B.M. Haisch and M. Rodonò eds.), p. 87.
Kodaira, K.: 1977, *Astron. Astrophys.* **61**, 625.
Kopp, R.A. and Pneuman, G.W.: 1976, *Solar Phys.* **50**, 85.
Kopp, R.A. and Poletto, G.: 1984, *Solar Phys.* **93**, 351.
Kostyuk, N.D. and Pikel'ner, S.B.: 1974, *Astron. J. of USSR* **51**, 1002.
Kuipers, J.: 1989, *Solar Phys.* **121**, 163.
Kundu, M.R., Jackson, P.D., White, S.M. and Melozzi, M.: 1987, *Astrophys. J.* **312**, 822.
Kundu, M.R., Pallavicini, R., White, S.M. and Jackson, P.D.: 1988, *Astron. Astrophys.* **195**, 159.
Kundu, M.R., Woodgate, B. and Schmahl, E.J. (eds.): 1989, *Energetic Phenomena on the Sun*, Kluwer Acad. Publ., Dordrecht.
Lacy, C.H., Moffett, T.J. and Evans, D.S.: 1976, *Astrophys. J. Suppl.* **30**, 85.
Lang, K.R.: 1989, in *Flare Stars in Star Clusters, Associations and Solar Vicinity* (L.V. Mirzoyan ed.), in press.
Livshits, M.A., Badalyan, O.G., Kosovichev, A.G. and Katsova, M.M.: 1981, *Solar Phys.* **73**, 269.
MacNeice, P., McWhirter, R.W.P., Spicer, D.S. and Burgess, A.: 1984, *Solar Phys.* **90**, 357.
Mirzoyan, L.V. (ed.): 1989, *Flare Stars in Star Clusters, Associations and Solar Vicinity*, Kluwer Acad. Press, Dordrecht (in press).
Mullan, D.J.: 1976, *Astrophys. J.* **207**, 289.
Mullan, D.J.: 1977, *Solar Phys.* **54**, 183.
Mullan, D.J.: 1984, *Astrophys. J.* **279**, 746.
Mullan, D.J.: 1985, in *Radio Stars* (R. M. Hjellming and D.M. Gibson eds.), p. 173.
Mullan, D.J.: 1989, *Solar Phys.* **121**, 239.
Nagai, F.: 1980, *Solar Phys.* **68**, 351.
Nagai, F. and Emslie, A.G.: 1984, *Astrophys. J.* **279**, 896.
Neiding, D.F.: 1989, *Solar Phys.* **121**, 261.
Nelson, G.J., Robinson, R.D., Slee, O.B., Ashley, M.C.B., Hyland, A.R., Touhy, I.R., Nikoloff, I. and Vaughan, A.E.: 1986, *Montly Not. Roy Astron. Soc.* **220**, 91.
Pallavicini, R., Peres, G., Serio, S., Vaiana, G.S., Acton, L., Leibacher, J. and Rosner, R.: 1983, *Astrophys. J.* **270**, 270.
Pallavicini, R., Serio, S. and Vaiana, G.S.: 1977, *Astrophys. J.* **216**, 108.
Pallavicini, R. and Tagliaferri, G.:1989, in *Solar and Stellar Flares - Poster Papers* (B.M. Haisch and Rodonò eds.), p. 17.
Pallavicini, R., Tagliaferri, G. and Stella, L.: 1989, *Astron. Astrophys.*, in press.
Peres, G.: 1989, *Solar Phys.* **121**, 289.
Peres, G., Reale, F., Serio, S. and Pallavicini, R.: 1987, *Astrophys. J.* **312**, 895.
Pettersen, B.R.: 1988, in *Activity in Cool Star Envelopes* (O. Havnes et al. eds.), p. 49.
Pettersen, B.R.: 1989, *Solar Phys.* **121**, 299.
Pettersen, B.R., Coleman, L. A. and Evans, D.S.: 1984, *Astrophys. J. Suppl.* **54**, 375.
Phillips, K.J.H.: 1989, in *Solar and Stellar Flares - Poster Papers* (B.M. Haisch and M. Rodonò eds.), p. 45.
Poletto, G.: 1989, *Solar Phys.* **121**, 313.
Poletto, G., Pallavicini, R. and Kopp, R.A.: 1988, *Astron. Astrophys.* **201**, 93.
Reale, F., Peres, G., Serio, S, Rosner, R. and Schmitt, J.H.M.M.: 1988, *Astrophys. J.* **328**, 256.
Robinson, R.D.: 1989, in *Solar and Stellar Flares - Poster Papers* (B.M. Haisch and M. Rodonò eds.), p. 83.
Rodonò, M. and other 9 authors: 1989, in *Solar and Stellar Flares - Poster Papers* (B.M. Haisch and M. Rodonò eds.), p. 53.
Shakhovskaya, N.I.: 1989, *Solar Phys.* **121**, 375.
Simnett, G.M.: 1989, in *Solar and Stellar Flares - Poster Papers* (B.M. Haisch and M. Rodonò eds.), p. 357.
Somov, B.V.: 1975, *Solar Phys.* **42**, 235.
Sturrock, P.A. (ed).: 1980, *Solar Flares*, Colorado Assoc. Univ. Press, Boulder, Co.
Svestka, Z.: 1976, *Solar Flares*, Reidel Publ. Co., Dordrecht.
van den Oord, G.H.J.: 1988, *Astron. Astrophys.* **207**, 101.
Worden, S.P.: 1983, in *Activity in Red-dwarf Stars* (P.B.Byrne and M. Rodonò eds.), p. 207.

DISCUSSION

SIVARAM: (i) Have there been observations of flares from red giants?
(ii) What about the durations and periodicities of stellar flares? Are they consistent with a Poisson process?

PALLAVICINI: (i) Flares have been occasionally reported from a few late-type giants and supergiants, but have not been confirmed (Pettersen 1989, Solar Phys. 121,299). The only evolved stars that have been confirmed to flare regularly are RSCVn and Algol-type binaries, WUMa binaries and FK Com stars.
(ii) As far as we know, there are no regularities observed in the periodicities and durations of stellar flares. Both are consistent with Poisson statistics.

KUIJPERS: In my view the important issue is not if the solar flare can be scaled up to the stellar case. Rather, it is if the various kinds of explosions on stars are also of magnetic origin, but under different conditions. For example, in the case of an accretion disk around a neutron star magnetic flares may be important in a radiation-dominated plasma; or, in the case of active galactic nuclei, flares may occur in an electron-positron plasma. So I would like to stimulate my colleagues very much to think of interesting physics for magnetic flares under conditions different from the Sun or dMe stars.

PALLAVICINI: Your point is well-taken. Certainly, we should address the flare problem with an open mind, especially in view of the fact that there are at least some differences between the solar and stellar flare, especially with regard to radio emission. Nevertheless, the analogy between solar flares and flares on dM stars is so strong that there are good reasons to investigate further the application of solar-type models to stars. This may not be well justified for other astrophysical objects, such as those you mentioned.

LANG: Scaled-up version of solar flares are *not* consistent with the radio data for dwarf m stars. The radio emission is often not correlated with the X-ray or optical emission from flare stars, whereas they are correlated on the Sun. Rise times of about 10 milliseconds for radio flares indicate that the size is relatively small(~ 3000km), not a scaled-up larger size. Coherent radiation mechanisms are required, and, if cyclotion maser emission applies, the magnetic fields are not exceptionally strong (~ a few hundred gauss).

PALLAVICINI: I agree with you that radio observations indicate the existence of fundamental differences between solar and stellar flares and, therefore, that scaled-up versions of solar flares may not be sufficient to explain stellar flares. On the other hand, there are also many similarities at other wavelengths between solar and stellar flares which argue for the common mechanism. The large sizes and/or higher magnetic fields that I have inferred for stellar flares derive from the *assumption* (which I have not attemped to justify) that the energy of stellar flares derives from magnetic fields as for the Sun. On the other hand, size and magnetic fields derived from radio observations are themselves model-dependent and I do not think you can settle the issue of size and magnetic field strengths from these data alone. The problem is that we still do not know whether solar and stellar flares derive from a common mechanism.

GANGADHONA: (i) What is the typical time scale over which the flare energy is released?
(ii) What is the frequency range over which energy is released?

PALLAVICINI: (i) The typical time scales for energy release in UV ceti-type flare stars are similar to those of solar flares. It is often mentioned that the time-scales of stellar flares are shorter than for solar flares, but I think this is a selection effect since stellar flares have been observed mostly in continuum optical emission which is typically more impulsive. In the Balmore lines and in soft X-rays there is virtually no difference in time scales between solar and stellar flares.
(ii) Energy is released in stellar flares at all frequencies that have been observed so far, *i.e.* optical, UV, soft X-ray, radio. We do not yet have observations of stellar flares in hard X-rays and γ-rays.

ALURKAR: Was the negative flare an isolated event? Is it specific to a particular class of stars or is it an absorption line?

PALLAVICINI: Infrared stellar flares have been observed so far in only two cases, but this may be due to the fact that there were no previous observations in the infrared. We cannot say, therefore, if this is a general property of flare stars. The negative dip, observed in the k-band, is certainly not an absorption line, but we do not have yet any plausible mechanism to explain it, except that it is probably due to modifications of atmospheric structure and opacity effects. Observations are being planned to see whether negative IR flares occur on the Sun. So far we simply do not know if the same phenomenon occurs also on the Sun.

SHEVGAONKAR: (i) As we carry out more and more observations at longer radio wavelengths we find higher brightness temperatures ($\sim 10^{12} - 10^{14}$K). These can only be explained by a coherent mechanism or by a large source size ($\sim 100R_*$).
(ii) As you have mentioned, on flare stars the magnetic field and density are large, so the sight for coherent emission must be above the gyro-resonance levels. At that height the magnetic fields, if they are similar to that on the Sun, must have a simple configuration. Why then are there frequent coherent flares on the flare stars but not on the Sun?

PALLAVICINI: (i) Your point is well taken. (ii) I do not have any simple answer. We do not know the strength and topology of magnetic fields in flare stars. At this stage we can only make conjectures, but this will not help much in answering your question.

DAVILA: We know that on the Sun flares exhibit 155-day periodicity. Are there stellar observations with sufficient time coverage to determine whether a similar phenomenon occurs on other stars?

PALLAVICINI: Flare stars have been observed at optimal wave lengths for a long-time and periodicities have been sometimes reported. None of these periodicities, however, have been confirmed and the general belief at present is that stellar flares occur randomly in time. Thus no 155-day periodicity has been found so far for stellar flares, but certainly it would be useful to look again at the data or at future data in search of such a periodicity.

VAHIA: Pye and McHardy have explained flares on RSCVn binaries as magnetic loops of interbinary sizes. Does the fact that most of our flare stars are in binaries have any bearing on the flare characteristics of flare stars?

PALLAVICINI: I purposely excluded RSCVn binaries and PMS stars from my talk, because there are many indications that flares on these stars are fundamentally different from solar flares and from flares on UV Ceti-type stars. It is well possible that interconnecting

loops are involved in flares in RSCVn binaries. However, I do not think that binarity has any effect on flares on UV Ceti-type stars, since these stars are not close binaries. The typical distances between components of UV Ceti-type binaries are orders of magnitude larger than the stellar radius, contrary to what occurs in RSCVn binaries.

STOKES PARAMETERS FOR THOMSON SCATTERING IN MAGNETIZED PLASMA

CHIH-KANG CHOU and HUI-HWA CHEN
Institute of Physics and Astronomy
National Central University
Chung-Li, Taiwan 32054
Republic of China

The effect of a superstrong magnetic field on neutron stars or white dwarfs is studied for Thomson scattering in a fully ionized collisionless plasma. The equation of motion for an electron in the presence of both the induced electric field of the plasma and a static uniform external magnetic field is used to determine the acceleration of the electron. The collective plasma effects due to the field and density fluctuations are investigated by using the test-particle picture. The scattering of a photon by a plasma is a function of the acceleration of particles by the electric field of the incident wave and the static external magnetic field $\vec{B}_0$. Assuming that the electrons are distributed with density $n_e(\vec{r}',t')$, the radiation field far from the scattering center is

$$\vec{E}_{sc.} = \frac{r_0}{R(1-u)} \int n_e(\vec{r}',t')\, \hat{k}' \times \hat{k}' \times \begin{pmatrix} E_x + i\sqrt{u}\, E_y \\ E_y - i\sqrt{u}\, E_x \\ (1-u)\, E_z \end{pmatrix} e^{-i\vec{k}'\cdot\vec{r}'}\, \delta(t'-t+\frac{R}{c}) \cos\delta_\lambda \, d\vec{r}' ,$$

$$\hat{k}' = \vec{k}'/k', \quad R = |\vec{r} - \vec{r}'|, \quad u = \omega_c^2/\omega^2, \quad v = \omega_p^2/\omega^2, \quad r_0 = e^2/m c^2.$$

where the delta function indicates that all quantities are to be evaluated at the retarded time $t' = t - |\vec{r} - \vec{r}'|/c$, and δ_λ is the angle between the wave vector $\vec{k}$ and the Poynting vector, which is given by

$$\tan\delta_\lambda = \frac{-1}{2 n_\lambda^2(\alpha)} \left(\frac{\partial n_\lambda^2(\alpha)}{\partial \alpha} \right) .$$

The two indices of refraction, $n_\lambda(\alpha)$ are defined by

$$n_\lambda^2 = 1 - \frac{2v(1-v)}{2(1-v) - u\sin^2\alpha + (-1)^\lambda \sqrt{u^2 \sin^4\alpha + 4u(1-v)^2 \cos^2\alpha}}$$

where α is the angle between the wave vector and the static magnetic $B_0\,\hat{z}$.

E. R. Priest and V. Krishan (eds.), Basic Plasma Processes on the Sun, 93–94.

The Stokes parameters for the scattered radiation are computed explicitly in terms of the state of polarization of the incident wave, the electron-cyclotron frequency, the plasma frequency, the angle of incidence, and the angle of scattering. The effect of the magnetoactive plasma on the polarization parameters of the scattered radiation is considered in detail for incoming radiation propagating along or across the static magnetic field.

For a cold plasma with uniform particle density, the scattered field is reduced essentially to the vacuum radiation expression (in the single-particle limit). The collective effects of the magnetoactive plasma on the scattered radiation is then determined entirely by the acceleration of the particles in the plasma. For a hot inhomogeneous plasma with a strong static magnetic field, thermal effects must be included, and it is expected that magnetic Thomson scattering on thermal electrons is partially or completely cancelled by nonlinear scattering on the polarization clouds of the electrons. Such joint thermal and nonlinear effects on Thomson scattering will be considered in future publications.

The most distinguishing feature is the fact that the superstrong magnetic field ($B \approx 10^{12}$ G) greatly affects the Thomson scattering process, resulting in resonances in the scattering cross section. The important consequences of these cyclotron resonances are the increase in the photon mean free path in the scattering regions, strongly affecting the angular distribution and polarization properties of the scattered photons.

The effects of the plasma are very insenitive to the specific values of v ($v = \omega_p^2 / \omega^2$, ω_p denotes the electron plasma frequency) so long as $v \ll 1$, whereas the criterion for the magnetic field to substantially affect the Stokes parameters is that the photon frequency be less than the electron-cyclotron frequency. The effects of classical radiation damping (and natural line broadening) are briefly discussed. The corresponding quantum mechanical treatment for Compton scattering in a relativistic plasma is now under investigation.

REFERENCES

Canuto, V., Lodenquai, J., and Ruderman, M. : 1971, phys. Rev. D3, 2303.

Chou, C. K. : 1986, Astrophys. and Sp. Sci. 121, 333

Chou, C. K., Chen, H. H. : 1989, International Conference on Plasma Physics, New Delhi, India (November 22-28 1989).

Pierre PAILLERE
Centre d'Etudes de Limeil-Valenton
BP 27 - 94190 - Villeneuve-St-Georges - France

CLASSICAL TREATEMENT OF THE COMPTON COLLISION IN GENERAL RELATIVITY

A systematic development of the photon 4-momentum in the non-quantic perspective of General Relativity is very difficult to find in the scientific litterature. However, the photon 4-momentum is a fundamental concept if one wants to study relativistic transport theory or Compton collision formulae. For these reasons, a coherent formalism has been elaborated. It sysmetically allows one to formulate these problems the criterion of validity being the fact that in the case of the Minkowski metric, the general formulae give the classical results of the Special Relativity.

This paper based on previous work (1) which contains detailed calculations, deals with the classical treatment of the Compton collision in a non diagonal Riemannian metric under the hypothesis of conservation of the global 4-momentum. Following a deep analogy with the non-zero rest mass particle, one defines the 4-momentum and the apparent mass of the photon, M. Then one writes the generalized Doppler effect and defines a new scalar parameter ξ to describe the trajectory of the photon and one calculates the frequency and velocity of the emerging photon and electron in a Compton collision, the collision parameters being the frequency and the direction of the incident photon, the velocity of the incident electron and the direction of the emerging photon. We have :

---

(1) Pierre PAILLERE :"Le corpuscule photon en espace de Riemann" Note CEA n° 2401 (March 1984)

E. R. Priest and V. Krishan (eds.), Basic Plasma Processes on the Sun, 95–96.

$q^\alpha = h\nu\Omega^\alpha/\sqrt{goo}$, $\Omega\alpha\Omega^\alpha = 0$, $\gamma ij\Omega^i\Omega^j = 1$, $\gamma ij = -gij + \gamma i\gamma j$, $\gamma i = goi/\sqrt{goo}$,

$H = cq_o = h\nu$, $a = \gamma i\Omega^i$, $a' = \gamma i\Omega^{i'}$, $M = h\nu(1-a)/c^2 goo$,

$q^\alpha = M_o c\, dx^\alpha/cd\zeta$, with $M_o c = q_\alpha u_s^\alpha$, u_s^α : source velocity ,

and (R) : $d\lambda = \frac{cd\zeta}{M_o} = \frac{cdt}{M}$ = relativistic invariant, so that :

$$\frac{\nu'}{\nu} = \frac{\sqrt{goo} + \gamma i\beta^i - \gamma ij\Omega^j\beta^i}{\sqrt{goo} + \gamma i\beta^i - \gamma ij\Omega^{j'}\beta^i + \Gamma\Lambda(1-\vec{\Omega}\vec{\Omega}')}, \quad \Gamma = \frac{h\nu}{moc^2\sqrt{goo}}$$

$$\frac{\beta^{i'}}{\sqrt{goo}} = \frac{\beta^i + \Gamma\Lambda(\Omega^i - \nu'\Omega^{i'}/\nu)}{\sqrt{goo} + \Gamma\Lambda\,[1-a-(1-a')\nu'/\nu]}, \quad \Lambda = [(\sqrt{goo} + \gamma i\beta^i)^2 - \beta^2]^{\frac{1}{2}}.$$

In the Schwarzschild's case one has : $a = a' = 0$ and

$$\frac{\nu'}{\nu} = \frac{\sqrt{1-2m/r} - (\vec{\Omega}.\vec{\beta})}{\sqrt{1-2m/r} - (\vec{\Omega}.\vec{\beta}) + \Gamma\Lambda(1-\vec{\Omega}\vec{\Omega}')}, \quad \Lambda = [1-2m/r - \beta^2]^{\frac{1}{2}} \quad , \alpha = 1,665.10^5 CG$$

while, with the help of the (R) relation we obtain by the geodesic equati $Dq_o = 0$, $h\nu$ = constant along the trajectory that we can write, with $u = 1/r$

$$\frac{du}{dct} = -u^2(1-2mu)\,[1-(Ru)^2\,(1-2mu)]^{\frac{1}{2}}, \quad \frac{d\varphi}{Dct} = Ru^2(1-2mu),$$

whence the transit time Δt from Sun to Earth :

$$\frac{c\Delta t}{R} = \int_{R/DSE}^{1} dx/[x^2\cdot(1-\alpha x)][1-x^2(1-\alpha x)]^{\frac{1}{2}}; \quad (\alpha = 2m/R \quad , x = Ru)$$

(DSE = distance of Sun to Earth/R = Sun radius)

We find Δt = 498,99669 s while the quotient of DSE by c gives $\Delta t'$ = 499,006813 s !

Notice that for goo = 0 we get : $\beta = 0$, $\beta' = 0$, $\nu = 0$, $\nu' = 0$, $1 = 0$, $\frac{du}{dt} = 0$ and $\frac{d}{dt} = 0$ (R = 6,9598 10^{10} cm, DSE = 1,495985 10^{13} cm).

CHROMOSPHERIC AND CORONAL HEATING MECHANISMS

P.Ulmschneider[1] and U.Narain[2]
1 Insti. f. Theor. Astrophys., Im Neuenheimer Feld 561, D-6900 Heidelberg, W.Germany
2 Astrophysics Research Group, Meerut College, Meerut-250003, India

Stellar chromospheres lose energy, predominantly, via radiation and coronae via conduction, radiation and stellar wind flows. To prevent these hot layers from rapidly cooling down to photopsheric temperature some source of heating must be present.

Various heating mechanisms have been proposed to date. For details see Narain and Ulmschneider (1989) and references contained therein.

This study leads us to conclude that in late-type stars short-period waves seem to contribute a basic but weak background heating, independent of rotation. In rotating late-type stars magnetic field related heating is dominant. Slow mode and longitudinal tube waves seem to heat low and middle chromosphere via shocks readily. Locally produced fast mode waves may heat coronal loops via shocks and Landau damping. Transverse and torsional Alfven waves can also heat coronal loops via resonant absorption, phase mixing, nonlinear mode coupling and Landau damping. Alfvenic surface waves can heat these loops via resonant absorption in thin sheaths. The loops can also be heated via current (magnetic field) dissipation through anomalous Joule heating and reconnection.

Coronal holes are heated, primarily, by dissipation of shear Alfven waves via phase mixing, nonlinear mode coupling turbulent and Landau damping.

One of the authors (U.N.) is very much thankful to Prof.Vinod Krishan and other organisers for their financial help and encouragement.

Narain,U. and Ulmschneider,P. : 1989 Space Sci. Rev. (In Press).

E. R. Priest and V. Krishan (eds.), Basic Plasma Processes on the Sun, 97.

IV. PHOTOSPHERIC FLOWS and MAGNETIC FIELDS

LARGE-SCALE FLOW PATTERNS IN THE SOLAR ATMOSPHERE

K.R.Sivaraman
Indian Institute of Astrophysics
Bangalore 560034
INDIA

INTRODUCTION: It is now well established (McIntosh 1979; Makarov and Sivaraman 1983) that the filament and filament channels seen in the H-alpha spectroheliograms (or filtergrams) can be used as reliable tracers for studying the time evolution of large-scale magnetic fields on the Sun. These features represent the neutral lines between the unipolar regions of opposite polarity. Comparison of the synoptic charts compiled from H-alpha pictures with those from full-disc magnetograms for the same period shows very good agreement and hence the former can be used with confidence for time evolution studies of large-scale unipolar regions for those periods when the magnetographs did not even exist. In this paper we shall present one of the results of our study (Makarov and Sivaraman 1983, 1989) on the migration of H-alpha filaments, namely, the existence of the meridional flow on the Sun. We shall extend it further to show the participation of this meridional flow in the solar cycle variation.

DATA AND ANALYSIS: Starting from the H-alpha and K_{232} spectroheliograms of the Kodaikanal and Kislovodsk stations we constructed the H-alpha synoptic charts for the period 1904-1982. For the period 1964-1973 we used McIntosh's (1979) synoptic charts themselves. These charts show clearly the continous boundaries of the large-scale unipolar regions and it is possible to assign a mean latitude to every filament band over each solar rotation. We then plot such mean latitude values for the polemost filament band as well as for other filament bands as a function of time. The filament migration derived thus for the period 1910-1982 (Makarov and Sivaraman, 1989) is shown in Fig.1.

RESULTS: The filament bands start their poleward migration with small speeds ranging from 5 to 20 m sec^{-1} synchronously with the rising phase of the solar cycle (Fig.1) Around the peak of activity, these filament bands, which previously move slowly, experience a sudden acceleration resulting in velocities as high as 40-m sec^{-1} or more and on reaching the pole they cause the reversal of the polar fields.

E. R. Priest and V. Krishan (eds.), Basic Plasma Processes on the Sun, 101–105.

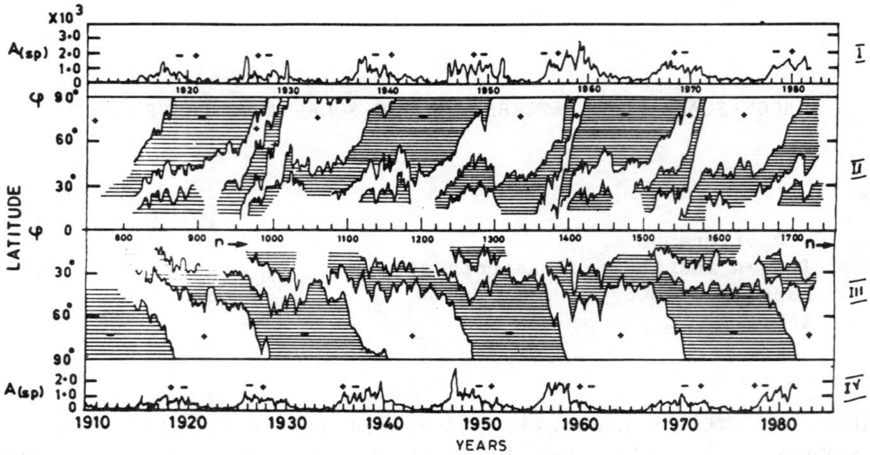

Fig.1. Boxes II and III show the migration trajectories of magnetic neutral lines (filament bands) in the North and South hemispheres for the period 1910-1982. Hatched areas: negative polarities and clear areas: positive polarities. Boxes I and IV: continous curve represents the run of the mean daily areas of sunspots [A(Sp)] in millionths of the visible hemisphere averaged over one rotation with 3 point smoothing.

Topka et al. (1982) have done a similar analysis for the period 1964-80 and there is perfect agreement between their results and ours for this period. We interpret that the motive force that causes the poleward migration of the filaments (which is also the transportation of unipolar magnetic regions) is a meridional flow towards the poles in the solar photosphere that has a mean velocity of 25-40 m s^{-1}. Topka et al. (1982) have shown that the process of diffusion alone is inadequate while a meridional flow can acount for the observed poleward transport. After the polar reversal, the filament bands which are to cause the next reversal are still at low latitudes (20°-30°) meandering with a quasi-oscillatory motion and they start the poleward journey only after the sunspot activity has increased.

DISCUSSIONS: Let us look at the Doppler measurements which show the meridional motion and compare these with the filament migration diagram. With the Doppler Zeeman Analyser, Beckers (1977) detected the meridional flow with velocity of 42 $\pm$ 9 ms $^{-1}$. Based on measurements on 89 days in the period July 1977-March 1978, Duvall (1979) detected in the latitude zone 10°-50° a meridional flow of 20 m s^{-1} from the Stanford magnetograph data. Howard (1979) from the analysis of Mt.Wilson Doppler measures for the period 1974-1977 also detected a poleward meridional flow with velocities of 20 m s^{-1}. Snodgrass reanalysing the Mt.Wilson data derived a latitude dependent polynomial fit to this flow and obtained a mean value of 10 m s^{-1} for the

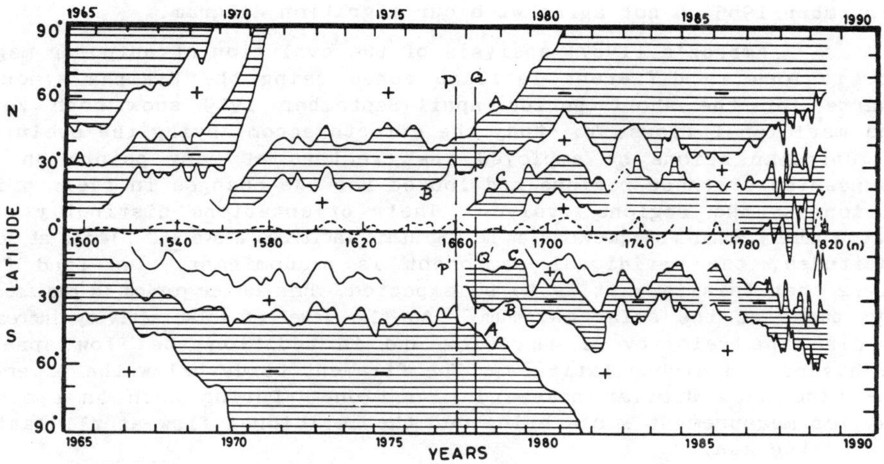

Fig.2. Migration trajectories for the period 1965-1988. The period July 1977 to March 1978 of Beckers' Doppler measurements (1978) is shown by the two vertical broken lines PP' and QQ'. Notice the filament bands, A, B and C in both the hemispheres show poleward migration during this period and are confined to the latitude zone 10°-40° where Beckers also detected the meridional flow.

period May 1982 to March 1984.

In Fig.2 we have reproduced the filament migration for the years 1965-1982 from Fig.1 and extended this to 1988 on a bigger scale to help comparison with the Doppler measurements. For example, Beckers' measures (1978) were obtained during July 1977 to March 1978. We have marked this period on Fig.2 in the north and south hemispheres. The meridional motion measured by him is reflected by the poleward moving filament bands A, B, & C (Fig.2). Similarly, the measurements of Duvall (1979) and of Snodgrass (1984) also fit well with the filament migration. Snodgrass's (1984) determinations show that the maximum of the velocity occurs around 30° latitude and this is what is to be expected from the migration diagram (Fig.2).

Another direction from which there is additional support is the meridional motion derived from line asymmetries of photospheric lines, but their agreement with the results of the filament migration studies have not been generally good. A possible reason could be that the Doppler measurement of meridional flow are in general complicated by a possible confusion with the latitude variation of the limb shift effect. In this respect the meridional motion determined from tracer measurements are superior and reliable. Cavallini et al. (1985, 1986) find a residual in the differences between the meridional and equatorial line shifts in the Fe I lines 6302.5A in July-August 1984 and July-September 1985. Their measurements on 6302.5A show a maximum at 30°-40° latitude and agree with Fig.2 although their velocities of 50 m s^{-1} are higher. Their measurements using 5576A done in July-

September 1985 do not agree with our migration diagram.

Ambroz's (1987) analysis of the evolution of unipolar magnetic regions in different latitude zones using the H-alpha synoptic charts for the short period April-September 1979 show both zonal and meridional flows. To study the effects accompanying the evolution of unipolar reions he employed test vectors at many points on the boundaries of these regions and looked for the changes in their orientation as the regions evolve. Their orientations distinctly show that a purely zonal flow prevails at heliographic latitudes above $\pm 50°$, while at lower latitudes, the meridional component is predominant. We find from Fig.2 that this is what is to be expected. This latter period is the ascending phase of the solar maximum. As the level of solar activity increases so also the velocity of the flow and in addition the flow spreads to higher and higher latitudes. The filament bands below the polemost one also show similar poleward migration. During such an epoch a Doppler measurement would bring out the meridional flow at all latitudes on the sun.

It is thus clear from the migration diagrams (Figs 1 & 2) that during the years of the solar minimum, the filament bands do not show signs of a continous migration towards the poles. They remain at low latitudes with a quasi-oscillatory motion. But with the commencement of the solar activity, they start all of a sudden moving polewards suggesting that the motive force driving them which is absent or very weak during the solar minimum starts operating from this epoch onwards. This behaviour is unmistakably seen cycle after cycle (1910-1982) in both hemispheres of the Sun.

REFERENCES

Ambroz, P., 1987. Bull. Astron. Inst. Czechosl. 38, 110.

Beckers, J.M., 1978, in G.Belvedere and L.Paterno (eds) Workshop on Solar Rotation, University of Catania, p.166.

Cavallini, F., Ceppatelli, G., Righini, A., 1985 Astron. Astrophys. 150, 256.

Cavallini, F., Ceppatelli., Righini, A., 1986, Astron, Astrophys. 163, 219.

Duvall, T.L., 1979, Solar Phys. 63, 3.

Howard, R., 1979, Astrophys. J. 229, L.45.

Makarov, V.I. and Sivaraman, K.R., 1983, Solar Phys. 85, 215 & 227.

Makarov, V.I. and Sivaraman, K.R., 1989, Solar Phys. 123, 367.

McIntosh, P.S., 1979, Annotated Atlas of H-alpha Synoptic charts, World Data Centre A for Solar Terrestrial Physics, NOAA.

Snodgrass, H.E., 1984, Solar Phys. 94, 13.

Topka, K., Moore, R., LaBonte, B.J. and Howard, R., 1982, Solar Phys. 79, 231.

DISCUSSION

VAN BALLEGOOIJEN: How did you define the latitude of a filament band?

SIVARAMAN: This is done by measuring the latitudes of the filament band in every 20° longitude zone and averaging them over one solar rotation. This mean value forms one data point on the migration trajectory curve.

PRABHAKARAN: The poleward migration of filaments does not have the same velocity throughout the cycle. How well does it agree with other measurements?

SIVARAMAN: The agreement between the Doppler measurements showing the meridional motion and our filament trajectory diagram is quite good. The fact that the poleward migration does not have the same velocity throughout the cycle is exactly the point I am emphasising by saying that the meridional motion participates in the solar cycle variations.

SUNSPOT MOTIONS FROM A STUDY OF KODAIKANAL AND MOUNT WILSON OBSERVATIONS

ROBERT F. HOWARD
National Solar Observatory
National Optical Astronomy Observatories †
Tucson, Arizona 85726, USA

K. R. SIVARAMAN
S. S. GUPTA
Indian Institute of Astrophysics
Bangalore 560 034, India

PAMELA I. GILMAN
Mount Wilson Observatory
And Department of Astronomy
University of California, Los Angeles
Los Angeles, California 90024, USA

ABSTRACT. A study of the daily motions of individual sunspots and of sunspot groups has been undertaken using Kodaikanal and Mount Wilson white-light observations. The Mount Wilson measurements were completed several years ago, and the Kodaikanal measurements have recently been started, using a technique identical to that used for the Mount Wilson data. Both data sets started early in this century, and both have continued to the present time. Preliminary results of a comparison of the two data sets are presented which show a good agreement between them in spot areas and motions. Analysis of the combined data set for the years that are available so far indicates that many more spots are available in such a reduction because the 12-hour time difference between the two sites allows for a more certain identification of individual spots from one observation to the next than in the case when data from only one site are used. This greatly increases the number of sunspots available for motion studies, particularly the smaller sunspots. Preliminary rotation and latitude drift reduction from the combined data set confirm earlier results from the Mount Wilson data alone.

1. Introduction

The important processes that produce the solar activity cycle take place beneath the solar surface. One method to advance our knowledge of the cycle mechanism is to study the motions at the surface of tracers that are rooted below. Considerable effort has gone into this activity in recent years (*e.g.* Schröter, 1985).

† Operated by the Association of Universities for Research in Astronomy, Inc., under Contract with the National Science Foundation.

E. R. Priest and V. Krishan (eds.), Basic Plasma Processes on the Sun, 107–111.

Several years ago the Mount Wilson white-light photoheliogram data set was measured for positions and areas of individual sunspots. This program resulted in a number of studies of properties of spots and spot groups (Howard, Gilman, and Gilman, 1984; Gilman and Howard, 1984a; 1984b; Gilman and Howard, 1985; 1986: Howard and Gilman, 1986; Bogdan, et al., 1988). A similar data set has been accumulated at the Kodaikanal site of the Indian Institute of Astrophysics. This set of plates extends back to 1905 and is in excellent condition.

Recently a digitizing instrument similar to that used for the Mount Wilson data was installed at Kodaikanal, and measurements of sunspots were begun. It is planned to measure every one of the plates at Kodaikanal from 1905 to the present. The technique being used is as nearly identical as possible to that used for the Mount Wilson plates and described in the first of the series of papers that came from this work (Howard, Gilman, and Gilman, 1984).

The measurement of the Kodaikanal white-light plates will soon give us a valuable data set with which to study the motions of sunspots at the solar surface and the radius of the sun during this century. This will be the largest collection of individual sunspot position measures in existence. The Greenwich data set, which has been much used in studies of sunspot motions, is an invaluable set of excellent data covering spot group areas and positions, but it does not have a complete set of individual spot area results. It has, however, been used to study the motions of single-spot groups (*e.g.* Wöhl, 1988).

The work is now in progress and so far several years of data have been measured. In this paper we discuss the measurements and present some of the first results from the Kodaikanal measures and from the combined Kodaikanal-Mount Wilson data set.

2. Observations, Measurements, and Reduction

Daily white-light photographs of the full-disk sun have been taken at Kodaikanal since 1905. These observations have been made with the same instrument and using the same observing procedures during the entire interval. The image size is approximately 20 cm, which is slightly larger than the Mount Wilson images.

The measuring technique for the Kodaikanal plates is quite similar to that used for the Mount Wilson plates. This is done in order to provide a uniform data set for comparison of the Kodaikanal results with the Mount Wilson results and for ease of combining the two data sets. A digitizing pad which is operated by a cursor containing a crosshair is used, as has been described for the Mount Wilson measurements (Howard, Gilman and Gilman, 1984). The accuracy of the digitizer is several times better than that of the best seeing on any of the solar images, and is therefore not a factor in the reduction process.

The measurement of a plate for a day consists of making 8 evenly spaced determinations of the limb position and then two position measures for each sunspot on the disk within 60 deg of the central meridian.

Each year is reduced separately as a whole. Accurate corrections to the positions of the limb points and the spot measures are made for atmospheric refraction using the time and date of the observation. The radius and central position of the solar disk are derived in digitizing board coordinates. Then solar coordinates are derived for each spot measurement point. Areas and positions of the spots are derived. Areas are corrected for projection effects.

With the data for each day tabulated, a search for consecutive day's observations is made. In the case of the combined data set from both observatories, the next available observation is used. The majority of the time differences in the combined data set cluster around 1/2 day. In practice, an upper limit of 1.5 days is used for the determination of 'daily' motions.

For each consecutive pair of observations, differences in latitude and CMD are calculated for each spot group and for each spot. In the case of groups the positions are area-weighted. The technique for determining the identification of individual sunspots is described (Howard, Gilman, and Gilman, 1984).

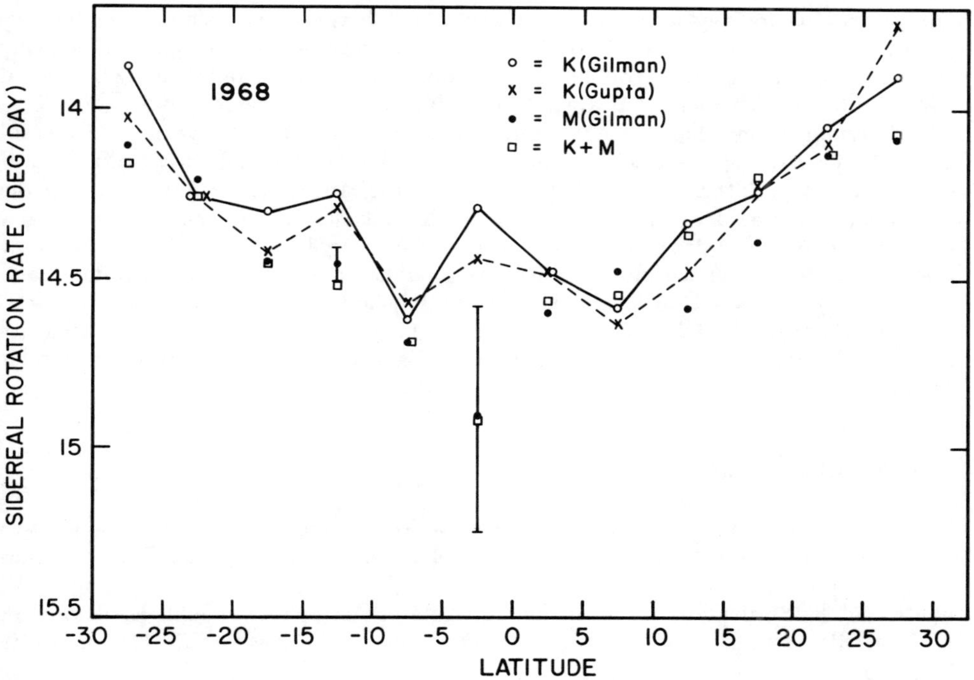

Figure 1. Sidereal rotation rate as a function of latitude for sunspots for the first 9 months of 1968. Circles are Kodaikanal data measured by Gilman; crosses are Kodaikanal data measured by Gupta; solid circles are Mount Wilson data; and squares are combined Kodaikanal-Mount Wilson data (Kodaikanal data measured by Gupta). Typical error bars are shown for low and intermediate latitudes. The errors at high latitudes are similar in magnitude to those at low latitudes.

3. Results

Figure 1 shows a comparison of solar rotation results for various data and various measurements for the first 9 months of 1968. Only part of the year was used because only that part of the year was available for the measurements of Kodaikanal data by PG. The

Kodaikanal data are represented by open circles and crosses. The agreement between the measurements made by the two measurers is seen to be well within the limits of the errors. Error bars (± one standard deviation) are shown for the Mount Wilson measures for low latitudes at 0 to −5 deg (typical also for high latitudes) and for intermediate latitudes at −10 to −15 deg, where there are more spots. The low latitude error bars refer to an average over 11 sunspots, and the intermediate latitude error bars refer to an average over 264 sunspots. These errors are of similar magnitude for Kodaikanal measures and slightly lower for the combined data set.

The differential rotation curve in Figure 1 representing the combined data set is indicated by squares. For the combined data set we used the Kodaikanal measures of SSG. The combined results for any latitude do not always fall between the results from the two separate measurements because it is not a simple average of the two sets. With many 1/2-day determinations in the combined result, this means that there are many individual determinations in this result that are not in the individual observatory results.

We determined the value of A, the sidereal equatorial rotation velocity in the representation $\omega = A + Bsin^2\phi$, where ϕ is the solar latitude. For the full disk data averaged over the 9-month interval, this value is 14.39 deg/day for the Mount Wilson data, 14.52 for the Kodaikanal data measured by PG, 14.62 for the Kodaikanal data measured by SSG, and 14.64 for the combined data set (Kodaikanal data measured by SSG). For the full 1921 through 1982 interval of the Mount Wilson data published earlier (Howard, Gilman, and Gilman, 1984) A is 14.522, although there were significant variations from year to year.

The results of the Kodaikanal measurements so far are encouraging, and we anticipate that we will have an opportunity soon to add significantly to our knowledge of solar dynamics.

4. Acknowlegements

This research has been made possible through a Grant from the US-India Cooperative Science Program of the Division of International Programs, US National Science Foundation. We gratefully acknowledge assistance from a number of people, including A. V. Ananth and R. Srinivasan, of the Indian Institute of Astrophysics, John E. Boyden of the University of California, Los Angeles, and Stephen A. Colley of the National Solar Observatory.

5. References

Bogdan, T. J., Gilman, P. A., Lerche, I., and Howard, R.: 1988, Astrophys. J. **327,** 451.
Gilman, P. A., and Howard, R.: 1984a, Astrophys. J. **283,** 385.
Gilman, P. A., and Howard, R.: 1984b, Solar Phys. **93,** 171.
Gilman, P. A., and Howard, R.: 1985, Astrophys. J. **295,** 233.
Gilman, P. A., and Howard, R.: 1986, Astrophys. J. **303,** 480.
Howard, R., Gilman, P. A., and Gilman, P. I.: 1984, Astrophys. J. **283,** 373.
Howard, R., and Gilman, P. A.: 1986, Astrophys. J. **307,** 389.
Schröter, E. H.: 1985, Solar Phys. **100,** 141.
Wöhl, H.: 1988, Solar Phys. **116,** 199.

DISCUSSION

PRIEST: (i) What do you believe is the cause for small spots to rotate faster than large ones? Is it because the small ones are affected more by the photospheric plasma?
(ii) Are there implications for the validity or otherwise of Piddington's tree-trunk model of the subphotospheric structure?

HOWARD: (i) I believe that the important factor is probably the age of the spot. This is not really an answer to this question, but one can imagine a weakening of the connection between the lower and surface layers as the spot ages and grows. Thus the spot would tend to rotate closer to the surface plasma rate as it aged.
(ii) It was this model that led me to assume that each spot group rotated as a whole *i.e.* showed no differential rotation. The fact that this is not the case implies either that the tree trunk model is not valid or that the connection of the spots to the subsurface layer is weak.

OBSERVATIONS OF MAGNETIC FEATURES WITH THE GERMAN SOLAR TELESCOPES AT THE OBSERVATORIO DEL TEIDE/TENERIFE

F. KNEER[1], D. SOLTAU[2], E. WIEHR[1]

[1] Universitäts-Sternwarte
Geismarlandstr. 11, D-3400 Göttingen, Fed. Rep. of Germany

[2] Kiepenheuer-Institut für Sonnenphysik
Schöneckstr. 6, D-7800 Freiburg, Fed. Rep. of Germany

ABSTRACT. The German solar facilities at the Obsrvatorio del Teide are described first. Then, a few examples of recent results from magnetic features are given: spatial variation and velocity fluctuation of small-scale magnetic fluxtubes in the quiet Sun, Evershed flow and magnetic field in connection with penumbral fine structure, and magnetic field variation in sunspot umbrae.

1. OBSERVING FACILITIES

The telescopes and their postfocus equipment are described extensively elsewhere (Wiehr, 1987, Kneer et al., 1987, Schmidt and Soltau, 1987, Kneer and Wiehr, 1989, Soltau 1989). Here, only a short summary is given.

1.1. GREGORY COUDÉ TELESCOPE (GCT)

At an altitude of 2400m, the evacuated GCT is housed in a dome on top of a 20m high tower. The 45cm parabolic primary mirror has a focal length of 2.40m which is enlarged to f_{eff} = 25.0m by the Gregorian. Instrumental polarisation is low and constant for a period of 2-3 days (Wiehr, 1971). The telescope feeds a) a slit-jaw camera for simultaneous observations in white light, Ca K, and Hα and b) a horizontal Czerny-Turner spectrograph (focal length 10m, grating: 316 grooves/mm, ruled area 180mmx360mm, blaze at 63.5 deg.). Computers perform positioning, scanning, data acquisition, and dome control. Exposure times in the spectrum (quiet sun centre) are 1/2-1s on film and few 10ms on CCDs. Operation of the GCT started 1986.

1.2. VACUUM TOWER TELESCOPE (VTT)

The VTT is located at the same site. The main mirror (diameter 70cm, focal length 46m) is fed by a coelostat. The VTT is equipped with a slit-jaw camera for simultaneous observation (photographically and on TV recording) in white light, Ca K, and Hα. The spectrograph consists of a low dispersion predisperser and the main vertical spectrograph (focal length 15m, echelle grating: 79 grooves/mm, ruled area 220mmx440mm, blaze at 63.5 deg.). Exposure times are similar as in the GCT. The spectrograph can also run in a multi-channel subtractive double pass mode (MSDP) supplied by colleagues from Meudon/France (Mein and Blondel, 1972). In addition, the telescope can feed optical laboratories for small-band imaging. Computer control exists as well. The VTT has operated since 1989.

E. R. Priest and V. Krishan (eds.), Basic Plasma Processes on the Sun, 113–117.

1.3. NEWTON TELESCOPE

This evacuated telescope with a 40cm parabolic main mirror on an equatorial mounting has an effective focal length of 37.5m. Filters, film cassettes, and photometers can be fixed to it. It has operated since 1972.

1.4. POLARIMETRY

Information about magnetic fields is obtained from the Zeeman splitting of spectral lines and from their polarisation properties. Several analysers for left and right circular polarized light, i.e. for the Stokes I and V intensities can presently be mounted at the entrances or exits of the spectrographs. An analyser for the complete Stokes vector $\underline{S}$ = (I, Q, U, V) is under construction.

2. MAGNETIC FEATURE OBSERVATIONS

From the many investigations concerning the observation of magnetic features only a few examples can be touched upon here. More detailed analyses are or will be published elsewhere.

2.1. SMALL FLUXTUBES

The motion of a small magnetic element beneath a Ca+ flocculum was investigated by Wiehr (1988). The zero-crossing of the Stokes V profile of the Fe I 6302 Å line shows wavelength shifts corresponding to peak-to-peak velocities of 750 m s^{-1} on timescales of 5min. Recent observations of magnetic elements in the quiet sun Ca network show rapid changes (within few arcsec) of the magnetic polarity.

2.2. SUNSPOT FINE STRUCTURE

2.2.1. *Penumbra*. Continuum intensity pattern, Evershed-flow, and magnetic field fluctuations are uncorrelated among each other (Wiehr and Stellmacher, 1989).

2.2.2. *Umbra.* Figure 1 shows a white light slit-jaw picture of umbral fine structures and the spectrogram containing the spectrum of the umbral dot to the left of the light bridge. Some densitometer recordings are collected in Figure 2. The Zeeman splitting of spectral lines yields magnetic field strengths given in Table 1. In the umbral dot, the field is weaker than in the surrounding penumbra, but not as much as suggested earlier (Kneer, 1973). Similar data are presently under analysis and the implications for atmospheric and magnetic field structure have yet to be investigated.

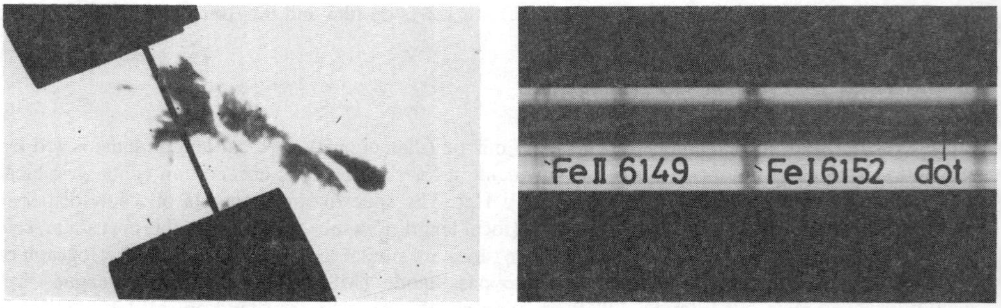

Fig. 1. White light picture and spectrogram of umbral fine structure.

TABLE 1. Magnetic field strengths (in gauss) in sunspot features.

	FeII 6149.2	FeI 6151.6	CaI 6156.0
inner penumbra	1900	1700	1900
umbra	-	2750	3050
umbral dot	2460	2330	2750

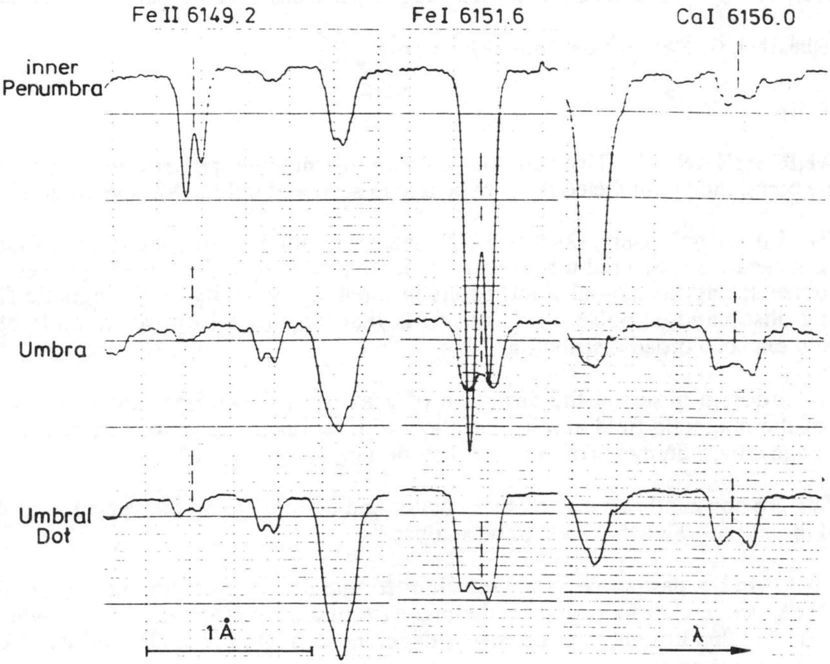

Fig. 2. Densitometer recordings of spectral lines from sunspot features

ACKNOWLEDGEMENTS. The "Hilfseinrichtung Sonnenphysik" is financed by the Deutsche Forschungsgemeinschaft.

REFERENCES

Kneer, F.: 1973, Solar Phys. **28**, 361

Kneer, F., Schmidt, W., Wiehr, E., Wittmann, A.D.: 1987, Mitt. Astron. Ges. **68**, 181

Kneer, F., Wiehr, E.: 1989, Proc. Workshop on Solar and Stellar Granulation, R.J. Rutten and G. Severino (eds.), Kluwer/Dordrecht, p. 13

Mein, P., Blondel, M.: 1972, Solar Phys. **27**, 489

Schmidt, W., Soltau, D.: 1987, Mitt. Astron. Ges. **68**, 184

Soltau, D.: 1989, Proc. Workshop on Solar and Stellar Granulation, R.J. Rutten and G. Severino (eds.), Kluwer/Dordrecht, p. 17

Wiehr, E.: 1971, Solar Phys. **18**, 226

Wiehr, E.: 1987, Proc. Workshop on the Role of Fine-Scale Magnetic Fields on the Structure of the Solar Atmosphere, E.-H. Schröter, M. Vázquez, A.A. Wyller (eds.), Cambridge Univ. Press, p. 354

Wiehr, E.: 1988, Proc. Symp. Seismology of the Sun and Sun-like Stars, Tenerife, Spain, ESA SP-286, p. 269

Wiehr, E., Stellmacher, G.: 1989, Astron. Astrophys., in press

DISCUSSION

VENKATAKRISHNAN (i) How do you measure instrumental polarisation? (ii) Can the weak line correlation with the continuum be due to scattered light in the spectrograph?

KNEER: (i) Linear polarisers, ($\lambda/4$) and $\lambda/2$ sheets are mounted to the entrance window and the Stokes vector is measured for several rotational positions of the linear polariser.
(ii) We do see strong horizontal fluctuations in intensity, velocity, and magnetic field, much beyond observational noise. I do not think that they arise from scattered light for which I would expect a broad distribution.

WEISS: In order to describe the structure of a sunspot it is important to know the inclination of the magnetic field and its variation with radius in the outer penumbra. Do You have any new measurements of the field direction in the penumbra?

KNEER: We are constantly trying to obtain better and better data on the variation of field strength and direction. They have yet to be analysed.

CHOUDHURI: In the past some observers claimed that umbral dots are merely granules which have been somewhat modified by the presence of a strong magnetic field, whereas other observers felt that umbral dots are totally different from granules. Would you like to comment on this controversy?

KNEER: The spectroscopic characteristics of umbral dots are very different from those in the dark umbra. FeII is certainly present in dots and the magnetic splitting of lines measured in dots is different from both umbra and penumbra. Thus temperature fluctuations are much higher than in ordinary quiet-Sun "granulation". I think dots need a different name!

SIVARAM: What is the range of magnetic fields you can measure?

KNEER: The magnetic flux densities given here are measured from the Zeeman splitting. The lower limit for such measurements is about 1200-1500 Gauss, due to thermal broadening of the spectral lines. The upper limit we see in sunspot umbrae is about 3500 Gauss.

VAN BALLEGOOIJEN: Could the lack of correlation between intensity and velocity fluctuations in sunspots be explained in terms of an upward motion followed by a downward motion for each bright point?

KNEER: It is a possibility. To my knowledge, such an analysis of time sequences has not yet been performed. Certainly it should be done.

INTERPRETATION OF THE "THIRD HARMONIC" OF THE SOLAR MAGNETIC CYCLE

M.H.GOKHALE and J.JAVARAIAH
Indian Institute of Astrophysics
Bangalore 560 034
India

ABSTRACT. The peak at $\sim 3/22\ y^{-1}$ seen for all odd-degree axisymmetric modes in the SHF analysis of 'nominal magnetic field' seems to be an artifact of some non-linearity between the field magnitude and the sunspot occurence probability.

1. Introduction

We have an on-going project of analysing in various ways the distribution of the sunspot occurence probability (and a 'nominal' magnetic field defined therefrom), on the sun's surface and in time, using Greenwich data for sunspot groups. We have shown earlier that the spherical harmonic Fourier (SHF) analysis of these quantities yields approximately constant amplitudes and constant phases for their odd-degree axisymmetric SHF components during 1874-1976 (over which the data is available on a magnetic tape provided by H.Balthasar). Detailed results of the SHF analysis are discussed in earlier papers (Gokhale, Javaraiah and Hiremath, 1989, Gokhale and Javaraiah 1990).

In this paper we investigate the behaviour and significance of a third harmonic peak at $\sim 3/22\ y^{-1}$ which is quite significantly present in the spectra of all the odd degree axisymmetric modes with respect to the temporal frequency. These peaks are unlikely to be caused by the asymmetry in the 11 year cycle of the data size, since the magnitude of the nominal field is normalised with respect to the data size in each interval of analysis. Similar periodicity is seen in the SHF analysis of magnetogram data (Stenflo and Vogel 1986) and the hydrogen-alpha spectroheliogram data 1917-1985(Gokhale, et al 1990). It is also seen in the Fourier analysis of solar rotation (Singh and Prabhu, 1985). It is therefore important to see if these 'third harmonic' peaks represent independent periodicity in the nominal magnetic field. For this purpose we have determined the correlation between the amplitudes and phases of the third harmonic components with those of the corresponding first harmonic components. The presence of high corelation between the amplitudes (and a stronger correlation between the phases) of the first and the third harmonics show that the third harmonic peaks do not represent any independent periodicity. It seems they result from some residual nonlinearity between the nominal field and the data size.

E. R. Priest and V. Krishan (eds.), Basic Plasma Processes on the Sun, 119–123.

2. Data and Method of Analysis

The data used consists of the heliographic latitudes b_i, longitudes Φ_i and times of observation t_i' of all the sunspot groups observed during 1874-1976 which are included in the ledgers I and II of the Greenwhich Photoheliographic Results. The t_i' are converted to values t_i of the time measured from the zero hour of 1 January 1874. The sunspot occurence probability $p(\Theta, \Phi, t)$ during any sufficiently long time interval (T_1, T_2) is defined as

$$p(\Theta, \Phi, t) = \frac{1}{N(T_1, T_2)} \delta(\mu - \mu_i, \Phi - \Phi_i, \tau - \tau_i)$$

in the neighbourhood of the points (μ_i, Φ_i, τ_i) in the (μ, Φ, t) space, and

$$p(\Theta, \Phi, t) = 0 \qquad \text{elsewhere}$$

where $\mu = \cos\Theta$, $\tau = (t-T_1)/(T_2-T_1)$, δ represents a delta function, $N(T_1, T_2)$ is the number of data points given by sunspot groups during in the time interval (T_1, T_2). It may be noted that this way the data of each sunspot group gets weighted by the number of days it is observed. The 'nominal magnetic field' 'B' (Θ, Φ, t) is defined as

$$'B'(\Theta, \Phi, t) = \pm p(\Theta, \Phi, t)$$

where the signs $\pm$ are chosen according to Hale's laws of magnetic polarities, taking care to separate the data from the old and the new cycles during the overlaps of successive cycles.

3. Presence of the "Third Harmonics" with constant amplitudes and phases

In an earlier paper (Gokhale and Javaraiah, 1990) we have plotted the SHF amplitudes of 'B' (Θ, Φ, t) during the interval 1902-1954, as functions of temporal frequency ν, for axisymmetric (m=0) modes of odd degrees of $\ell = 23$. All these plots show a very high 'primary' peak at $\nu = 1/22y^{-1}$ as expected. But all of them also show a smaller "third harmonic" peak at $\sim 3/22\ y^{-1}$, which, in the case of modes $\ell \geq 9$, is comparable to the primary peak. For $\ell = 13$, 15 it is as high as the primary peak. Similar concentration of SHF power at the third harmonic is also seen in the results obtained by Stenflo and Vogel (1986) from the magnetogram data and by Gokhale et al (1990) from the hydrogen-alpha spectroheliogram data 1915-1985. The '7 year' periodicity is present even in the analysis of solar rotation using CaII K spectroheliograms (Singh and Prabhu 1985). We find that for $\ell \geq 13$ the amplitudes and the phases of the third harmonic mode are approximately constant during all the 82 intervals of 22 y length, from 1874-1895 to 1955-1976, each of which is shifted with respective to the previous intervals by one year (Fig.1).

4. Interpretation

It may be thought that the 'third harmonic' peaks result from the asymmetry of the 11 year half cycles of the data size. However this is not

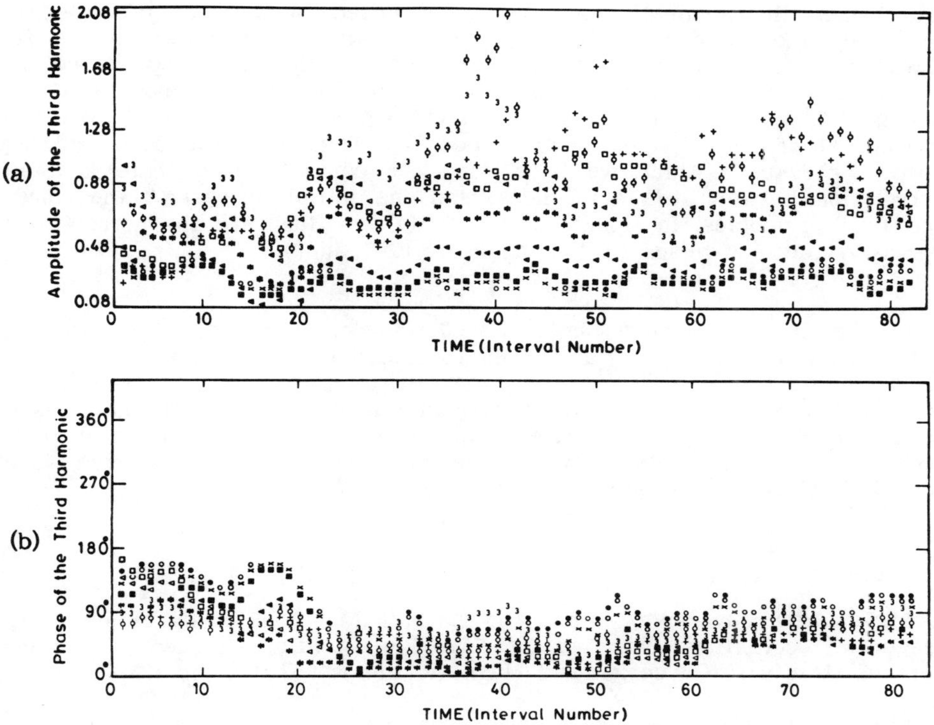

Figure 1 (a) Amplitude ratio A3/A1 and (b) the phase difference $\delta 3- \delta 1$ as functions of time represented by the interval number. Symbols: $\ell=1$: filled circle, $\ell=3$: open circle,.......... $\ell=21$: letter 'phi'.

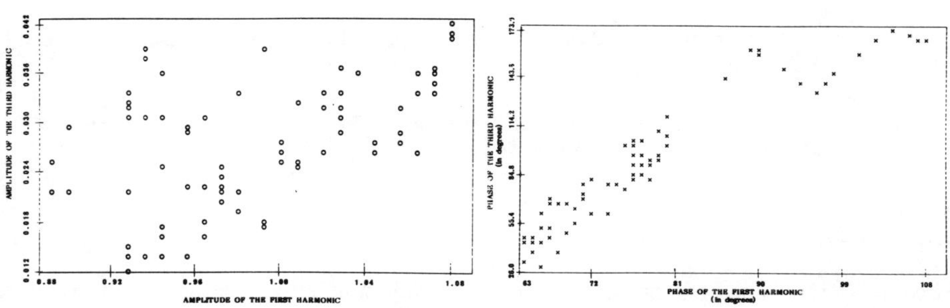

(a) (b)

Figure. 2 (a) Amplitude A3 of the third harmonic versus amplitude A1 of the first harmonic, and (b) phase $\delta 3$ of the third harmonic versus phase $\delta 1$ of the first harmonic, for the mode $\ell=1$, m=0.

true, since the nominal field 'B' is normalised to the size of the data as shown by the factor $1/N\ (T_1, T_2)$ in equation (1). In order to see if the third harmonic is an independent SHF mode in the time dependence of 'B' we plotted for each odd degree ℓ , the amplitudes A3 (and the 'initial' phases δ 3) of the third harmonic mode against the amplitude A1 (initial phase δ 1) of the first harmonic mode. We find, that for each odd degree upto ℓ =13 (upto which we have done our computations), there is a strong correlation between the amplitudes (and the initial phases) of the third and the first harmonics. In figures 2(a) and 2(b) we illustrate the correlations A3 versus A1 and δ 3 versus δ1 for the mode (ℓ=1, m=0).

For all the odd modes upto ℓ= 13, the correlation coefficients are ~ 0.49 between A1 and A3 and ~0.85 to 0.95 between δ_1 and δ_3.

This shows that the third harmonic peaks in the SHF power are unlikely to represent an independent set of modes. They result most probably from some non-linearity between the magnitude of the analyzed function 'B' (Θ, Φ, t) and the true sunspot occurence probability.

References

Gokhale M.H., Javaraiah, J. and Hiremath, K.M., 1990,"Study of sun's 'hydromagentic' oscillations using sunspot data" in Stenflo, J.O. (ed) "Solar Photosphere: Structure, Covection and Magnetic Fields", IAU Symp. **138**, (in press).

Gokhale M.H. and Javaraiah, J., 1990 "Sunspot activity as originating in interference of sun's global MHD oscillations", Mon.Not.Roy.Astron.Soc. (in press).

Gokhale, M.H., Makarov, V.I., Sivaraman, K.R., Javaraiah, J., Tavastsherna, K.S., Vasileva, V.V., 1990 "Global modes of solar magnetic fields derived from H-alpha synoptic charts", (in preparation).

Singh, J. and Prabhu, T.P., 1985 "Variations of the solar rotation rate derived from Ca+K plage areas", Solar Phys. **97**, 302.

Stenflo, J.O. and Vogel, M 1986 "Global resonances in the evolution of solar magnetic fields" Nature, **319**, 285.

DISCUSSION

DALSGAARD: The presence of the third harmonic may simply reflect the non-sinusoidal shape of the sunspot index as function of time. This is supported by the constancy of the amplitude ratios and phase difference. The situation is equivalent to Fourier analysis of Cepheid light curves.

GOKHALE: Certainly one does expect a third harmonic due to the non-sinusoidal nature of the sunspot cycle, or due to a non-linear dependence of the 'data' variable (P) on a basic variable such as the magnetic field (B). That is why we have tested the correlation of the amplitudes and phases of the third harmonic with respect to those of the first harmonic. However, we have *normalized* the sunspot probability (p) to the *total* number of data points in each 22-year sample, thereby filtering out the time-dependence of the data size (if p depends linearly on B). Since we still see the third harmonic, with different amplitudes for different odd ℓ (≤ 9), but the amplitude and phase for each ℓ correlates with those of the first harmonic for the same ℓ, we conclude that there must be some non-linearity between p and B. As for the independence of the third and first harmonics for $\ell > 9$, further analysis will be needed.

SELF-ORGANIZATION PROCESSES ON THE SUN : THE HELIOSYNERGETICS

V.Krishan
Indian Institute of Astrophysics, Bangalore 560034, India

and

E.I.Mogilevskij
IZMIRAN 142092, TROISTSK, Moscow Region, Moscow, USSR

Abstract Non-linear interactions between small fluid elements, magnetized or otherwise, in an energetically open nonlinear system facilitate the formation of large coherent stable structures. This is knwon as self-organization. We interpret solar granulation on all scales and the formation and evolution of some structures in solar active regions to be the result of self-organization processes occur ing in a turbulent medium.

Introduction

In an isolated turbulent medium, decay of large eddies into small eddies is a common occurrence. The converse i.e. small eddies coalescing into large ones can be accomplished in an energetically open system with very special properties. Generally, such a system must be dissipative and describable by a nonlinear partial differential equation. In the absence of dissipation, the equation should have two or more quadratic or higher order conversed quantities called invariants. On the inclusion of dissipation, the invariants decay differentially. The nature of the nonlinear interaction between the fluid elements is such that the slow decaying invariant (say B) cascades towards large spatial scale and the fast decaying invariant (say A) cascades towards small spatial scales. The intial random field of velocities and magnetic fields can then be described through a variational principle in which the invariant A is minimized keeping the invariant B constant i.e.

$$\delta A - \lambda \delta B = 0$$

where $\delta A, \delta B$ are small variations and λ is the Lagrange multiplier Hasegawa (1985). Kraichnan (1967) found that in two-dimensional hydrodynamic turbulence, the energy invariant cascades towards large spatial scales and the enstrophy invariant towards small spatial scales where it suffers heavy dissipation. It is this property of selective decay that facilitates the formation of large structures. But can one use the results of 2-D turbulence theory to explain phenomena on a real atmosphere which is a 3-D system? What one needs is the process of inverse cascade to occur in a 3-D system and Levich and Coworkers (Levich and Tzvetkov 1985)

E. R. Priest and V. Krishan (eds.), Basic Plasma Processes on the Sun, 125–133.

have shown such a possibility. We will use the ideas developed in this work to construct a model of solar granulation in its entirety from granules to giant cells.

One can also study the dynamics of a turbulent magnetoactive medium consisting of small scale strongly magnetized elements embedded in a large spatial scale weakly magnetized ambient medium. It can be shown that a stimulating disturbance of the ambient magnetic field enhances the nonlinear interaction between the small elements and the complete system is describable by a Schrodinger type equation that gives a full spectrum of discrete structures depending on the available free energy. This approach will be used to search sources of solar flares.

Modeling Solar Granulation

The cellular velocity patterns observed on the solar surface are believed to be manifestations of convective phenomenon occuring in the sub-photospheric layers. (Antia et al (1984) Bogart et al (1980) Brandt et al (1988) Schwarzschild (1975) and Simon and Leighton (1964). Here, we explore if the excitation of random small scale motions can lead to large organized structures which are observed in the form of granules, mesogranules, supergranules and giant cells. In this picture, large helicity fluctuations present in a turbulent medium play an essential role in the inverse cascading process. The helicity density γ, a measure of the knottedness of the vorticity field is defined as $\gamma = \vec{V}\cdot(\vec{\nabla}\times\vec{V})$. It is found that the quantity I defined as $I = \int \langle \gamma(x)\,\gamma(x+r)\rangle d^3r$ is also an invariant of an ideal 3-D hydrodynamic system in addition to the total energy. Assuming a quasinormal distribution of helicities, the invariant I can be expressed as:

$$I = C\int [E(K)]^2\, dK \tag{1}$$

where C is a constant and $E = \int E(k)\, dk$ is the total energy density. Using Kolmogorovic arguments one finds inertial range for energy invariant to be:

$$E(K) \propto K^{-5/3} \quad \text{and} \quad E \propto L^{2/3} \tag{2}$$

and for 'I' invariant to be:

$$E(K) \propto K^{-1} \quad \text{and} \quad E \propto \log \frac{L(t)}{\ell} \tag{3}$$

where L(t) is the largest length scale excited at time t Levich and Tzvetkov (1985). In analogy to the 2-D case (Hasegawa 1985) one expects that in 3-D the invarant 'I' would cascade towards large spatial scales and the energy towards small spatial scales. The cascading of 'I' towards large spatial scales essentially enhances the correlation length of helicity fluctuations, without much increase in energy associated with it, eq.(3). We propose that this regime of turbulence favours the formation of solar granulation at the smallest scales. One can ask if there is an upper limit to the size of granules, the lower limit of course is determined by dissipation. We recall that all atmospheres are restricted in the vertical direction due to gravity. The largest dimension of fully 3-D structures is given by the ratio I/E = L = L_Z where L_Z is the characteristic vertical scale. We identify this scale with the size of the region with superadiabatic temperature gradient, since this is the region that provides energy in the vertical velocity field which then drives the horizontal flow. This limit (Nelson and Musman 1978) of 1000 km puts the 3-D granules right at the top of the convection zone. When the correlation length of helcity fluctuations reaches the limit L_Z, it can only grow in the horizontal plane. Another consequence of the growth of correlation length is that the velocity and vorticity get aligned which reduces the nonlinear term $(\vec{V}\cdot\vec{\nabla})\vec{V} = \vec{\nabla}\times(\vec{V}\times\vec{\omega})$ of the Navier-Stokes equation and thus retards the flow of energy to small spatial scales. With the growth of correlation length only in the horizontal plane, the system becomes more and more anisotropic. Under these circumstances, the vertical component of velocity V_Z becomes independent of (x,y,z) and the horizontal components V_x and V_y independent of Z, leading to $W_{x,y} = (\vec{\nabla}\times\vec{V})_{x,y} = 0$. The invariant I becomes

$$I = \int \langle (V_z \omega_z)^2 \rangle dx\, dy\, dz$$

$$\simeq L_z \langle V_z^2 \rangle k^2 V_k^2 k^{-2} \propto V_k^2 = k E(k) \propto L^{2/3} \tag{4}$$

and from $I = \int I(k) dk$, one finds $I(k) \propto k^{-5/3}$ (5)

Here L is the largest length scale in the horizontal plane. Thus I(k) spectrum coincides with the energy spectrum of 2-D turbulence $E(k) \propto k^{-5/3}$ corresponding to the inverse cascade. One expects that an increasing fraction of energy is transferred to large spatial scales as the anisotropy in the system increases. We propose this part of the turbulence spectrum to be conducive to the formation of supergranules which appear predominantly on large horizontal scales with energy spectrum given by eq.(4). The intermediate region where a 3-D system is developing anisotropy, the energy spectrum being given by eq.(3) can be identified with mesogranulation or the gap between granulation and supergranulation. One can again ask if there is a limit to the size of supergranules. The growth of large structures in a highly anisotropic turbulence can again be interrupted as a result of symmetry breaking caused by the Coriolis force. The length scale L_c where the nonlinear term of Navier-Stokes equation becomes comparable to the coriolis force, can be determined from

$$(\vec{V}\cdot\vec{\nabla})\vec{V} = 2(\vec{V}\times\vec{\Omega}) - \vec{\Omega}\times(\vec{\Omega}\times\vec{R})$$

or

$$L_c \simeq V/\Omega \tag{6}$$

where $\vec{\Omega}$ is the angular velocity. Given sufficient energy, structures of size L_c must form. At these large spatial scales, the system simulates 2-D behaviour and enstrophy conservation begins to play its role. One may consider scales $L \gtrsim L_c$ as a source of vorticity injection into the system. The enstrophy then cascades towards small scales with a power law spectrum given by

$$E(k) \propto k^{-3} \quad and \quad E \propto L^{2} \tag{7}$$

Thus there is a break in the energy spectrum as energy must cascade to larger spatial scales as $L^{2/3}$ and to small scales as L^2. Therefore the energy must accumulate at $L \sim L_c$ and eventually pass on to the highest possible scales of the general circulation of the atmosphere. It is tempting to identify this region of turbulence with the excitation of giant cells. This perhaps is the complete story of the energy spectrum, Figure (1) in the turbulent medium of the solar atmosphere.

Energetics

According to the picture (Levich and Tzvetkov 1985) presented here, the energy in the larger structures has been inverse cascaded from the smaller structures, if so, then the energy density per unit gram E(L) in the large scale L should not exceed that in the small scale (l). From the energy spectrum $E \propto L^{2/3}$, it follows that

$$E(L) \leq \left(\frac{E_o(\ell)}{\tau}\right)^{2/3} L^{2/3} \simeq E_o(\ell) \tag{8}$$

where τ is thetime for which energy injection must occur. If we take τ to be the lifetime of the larger structure then for $\tau \sim 20$ hrs for supergranules and $E_o(\ell) \sim (0.5)^2$ km^2/Sec^2 for energy density/gm in the granules, one gets $L \sim L_{SG} \sim 36000$ Km which is the typical size of a supergranule. The size of a giant cell can be determined from eq.(6) for $\Omega = (2\pi/27)$ day^{-1} and assuming $V \sim 0.3$ Km/Sec for supergranular velocity (since they provide the stirring force for turbulence that organizes itself into giant cells) one gets $L_c = L_{GC} \sim 1.17 \times 10^5$ Km. Again the energy content of giant cells should not exceed that of supergranules. Using equation (7) for the energy spectrum in this region, one gets:

$$E_{GC} \leq \left(E(L_{SG})/L_{SG}^2\, \tau\right)^{2/3} L_{GC}^2 \lesssim E(L_{SG}) \tag{9}$$

where $(E(L_{SG})/L_{SG}^2 \tau)$ is the enstrophy injection rate. For $L_{GC} \sim 10^5$ Km, $\tau = 30$ days = lifetime of giant cell, $L_{SG} \sim 30{,}000$ Km, and $E_{SG} \sim (0.3)^2$ Km^2/Sec^2. One finds that eq.(9) can be barely satisfied. Furthermore, in the presence of coriolis force, the pressure balance condition becomes:

$$\frac{1}{\rho}|\vec{\nabla} P| \simeq V^2/L_c + F_c \simeq 2V^2/L_c \tag{10}$$

in contrast to the case with no coriolis force where $\frac{1}{\rho}|\vec{\nabla}P| \simeq V^2/L_c$. Thus one concludes that larger energy density is required to maintain structures at scale L_c. This may be the reason for their rare observability. The appearance of structures at Lc must be accompanied by a corresponding increase in the convective flux and therefore probably of total solar flux. Total solar luminosity changes of 1% have been observed. If we attribute all of this 1% to increase in the convective flux, equation (10) can be satisfied and structures of size Lc can be excited. The differential rotation of the sun favours the formation of larger structures at the polar regions in comparison to the equatorial regions. This is further substantiated by the fact that the dominantly open magnetic fields in polar regions do not inhibit flow of convective flux. Thus one may look for probable correlation between polar phenomena and solar luminosity enhancements with the appearance of giant cells. A very steep spectrum eq.(7) practically forbids further organization of turbulence into structures larger than Lc. The simultaneous organization of the velocity and magnetic field is discussed elsewhere in these proceedings.

Energy injection in solar active regions

The observations show that magnetic field variations in the flaring active regions either are of the same order as without flares, or have a reversible character. Therefore the currents associated with the evolutionary changes in the active regions though play an active role (like accelerating the particles) in the flare process are not the primary sources of flare energy. It is proposed here that the current changes that are induced by the external stirring provided by the active region subphotospheric zone in the form of non-linear impulses, are responsible for the flaring of the plasma.

The solar convection zone can be regarded as a kinetic active medium where turbulent dynamo generates most probably, small scale (cross section 10^7 cm; length 10^8 cm) magnetoplasma elements with a lifetime of 10^3 sec a magnetic field strength of 2-3 KG (Parker 1979). If the energy supply is sufficient, the concentration of these magnetic elements reaches a threshold (and then exceeds) value at which pairwise non-linear interactions become effective. This is because each magnetic element absorbs energy from the surrounding turbulent medium of the convection zone and emits sound waves in the nearby zone, which primarily results in a close coupling between magnetic elements. Such a process of non-linear wave interaction repeated more than once in element pairs and other combinations leads to the formation of large scale structures such as pores, sunspots, arches, etc. This is an equivalent description of the inverse cascade of energy where it tends to accumulate at the largest scales. A dispersive disturbance propagating in such a non-linear medium can develop into a soliton which retains its characteristics over long distances of propagation. We propose that this is the form in which flare energy is transported from subphotospheric layers to the flaring region, Mogilevskij (1984).

The common occurrence of helical structures in a flare active region is evidence of the presence of helicity in the magnetic and velocity fields at the photospheric and subphotospheric layers. As discussed in Sec. III, the formation of large structures is favoured in a medium possessing large correlated helicity fluctuations. The helicity density h of magnetic field B is defined as $h = \vec{A}\cdot\vec{B} = \vec{A}\cdot(\vec{\nabla}\times\vec{A})$ where $\vec{A}$ is the vector potential such that $\vec{B} = \vec{\nabla}\times\vec{A}$. The inverse cascade of energy to large scales follows the increase in the correlation length of the helicity fluctuations, which happens in a Gromeko-Beltramy flow given by $\vec{\nabla}\times\vec{V} = \alpha\vec{V}$. Analogous to this situation, the magnetic field is also expected to attain this form since the increase of magnetic helicity leads to the alignment of the magnetic field $\vec{B}$ and the vector potential $\vec{A}$ such that $\vec{\nabla}\times\vec{A} = \alpha\vec{A}$ and $\vec{\nabla}\times\vec{B} = \alpha\vec{B}$ where α is a constant. If we take the model of convection zone suggested by Spruit (1978), then for subphotospheric layers $\beta = (E_{kin}/E_{mag}) \approx 1$. In this case, the kinematics of magnetoplasma in the active region will be combined with Gromeko-Beltramy motion i.e. $\vec{\nabla}\times\vec{V} = \alpha\vec{V}$, $\vec{\nabla}\cdot\vec{V} = 0$, $V < V_s$ where V_s is the sound speed. Under these circumstances the magnetic and velocity fields are transported such that

$$\vec{\nabla}\times(\vec{B}\cdot\vec{\nabla})\vec{V} = \alpha(\vec{B}\cdot\vec{\nabla})\vec{V}$$

and

$$\vec{\nabla}\times(\vec{V}\cdot\vec{\nabla})\vec{B} = \alpha(\vec{V}\cdot\vec{\nabla})\vec{B} \tag{11}$$

The term $(\vec{B}\cdot\vec{\nabla})\vec{V}$ describes velocity transport at the magnetic field gradient and the term $(\vec{V}\cdot\vec{\nabla})\vec{B}$ describes magnetic field transport at the velocity gradient. Then operator rotor over these convection terms will define the currents and Eqs.(11) will represent the force free character of these currents. This is the energy state to which an active region will evolve under the absence of external disturbances. We shall consider the problem of a large wavelength disturbance ($\lambda \gtrsim 10^7$ cm) propagating vertically across the magnetic field near the neutral line (at B = 0) at the photospheric level. The origin of this disturbance could be traced to the restricted zones of enhanced energy release (e.g. due to He^3 abundance) at the core convection zone boundary. This energy could be transported towards the photosphere efficiently in the form of long non-linear waves in a medium of decreasing pressure and density. These upward propagating waves in a dispersive medium like the convection zone may develop into solitons. The non-linear Eqs.(11) in a dispersive medium can be shown to attain the form

$$\frac{\partial U}{\partial t} + (U+1)\frac{\partial U}{\partial z} + \frac{(U+1)}{\alpha^2}\frac{\partial^3 U}{\partial z^3} = 0 \tag{12}$$

where U represents the velocity or magnetic field disturbance. If propagation takes place in a non-linear medium with a dispersion $(\omega/U+1) = k + \epsilon k^3$ where $\epsilon = 1/\alpha^2$ is the dispersion length, the intial disturbance $U = U_0 e^{i(k\cdot z - \omega t)}$ is transformed in accordance with Eq.(12) into an MHD soliton of the type

$$U = U_0 \operatorname{Sech}^2 \xi \,, \quad \xi = \left(\frac{U_0\alpha^2}{12}\right)^{1/2}\left(z - \frac{U_0}{3}\right) \tag{13}$$

For such a disturbance, the energy conservation law may be written as

$$W\lambda = \text{Constant} \,, \quad \lambda \simeq z \leq \delta$$

where $W \simeq U_0^{3/2}$ is the energy in a unit wavelength of the disturbance λ. The total energy that reaches the photosphere in one MHD-soliton propagating at speed $V \gtrsim V_A$, the Alfven speed can be estimated as

$$E_m \simeq \epsilon_m V_A \ell^2 \Delta t \tag{14}$$

where $\epsilon_m = b^2/8\pi$ is the average magnetic energy density in the soliton, t is the lifetime of the soliton at the photospheric level. If we assume $b \sim 4 \times 10^2$ Gauss, $V_A \sim 6 \times 10^6$ cm/sec, $\ell = 7 \times 10^8$ cm and $t = 6 \times 10^2$ sec, then $E_m \simeq 10^{31}$ ergs. Thus either a series of successive solitons or a distribution of solitons in the vicinity of the neutral line may account for the total energy of a large flare. In fact it can be shown that a single large soliton can be formed from the many interacting solitons through self-organization processes, Hasegawa (1985). Dissipationless KDV equation has an infinite number of conserved quantities; its solution permits n solitons and dispersive wave for arbitrary initial conditions. In the presence of dissipation, the differential dissipation of the conserved quantities leads to an inverse cascade in the wave number spectrum and the soliton gas is expected to condense into one big soliton, Hasegawa et al. (1981). Since the conserved quantities of KDV equations are of order higher than quandratic, the organized state obtained from variational principle is non-linear localized state.

Conclusion

Formation of large coherent structures in an otherwise turbulent medium occurs if it is injected with enough energy. A Navier-Stokes system has been investigated to explore the possibility of organizing energy associated with solar granules into supergranules and giant cells. The suggested correlation between the solar luminosity increases and the appearance of the giant cells should be looked for. In addition, with the availability of extremely high resolution observations of the photospheric velocity fields (Brandt et al. 1988) it should be possible to estimate the quantity 'I' at various heights on the photosphere. The general overall spectrum except at the breaks is the well-known Kolmogorov law $K^{-5/3}$.

In the second example trying to understand the nature of the initial source of flare energy,it is found that a magnetoactive medium possessing helicity will submit itself to the formation of solitons which themselves could be organized state of a non-linear medium describable by KDV equation. A gas composed of many solitons can get further organized into one big soliton which would carry energy enough for a big flare. Observationally, this phenomenon could be identified with magnetic transients, multiple flares and multiple electron beams accelerated during the collision of this soliton with the pre-existing current carrying plasma. In fact, one could equally well give a 'soliton interpretation' for the supergranular and the giant cells. We hope to accomplish this in the near future.

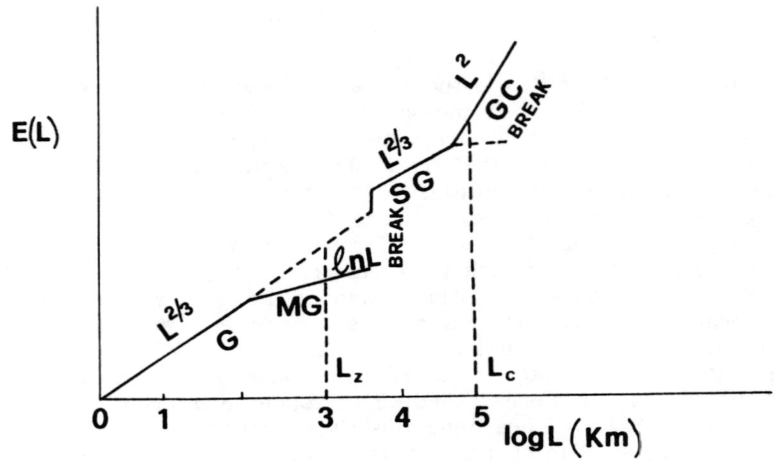

Fig. 1. Turbulent energy spectrum. L_Z = scale of the first break due to anisotropy, L_C = scale of the second break due to the Coriolis force, G = granule, MG = mesogranule, SG = supergranule and GC = giant cell.

References

Antia,H.M., Chitre,S.M. and Narasimha,D. 1984, Ap. J. 282, 574-583.

Bogart,R.S., Gierasch,P.J. and MacAuslan,J.M. 1980, Ap. J. 236, 285-293.

Brandt,P.N. et al. 1988, Nature 335, 238-240.

Hasegawa,A. 1985, Advances in Physics 34, 1-42.

Hasegawa,A., Kodama,Y. and Watanabe,K. 1981, Phys. Rev. Lett. 47,1525.

Kraichnan,R.H. 1967, Phys. Fluids 10, 1417-1423.

Levich,E. and Tzvetkov,E. 1985, Physics Reports 128, No.1, 1-37.

Mogilevskij, E.I. 1984 Preprint No.43a (517).

Nelson,G.D. and Musman,S. 1978, Ap.J. 222, L69.

Parker,E.N. 1979, Cosmical Magnetic Fields, Clarendon Press, Oxford.

Schwarzchild,M. 1975, Ap. J. 195, 137-144.

Simon,G.W. and Leighton,R..B. 1965, Ap. J. 140, 1120-1147.

Spruit,H. 1978, in Magnetic Flux Tubes and Transport of Heat in the Convection Zone of the Sun, Utrecht, Holland.

DISCUSSION

CHITRE: I am just curious to know how the energy content in granulation compares with that in super-granulation in your inverse-cascade picture.

KRISHAN: Since the energy in the larger structures has been inverse-cascaded from smaller structures, the energy density per unit gram E(L) in the large scale (L) should not exceed that in the small scale (ℓ). This consideration determines the sizes of super-granules and giant cells if the energy injection at small scales takes place at least as long as the lifetime of the larger structure.

WEISS: Recent high-resolution observations at La Palma and the Pic du Midi suggest that granules emerge at the centres of meso-granules and move outwards. Some large granules explode, generating rapid outward motion and it is not clear how much of the meso-granular flow is produced by these exploding granules. Is this mechanistic description compatible with your statistical model?

KRISHAN: In my statistical model, meso- and super-granules originate from a reorganization of the energy contained in granules. In this sense, it may be compatible with the observations that link meso- and super-granules with the agglomeration and explosion of granules. More can be said only after the dynamics of these structures has built up.

LINEAR AND NONLINEAR CONVECTION WITH AN ALIGNED MAGNETIC FIELD

N Rudraiah, I S Shivakumara and P Geetavani
UGC-DSA Centre in Fluid Mechanics,
Department of Mathematics, Central College,
Bangalore University, Bangalore-560001, INDIA

Abstract: The effect of horizontal magnetic field on the onset of three-dimensional convection in a horizontal fluid layer is studied. It is found that the two-dimensional solutions are unstable to three-dimensional disturbances. A detailed bifurcation study is reported.

Introduction: Convection in the presence of a vertical magnetic field has been extensively investigated because of its relevance in astrophysics and geophysics [1]. It is known that the penumbra of the sunspot is a well developed region for temperature about 5000°K where the magnetic field lines expand rapidly to become more horizontal of strength about 2 KG. Though small, they are important in providing a mechanism to cool the photosphere and upper convection zone [1-3]. Therefore, the effect of horizontal magnetic field on convection has been studied in [2-3] by considering only two-dimensional disturbances. Its effect on three-dimensional convection has not been given attention. This is considered in this paper with the object of showing that the two-dimensional solutions may become unstable to three-dimensional disturbances which possess added degree of freedom in the study of heat transfer. We have investigated both linear and nonlinear convection in the presence of horizontal magnetic field. Due to want of space we present here only the results pertaining to linear theory.

Mathematical Formulation: We consider a Boussinesq electrically conducting fluid layer bounded by stress-free surfaces. The basic equations of motion are conservation of mass, momentum, energy and magnetic field as given in [4]. The usual process of linearization and employing normal mode analysis, we get

$$R = k^2[k^4 - p^2/\sigma + (p^2 + Bk^4)G/(p^2 + B^2k^4)]/\pi^2\alpha^2 + ipk^2N, \tag{1}$$

with $N = k^2(\sigma+1)/\pi^2\alpha^2\sigma + (B-1)G/(p^2+B^2k^4)$. Here $R = \beta g\Delta T d^3/\nu\kappa$, $Q = \mu H_0^2 d^2/\rho_0 \nu\nu_m$, $B = \nu_m/\kappa$, $\sigma = \nu/\kappa$, $k^2 = \pi^2(\alpha^2+1)$, $\alpha^2 = l^2 + m^2$, $G = QB\pi^2\ell^2$. Where ℓ and m are wavenumbers in x and y directions and p is the frequency. R is a physical quantity, therefore (1) demands that either p = 0 or N = 0. These conditions help us to study the types of bifurcations.

Direct bifurcation (p = 0): Now from (2), we get

$$R^{(s)} = k^2[k^4 + Q\pi^2\ell^2]/\pi^2\alpha^2. \tag{2}$$

E. R. Priest and V. Krishan (eds.), Basic Plasma Processes on the Sun, 135–136.

This becomes minimum, for a fixed value of ℓ, when $\alpha = \alpha_c$ is a root of $2\alpha_c^6 + 3\alpha_c^4 =$ $1+Q\ell^2/\pi^2$. As $Q \to \infty$, $\alpha_c \sim \sqrt[6]{Q\ell^2/2\pi^2}$ and $R^{(s)} \sim Q\pi^2\ell^2$. As $Q \to 0$, $\alpha_c \sim 1/\sqrt{2}$, $R^{(s)} \sim$ $27\pi^4/4$.

Hopf bifurcation ($p \neq 0$, $N = 0$): There is a Hopf bifurcation at

$$R^{(0)} = k^6(\sigma+B)(1+B)/\sigma\pi^2\alpha^2 + k^2(\sigma+B)G/\pi^2\alpha^2(1+\sigma), \tag{3}$$

provided $p^2 = \alpha^2\sigma B(R^{(s)} - R^{(0)})/(\alpha^2+1)(1+\sigma+B) = -B^2k^4 + (1-B)\sigma G/(1+\sigma) > 0$. So the necessary condition for the existence of a Hopf bifurcation is $B < 1$, $Q > B(1+\sigma)k^4/\pi^2\ell^2\sigma(1-B)$. We note that when $\ell = \alpha$ (i.e. $m = 0$), (2) and (3) tend to those given in [2] for two-dimensional disturbances.

Conclusions: The critical Rayleigh numbers for direct and oscillatory motions are computed for different values of B and Q and the results are compared with those of two-dimensional disturbances in Fig. 1. This shows that the two-dimensional solutions become unstable to three-dimensional disturbances. The nature of bifurcation, for different values of σ, is depicted in Fig. 2. The bifurcation from steady to oscillatory solutions occurs earlier for small values of σ compared to the large values of σ. These conclusions are used in the study of nonlinear theory.

References:

[1] Proctor, M.R.E. & Weiss, N.O. (1982), Rep. Prog. Phys. 45, 1317-1379.
[2] Arter, W. (1983), Geophys. Astrophys. Fluid Dynamics, 25, 259-292.
[3] Knobloch, E. (1986), Geophys. Astrophys. Fluid Dynamics, 36, 161-177.
[4] Rudraiah, N. (1986), Praman - J. Phys. 27, 233-266.

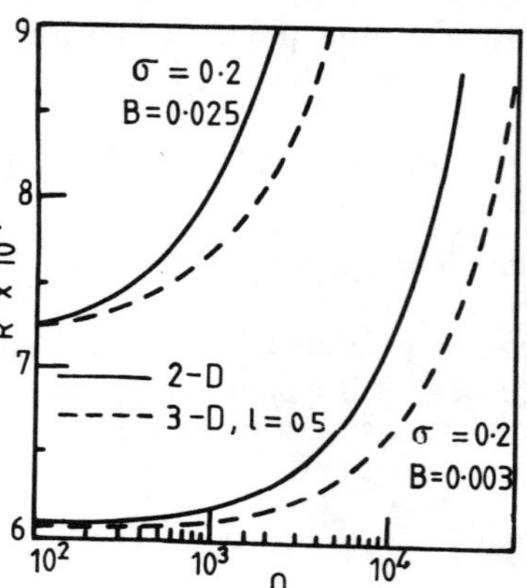

Fig.1 Critical Rayleigh number Vs.Q

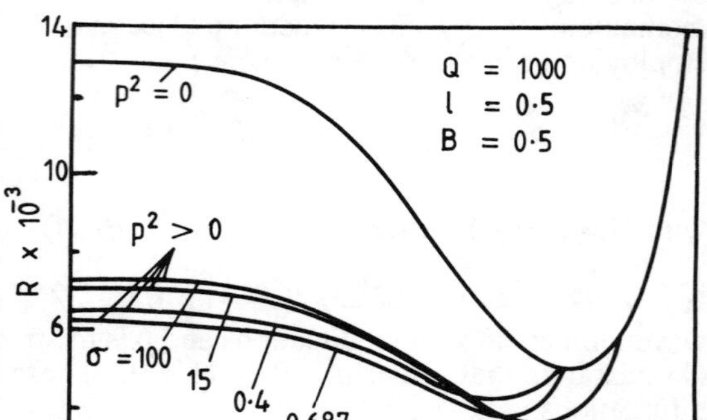

Fig.2 Curves of neutral stability

V. PHOTOSPHERIC FLUX TUBES

MAGNETOHYDRODYNAMICS OF SUNSPOTS

N.O. WEISS
Department of Applied Mathematics and Theoretical Physics
University of Cambridge, Cambridge CB3 9EW, UK

ABSTRACT. Recent numerical investigations of fully compressible nonlinear magnetoconvection have clarified the nature of convection in sunspot umbrae. In a shallow layer with a strong vertical magnetic field oscillations give way to travelling waves as the Rayleigh number is increased but in a deep stratified layer oscillatory behaviour only appears after secondary bifurcations. This behaviour leads to a model that explains the formation of umbral dots. Penumbral structure is more difficult to explain owing to the apparent presence of adjacent horizontal and inclined fields in dark and bright filaments. The inner penumbra lies above a transition zone where volume currents are needed to maintain an overall static equilibrium; instabilities in this region may be responsible for filamentary structure in the penumbra as well as for fine structure at the umbral-penumbral boundary.

1. Introduction

Sunspots have been observed for centuries and the ideal circular spot with a filamentary penumbra is a fairly good approximation to reality. Yet this apparently simple structure remains surprisingly difficult to explain. Our inability to provide a proper description of a sunspot is an obvious embarrassment to theoreticians. A few days ago the "Navhind Times" referred to "sunspots, *mysterious* areas on the solar surface that are cooler and darker than the surrounding area". In this review I hope to explain why they continue to be so mysterious.

One sign of our ignorance is the lack of any consensus on the overall structure of a flux-tube beneath a sunspot. Early discussions contrasted a monolithic plug extending deep into the convection zone with a jellyfish that split into innumerable tentacles below the photosphere. Now we know that convection must lead to an irregular structure which still remains sufficiently coherent to support global oscillations. But disagreement persists between those who favour an inhomogeneous plug and those who favour a tight cluster of independent flux tubes. Moreover, surface observations do not allow us to decide between these models.

In what follows I shall give a selective discussion of some major problems; others, such as waves, are treated elsewhere in these Proceedings. First I shall describe recent progress in modelling convection in the umbra. Then I shall outline the

E. R. Priest and V. Krishan (eds.), Basic Plasma Processes on the Sun, 139–148.

difficulties posed by the field configuration in the outer penumbra and finally I shall comment on the global structure of a sunspot.

2. Energy Transport in Sunspot Umbrae.

It has been accepted for almost 50 years that sunspots are cool owing to partial suppression of convection by a strong magnetic field (Biermann 1941; Cowling 1976). As more detailed sunspot models were computed (Schlüter & Temesvary 1957; Chitre 1963; Deinzer 1965; Chitre & Shaviv 1967; Yun 1970) it became apparent that even the reduced energy flux emitted by the umbra cannot be supplied by radiative transport alone. Hence it is essential to establish the nature of convection in an imposed vertical magnetic field – a topic to which Indian astrophysicists have made important contributions.

Boussinesq magnetoconvection has been studied in considerable detail. For a fluid with magnetic diffusivity η and thermal diffusivity κ behaviour depends critically on the ratio

$$\zeta = \eta/\kappa \; ;$$

in a star the radiative diffusivity κ is typically large ($\zeta \approx 10^{-6}$ in the solar photosphere). Linear theory shows that for $\zeta < 1$ and a sufficiently strong magnetic field convection sets in at an oscillatory (Hopf) bifurcation (Chandrasekhar 1961) and the development of nonlinear oscillations has been followed with analytical and numerical techniques (Proctor & Weiss 1982; see also Rudraiah, Kumudini & Unno 1985a, b).

Recent work has been more concerned with compressible magnetoconvection (Hughes & Proctor 1988). The linear instability of a polytropic atmosphere in a vertical magnetic field has been investigated by Antia & Chitre (1979) and Cattaneo (1984). In the regime of interest there is a Hopf bifurcation from the static solution, giving rise to branches of standing wave and travelling wave solutions, followed eventually by a stationary (pitchfork) bifurcation which gives rise to a branch of steady solutions. The steady solutions are initially unstable but may ultimately gain stability. The effects of compressibility and stratification on steady nonlinear magnetoconvection were studied by Hurlburt & Toomre (1988). Subsequently, Hurlburt *et al.* (1989) investigated time-dependent convection in a weakly stratified layer. The behaviour found depends on the parameter $\beta = 2\mu_0 p/B_0^2$, where B_0 is the mean field strength and p is the gas pressure; in a sunspot β increases from unity at the photosphere to 30 at a depth of 1000 km. Although convection sets in as periodic oscillations, stability is transferred to travelling waves for $32 \geq \beta \geq 6$ if $\zeta < 0.3$. The travelling waves have strong prograde velocities but carry very little heat. If they were excited in the umbra we might expect to observe them but no behaviour of this type has yet been seen (except perhaps at the umbral-penumbral boundary). Convection in a deep stratified layer apparently takes a different form, though locally generated travelling waves, at depths of only 100 km below the photosphere, may still be able to excite umbral oscillations (e.g. Evans & Roberts

1990).

Within a sunspot the density increases rapidly with depth. Moreno Insertis & Spruit (1989) have studied the stability of a model atmosphere in the absence of diffusion. In the sun the opacity increases beneath the photosphere owing to ionization and the radiative diffusivity consequently falls so that $\zeta > 1$ at depths between 2000 and 20000 km (Meyer *et al.* 1974). This raises the following question: what happens in a layer where $\zeta > 1$ at the bottom, favouring steady overturning, but $\zeta < 1$ at the top, where oscillations are locally preferred? In sunspot umbrae magnetic fields are most effective in hindering convection immediately below the photosphere. Umbral dots have been ascribed to overshooting from below (Obridko 1974; Parker 1979a). Knobloch & Weiss (1984), extrapolating from Boussinesq convection, suggested that oscillations would be decoupled from overturning motion, while Parker (1979a, b) and Choudhuri (1986) favoured oscillatory convection in field-free inclusions.

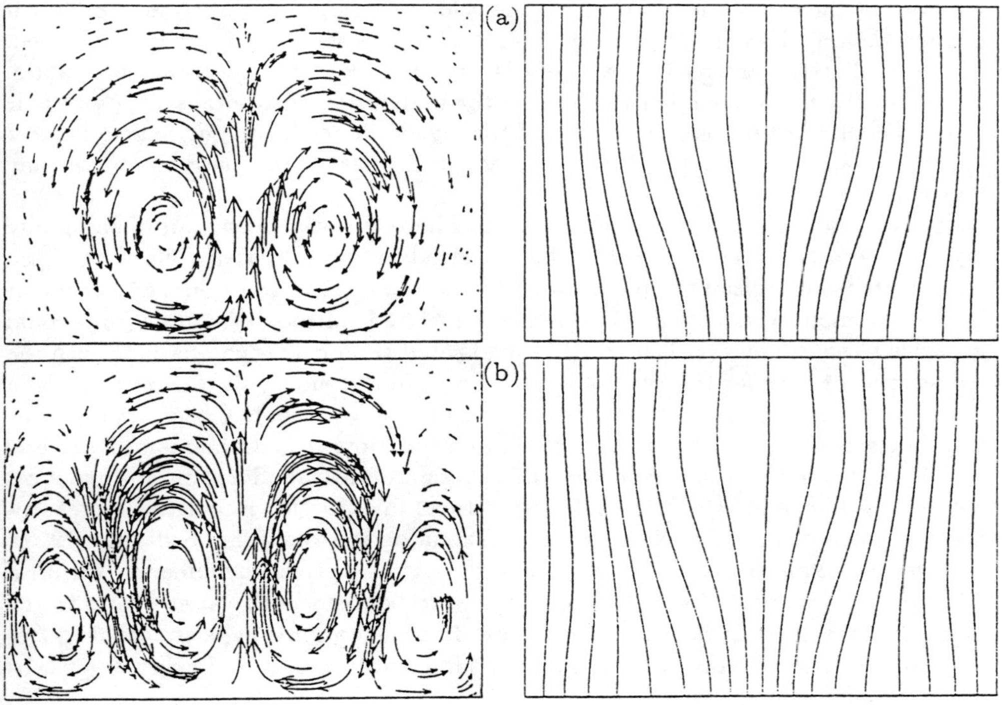

Figure 1. Nonlinear convection in a stratified compressible layer ($0.2 \leq \zeta \leq 2.2$). Streaklines (left) and fieldlines (right) for steady solutions with (a) $R = 1.07\ R_0$ and (b) $R = 1.39\ R_0$. Note the development of countercells in the corners.

Weiss *et al.* (1990) have recently investigated fully compressible nonlinear magnetoconvection in a stratified two-dimensional model, where $\zeta \propto \rho$ and increases continuously with depth. The most interesting results are obtained when $0.2 \leq \zeta \leq 2.2$, so that oscillations are favoured in the upper half of the layer. A linear stability analysis shows, however, that convection sets in at a stationary (pitchfork) bifurcation when the Rayleigh number $R = R_0$, giving rise to overturning motion. Numerical experiments yield weakly nonlinear solutions that are steady. At first motion barely penetrates to the top of the layer, as shown in Figure 1(a) for $R = 1.07\ R_0$. As R is increased countercells develop in the region where the magnetic field is strong, as illustrated in Figure 1(b) for $R = 1.39\ R_0$. These steady solutions eventually lose stability in a second Hopf bifurcation at $R = 1.47\ R_0$ and (after various vicissitudes) all trajectories are attracted to a branch of periodic solutions that exists for $R \geq 1.44\ R_0$. Figure 2 shows the behaviour of the velocity and magnetic field for an oscillatory solution at $\mathrm{R} = 2.59\ \mathrm{R_0}$. The horizontal scale has changed so that there are four rolls with rising plumes at the centres and sides of the region. These plumes are modulated periodically so that they alternate in vigour. Motion is now more vigorous at the top of the layer, where the velocity reverses its direction in each half of the domain, though the sense of motion at the base of the layer is unchanged. These oscillations are robust and persist over a wide range of parameter values. Note that the time-dependence as well as the underlying horizontal scale (with four rolls) is controlled by the upper part of the layer, where the magnetic field is most effective. On the other hand, the increase in ζ towards the base eliminates any travelling waves.

It is easy to generalize from the results in Figure 2 to the behaviour of an infinite layer. There one might expect to find aperiodically modulated rolls with rising plumes that sporadically burst through to the top of the layer. Similar behaviour in three dimensions, with an irregularly tesselated pattern of convection, would give rise to spatiotemporal chaos. The associated timescale is comparable with the Alfvén transit time for the layer, around 2 hours in the sun. This process provides a natural explanation for umbral dots in the centre of a sunspot. Observers have distinguished between central and peripheral umbral dots (e.g. Grossmann-Doerth *et al.* 1986). The latter occur near the umbral-penumbral boundary and may connect with bright filaments or even form linear features that develop into light bridges separating different portions of the umbra (García de la Rosa 1987; Scharmer 1989). The central umbral dots can be regarded as a convective phenomenon within a more or less homogeneous flux rope, where the field strength varies owing to the motion, but behaviour near the inner edge of the penumbra suggests that field-free plasma is intruding from outside the flux tube, as proposed by Parker (1979a, b) and Choudhuri (1986).

Figure 2. Oscillatory convection in a stratified layer for $R = 2.59\ R_0$. Streaklines and fieldlines at successive intervals during half a period of the oscillation.

3. Penumbral Field Configurations

The field becomes increasingly inclined to the vertical towards the outer boundary of the penumbra. Indeed the Evershed flow in dark penumbral filaments is virtually horizontal. Thus it is tempting to suppose that the magnetic field in the outer penumbra is likewise almost horizontal, as implied by Wittmann's (1974) measurements. Now the transition between the magnetic flux tube beneath the sunspot and the external plasma is observed to be sharp. If the field is horizontal this interface reduces the efficiency of convection so drastically that only shallow penumbral models (about 100 km thick) can be constructed (Schmidt, Spruit & Weiss 1986). These models allow one to develop an attractive and consistent picture of the outer penumbra but they have several fatal flaws. First of all, the magnetic flux emerging through the umbra and penumbra is insufficient for a monopole field with a strength of 1500G at the outer edge of the penumbra (Schmidt 1987). Secondly, the most recent precise measurements show that the mean field has an inclination of about 70° at the penumbral boundary (Adam 1990). Finally, high resolution measurements of the line-of-sight field in a spot as it rotates across the disc show that fields in bright and dark filaments have different directions, with the former inclined at $60 - 70°$ at the edge of the spot (Title *et al.* 1989). Thus one is forced back to considering a two-component field in the penumbra. Unfortunately no-one has yet produced a convincing description of such a configuration.

Some isolated features are understood: for example, the Evershed outflow (discovered at Kodaikanal) can be explained as a siphon effect driven by pressure differences along slightly inclined flux tubes (Meyer & Schmidt 1968; Thomas 1988; Montesinos & Thomas 1989). The inward motion of bright grains (Muller 1973) suggests that the filamentary structure is essentially time-dependent and that the intersection of a bright filament with the photosphere moves towards the umbra as the filament rises (Spruit 1981). The alternation between bright and dark filaments could then be related to some undular instability but it is not clear how this process could be represented by a local model without involving the deep structure of the sunspot.

4. Global Structure

The simplest models of a flux tube in magnetohydrostatic equilibrium assume an axisymmetric force-free poloidal field. This requires a potential field within the flux tube, separated by a current sheet from the external field-free plasma. The boundary condition is that the total pressure must be continuous across the interface, whose position is unknown. Sophisticated procedures for solving the resulting free boundary problem have now been developed (Schmidt & Wegmann 1983; Jahn 1989; Pizzo 1990). These potential field models provide valid descriptions of small flux tubes and pores but fail for sunspots. In a spot with a Wilson depression of 600 km the difference between the gas pressures at the centre of the umbra and in the external plasma is about 10^6 dyne cm^{-2}. This corresponds to a field of 5000 G, much greater than anything observed in normal sunspots. Hence it is necessary to invoke volume currents within the flux tube in order to explain the equilibrium of a sunspot. Jahn (1989) has made a great step forward in modelling the static structure of an (azimuthally averaged) axisymmetric sunspot. Figure 3 shows details of one of his solutions, which is matched to the observed surface field distribution. The currents are confined to a thick sheath below the penumbra, enclosing a central core with a potential field. This equilibrium model, with a gradually tapering flux tube, seems convincing.

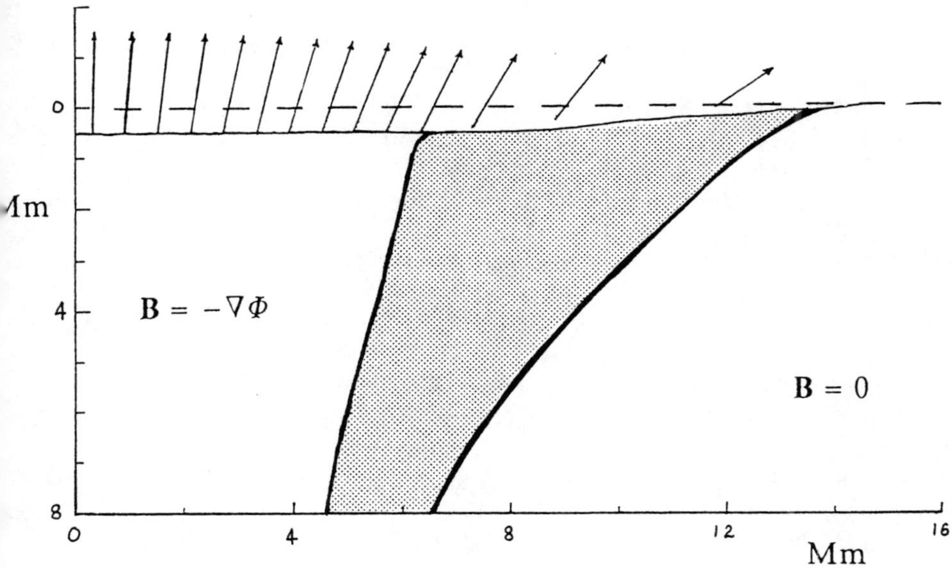

Figure 3. Cross-section through an axisymmetric model of a sunspot in magnetohydrostatic equilibrium (after Jahn 1989). The current sheath (shaded) encloses the potential field underneath the umbra. Arrows indicate the magnitude and direction of the magnetic field at the surface.

The difficulties come when we consider non-axisymmetric fine structure and energy transport mechanisms in the penumbra. Then we must describe dynamical processes in the flux tube. The region below the penumbra has an inclined magnetic field. It is convectively unstable and may also be unstable to interchanges, owing to the curvature of the boundary. Hence we may expect undular instabilities which could lead to the formation of penumbral filaments. In addition they might give rise to intrusions of field-free plasma into the umbra, as suggested by observations made at Tenerife and La Palma. So we are faced with the problem of describing the nonlinear development of instabilities in the current sheath below the penumbra. Here the geometry is awkward, the boundary conditions are uncertain and several effects are simultaneously involved. It is not obvious how to isolate individual processes, while a full simulation would be complicated and difficult to interpret. Yet we cannot claim to understand a sunspot until we are able to analyse this structure.

References

Adam, M.G. 1990. *Solar Phys.* **125**, 37.

Antia, H.M. & Chitre, S.M. 1979. *Solar Phys.* **63**, 67.

Biermann, L. 1941. *Vierteljahrschr. Astron. Ges.* **76**, 194.

Cattaneo, F. 1984. In *The hydromagnetics of the sun*, p. 47. ESA SP-220, Noordwijk.

Chandrasekhar, S. 1961. *Hydrodynamic and hydromagnetic stability.* Clarendon Press, Oxford.

Chitre, S.M. 1963. *Mon. Not. Roy. Astron. Soc.* **126**, 431.

Chitre, S.M. & Shaviv, G. 1967. *Solar Phys.* **2**, 150.

Choudhuri, A.R. 1986. *Astrophys. J.* **302**, 809.

Cowling, T.G. 1976. *Magnetohydrodynamics.* Hilger, Bristol.

Deinzer, W. 1965. *Astrophys J.* **141**, 548.

Evans, D.J. & Roberts, B. 1990. *Astrophys. J.* **348**, 346.

García de la Rosa, J.I. 1987. *Solar Phys.* **112**, 49.

Grossmann-Doerth, U., Schmidt, W. & Schröter, E.H. 1986. *Astron. Astrophys.* **156**, 347.

Hughes, D.W. & Proctor, M.R.E. 1988. *Ann. Rev. Fluid Mech.* **20**, 187.

Hurlburt, N.E., Proctor, M.R.E., Weiss, N.O. & Brownjohn, D.P. 1989. *J. Fluid Mech.* **207**, 587.

Hurlburt, N.E. & Toomre, J. 1988. *Astrophys. J.* **327**, 920.

Jahn, K. 1989. *Astron. Astrophys.* **222**, 264.

Knobloch, E. & Weiss, N.O. 1984. *Mon. Not. Roy. Astron. Soc.* **207**, 203.

Meyer, F. & Schmidt, H.U. 1968. *Zs. Angew. Math. Mech.* **48**, 218.

Meyer, F., Schmidt, H.U., Weiss, N.O. & Wilson, P.R. 1974. *Mon. Not. Roy. Astron. Soc.* **169**, 35.

Montesinos, B. & Thomas, J.H. 1989. *Astrophys. J.* **337**, 977.

Moreno Insertis, F. & Spruit, H.C. 1989 *Astrophys. J.* **342**, 1158.

Muller, R. 1973. *Solar Phys.* **29**, 55.

Obridko, V.N. 1974. *Astr. Zh.* **51**, 1272; *Sov. Astron.* **18**, 758.

Parker, E.N. 1979a. *Astrophys. J.* **230**, 905.

Parker, E.N. 1979b. *Astrophys. J.* **234**, 333.

Pizzo, V.J. 1990. *Astrophys. J.*, submitted.

Proctor, M.R.E. & Weiss, N.O. 1982. *Rep. Prog. Phys.* **45**, 1317.

Rudraiah, N., Kumudini, V. & Unno, W. 1985a. *Publ. Astron. Soc. Japan* **37**, 183.

Rudraiah, N., Kumudini, V. & Unno, W. 1985b. *Publ. Astron. Soc. Japan* **37**, 207.

Scharmer, G. 1989. Private communication.

Schlüter, A. & Temesvary, S. 1958. In *Electromagnetic phenomena in cosmical physics* (ed. B. Lehnert), p. 263. Cambridge University Press.
Schmidt, H.U. 1987. In *The role of fine-scale magnetic fields on the structure of the solar atmosphere* (ed. E.-H. Schröter, M. Vázquez & A.A. Wyller), p. 219. Cambridge University Press.
Schmidt, H.U., Spruit, H.C. & Weiss, N.O. 1986. *Astron. Astrophys.* **158**, 351.
Schmidt, H.U. & Wegmann, R. 1983. In *Dynamical Problems in mathematical physics* (ed. B. Bronowski & E. Martensen), p. 137. Lang, Frankfurt.
Spruit, H.C. 1981. In *The physics of sunspots* (ed. L. H. Cram & J. H. Thomas), p. 359. Sacramento Peak Observatory, Sunspot.
Thomas, J.H. 1988. *Astrophys. J.* **333**, 407.
Title, A.M., Frank, Z.A., Shine, R.A., Tarbell, T.D. & Scharmer, G. 1989. *Bull. Am. Astron. Soc.* **21**, 837.
Weiss, N.O., Brownjohn, D.P., Hurlburt, N.E. & Proctor, M.R.E. 1990. *Mon. Not. Roy. Astron. Soc.*, in press.
Wittmann, A. 1974. *Solar Phys.* **36**, 29.
Yun, H.S. 1970. *Solar Phys.* **16**, 398.

SCUSSION

)NTGOMERY: In our MHD computations, when we have gone from two dimensions to :e, we have often seen large changes and qualitatively new effects. Do you expect the same l happen with your convective MHD computations? What can one learn in two dimensions t is likely to remain true in three?

:ISS: Two-dimensional models are obviously over-simplified, though they enable us to ntify bifurcations and to investigate transitions from one form of solution to another. ee-dimensional simulations of compressible convection in the absence of a magnetic field w that sinking sheets are focussed into plumes and allow a richer structure. It is likely that two-dimensional solutions are unstable to three-dimensional perturbations. Nevertheless, expect travelling waves to persist in the regime where $1 \lesssim \beta \lesssim 10$ and secondary oscillations urvive in stratified layers. We plan to carry out three-dimensional numerical experiments he near future.

'EER: Could you make predictions on the timescale, temperature- and magnetic field :tuations, and velocities involved with the oscillatory convection you simulated?

:ISS: Our model is too idealized to predict the magnitudes of temperature and field turbations at the umbral photosphere. The timescale is determined by the Alfvén speed r a layer about 3000 Km deep, which is compatible with the observed lifetime of umbral s.

ITRE: Is the time ripe now to construct two-dimensional models of sunspots in order to erstand problems concerning the depth of a sunspot (*i.e.* location of the base of the spot)?

:ISS: I think there is scope for more detailed models extending the magnetohydrostatic lels developed recently by K. Jahn. There is still a difficulty in describing thermodynamic perties of the region with volume currents, which makes it hard to formulate a consistent tment.

RUDRAIAH: (i) Are there observations to reveal the existence of horizontal magnetic fi comparable in magnitude to that of vertical magnetic field?
(ii) How dark is the Sunspot? Is your model capable of explaining this?

WEISS: (i) There is some evidence for azimuthal fields but they do not seem to significant. The radial field increases from zero at the centre of the umbra to the edge of spot where the field is almost horizontal. Small-scale variations in field directions are v difficult to measure.
(ii) The mean umbral heat flux is about 20% of the photospheric flux. This is consistent w our model but any quantitative prediction will need detailed simulations like that being carr out by Nordlund and Stein.

AL-KHASHLAN: How dark is the sunspot?

WEISS: The umbra is about 2000K cooler than the normal photosphere so the intensity reduced by a factor of 6. The isolated umbral dots are small and have photosphe brightness. The mean penumbral intensity is about 25% less than the normal photosphe value but dark filaments are 40% less bright while bright filaments are only slightly cooler.

SUNSPOT SEISMOLOGY THEORY

JOSEPH M. DAVILA
Laboratory for Astronomy and Solar Physics
NASA-Goddard Space Flight Center
Greenbelt, MD 20771

ABSTRACT. Recent observations have demonstrated that sunspots can absorb significant p-mode wave power, and that this absorption has a well defined wavelength dependence. These observations open, for the first time, the potential to deduce semi-empirically the subsurface structure of sunspots and active regions. Several physical mechanisms have been proposed to explain the absorption. In this paper, the proposed mechanisms are reviewed and their viability in view of the current knowledge of the scattering process is discussed.

1. Introduction

Below the solar photosphere waves are energetically unimportant. Most of the energy is carried by the convective or radiative flux. Sub-photospheric waves can however serve as powerful probes of the internal structure of the sun. Most of these studies have concentrated on the large scale properties of the solar interior. Recently however observations of the scattering of solar p-modes from sunspots have been obtained.

During the first half of this century physicists were able to deduce the structure of the atomic nucleus by observing the scattering of electromagnetic radiation. One might expect, by analogy, that the process of scattering of p-modes by sunspots will ultimately yield significant new insights into the subsurface structure of solar active regions, and ultimately to new ideas regarding the formation of active regions, sunspot structure and the influence of the dynamo on the solar activity cycle.

2. Observations

Previous attempts to use oscillations as a probe of sunspot structure have concentrated on velocities measured within the sunspots. These observations enabled a test of the hypothesis that the magnetic field of the spot acts as a selective filter of external wave energy (Thomas et al., 1982). There is also an apparent shift in wavenumbers in the spot relative to the surrounding photosphere, possibly related to the difference between

E. R. Priest and V. Krishan (eds.), Basic Plasma Processes on the Sun, 149–157.

sunspot magnetoacoustic speed and sound speed in the surrounding atmosphere (Abdelatif et al., 1986).

More recent work has taken a different point of view. Braun, Duvall and LaBonte (1987) have studied how the presence of the sunspot affects externally impinging waves, much as one would study the structure of an atom by bombarding it with electromagnetic radiation and observing the scattered radiation field. The high-degree acoustic oscillations are vertically trapped in a spherical shell with the surface as the upper boundary and the lower boundary depending on the frequency and horizontal wavenumber in a known way. It is likely that the high-degree waves are not global modes, that is, they do not remain coherent while travelling over a circumference to interfere with themselves. So locally they can be considered as horizontally travelling, vertically trapped waves. The waves are normally observed by their vertical motions near the surface, which we see as Doppler shifts of spectrum lines. In the classical analysis, waves in one fixed direction are isolated by spatial filtering of two-dimensional Doppler shift images and then a two-dimensional power spectrum isolates the behavior in horizontal wavenumber and temporal frequency. The two independent quadrants of this power spectrum correspond to waves travelling in opposite directions on the solar surface.

Braun, Duvall and LaBonte (1987) replaced the classical analysis by one in a cylindrical coordinate system centered on the sunspot. Radially propagating waves are isolated by a Hankel function decomposition from which the sunspot itself is excluded. The resulting two quadrants of the power spectrum correspond to inward and outward propagating waves. It is immediately obvious from one of these spectra that there is a large excess of power in the incoming waves over the outgoing waves. This excess is interpreted as being due to absorption by the sunspot. The absorption appears to be a function only of the horizontal wavenumber k and not of the frequency ω or the phase velocity ω/k which would be expected for a purely depth dependent effect (Duvall, 1982). The absorption is small at low k and increases to a maximum of 50% absorption at $k \sim 1\ Mm^{-1}$ and remains at that level to the maximum observed wavenumber $k = 1.5\ Mm^{-1}$. This result was found for spots with 10 Mm umbral diameter. For a smaller spot a reduced magnitude of absorption was found (Braun et al., 1988).

3. Qualitative Discussion of the Scattering Process

What is the physical nature of the mechanism responsible for this absorption? There are two possibilities, both illustrated in figure . The incoming wave flux, P_{in} is greater than the outgoing power, P_{out}. The difference between these two quantities is defined as the absorbed power. The excess energy could be converted into heat within the sunspot by dissipative (irreversible) processes. The mechanism most frequently suggested is resonance absorption, a process which has been well studied in the context of heating coronal loops. It is also possible that the incoming sound wave shakes the magnetic field of the sunspot flux tube, and energy which is originally in sound waves is transferred into magnetic wave modes propagating more or less parallel to the magnetic field of the sunspot flux tube. If these waves can then leak from the system, say upward through the chromosphere or downward through the base of the sunspot (the flux is denoted as S_{mag}

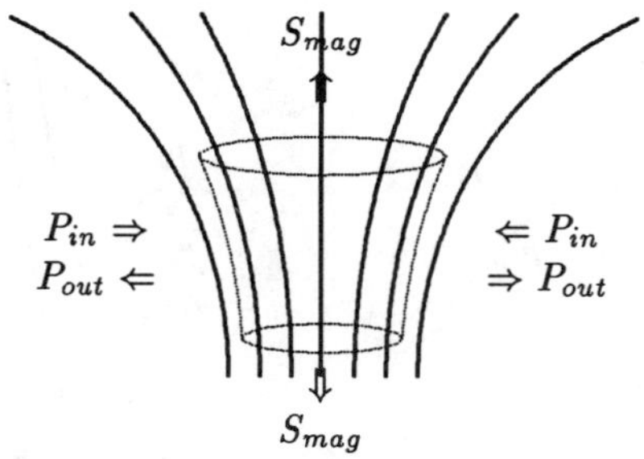

Figure 1: *Physical processes which can result in the absorption of p-modes. The net power absorbed $P_{in} - P_{out}$ can be converted into magnetic modes which then propagate out through one of the ends of the sunspot flux tube, S_{mag}, or the energy can be dissipated in a thin hollow cylindrical region.*

in figure 1), one could explain the observed absorption by a completely non-dissipative process involving mode conversion in the field of the sunspot. Finally, it should be noted that these processes are not mutually exclusive. It is indeed possible that both of these basic mechanisms, i.e. dissipation and mode conversion, are responsible for the absorption seen in a real spot. At this point there is no observational way to distinguish between these possibilities.

From relatively simple considerations, however, one can make some educated guesses about what might be learned about the structure of sunspots by comparing theory and observational descriptions of the scattering process. For example if the magnetic field of the sunspot is converging with depth, as shown in figure 2a, one can see that the propagation direction of the outgoing p-modes is on the average directed more upward toward the solar surface than for the incoming waves. If on the other hand, the magnetic field diverges with depth, as shown in figure 2b, the average propagation direction of the p-modes is directed more downward toward the interior of the sun. It seems reasonable to hope that these two situations could be distinguished observationally once models are available to predict the magnitude and exact signature of the effect.

It seems unreasonable to imagine that a sunspot would remain as a vertical monolithic structure in the solar atmosphere. It is likely to be displaced from its vertical position by convective motions in the surrounding fluid and/or by magnetic buoyancy. By considering the simple structure shown in figure 2c, it is clear that the scattering of sound from such a structure would not be axisymmetric. The p-mode flux on one side of the spot is reflected upward on the average, and the flux on the other side is reflected downward. Asymmetries in the scattering rate have been observed. However it seems that they can be explained primarily by the "impact parameter" effect. (Braun et al., 1988) Additional modeling must be completed before one can hope to separate any additional asymmetry due to a non-vertical sunspot flux-tube from the asymmetry caused by this basic mechanism.

Finally, as indicated in figure 2d, one might hope to determine whether the sunspot

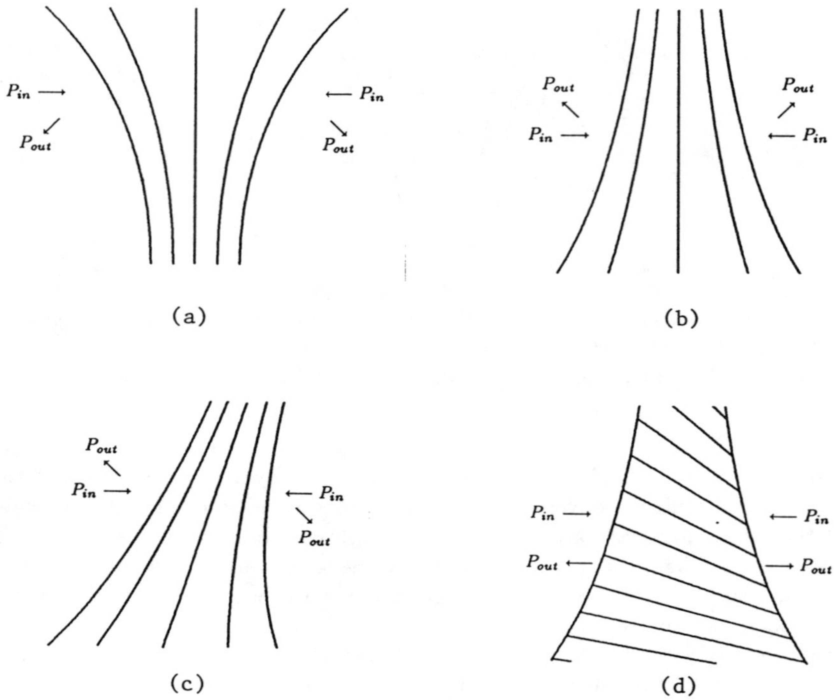

Figure 2: *The geometry of the subsurface sunspot flux tube can affect the amplitude of the scattered wave observed at the surface. The panels represent diverging flux tubes (a), a converging flux tube (b), a non-vertical sunspot (c) and a twisted flux tube (d). Potential effects of these different configurations are discussed in the text.*

flux tube carries a significant current, i.e. whether the flux tube is twisted. One might suspect that the presence of twist could significantly stiffen the flux tube and thereby change the predicted amplitude for waves reflected from the spot interface. Alternately as discussed below the presence of twist seems to be necessary for the efficient dissipation of axisymmetric modes by the resonance absorption process.

As suggested in the discussion above, the physical models proposed thus far can be divided into two broad classes; (1) models which incorporate dissipation of the p-mode energy within the spot (i.e. the irreversible transfer of energy into heat), and (2) dissipationless models which rely on conversion of the p-mode flux into other wave modes. Let us begin with a discussion of the dissipative models.

4. Dissipative Models

Two dissipative mechanisms have been proposed. The first, resonance absorption, is an extension of a wave dissipation model developed for the corona. The second is radiative damping within the sunspot. These are discussed below.

4.1 RESONANCE ABSORPTION

Hollweg (1988) was the first to propose a theoretical explanation for the absorption of p-modes by sunspots in the context of a simplified, planar resonance absorption model. He found that resonance absorption could be important under certain circumstances relating to the angle of incidence of the wave and the ratio of specific heats inside and outside of the spot. But he concluded that the resonance absorption process, on the whole, probably could not explain the large amount of absorption seen observationally.

Lou (1989) investigated the resonant absorption of MHD waves in a viscous medium. The linearized MHD wave equations including both shear and compressive viscosity terms were solved numerically using a standard algorithm. He concluded that for non-axisymmetric perturbations absorption at the Alfvenic singular layer could be quite effective in absorbing wave energy. Typical values for the absorption coefficient obtained in these models is on the order of $40 - 45\%$, in reasonable agreement with the observed value.

For axisymmetric perturbations the story is quite different. Lou showed that when axisymmetric waves impinge on a perfectly cylindrical sunspot, there is no singularity in the ideal MHD equations. The result is that there is no resonance layer set up in the spot, and the dissipation is therefore drastically reduced within the sunspot. The analogous result for resonant dissipation in a low β plasma was noted by Davila (1985). Whenever the transverse wavenumber ($k_{\perp}$) vanishes the two transverse components of the momentum equation decouple and the solution changes character dramatically. Lou showed that to dissipate 50% of the amplitude of an incoming axisymmetric wave the viscosity within the spot would have to be of order $10^5\ g - cm^{-1} - s^{-1}$. This value is much larger than the classical value for the viscosity, implying turbulent processes are important. However the turbulent viscosity needed to explain the observations is roughly 100 times larger than the upper limit on the spatially averaged turbulent viscosity obtained by considering the observed width of the p-mode ridges themselves. Lou also points out that it is somewhat counter intuitive to expect a large increase in the turbulent viscosity within a sunspot where the magnetic field is presumed to inhibit convection.

Chitre and Davila (1989, in these proceedings) considered an analytic model for resonance absorption in a sunspot flux tube. The model was basically an extension of the calculation of Davila (1985) to incorporate cylindrical geometry and a finite sound speed. The absorption rate was calculated from the ideal portion of the solution by integrating the Poynting flux over the surface of the resonance layer. It was found that 50% of the incoming wave energy could be absorbed with rather modest wave amplitudes ($\approx .3 - .5$ of the incoming wave) within the spot. The expression for the energy absorbed was seen to be proportional to m^{-2} in rough agreement with the calculation of Lou.

It should be emphasized that the absorption rate obtained in this calculation is independent of the details of the dissipation mechanism (magnitude or physical process) assumed. Therefore the results are very general, and can be applied for viscous, ohmic, radiation or any other damping mechanism. The analytic example presented by Chitre and Davila assumed radiative damping, but the same absorption rate would be obtained for viscous or ohmic dissipation. There are however two important observables which *do* depend on the dissipation process; (1) the amplitude of the wave within the dissipation region, and (2) the thickness of the dissipation region. Estimates for these quantities have not yet been obtained.

The problem with the absorption of axisymmetric modes remains however. Chitre and Davila suggest that if one considers more realistic spot models this problem may disappear. For example if one considers a spot which is not exactly cylindrical, perhaps it is elliptic or it has a nearly circular but fluted surface it is clear that incoming waves which are axisymmetric would be coupled because of the asymmetric geometry into asymmetric internal modes within the spot. These modes could then be efficiently dissipated by the resonant absorption process. These same arguments would hold if the sunspot were not precisely vertical within the atmosphere. There have been no calculations of the strength of these effects, however it is clear that at least in principle non-axisymmetric geometry of the spot itself may be able to account for the observed absorption of symmetric waves.

In addition, preliminary analysis of twisted flux tubes indicates that the absorption of m=0 modes is possible, even in the context of an otherwise perfectly symmetric sunspot.

4.2 RADIATIVE DAMPING

In a recent paper Bogdan and Knolker (1989) consider the propagation of linearized MHD waves in a uniform radiating plasma. They conclude that in the uniform limit, radiation damping is not likely to contribute significantly to the damping of p-modes within sunspots. This conclusion can be significantly modified when cross field gradients within the sunspot flux tube are accounted for (see the discussion above).

5. Non-dissipative Models

There have been two non-dissipative models proposed in the literature. Neither seems to be able to explain the observed absorption of p-mode flux in their present form.

5.1 CHROMOSPHERIC WINDOW EFFECT

Campbell and Roberts considered the effect of horizontal chromospheric magnetic fields on the reflection of p-modes. They found that the efficiency of the reflection was a function of the horizontal wavelength of the disturbance (the results of their calculation was summarized in figure 1 of that paper). They suggested that the observed absorption of p-modes could be due to enhanced leakage of the wave flux through the chromosphere. If however one examines the calculated result carefully one finds that as ℓ (i.e. $k_\perp$) increases more and more frequencies are trapped in the photosphere. This wavelength dependence seems to be contrary to the observed result that the sunspot absorption increases with increasing $k_\perp$.

5.2 RESONANT SCATTERING

Bogdan (1989) considers the scattering of plane sound waves by sunspots. He finds that the transmission coefficient has resonant peaks, suggesting increase absorption at discrete frequencies. Bogdan suggests that this effect could result in the absorption observed by Braun, Duvall and LaBonte.

It seems clear however that this mechanism alone cannot explain the absorbed p-mode flux because the total energy in the system is conserved. The model must be modified to include dissipation or leakage at the boundary to obtain agreement with the observations.

6. Conclusions

There is no satisfactory theoretical model for the absorption of p-modes by sunspots available at this time. The resonance absorption model seems to be able to obtain the large observed absorption coefficients observed for non-axisymmetric perturbations. For axisymmetric perturbations, departures from perfect cylindrical symmetry or the inclusion of a slight twist in the sunspot flux tube may be able to resolve the problem with the absorption of $m = 0$ wave modes. A calculation of the magnitude of these effects is not available at this time. Other dissipative models, which do not incorporate the background gradient effects inherent in the resonance absorption mechanism, require uncomfortably large dissipation coefficients within the sunspot.

The chromospheric window effect seems to have a wavelength dependence which is contrary to observation, and the resonant scattering effect proposed by Bogdan cannot explain the observed p-mode absorption without the addition of an energy loss mechanism to the model.

Acknowledgement: *The author wishes to Drs. S. M. Chitre and T. L. Duvall for constructive comments on the initial draft of this paper.*

References

Abdelatif, B. W., B. W. Lites, and J. H. Thomas, 1986, *Ap. J.,* **311,** 1015.

Bogdan, T. J., 1989, *Ap. J.,* **345,** 1042.

Bogdan, T. J. and M. Knolker, 1989, *Ap. J.,* **339,** 579.

Braun, D. C., T. L. Duvall, and B. J. LaBonte, 1987, *Ap. J.,* **319,** L27.

Braun, D. C., T. L. Duvall, and B. J. LaBonte, 1988, *Ap. J.,* **335,** 1015.

Campbell, W. R. and B. Roberts, 1989, *Ap. J.,* **338,** 538.

Davila, J. M., 1985, *Ap. J.,* **291,** 328.

Duvall, T. L., 1982, *Nature,* **300,** 242.

Hollweg, J. V., 1988, *Ap. J.,* **335,** 1005.

Lou, Y., 1989, Viscous Magnetohydrodynamic Modes and p-Mode Absorption by Sunspots, in press.

Thomas, J. H., L. E. Cram, and A. H. Nye, 1982, *Nature,* **297,** 485.

DISCUSSION

PRIEST: (i) What is the wavelength of the observed modes compared with the spot radius?
(ii) If a wavetrain is locally incident from one direction rather than from all directions, why should the non-axisymmetric modes not be more important than the axisymmetric ones?
(iii) What has been learnt so far from sunspot seismology about the subsurface nature of sunspots?

DAVILA: (i) A range of horizontal wavelengths is observed but typically the wavelengths observed are on the order of the spot diameter.
(ii) The p-modes in the vicinity of the spot are assumed to be nearly isotropic. For that reason the velocity expansion in terms of cylindrical waves is probably more appropriate than the plane wave expansion that you suggest.
(iii) We have learned quite a lot observationally about the interaction of p-modes and sunspots. Converting this information into inferences regarding the physical nature of the sunspot below the surface, one must have a theory for the interaction. A generally accepted theory does not exist at this time.

ROBERTS: I think it may be important that a spot has both a vertical structure (the umbra) and a horizontal one (in the penumbra). So *two* effects may be at work in the absorption of p-modes, one due to the vertical plug of field and the other due to the horizontal penumbral canopy. Is there any evidence for two aspects in the data of Braun *et al.*?

DAVILA: The absorption seen in the observations is clearly not confined to the umbra of the spot. Typically absorption seems to occur in a region 2-3 times larger than the sunspot umbra.

MOGILEVSKIJ: We have measured the spectrum of oscillations in sunspot umbra and found that in the low sound spectrum there are some discrete absorption lines with the frequences $\omega = n\omega_p$ $(n = 1,2,...)$. These lines give information on subphotospheric dynamics down to 10^9cm.

UBEROI: (i) Why do you rule out the additional singularity arising due to compressibility for the m = 0 mode?
(ii) What are the magnetic field ranges in which sound wave conversion takes place?

DAVILA: (i) Any singularity in the equation can in principle provide a site for dissipation in the plasma.

(ii) p-modes are absorbed in regions with surface field strengths which range from 100 to a few thousand Gauss.

HOLLWEG: (i) In response to Dr Uberoi's question, I have looked at the cusp singularity and concluded that it exists only under *very* extreme parameters. (ii) If the spot "surface" is corrugated, it loses its meaning and the m-difficulties go away. The talk by Nigel Weiss suggests evidence for fluting so we may indeed have corrugation.

DAVILA: (ii) The coupling between m = 0 and m > 0 modes caused by surface corrugations on the sunspot flux tube will have a strength proportional to the amplitude of the corrugation We need to calculate what amplitude is necessary to provide the observed coupling. The initial sunspots analyzed by Braun, Duvall and La Bonte were chosen for their circular symmetry. Of course the sub-surface structure of the tube is not directly observed.

WAVES IN MAGNETIC FLUX TUBES

B Roberts
Mathematical Sciences Dept, The University, St Andrews KY16 9SS, Scotland

ABSTRACT. The basic aspects of wave propagation in a magnetic flux tube are reviewed, with particular emphasis on the types of flux tube that occur in the solar atmosphere. Two fundamental speeds arise naturally in a description of wave propagation in a flux tube: the slow magnetoacoustic (cusp) speed c_T, which is both subsonic and sub-Alfvénic, and a mean Alfvén speed c_k. Both surface and body modes are supported by a tube. It is stressed that a flux tube may act as a wave guide, similar to the guidance of light by a fibre optic, or sound in an ocean layer, or seismic waves in the Earth's crust.

1. INTRODUCTION

The evidence for the occurrence of magnetic flux tubes in astrophysical plasmas is increasing all the time. Flux tubes (or flux ropes) are believed to occur in extragalactic jets, in the magnetospheres of some planets, and in the Sun. Indeed, in the case of the Sun almost all of the magnetic flux emerging through the photospheric surface is found to occur in concentrated forms, ranging in scale from the visible sunspot to the sub-telescopic intense flux tubes. Thus, the source of coronal and solar wind magnetism is to be found in concentrated roots of magnetic field emerging from the deep interior. The field strengths in these sources, where they can be directly measured or indirectly inferred, are generally high: in a photospheric tube the field is as high as 0.2T (= 2kG), confined to a radius of about 100 km, whereas in a sunspot the field is typically 0.2-0.3T (= 2-3kG) and covers a wide area of the solar surface.

The concept of a magnetic flux tube goes back to the work of Michael Faraday (1791-1867) and James Clerk Maxwell (1831-1879). Faraday, the intuitive scientist with no formal training in mathematics, used the concept as a useful guide to a physical understanding of electrical and magnetic phenomena. Maxwell, with his more mathematical bent, gave the idea a more formal development and substance. Faraday was a little surprised, and no doubt pleasantly, to see the concept of a flux tube responding to Maxwell's mathematical overtures without exhibiting any undue signs of strain. It is amusing to note the manner in which Faraday, writing on the 4th May 1885 to the engineer Isambard Kingdom Brunel, puts his feelings: "I sometimes feel glad that I am not a mathematician for though Mathematical science is sure, I do not find that the conclusions of its professors can be trusted much more than those of other Professors" (Faraday 1885). Good advice indeed!

E. R. Priest and V. Krishan (eds.), Basic Plasma Processes on the Sun, 159–174.

The mathematical substance that Maxwell gave to the concept of a magnetic flux tube rested on a very general definition. In his paper 'On Faraday's Lines of Force', Maxwell puts it thus: "If upon any surface which cuts the lines of fluid motion we draw a closed curve, and if from every point of this curve we draw a line of motion, these lines of motion will generate a tubular surface which we may call a *tube of fluid motion*" (Maxwell, 1855/6). By the term 'fluid' Maxwell means a purely imaginary substance and by 'lines of fluid motion' we would now say field lines.

The types of flux tubes that we are interested in astrophysical plasmas are more specific than those resting upon the rather general definition Maxwell gives us, for the interior and exterior of a tube are commonly distinct from one another by reason of physical differences additional to magnetism, though magnetism may well be the *cause* of such differences. For example, the field outside of sunspots and intense photospheric tubes is negligible in comparison with the interior field; so these objects are *isolated* flux tubes (see Roberts 1990a for a recent review). Such tubes are in magnetostatic pressure balance with their non-magnetic surroundings, so there is a deficit in gas pressure inside the tube: the tube is partially evacuated. For sunspots, the tube is also cooler than surroundings so there is a temperature deficit as well. The picture is more complicated for intense tubes, the temperature structure varying with height. (For recent reviews of solar magnetism and magnetic flux tubes see Schüssler (1987, 1990), Solanki (1987), Stenflo (1989) and Roberts (1990a).)

The tubes of the solar corona are different from those of the photosphere, for the corona is dominated by magnetic field and so there are no isolated tubes there. Instead, inhomogeneities in density and temperature give rise to magnetic flux tubes in which the magnetic field fills all of space; the interior of a coronal tube is generally denser than surroundings. Such *coronal loops* make up the coronal plasma in *active regions*.

What can magnetic flux tubes do? Firstly, we note that a tube is a *wave guide*. It permits waves to propagate without spatial attenuation. Thus a tube is likely to be *a good communication channel* between one region of a plasma and another, perhaps providing a connection between an energy source and an energy sink. For example, photospheric flux tubes connect the convection zone - an ample source of energy, especially in granules - with the chromosphere and corona, and so are almost certainly central to the question of heating of those regions. Also, they guide spicular material that produces the spicule jets in the chromosphere. Secondly, we observe that a magnetic flux tube is an *elastic* object (and so an elastic, not rigid, wave guide) and as such is likely to respond to sudden changes by guiding *waves*. Sunspots, for example, are known to support a variety of wave phenomena (see Evans and Roberts (1990) for a recent discussion). Finally, we point out that flux tubes may both *leak* waves to and *absorb* waves from their environment (see Roberts and Webb 1979; Spruit 1982; Cally 1985, 1986; Davila 1985). Braun, Duvall and LaBonte (1987, 1989), for example, have discovered that sunspots may be absorbing over 50% of the sound waves (p-modes) in their vicinity.

2. OSCILLATIONS OF AN ISOLATED TUBE

2.1 Propagation Speeds

We consider the modes of oscillation of an isolated magnetic flux tube of strength B_0 embedded in a field-free gas of pressure p_e and density ρ_e. If such a tube is subjected to a sudden twisting motion it will respond by producing a *torsional oscillation* of the tube; such a torsional motion may propagate as an *Alfvén wave* with speed $v_A = B_0/(\mu\rho_0)^{1/2}$, for a tube with gas density ρ_0 and magnetic permeability μ. On the other hand, a symmetric squeezing of the tube may produce the *sausage mode*(see Figure 1). Here both the gas and magnetic field within the tube are expanded and contracted in the motion. Compressions of the gas are characterized by its sound speed $c_s = (\gamma p_0/\rho_0)^{1/2}$, for a tube with gas pressure p_0, whilst magnetic compressions are represented by the Alfvén speed v_A. The two speeds c_s and v_A combine in the form

$$\left(\frac{1}{c_s^2} + \frac{1}{v_A^2}\right)^{1/2} \equiv \frac{1}{c_T} \tag{1}$$

to produce the basic speed c_T. The tube speed c_T is both sub-sonic and sub-Alfvénic and therefore relates to the *slow* magnetoacoustic wave; see Roberts (1985a, 1990b) and Hollweg (1986) for a general discussion of magnetohydrodynamic waves.

It is interesting to note that any elastic tube will have a characteristic propagation speed of the form given by equation (1) for one dimensional waves (see Lighthill 1978). For example, a blood vessel is an elastic tube with a speed v_A arising from the elasticity of the vessel's membrane and c_s being the speed of sound in blood. In this example, the speed c_s is much greater than the 'elastic' speed v_A and so the propagation speed c_T is close to the elastic speed. Another example is provided by water in a hose-pipe and the phenomenon of 'water-hammer'. If the pipe is made of metal, then the elastic speed v_A is much greater than the sound speed in water, and so the propagation speed c_T is approximately the sound speed in water (about 1.4 km s^{-1}). On the other hand, for water in a plastic pipe the ordering in the speeds is reversed and now the propagation speed c_T is close to the elastic speed and so much lower (about 10 ms^{-1}). In the case of a magnetic flux tube under photospheric conditions, the sound and Alfvén speeds are roughly comparable at about 10 km s^{-1} and so the tube speed c_T is reduced to about 7 km s^{-1}.

Asymmetric disturbances of the tube produce *kink* (or serpentine) modes. This may be viewed as similar to the waves on an elastic string, producing a propagation speed of $(T/\rho)^{1/2}$. The tension T in the string is clearly here due to magnetic tension, so $T = B_0^2/\mu_0$. The density ρ is here taken to be the sum of the gas density ρ_0 within the tube and the gas density ρ_e in the environment, reflecting the fact that the vibrating tube displaces about the same amount of material in its surroundings as that in the tube itself. Thus, with $\rho = \rho_0 + \rho_e$ we obtain a characteristic kink speed c_k given by

$$c_k = \left(\frac{\rho_0}{\rho_0 + \rho_e}\right)^{1/2} v_A. \tag{2}$$

The kink speed c_k is sub-Alfvénic but not necessarily sub-sonic.

2.2 Modes of Oscillation

To describe the modes of oscillation of an isolated tube in greater detail, we consider the usual equations of ideal magnetohydrodynamics as applied to a tube. The tube is taken to have a field strength B_0, gas pressure p_0, density ρ_0, and sound speed c_0 ($\equiv (\gamma p_0/\rho_0)^{1/2}$), confined to a radius a by a field-free environment of gas pressure p_e, density ρ_e, and sound speed c_e ($\equiv (\gamma p_e/\rho_e)^{1/2}$). Then pressure balance across the boundary of the tube implies that

$$p_e = p_0 + \frac{B_0^2}{2\mu}, \tag{3}$$

which combined with the ideal gas law implies that

$$\frac{\rho_e}{\rho_0} = \frac{c_0^2 + \frac{1}{2}\gamma v_A^2}{c_e^2}. \tag{4}$$

Modes of oscillation of this equilibrium configuration may be sought by requiring in cylindrical coordinates (r,θ,z) that perturbations in flow, pressure and field be of the form

$$f(\mathbf{r},t) = f(r)\, e^{i(\omega t + n\theta - kz)}, \tag{5}$$

representing a wave of frequency ω, longitudinal wave number k, azimuthal wave number n, and amplitude f(r). The geometry of the modes is then described in terms of the integer n: sausage modes correspond to $n = 0$; kink modes to $n = 1$; and fluting modes to $n \geq 2$. Additionally, the radial behaviour of the amplitude f(r) within $r \leq a$ may be either oscillatory or decaying. Oscillating modes are classified as *body* waves; decaying or evanescent modes as *surface* waves (Roberts 1981 a,b). See Figure 1.

It proves convenient to work in terms of the total pressure perturbation p_T, made up of the sum of the gas pressure perturbation and the magnetic pressure perturbation. For p_T is readily shown (see Edwin and Roberts 1983) to satisfy Bessel's equation:

$$\frac{d^2 p_T}{dr^2} + \frac{1}{r}\frac{dp_T}{dr} - \left(m_0^2 + \frac{n^2}{r^2}\right) p_T = 0, \quad r < a, \tag{6}$$

where the effective transverse wavenumber (squared) is given by

$$m_0^2 = \frac{(k^2c_0^2 - \omega^2)(k^2v_A^2 - \omega^2)}{(c_0^2 + v_A^2)(k^2c_T^2 - \omega^2)}. \tag{7}$$

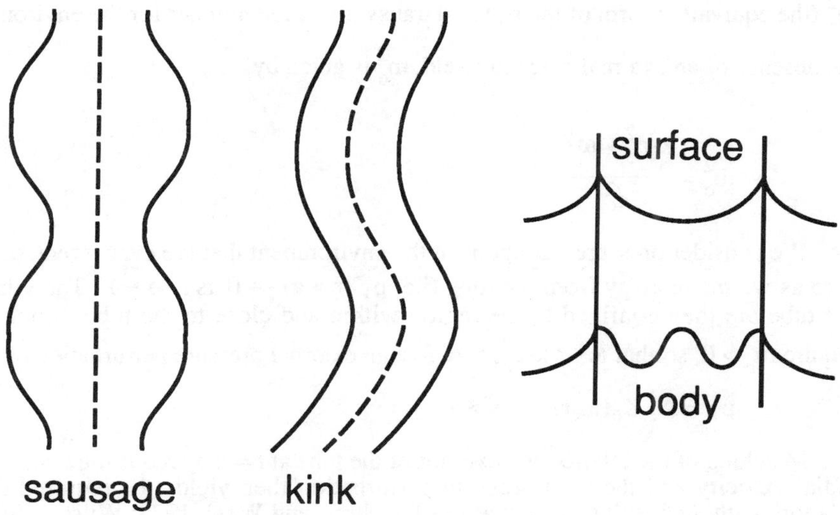

Fig. 1. Surface and body waves in a flux tube, showing the sausage (n=0) and kink (n=1) modes.

One solution of equation (6) which we should not overlook is the apparently uninteresting case of $p_T = 0$. In fact, with $p_T = 0$ the motions are torsional (i.e. $v_\theta \neq 0$, $v_r = v_z = 0$) and incompressible (i.e. div $\mathbf{v} = 0$); these are the torsional Alfvén waves satisfying the wave equation

$$\frac{\partial^2 v_\theta}{\partial t^2} = v_A^2 \frac{\partial^2 v_\theta}{\partial z^2} . \tag{8}$$

Returning to equation (6) we may construct its general solution in terms of Bessel functions. Requiring p_T to be bounded at the centre of the tube gives

$$p_T = \begin{cases} A_0\, I_n(m_0 r), & m_0^2 > 0, \\ A_0\, J_n\, (n_0 r), & m_0^2 < 0, \end{cases} \qquad (r < a), \tag{9}$$

where I_n is the modified Bessel function (of order n) and J_n is the Bessel function. We use the form $I_n\,(m_0 r)$ for *surface* waves ($m_0^2 > 0$), giving exponential decay from the boundary at r = a, and the form $J_n\,(n_0 r)$, giving oscillatory behaviour, for *body* waves ($m_0^2 < 0$). Here $n_0^2 = -m_0^2$.

The region outside the tube may be treated in a similar fashion. Pressure perturbations in r > a are governed by an equation of the form (6), with m_0^2 replaced by

m_e^2 (the equivalent form of the squared transverse wave number for the environment). In the absence of an external magnetic field, m_e^2 is given by

$$m_e^2 = \frac{k^2c_e^2 - \omega^2}{c_e^2}. \tag{10}$$

We consider pressure variations in the environment that are evanescent, declining to zero as we move away from the tube (i.e. $p_T\ (r > a) \rightarrow 0$ as $r \rightarrow \infty$). The vibrations of the tube are then confined to the region within and close to the tube. Such solutions require $m_e^2 > 0$, so that $\omega^2 < k^2c_e^2$, and give an external pressure perturbation of the form

$$p_T = A_e\, K_n(m_e r), \qquad r > a. \tag{11}$$

Matching of the interior and exterior of the tube at $r = a$ by requiring continuity of the radial velocity and the total pressure perturbation then yields the required dispersion relation, with the result (McKenzie 1970; Roberts and Webb 1978; Wilson 1980; Spruit 1982; Edwin and Roberts 1983; Abdelatif 1988; Evans and Roberts 1990)

$$\rho_e\omega^2 m_0 \frac{I_n'(m_0 a)}{I_n(m_0 a)} + \rho_0(k^2v_A^2 - \omega^2)m_e \frac{K_n'(m_e a)}{K_n(m_e a)} = 0, \tag{12}$$

where the dash denotes the derivative of the Bessel function. Equation (12) is valid for $m_e^2 > 0$. The derivation of the dispersion relation (12) given in Roberts and Webb (1978) was for $n = 0$ (sausage) modes only. It is interesting to note that a structurally similar relation was obtained by Tayler (1957) in a stability analysis of a twisted tube with a vacuum environment.

The dispersion relation (12) is transcendental and so obtaining its solutions is a complicated and largely numerical process, though certain cases (e.g. thin tubes, $|k|a << 1$) may be investigated analytically. In the special case of an *incompressible* medium ($c_0 \rightarrow \infty$, $c_e \rightarrow \infty$), for which $m_0^2 \rightarrow k^2$, $m_e^2 \rightarrow k^2$ and $c_T \rightarrow v_A$, the transcendental nature of equation (12) is removed and the solutions become transparent (Dungey and Loughhead 1954; Roberts and Webb 1978; Uberoi and Somasundaram 1980, 1982; Edwin and Roberts 1983)

$$\frac{k^2v_A^2}{\omega^2} = 1 - \left(\frac{\rho_e}{\rho_0}\right)\frac{I_n'(|k|a)K_n(|k|a)}{I_n(|k|a)K_n'(|k|a)}. \tag{13}$$

The modes described by dispersion relation (13) are *surface* waves (since $m_0^2 = k^2 > 0$); they propagate dispersively, a reflection of the fact that the tube provides a lengthscale, namely its radius, against which the waves may be measured. In the *thin* tube limit ($|k|a << 1$), the sausage mode (n=0) propagates with a speed close to v_A ($\equiv c_T$, since the medium is incompressible) and the kink (n=1) mode with a speed close to c_k.

The modes of the *compressible* case (equation (12)) are more complicated to describe, their character depending upon the relative orderings between the speeds c_e, c_0, and v_A (Edwin and Roberts 1983; Cally 1985, 1986; Davila 1985; Abdelatif 1988; Evans and Roberts 1990). For example, a magnetic flux tube with $c_e > c_0 > v_A$, typical of conditions below the surface layers of a sunspot, supports a slow surface wave, with speed close to c_T for $|k|a << 1$, as well as the harmonics of slow and fast body waves; a fast surface is unable to propagate under these conditions. The slow body waves have phase speeds between c_T and v_A. The fast body waves have phase speeds between the two sound speeds and are particularly interesting in that, with the exception of the fundamental sausage mode, they possess propagation cutoffs, corresponding to total internal reflection of the fast modes on the boundary of the tube.

By contrast, for a flux tube with speeds ordered as $v_A > c_e > c_0$, typical of conditions in an intense flux tube or in the upper layers of a spot, the fast body waves no longer arise and instead fast surface waves may propagate; the slow body waves remain much as before and there is a slow sausage surface mode. See diagrams in Edwin and Roberts (1983) and Evans and Roberts (1990).

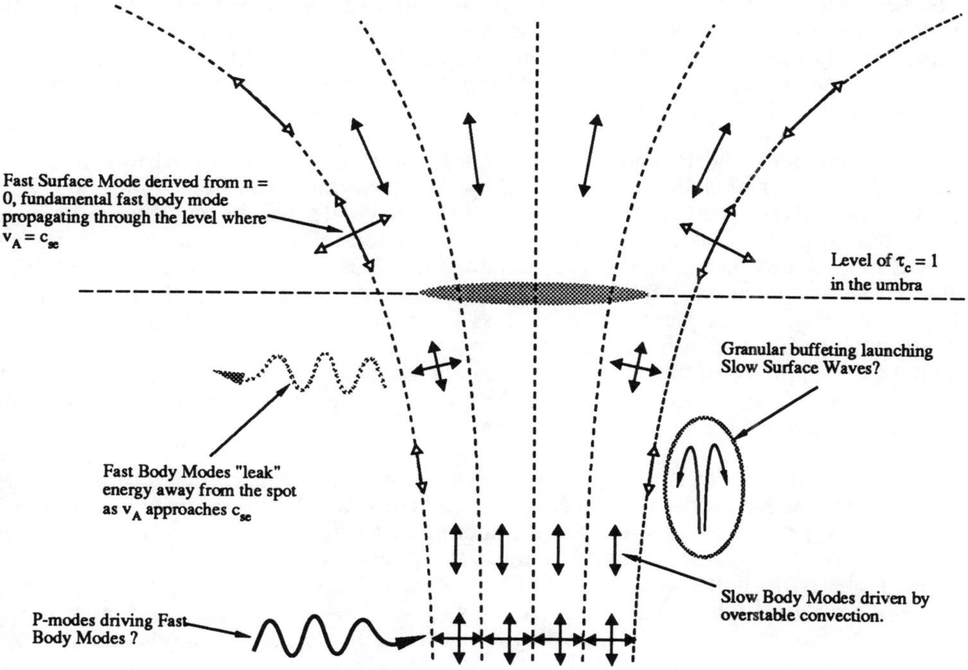

Fig. 2. A sketch of the various waves possible in a sunspot on the basis of a simple flux tube model. After Evans and Roberts 1990.

The fact that fast body waves in the deep layers of a spot, regarded as a simple flux tube, give way to fast surface waves in its upper layers has lead Evans and Roberts (1990) to suggest that p-modes in the environment of a spot will generate fast body modes within the spot, and these in turn will be ducted up or down the spot. Consequently, sunspots will act as sinks for p-modes. This is consistent with the observations by Braun *et al.* (1987, 1988), of sunspots absorbing p-modes, though a detailed theory of the effect has yet to be worked out. The fast body modes thus generated in the spot will then convert to fast surface waves in the upper layers of the spot. Such surface modes may in fact be the *running penumbral wave* (see Small and Roberts 1984; Miles and Roberts 1989). A schematic of the complex structure of modes possible in a sunspot flux tube is given in Figure 2.

3. CORONAL LOOPS AS COMMUNICATION CHANNELS

The isolated magnetic flux tubes of the photosphere act as communication channels between the dense energy reservoir of the convection zone, with its granules, supergranules and p-modes, and the relatively tenuous atmosphere of the chromosphere and corona. In much the same way the loops of magnetic field in the coronal atmosphere may act as communication channels there, covering distances of the order of 10^5km. Of course, such flux tubes are quite different from photospheric tubes, since the magnetic field of the corona fills all of the coronal gas and in general dominates the plasma both mechanically (the plasma beta is low) and thermally (heat conduction across the field is severely limited, allowing regions of different temperature to exist in close proximity to one another).

The magnetoacoustic modes of such a situation can be modelled in terms of a magnetic cylinder of radius a surrounded by a uniform magnetic field. For a low beta plasma, the Alfvén speeds within and external to the cylinder will be much larger than the corresponding sound speeds c_0 and c_e. The dispersion relation for such circumstances (see Edwin and Roberts 1983) can be obtained much as for the isolated tube discussed above. There are basically two distinct cases, depending upon whether the Alfvén speed v_A within the tube is greater than or less than the Alfvén speed v_{Ae} in the tube's environment. If the tube has a gas density that is *higher* than that in the surroundings, then $v_A < v_{Ae}$, provided the magnetic field is not too different from a uniform one; this is the "top-hat" profile of Alfvén speed. The opposite case, corresponding to a region of low density, gives $v_A > v_{Ae}$ and looks like an "inverted top hat"; see Roberts (1990b).

The coronal plasma provides examples of both structures, with coronal loops matching the top hat profile (in Alfvén speed) and coronal holes the inverted hat. Coronal loops with $v_A < v_{Ae}$ act as wave guides, ducting fast magnetoacoustic waves along the structure, much the same as light is guided along an optical fibre (Edwin and Roberts 1983; Roberts, Edwin and Benz 1984). Coronal holes or indeed any regions of the atmosphere with high Alfvén speed, $v_A > v_{Ae}$, are *leaky* wave guides (Spruit 1982; Edwin and Roberts 1983; Cally 1985,1986; Davila 1985) for fast modes. The slow mode is ducted irrespective of whether $v_A > v_{Ae}$ or $v_A < v_{Ae}$.

The guided fast body modes are analogous not only to light in a fibre optic but also to Love waves in the Earth's crust and Pekeris waves in an internal ocean layer (Edwin and Roberts 1983; Roberts *et al.* 1984). This rather pleasing mathematical analogy implies that coronal loops are likely to produce seismic signatures when fast waves are impulsively generated within them (Roberts *et al.* 1984). Such signatures could provide a diagnostic means of determining magnetic conditions in the corona (Roberts 1986b; Edwin and Roberts 1986). Also, because the waves propagate with high phase speeds, between v_A and v_{Ae}, the pulsations they produce are likely to be rapid, of the order of the crossing time for the diameter of the structure. For example, a loop with diameter 10^3km and Alfvén speed of order 10^3km s^{-1} will sustain fast waves with timescales of the order of a second. Such rapid oscillations may in fact be the long-observed radio pulsations (see Aschwanden 1987 for a review).

4. THIN FLUX TUBE THEORY

The thin flux tubes of the photosphere provide important communication channels between the convection zone, photosphere, and chromosphere. This region of the solar atmosphere is the most strongly stratified part of the Sun, for here the pressure scale height $\Lambda(z)$ of the atmosphere is smallest: Λ is about 100 km at the temperature minimum, compared with some 10^5km in the high temperature corona. Additionally, the upper layers of the convection zone are strongly super-adiabatic. These effects of stratification have important consequences for the modes of oscillation of a tube.

Analyzing the modes of oscillation of a magnetic tube in a stratified atmosphere is fraught with difficulties, not least beause the very equilibrium profile of the tube is complicated and has largely to be determined numerically (e.g., Pizzo 1986; Pneuman *et al* 1986; Fiedler and Cally 1990). Analytical progress has, however, proved possible for thin tubes. The idea has been to utilize the fact that physical variables such as pressure, density and longitudinal velocity are not likely to vary strongly *across* the tube, their main variations being along the tube. Other variables, such as the radial component of velocity, may vary linearly about the axis of the tube. A systematic expansion (in a Taylor series) of the magnetohydrodynamic equations about the axis of a tube is possible and provides the thin tube equations for symmetric modes of the tube (Roberts and Webb 1978):

$$\frac{\partial}{\partial t}\rho A + \frac{\partial}{\partial z}\rho v A = 0, \tag{14}$$

$$\frac{\partial v}{\partial t} + v\frac{\partial v}{\partial z} = -\frac{1}{\rho}\frac{\partial p}{\partial z} - g, \tag{15}$$

$$\frac{\partial p}{\partial t} + v\frac{\partial p}{\partial z} = \frac{\gamma p}{\rho}\left(\frac{\partial \rho}{\partial t} + v\frac{\partial \rho}{\partial z}\right), \tag{16}$$

$$p + \frac{B^2}{2\mu} = p_e, \tag{17}$$

$$BA = \text{constant}. \tag{18}$$

Here B(z,t) is the longitudinal field strength, A(z,t) the cross-sectional area of the tube, v(z,t) the longitudinal flow speed, and p(z,t) and ρ(z,t) the gas pressure and density within the tube. The external gas pressure $p_e(z,t)$ is calculated on the boundary of the tube.

Equations (14)-(18) are the *thin tube equations* for the sausage (n=0) mode of oscillation. The equilibrium of the tube is one of hydrostatic stratification, with gas pressure $p_0(z)$ and density $\rho_0(z)$ satisfying

$$p_0'(z) = - g\rho_0(z) \tag{19}$$

for gravity g (= 274 m s^{-2}). The temperature structure may be arbitrary.

In general, the isolated tube expands with height z as it maintains pressure balance with its surroundings (eqn (17)). Specifically, the equilibrium magnetic pressure $p_m(z) \equiv B_0^2(z)/2\mu$, for a tube with longitudinal magnetic field strength $B_0(z)$, satisfies

$$p_m' = g(\rho_0(z) - \rho_e(z)). \tag{20}$$

Introducing the ideal gas law, viz.

$$p = \frac{k_B}{\hat{m}}\rho T, \tag{21}$$

for Boltzmann's constant k_B, mean particle mass $\hat{m}$, and temperature T, applied both inside and outside the tube, gives

$$p_m' = \frac{1}{\Lambda_0}\left[\left(1 - \frac{T_0}{T_e}\right)p_0 - \frac{T_0}{T_e}p_m\right], \tag{22}$$

where $\Lambda_0 \equiv p_0/g\rho_0$ is the pressure scaleheight inside the tube. (We have assumed that the levels of ionization, represented by $\hat{m}$, are the same inside and outside the tube.) If the temperatures inside and outside the tube are the same ($T_0(z) = T_e(z)$), then

$$p_m' = -\frac{1}{\Lambda_0}p_m. \tag{23}$$

Thus, the magnetic pressure declines at the same rate as the hydrostatic pressures p_0 and p_e, and so the magnetic field strength $B_0(z)$ declines at *half* the gas pressure rate. Consequently, the tube expands in height.

It should be noticed that the nonlinear set of equations (14)-(18) may be grouped into two sets, equations (14)-(16) describing longitudinal motions in *any* elastic tube and equations (17) and (18) defining that elasticity as specifically magnetic. This property affords a convenient way of identifying a variety of effects and attributing those effects appropriately to the tube's elasticity, shape or magnetism.

4.1 Linear Disturbances

The case of linear disturbances is readily treated by the thin tube equations, though the algebra is tedious. Linear motions satisfy an equation of the Klein-Gordon type for amplitude Q(z,t), related to velocity v(z,t) (see Rae and Roberts 1982):

$$\frac{\partial^2 Q}{\partial t^2} - c_T^2(z)\frac{\partial^2 Q}{\partial z^2} + \Omega_{saus}^2(z)Q = 0. \tag{24}$$

Here c_T is the tube speed (defined in eqn (1)) and $\Omega_{saus}^2(z)$ is a complicated expression involving the equilibrium profile of the tube and its atmosphere. In deriving equation (24) it was assumed that $p_e(z,t)$ is in fact the undisturbed external gas pressure field.

If the atmosphere is *isothermal* (Λ_0 = constant) the coefficients in the Klein-Gordon wave equation (24) become constants, with Ω_{saus} given by (Defouw 1976; Roberts and Webb 1978)

$$\Omega_{saus}^2 = \left(\frac{9}{4} - \frac{2}{\gamma}\right)\omega_a^2 - \left(\frac{3}{2} - \frac{2}{\gamma}\right)\left(\frac{\beta}{\beta + \frac{2}{\gamma}}\right)\omega_a^2, \tag{25}$$

where $\omega_a \equiv c_0/2\Lambda_0$ is the acoustic cutoff frequency of the atmosphere and $\beta \equiv 2\mu_0 p_0(z)/B_0^2(z)$ is the usual plasma beta (a constant). It is possible to identify the first term on the right of equation (25) as arising from the *geometry* (shape) of the expanding tube and the second term as arising from the *elasticity* of the tube. Accordingly, the frequency Ω_{saus} is reduced by the elasticity of the tube. A *rigid* tube corresponds to $\beta = 0$.

Observe that in the absence of stratification (g=0) $\Omega_{saus} = 0$, and we obtain the simple wave equation with dispersion relation $\omega^2 = k^2 c_T^2$. The thin tube equations, then, relate to the slow sausage mode; in fact, they relate to the slow surface wave.

Kink modes may also be described by thin tube equations, though of a somewhat different form from equations (14)-(18); see Spruit (1981). Indeed, Spruit's thin tube equations for the kink mode also lead to a Klein-Gordon equation (see also Roberts 1986a):

$$\frac{\partial^2 Q}{\partial t^2} - c_k^2\frac{\partial^2 Q}{\partial z^2} + \Omega_{kink}^2 Q = 0, \tag{26}$$

with

$$\Omega^2_{\text{kink}} = \frac{c_k^2}{4\Lambda_0^2}\left(\frac{1}{4} + \Lambda_0'\right). \tag{27}$$

As with equation (24), the external pressure field is assumed to be its unperturbed value. In the absence of stratification (g=0), equation (26) reduces to the simple wave equation, with dispersion relation $\omega^2 = k^2 c_T^2$.

What is the significance of the occurrence of the Klein-Gordon equation? In an isothermal atmosphere both Klein-Gordon equations have constant coefficients, and so both have dispersion relations of the form

$$\omega^2 = k^2 c^2 + \Omega^2 \tag{28}$$

for propagation speed c (= c_T or c_k) and frequency Ω (= Ω_{saus} or Ω_{kink}). A dispersion relation of the form (28) possesses *propagation cutoff*: high frequencies ($\omega > \Omega$) propagate vertically ($k^2 > 0$) whereas low frequencies ($\omega < \Omega$) are evanescent ($k^2 < 0$). (This is much the same as for the vertical propagation of sound waves in an isothermal atmosphere, with c there becoming the sound speed c_s and Ω the acoustic cutoff ω_a.) Impulsively generated disturbances lead to a wave front propagating with the speed c, behind which an oscillating *wake* arises (Rae and Roberts 1982).

It is of interest to compare the sausage and kink modes for an isothermal atmosphere (Spruit and Roberts 1983; Roberts 1986a). Taking photospheric values of Λ_0 = 125 km and $c_0 = v_A$ = 7.5 km s^{-1}, we find that both modes e-fold in a distance of 500 km (four scale heights), to be compared with a vertically propagating sound wave which e-folds in two scale heights (250 km). The propagation speeds of the two modes are similar, about 5.3 km s^{-1} for the sausage mode and 4.5 km s^{-1} for the kink mode. However, the frequencies $\Omega_{\text{saus}}/2\pi$ and $\Omega_{\text{kink}}/2\pi$ are quite different, being 4.8 mHz (period 208s) for the sausage mode but only 1.4 mHz (period 700 s) for the kink mode. (The sausage mode's frequency is close to the acoustic cutoff frequency.) This marked difference in the two frequencies has lead Spruit (1981) to suggest that kink modes are likely to be more readily generated and propagated in the solar atmosphere than sausage modes.

What observational evidence is there for the existence of tube waves in the photosphere? Solanki and Roberts (1990) have argued that the asymmetry observed in the Stokes profile formed in photospheric lines may be explained as a consequence of tube waves. The waves are of relatively strong amplitude, about 1 km s^{-1}, and contribute significantly to the heating of the chromosphere. Evidence for the existence of waves in flux tubes also comes from observations by Deubner and Fleck (1990), who report the existence of waves on the chromospheric magnetic canopy propagating along the canopy and ultimately *down* the converging funnels provided by the tubes! It would seem that photospheric flux tubes are both sources for tube modes and receptors for canopy modes.

4.2 Nonlinear Waves

The stratification in the upper photosphere and above makes it likely that tube waves will grow in amplitude and possibly form shocks. There are, however, a number of effects working against the tendency for wave amplitudes to grow. The photosphere-chromosphere is a region of the solar atmosphere where nonadiabatic effects (radiative losses) are likely to be significant, and such effects work to limit the amplitude of waves as their energy is irreversibly dissipated in the form of heat. Also, it is known that in the absence of gravity flux tubes can support solitons (Roberts and Mangeney 1982; Roberts 1985b; Weisshaar 1989), nonlinear waves in which the amplitude is in balance with dispersive effects. These dispersive effects arise from the inertia of the tube's surroundings, an effect evident in the linear theory (Section 2.2) and included in thin flux tube theory by retaining the effect of $p_e(z,t)$ in equation (17). (The derivations of the Klein-Gordon equations (24) and (26) for sausage and kink modes neglects this effect.) How important all these effects are likely to be in a fully stratified, radiative flux tube is presently unknown and will no doubt require detailed numerical simulations (e.g. Herbold *et al.* 1985; Musielak *et al.* 1989) to fully assess.

Nonlinear effects are reviewed in detail by Ryutova (1990a,b).

5. CONCLUSIONS

The magnetic flux tubes we have described here are basically simple structures, involving sharp changes in density and/or pressure. The role of the magnetic field is to provide a channel connecting one region of the solar atmosphere with another; such channels are elastic. Despite this simplicity the waves that such tubes support are not especially simple to describe. This complexity springs from the fact that in magnetohydrodynamics there are three basic waves in a magnetic medium and additionally that a tube can duct these modes in essentially two ways (as a surface wave or a body wave). In reality, of course, photospheric and coronal magnetic flux tubes will be more intricate objects than the simple models explored here. Nonetheless, we trust that some of our descriptions remain and that the simple concept of a tube, grounded in the work of Faraday, is also a guide to the reality of magnetism in the solar atmosphere.

6. ACKNOWLEDGEMENTS

I am grateful to the Royal Society for travel support and to Professor Vinod Krishan for the hospitality she and her colleagues extended to me in Bangalore.

References

Abdelatif, T E (1988) *Astrophys. J.* **333**, 395.
Aschwanden, M J (1987) *Solar Phys.* **111**, 113.
Braun, D C, Duvall, T L and Labonte, B J (1987) *Astrophys. J. (Letters)* **319**, L27.
Braun, D C, Duvall, T L and Labonte, B J (1988) *Astrophys. J.* **335**, 1015.
Cally, P S (1985) *Australian J. Phys.* **38**, 825.
Cally, P S (1986) *Solar Phys.* **103**, 27.
Davila, J M (1985) *Astrophys. J.* **291**, 328.
Defouw, R J (1976) *Astrophys. J.* **209**, 266.

Deubner, F-L and Fleck, B (1990) *Astron. Astrophys.*, preprint.
Dungey, J W and Loughhead, R E (1954) *Australian J. Phys.* **7**, 5.
Edwin, P M and Roberts, B (1983) *Solar Phys.* **88**, 179.
Edwin, P M and Roberts, B (1986) in *Rapid Fluctuations in Solar Flares* (eds. B Dennis, L Orwig and A Kiplinger), NASA-2449, p.347.
Evans, D J and Roberts, B (1990) *Astrophys. J.* **348**, 346.
Faraday, M (1885) *The Selected Correspondence of Michael Faraday* (ed. L Pearce Williams), Cambridge: Cambridge University Press, 1971, Vol. 1, p.292.
Fiedler, R A S and Cally, P S (1990) *Solar Phys.* **126**, 69.
Herbold, G, Ulmschneider, P, Spruit, H C and Rosner, R (1985) *Astron. Astrophys.* **145**, 157.
Hollweg, J V (1986) in *Advances in Space Plasma Physics* (ed. B Buti), Singapore: World Scientific, p.77.
Lighthill, J (1978) *Waves in Fluids*, Cambridge: Cambridge University Press.
Maxwell, J C (1855/6) *The Scientific Papers of James Clerk Maxwell* (ed. W D Niven), Cambridge: Cambridge University Press 1890 (reprinted by Dover (New York, 1965)), Vol. 1, Art VIII.
Miles, A J and Roberts, B (1989) *Solar Phys.* **119**, 257.
McKenzie, J F (1970) *J. Geophys. Research, Space Physics* **75**, 5331.
Musielak, Z A, Rosner, R and Ulmschneider, P (1989) *Astrophys. J.* **337**, 470.
Pizzo, V J (1986) *Astrophys. J.* **302**, 785.
Pneuman, G W, Solanki, S K and Stenflo, J O (1986) *Astron. Astrophys.* **154**, 231.
Rae, I C and Roberts, B (1982) *Astrophys. J.* **256**, 761.
Roberts, B (1981a) *Solar Phys.* **69**, 27.
Roberts, B (1981b) *Solar Phys.* **69**, 39.
Roberts, B (1985a) in *Solar System Magnetic Fields* (ed. E R Priest), Dordrecht: Reidel, Chap.3.
Roberts, B (1985b) *Phys. Fluids* **28**, 3280.
Roberts, B (1986a) in *Small Scale Magnetic Flux Concentrations in the Solar Photosphere* (ed. W Deinzer, M Knölker and H H Voigt), Gottingen: Vandenhoeck & Ruprecht, p.169.
Roberts, B (1986b) in *Coronal and Prominence Plasmas* (ed. A Poland), NASA-2442, p.325.
Roberts, B (1990a) in *Physics of Magnetic Flux Ropes,* Chapman Conference (ed. C T Russell, E R Priest and L C Lee), Washington: AGU, in press.
Roberts, B (1990b) in *Solar and Planetary Plasma Physics* (ed. B Buti), Singapore: World Scientific, in press.
Roberts, B, Edwin, P M and Benz, A O (1984) *Astrophys. J.* **279**, 857.
Roberts, B and Mangeney, A (1982) *Mon. Not. Roy. Astr. Soc.* **198**, 7p.
Roberts, B and Webb, A R (1978) *Solar Phys.* **56**, 5.
Roberts, B and Webb, A R (1979) *Solar Phys.* **64**, 77.
Ryutova, M P (1990a) in *The Solar Photosphere* (ed. J O Stenflo), IAU Symp. **138**, Dordrecht: Kluwer.
Ryutova, M P (1990b), these proceedings.
Schüssler, M (1987) in *The Role of Fine-Scale Magnetic Fields on the Structure of the Solar Atmosphere* (eds. E-H Schröter, M Vazquez and A A Wyller), Cambridge: Cambridge University Press, p.223
Schüssler, M (1990) in *The Solar Photosphere* (ed. J O Stenflo), IAU Symp. **138**, Dordrecht: Kluwer.
Small, L M and Roberts, B (1984) in *The Hydromagnetics of the Sun*, ESA SP-200, p.257.

Solanki, S K (1987) in *The Role of Fine-Scale Magnetic Fields on the Structure of the Solar Atmosphere* (eds. E-H Schröter, M Vazquez and A A Wyller), Cambridge: Cambridge University Press, p.67.
Solanki, S K and Roberts, B (1990) in *The Solar Photosphere* (ed. J O Stenflo), IAU Symp. **138**, Dordrecht: Kluwer, p.103.
Stenflo, J O (1989) *Astron. Astrophys. Rev.* **1**, 3.
Spruit, H C (1981) *Astron. Astrophys.* **98**, 155.
Spruit, H C (1982) *Solar Phys.* **75**, 3.
Spruit, H C and Roberts, B (1983) *Nature* **304**, 401.
Tayler, R J (1957) *Proc. Phys. Soc. (London)* **B70**, 1049.
Uberoi, C and Somasundaram, K (1980) *Plasma Phys.* **22**, 747.
Uberoi, C and Somasundaram, K (1982) *Plasma Phys.* **24**, 465.
Weisshaar, E (1989) *Phys. Fluids* **A1**, 1406.
Wilson, P R (1980) *Astron. Astrophys.* **76**, 20.

DISCUSSION

UBEROI: In your last part of the talk you pointed out an anology with Pekeris's work on surface waves in the ocean. The source in the Pekeris work is located inside the ocean bed (I think). In your case is the source inside or outside? If it is outside does this analogy hold good?

Comment: Regarding your existence of surface waves for $v_A > v_{Ae}$ I like to point out that I have published a paper in 1982 in Solar Physics studying the parametric analysis of the dispersion equation and giving the parameters (ρ,B) for which the surface waves will exist and if these parameters are not suitable surface waves will not exist.

ROBERTS: Concerning your question about Pekeris waves, Roberts *et al.* (1984) assumed that the source was located in the coronal loop wave-guide, say at the bottom of the loop. Locating the source off centre or outside the loop would of course give slightly different results. But one would expect a flare, say, to generate waves across a number of loops, and each loop would exhibit a dispersive wave-packet behaviour that is of the *general* form as given for a symmetric source. It would be interesting to see what differences do in fact arise but the aim in Roberts *et al.* was more to exhibit the impulsive signature of a rapid pulsation, and that is most simply done in the case of a symmetric disturbance.

Regarding your comment, thank you for the reference [Uberoi (1982), *Sol. Phys.* **78**, 351]. In my paper in 1981 [Roberts (1981a)] I pointed out that *two* magnetoacoustic waves may arise at a single interface, one fast and one slow surface mode. The slow surface wave is generally present but the fast surface wave exists only under certain conditions, as you have also noted. This topic is being investigated further (see Miles and Roberts 1989; Jain and Roberts, in preparation).

RUDERMAN: What do you say about the thin tube approximation conveniency in the solar atmosphere?

ROBERTS: I think that the thin tube approach is very valuable for photospheric tubes and gives a good guide to the behaviour of the slow surface (sausage) mode and the slow kink mode. Body modes, however, have a different lateral structure and so need to be treated differently; curiously, though, slow body waves have phase-speeds that are very close to the thin tube speed, c_T. It should also be kept in mind that the external pressure field on the boundary of the tube, sometimes neglected in analytical and numerical studies of stratified tubes, is important in some circumstances and so should be retained in the thin tube equations. Indeed, its retention gives dispersive corrections to wave speeds and these are important for nonlinear studies. In fact the thin tube equations, with the external pressure term included, lead to flux tube solitons, as shown, for example, in Roberts and Mangeney (1982) and Roberts (1985).

AL-KHASHLAN: What other things can magnetic flux tubes do besides supporting a variety of magnetohydrodynamic waves?

ROBERTS: I have of course concentrated on the wave aspects of a flux tube but a tube, in general terms, transmits stresses and strains and also may provide a duct for flows and for spicules. Coronal flux tubes are also liable to ideal mhd instabilities, ballooning modes, etc. - see, for example, work by A W Hood. Such instabilities of coronal tubes and coronal arcades may be important for prominence eruption.

NONLINEAR WAVES IN FLUX TUBES

M RYUTOVA
630090
Novosibirsk, 90
USSR

ABSTRACT. Some results of nonlinear studies of flux tube oscillations are summarized showing that each step made towards the real conditions of the solar atmosphere opens the existence of qualitatively new effects which form the basis of strongly nonlinear dynamics. Among these effects are, for example: explosive instability in the presence of negative energy waves which leads to an efficient and fast energy release; the generation of secondary plasma flows and current drive which strongly influence the evolution of magnetic flux tubes and their stability conditions; formation of shocks and solitary waves, temporal brightening of flux tubes and others.

1. Introduction

Here, in Bangalore, we could celebrate the 30th anniversary of the suggestion by Robert Howard (1959) that the solar magnetic field can be concentrated in bundles of field lines. After the first direct observations by Sheeley (1967) there has been no lack of new observational data; there were results of Beckers and Schröter (1968), Tanenbaum et al. (1969), Livingston and Harvey (1969), Frazier (1971), Grigorjev and Kuklin (1971), Gopasyuk and Tsap (1971), Howard and Stenflo (1972) and others (for more details see, for example, Stenflo (1989)). Even this very incomplete list of explorers shows the exciting and great activity of the observers, which was the beginning of a new period in Solar Physics - the physics of structured magnetic fields. For understanding the basic processes in the solar atmosphere, which are obviously governed by magnetic fields, the properties of separate magnetic flux tubes as well as the properties of their ensembles should be analysed. Beside its significance for different astrophysical objects where structured magnetic fields are met quite often, such a study is also of interest as a fundamental problem of plasma physics due to a rich picture of wave processes and instabilities in such structures. For the time being we have a good theoretical experience as well due to the essential contribution by P R Wilson, L E Cram, E Parker, W Unno, N Weiss, S Syrovatskii, J Uberoi, E R Priest and others (see Priest, 1982 and References therein).

Today there is a rather good understanding of the properties of linear oscillations of a single flux tube: eigenmodes, mechanisms of damping, effects of vertical stratification and so on. Even in the simplest model of a homogeneous tube the physics of flux tubes is very rich. Motions in the convective zone and, as will be shown below, strong shear flows excite different kinds of oscillations of a flux tube. Among these the most important are longwave oscillations (with $kR \ll 1$, where $k = 1/\lambda$ is the wave number and R is the tube radius) corresponding to a dipole mode (with an azimuthal wave number $m = \pm 1$) that has bending or kink oscillations and an axisymmetric sausage mode ($m = 0$). For both modes the frequency scales linearly with wave number.

The bending mode is an analogue of the Alfven wave with the phase velocity containing the density of the ambient plasma which is caused by the effect of «added mass» (Ryutov and Ryutova, 1976; Spruit, 1981):

$$c_b = \frac{\omega}{k_z} = \frac{a}{\sqrt{1 + 1/\eta}}\,, \tag{1}$$

where $\eta = \rho_i/\rho_e$, ρ is the plasma density, $a = b/4\pi\rho_i$ is the Alfven velocity. Here and below the subscripts i and e correspond to the regions inside and outside the flux tube. The phase velocity (1) is written assuming that there is no magnetic field outside the tube.

E. R. Priest and V. Krishan (eds.), Basic Plasma Processes on the Sun, 175–185.

The sausage mode is the analogue of the slow magnetosonic wave. In these oscillations the sum of gas-kinetic and magnetic pressures is almost not perturbed, so that the plasma parameters outside the tube have little influence on their dispersion relation (Defouw, 1976):

$$c_T = \frac{\omega}{k_z} = \frac{a s_i}{\sqrt{a^2 + s_i^2}}. \quad (2)$$

Here $s^2 = \gamma p/\rho$ is a sound speed squared (γ is a specific heat ratio). As to the oscillations with higher azimuthal mode numbers $m = \pm 2, \pm 3, \ldots$ they are weakly coupled with the larger-scale motions of the atmosphere (their amplitudes are proportional to $(kR)^m$) and are of relatively less interest.

Of course, in a magnetic flux tube as in any oscillating system there is present a standard set of nonlinear phenomena like shocks, solitons, collapse etc. Besides that, there are the specific effects caused by the peculiarity of the flux tube itself. Before proceeding to some results of nonlinear studies of flux tube oscillations we will describe in the next Section some features of flux tubes which are important for understanding more complicated nonlinear phenomena.

2. Nondissipative Damping Mechanisms

In a compressible medium an oscillating magnetic flux tube can be a source of secondary acoustic waves (Ryutov and Ryutova, 1976). In general, if outside the tube there is an external magnetic field the tube radiates all types of MHD-waves. The radiation takes place in the case when the phase velocity of oscillations is larger than the sound speed in the ambient plasma, that is when the radial wave number

$$k_r = \frac{1}{s_e}\left(\omega^2 - k^2 s_e^2\right)^{1/2}$$

is real. In the opposite case the radiated waves are evanescent and their presence leads only to a small change of phase velocity. The radiative damping rate caused by this effect for an arbitrary mode (except $m = 0$) is as follows:

$$\alpha_{rad}^{(m)} = \frac{\pi\omega}{|m|!\,(|m| - 1)!\,(1 + \eta)} \frac{(kR)^{2|m|}}{2} \left[\frac{2}{\gamma(1 + \eta) - 1}\right]^{|m|}.$$

Respectively, for the kink mode we have

$$\alpha_{rad}^{(1)} = \frac{\pi\omega}{1 + \eta} \frac{(kR)^2}{2} \left[\frac{2}{\gamma(1 + \eta)} - 1\right]. \quad (3)$$

Note, that the natural frequencies of tube oscillations are independent of azimuthal mode number and are of the same form as expression (1) for all $|m| > 1$.

The radiative damping works in the case of sausage oscillations as well. The corresponding damping rate is (Ryutova, 1981)

$$\alpha_{rad}^{(0)} = \frac{\pi}{2}\left(\frac{kR}{2}\right)^2 \frac{\omega s_i^6}{s_e^2 (a^2 + s_i^2)^2} \quad (4)$$

The damping rate contains for all cases the small parameter $(kR)^2$ and is typically small, which means that once excited these oscillations propagate along the flux tube. The situation becomes quite different in the more realistic case of smooth radial profiles of density and magnetic field inside the tube.

In the case of a radially inhomogeneous tube there exists another nondissipative mechanism of damping connected with the pumping of the oscillation energy into the resonance layer where the phase velocity of oscillations becomes equal to the local Alfven velocity. The mathematical form of the corresponding equation is of the Rayleigh equation in usual hydrodynamics with the coefficient of the higher derivative approaching zero at the singular point. The general features of such a type of equation and, in particular, Alfven resonances in laboratory plasmas were studied by many authors (Timofeev, 1970; Tataronis and Grossman, 1973; Chen and Hasegawa, 1974; Timofeev, 1989). It was explored as well in connection with the physics of coronal loops and heating processes in the solar atmosphere (Ionson, 1978; Hollweg, 1979; Heyvaerts and Priest, 1983; Sudan and Similon, 1988).

The resonance damping of bending oscillations of a magnetic flux tube was first studied in a model of a tube with diffuse boundaries (Ryutova, 1977). It was assumed that the magnetic field is almost constant across the whole tube except in a narrow region of width $l \ll R$ where it linearly decreases to zero. Respectively, the dependence of pressure and plasma density follows the magnetic field in accordance with the equilibrium condition. In this approach the resonance damping rate of bending oscillations is as follows:

$$\nu_{res} = \frac{\pi\omega}{4} \frac{\rho_i}{\rho_i + \rho_e} \frac{l}{R} . \tag{5}$$

For the slow (sausage) oscillations resonance absorption does not take place, but the radial inhomogeneity plays a crucial role: these oscillations appear to be very sensitive to inhomogeneities of plasma and magnetic field across the tube and experience a strong radiative damping (Ryutova, 1981). To be more precise, if the relative drop of plasma density and/or magnetic field over the tube radius exceeds the magnitude $(kR)^2$ (which is extremely small for longwave oscillations) the sausage oscillations vanish. In the case of a really smooth radial dependence of flux tube parameters the weakly damped wave with smoothly varying radial eigenfunction does not exist any more. And the question arises of whether under such conditions the oscillations of flux tubes can still be the agent responsible for energy transfer from the lower to upper layers of the atmosphere. The answer is positive (Ryutova and Khijakadze, 1989). This is connected with the fact that the resonance damping in a smoothly inhomogeneous tube gives rise to a pecular evolution of the radial mode structure of the perturbation: it becomes more and more spiky at higher altitudes. The characteristic radial scale length Δr diminishes inversely proportional to z:

$$\Delta r \sim aR/\omega z .$$

Respectively, due to the presence of the small-scale structure the usual dissipative processes (viscosity, thermal conductivity, ets.) become important at larger z. The longitudinal dependence of the energy flux of oscillations becomes nonexponential. For example, the volume density of power Q released by viscous dissipation is of the order of

$$Q \sim \rho\nu \frac{\omega^2\xi^2}{(\Delta r)^2}$$

(ν is a kinematic viscosity coefficient and ξ is the tube's displacement). As Δr diminishes with height Q is growing with z. At some altitude z^*, Q reaches a maximum, and then rapidly decreases. Thus, the heating power has a pronounced maximum at some altitude z^*, which can estimated as

$$z^* \sim \frac{a(r)}{\omega} \left(\frac{\omega R^2}{\nu}\right)^{1/3} .$$

In the observational data this effect can manifest itself by a temporal brightening of the flux tube region at a definite height.

3. Effects of Sheared Flows

According to the observational data there are plasma flows along magnetic structures with different velocities inside and outside them.

The presence of shear flows along the magnetic fields leads to the appearance of qualitatively new effects. First of all, to the existence of negative energy waves, and thus to a widening of the classes of instability. And second, to a linear hydrodynamic instability which is similar to a tangential discontinuity and which can be responsible for the excitation of flux tube oscillations in regions far from the convective zone.

The specific features of negative energy waves are that the energy of a system with these waves is lower than without them and losing the energy of negative energy waves due to dissipative effects (say, the interaction with the waves of positive energy, with particles and so on) leads to a growth of their amplitudes (Kadomtsev et al., 1964; Dikasov et al., 1965; Coppi et al., 1969; Weiland and Wilhelmson, 1977).

In a cylindrical coordinate system with the z-axis directed along the flux tube the linearized MHD-equations in the presence of shear flow with a velocity u(r) are as follows (Ryutova, 1988):

$$i\delta\mathcal{P} = \rho(r)\frac{\Omega^2(r)\,(s^2+a^2) - k^2s^2a^2}{\Omega^2 - k^2s^2}\left[\frac{1}{r}\frac{\partial}{\partial r}\, r\,\frac{v_r}{\Omega(r)} + \frac{i\,m}{r}\frac{v_\varphi}{\Omega(r)}\right],$$

$$\frac{\partial\delta\mathcal{P}}{\partial r} = i\rho(r)\,[\Omega^2(r) - k^2a^2]\,\frac{v_r}{\Omega(r)},$$

$$\frac{im}{r}\delta\mathcal{P} = i\rho(r)\,[\Omega^2(r) - k^2a^2]\,\frac{v_r}{\Omega(r)}. \tag{6}$$

Here $\delta\mathcal{P} = \delta p + b_zB/4\pi$ is the total pressure perturbation (all perturbed quantities are assumed to be proportional to $\exp(-i\omega t + ikz + im\varphi)$), and

$$\Omega(r) = \omega - ku(r).$$

The set (6) describes all the types of linear oscillations of magnetic flux tubes in the presence of arbitrary shear flows along the magnetic field.

As an example we present here the results for a kink mode in the simplest model of shear flow without a radial profile (taking into account a radial profile of flows near the tube boundary leads to very interesting physics and will be presented elsewhere). In this case the set (6) is reduced to the following equation for the tube's displacement vector $\vec{\xi}$:

$$\rho_i\frac{\partial^2\vec{\xi}}{\partial t^2} = -\rho_e\left(\frac{\partial}{\partial t} + \frac{\partial}{\partial z}\right)^2\vec{\xi} + \frac{B^2}{4\pi}\frac{\partial^2\vec{\xi}}{\partial z^2}. \tag{7}$$

The dispersion relation corresponding to (7) has a form:

$$\omega^2 + \frac{1}{\eta}(\omega - ku)^2 - k^2a^2 = 0. \tag{8}$$

From (8) we have

$$\frac{\omega}{k} = \frac{1}{1+\eta} \left\{ u \pm \sqrt{\eta \, [a^2(1+\eta) - u^2]} \right\} . \tag{9}$$

One can see from (9) that if the velocity of the shear flow exceeds the threshold

$$u > a\sqrt{1+\eta}$$

the system becomes unstable: this is a linear hydrodynamic instability similar to the instability of a tangential discontinuity. This instability may play an essential role in the excitation of oscillations in magnetic structures in the chromosphere and corona: if somewhere the magnetic flux tube is «blown over» by the upstream flow, the oscillations excited there will subsequently propagate upstream. The frequency of these oscillations is of course unrelated to time scales in the convection zone and may be much greater than the reciprocal of the time for granulation.

In the region $u < a\sqrt{1+\eta}$ where the system is stable with respect to TD-instability the instability of negative energy waves can still develop. The direct calculations for the energy density of waves corresponding to the lower branch («minus» sign) of the dispersion relation (9) give:

$$W = \pi R^2 k^2 \rho_e \xi^2 (\eta a^2 - u^2) \frac{x}{x+u} ,$$

where $x = \sqrt{\eta \, [a^2(1+\eta) - u^2]}$. So, at flow velocities exceeding the threshold

$$u > a\sqrt{\eta}$$

the system becomes unstable with respect to negative energy waves. The interval $a\sqrt{\eta} < u < a\sqrt{1+\eta}$ where the instability of negative energy waves can exist can be written in terms of $\beta = 8\pi p/B^2$:

$$a\sqrt{\frac{\beta}{1+\beta}} < u < a\sqrt{\frac{1+2\beta}{1+\beta}} . \tag{10}$$

For the development of the instability the dissipative processes must turn on. In other words, incorporating dissipative effects results in a transfer of energy away from the negative-energy waves and an increase in their amplitude. A remarkable property which a magnetic tube exhibits is that even in the absence of the usual dissipative processes this instability can develop due to the collisionless dissipation mechanisms described in Sec. 2, namely, due to the radiation of secondary acoustic waves and due to resonance damping.

In the case of radiative damping there are two possibilities: the amplitude of oscillation of a flux tube will grow if the energy of these oscillations and the energy of radiated waves have opposite signs, that is if tube oscillation has a positive energy, and the radiated sound wave has negative energy and vice versa. For example, the conditions for bending oscillations with positive energy to radiate a negative-energy sound wave have the form:

$$a > s_e/\sqrt{\eta} , \quad u > s_e + \sqrt{a^2 - s_e^2/\eta}$$

and the corresponding growth rate is as follows (Ryutova, 1988):

$$\frac{\nu^b_{rad}}{\omega} = \pi k^2 R^2 \frac{(c_b - u)^2 \left[(c_b - u)^2 - s_e^2 \right]}{2 s_e^2 c_b \left[(1+\eta) c_b - u \right]} .$$

In the case of sausage oscillations the instability occurs under the following condition:

$$u > c_T + s_e .$$

The corresponding growth rate has the form:

$$\frac{\nu^T_{rad}}{\omega} = \frac{\pi k^2 R^2}{4\eta} \frac{c_T^2 (c_T - u)^2}{a^4} .$$

Recently, similar results were obtained by Hollweg et al., 1989, who studied the effects of velocity shear on resonance absorption of incompressible MHD surface waves and indicated the importance of this effect for the development of turbulence in regions of strong velocity shear.

4. Explosive Instability of Negative Energy Waves

A specific nonlinear instability, an explosive instability, occurs in a system which contains waves with energies of different signs. A distinctive feature of an explosive instability is that the amplitudes of the interacting waves reach infinitely large values in a finite time. As an example, we present results for three wave interaction under the assumption that only one wave, say the sausage one is excited in the system at the origin, and the time scale and amplitudes of the two other waves (kink modes with m = -1 and m = 1) are determined by the thermal noise. The conditions for explosive instability for this three-wave process are the following matching conditions:

$$\omega_T + \omega_{b_+} + \omega_{b_-} = 0, \quad k_T + k_{b_+} + k_{b_-} = 0, \quad |m_T| \pm |m_{b_+}| \pm |m_{b_-}| = 0. \tag{11}$$

The - and + with the subscript b correspond to waves which are travelling downstream and upstream. The third condition in set (11) corresponding to the conservation of the angular momentum of the waves can be easily satisfied since for the sausage mode $m_T = 0$. The kink mode can be either linearly polarized, that is it can be a mixture of waves with m = + 1 and m = - 1, or circularly polarized in opposite directions. To satisfy the first condition in set (11) which corresponds to energy conservation it is obvious that at least one of the waves has to be the negative energy wave. As was mentioned above, the sausage mode is always a positive energy wave. Of the two kink modes the wave travelling downstream has a negative energy and negative wave number. Note, that in analyzing nonlinear processes the sign of the frequency is assumed to correspond to the sign of the energy, so $\omega_{b_-} < 0$. When this approach is taken, it is simple to verify that the matching conditions (11) are compatible if the following inequality holds:

$$u \geq u_{expl} = \frac{c_T}{1 + \eta} + \left[a^2\eta - c_T^2 \frac{\eta(\eta^2 + 3\eta + 3)}{1 + \eta^2} \right]^{1/2} . \tag{12}$$

The value of u_{expl} lies below the limit of hydrodynamic instability and just gets into the interval (10) corresponding to the existence of negative energy waves. Thus, under the condition

$$a\sqrt{\eta} < u < u_{expl}$$

the three wave interaction (note that two of them might be at the level of noise) can be explosive; in the initial stage of evolution of the explosive instability amplitudes of interacting waves grow exponentially, and the typical growth rate is $k_T v_{T\sim}$ in order of magnitude, where $v_{T\sim}$ is the velocity amplitude of the tube's boundary for sausage oscillations. After a time of the order of several times the reciprocal of the growth rate, at which the amplitudes of all three waves have become equal in order of magnitude, the amplitudes begin a power-law growth in accordance with

$$\nu_{\sim} \sim 1/(t - t_0),$$

where t_0 is the «explosion» time, which is also of the order of $(k_T v_{T\sim})^{-1}$. So, formally, in a time t_0 the amplitudes of all three waves achieve infinitely large values. At this strongly nonlinear stage higher-order nonlinear processes are, of course, involved. These processes can limit the growth of amplitudes at a finite level. The further development of explosive instability depends on the character of these nonlinear processes as well as the dispersive properties of the system (see, for example, Weiland and Wilhelmsson, 1977). The earliest noninear effect which turns on is the nonlinear phase shift which leads to a breaking of the synchrony in time. Note, that the inhomogeneities of the medium lead to a breaking of «synchrony» in space. The stabilization of the explosive instability by the nonlinear phase shift can result in the formation of a solitary wave. Very important is the influence of the second dispersive effects, which lead to rapid oscillations (ripplings) superimposed on the main nonlinear solution. The interaction between new oscillations can again give rise to explosive growth of their amplitudes. This can happen of the nonlinear frequency shift counteracts the stabilization by second-order dispersive effects. Thus, in this case an «explosion» or fast and efficient energy release takes place.

In the opposite case, where the nonlinear phase shift enhances the stabilization due to dispersive effects there appear new solitons. There are situations (for example, due to a weak imbalance between the two stabilizing effects) when solitons with explosively growing amplitudes appear.

A soliton solution in flux tubes (in the absence of shear flows) was found earlier by Roberts and Mangeney (1982) by the traditional way of using the dispersion relation of tube oscillations keeping higher-order terms with respect to $(kR)^2$. Today the problem of solitons in flux tubes is a current interest of many explorers.

5. «Magnetosonic Flows»

A specific nonlinear effect which can play an essential role in the dynamics of magnetic structures and in their evolution in time consists of the generation of secondary plasma flows by oscillating flux tubes (Ryutova, 1986). This effect is similar to the effect of «acoustic flows» or «quartz wind» (Eckart's flows) in usual gasdynamics (see, for example, Nyborg, 1965), but its picture in magnetohydrodynamics appears to be much richer. These effects are caused by the presence of the boundary between the plasma and magnetic field due to the appearance of additional tangential tensions at the boundary. There are mainly two reasons leading to the generation of «magnetosonic flows». First is the action of the ponderomotive force on a plasma. This mechanism is not connected with the absorption of oscillation or with any dissipative effect. The second mechanism is just connected with the absorption of momentum and angular momentum of oscillations propagating along the flux tube. The absorption of longitudinal momentum (nonzero angular momentum can be transferred by circularly polarized bending oscillations) causes stationary convection across the flux tube (rotational mass flows inside and outside the tube). Moreover, the action of this mechanism leads also to current drive. The geometry of currents corresponds to the geometry of secondary flows.

The equation for the stationary velocity field of generated secondary flows $\vec{u}$ has the form:

$$\rho\nu\nabla^2\vec{u} = -\nabla\left(p + \frac{B^2}{8\pi}\right) + \vec{f}\,, \tag{13}$$

where ν is the kinematic viscosity and $\vec{f}$ is the ponderomotive force acting on unit volume:

$$\vec{f} = - \langle \rho' \frac{d\vec{v}'}{dt} \rangle + \frac{1}{c} \langle [\vec{j}'\vec{B}'] \rangle. \tag{14}$$

Here a prime denotes perturbed quantities in linear oscillations and is defined from the linearized MHD-equations corresponding to one of the modes (kink or sausage).

One can see from Eq. (13) that convective motions can arise only if

$$\text{rot } \vec{f} \neq 0.$$

Otherwise, the ponderomotive force leads only to an insignificant redistribution of plasma parameters inside the flux tube. The magnitude of f becomes especially large in the Alfven resonance layer where quite strong convective motions can arise.

The most essential consequence of the effect of secondary flows is that it affects the evolution of flux tubes and actually determines their lifetime. There are two different regimes of the evolution of a flux tube. During the action of the ponderomotive force a levelling of density and magnetic field along the field lines of this force takes place. The field of forces as well as the character of secondary flows and hence the further evolution of the flux tube depend on the relation between the duration T of the primary wavetrain and the time of establishing a viscous flow $\tau_v \sim R^2/\nu$.

In the case of a long coherent wavetrain when $T > \tau_v$ the equalization of density and magnetic fields leads to the «splitting» of the flux tube into four independent flux tubes. Gravity affects in this process (in the regions of small pressure scale height) can cause the separation of new flux tubes from each other and thus the formation of a «magnetic tree».

In the opposite case of a sequence of mutually incoherent short wavetrains with $T < \tau_v$, a kind of stochastic motion is induced which results in a diffusive broadening of the tube and ends up in a complete «dissolving» of the flux tube in the ambient plasma.

As was mentioned above the second mechanism of the generation of plasma flows and current drive is connected with the absorption of oscillations. Formally, the expression of the force has the same form as Eq. (14), but now the terms directly connected with the absorption are taken into account. Note, that besides the absorption due to the usual dissipative effects nondissipative mechanisms peculiar to flux tubes (and described above) are very efficient. For example, in the case of resonance damping when the energy of bending oscillations is pumping into the resonance layer where the dissipation occurs, the whole momentum of oscillations is transferred to the plasma in a narrow layer, causing a strong upward mass flow along the magnetic field. The same holds in the case of absorption of angular momentum when strong convection across the flux tube takes place. The corresponding force can be estimated as follows:

$$f_z \sim \nu_{res} \frac{B^2}{4\pi} \frac{|v'|^2}{a^2 R}.$$

As was mentioned above, unlike the usual gasdynamics, in magnetic flux tubes the generation of plasma flows is accompanied by current drive. The quantitative solution of the problem is based on the two-fluid approach, where the current density is

$$\vec{j} = en(\vec{v}_i - \vec{v}_e).$$

Here v_i and v_e are the ion and electron velocities and n is the plasma concentration. Two-fluid MHD leads to the following equation for the current density:

$$\frac{\partial \vec{j}}{\partial t} = e \frac{m_e \vec{f}_i - m_i \vec{f}_e}{m_i m_e} - \frac{e^2 n}{\sigma} \frac{m_i + m_e}{m_i m_e} \vec{j}. \tag{15}$$

Here all the definitions are standard and $\vec{f_i}$ and $\vec{f_e}$ are the forces acting on ions and electrons, respectively. Note that the main contribtuion to the effect of current drive (as is clear from Eq. (15)) comes from the force $\vec{f_e}$ acting on the electrons. In the stationary case we have simply

$$\vec{j} = \sigma(\vec{E} - \vec{f_e}/e). \tag{16}$$

The detailed calculation of the force $\vec{f_e}$ needs a kinetic approach and is too complicated, but in the problem under consideration there are circumstances which make the problem easy. The point is that in the longwave limit the particle mean free path (in a highly collisional solar plasma) l is much less than the wavelength of flux tube oscillations λ: $l \ll \lambda\sqrt{m_e/m_i}$. This means that the perturbations of ion and electron temperatures as well as perturbations of ion and electron pressures in fluxtube oscillations are equal:

$$p_i = p_e.$$

So, from the momentum conservation law it is clear that half of the wave momentum is transferred to electrons and half to ions (Ryutov and Ryutova, 1989).

Therefore, for the mean force acting on the electron gas we can write:

$$\vec{f_e} = \frac{1}{2}\dot{\vec{\mathcal{P}}}.$$

Here $\vec{\mathcal{P}}$ is the longitudinal momentum of tube oscillations:

$$\dot{\mathcal{P}} = \frac{\gamma}{v_\varphi} W,$$

where γ, v_φ and W are the damping rate, phase velocity and energy density of the oscillations which causes the corresponding current drive. For example, an estimate of the current which can be generated due to the absorption of bending oscillations in the resonance layer is

$$j_z \sim v_{res} \frac{eW}{2vm_e} \frac{R}{l}.$$

We used here the model of a flux tube with a diffuse boundary of width l; $v_\varphi = c_b$ is determined by Eq. (1) and $\gamma = v_{res}$ by Eq. (5).

The same considerations hold for the oscillations with circular polarization. In this case besides momentum the wave carries angular momentum. Now, the absorption of angular momentum leads to the generation of azimuthal currents.

6. Summary

In the present paper an attempt is made to describe some basic plasma effects which can be responsible for particular active processes in the Sun. There is a scheme of these effects and the links between them. Many links in this picture have yet not been found. And, of course, the picture itself is far from complete. But the most important next step is to find the quantitative relevance of these effects to particular observational data.

References

Beckers, J. M. and Schröter, E. H., 1968, Solar Phys., **4**, 142.
Chen, L. and Hasegawa, A., 1974, J. Geophys. Res., **79**, 1024.
Coppi, B., Rosenbluth, M. N. and Sudan, R., 1969, Ann. Phys., **55**, 207.
Defouw, R. J., 1976, Ap. J., **209**, 266.
Dikasov, V. M., Rudakov, L. I. and Ryutov, D. D., 1965, JETR, **48**, 913.
Frazier, F. N., 1971, in Proc. IAU Symposium No. 43, 93, R. Howard , ed.
Grigorjev, V. M. and Kuklin, G. V., 1971, Ibid.
Gopasyuk, S. I. and Tsap, T. T., 1971, Ibid.
Heyvaerts, J. and Priest, E., 1983, Astr. Ap., **117**, 220.
Hollweg, J., 1979, Solar Phys. **62**, 227.
Hollweg, J., Yang, G., Cadez, V. M. and Gakovic, 1989, Ap. J.
Howard, R., 1959, Ap. J., **130**, 193.
Howard, R. and Stenflo, J., 1972, Solar Phys., **22**, 402.
Ionson, J. A., 1978, Ap. J. **226**, 650.
Kadomtsev, B. B., Mikhailovsky, A. B. and Timofeev, A. V., 1964, JETP, **47**, 2267.
Livingston, W. and Harvey, J., 1969, Solar Phys., **10**, 294.
Nyborg, W. L., 1965, in Physical Acoustics, Vol. 2, Part B, Properties of Polimers and Nonlinear Acoustics, Academic Press.
Priest, E. R., 1982, Solar Magnetohydrodynamics, Dodrecht, D. Reidel Publ. Co.

Discussion

UBEROI: What are the time scales involved in the nonlinear effects? How do these compare with the resonant and radiative damping time scales?

RYUTOVA: Several inverse damping rates, but to be more strict we have to specify which of the nonlinear effects we are dealing with. For example, in a case of secondary flows the duration of the incident wave train and the time of establishing viscous flows are involved as well.

PRIEST: (i) Do you think nonlinear effects are more important in the photosphere than the corona, where the Alfven Mach number is larger? (ii) Because the observed turbulent or wave flows are so small compared with the Alfven speed, do you think that coronal heating by waves is essentially a linear process? (iii) What effects do you expect to saturate your explosive instability in practice?

RYUTOVA: (i) No, and I mean no for all the different effects which I've considered today. Almost for all of them there is not only the question of the Alfven Mach number. For generation of secondary plasma flows and current drive the most important parameters are the duration of the incident wave train and the viscosity of the ambient plasma. The phenomena of splitting of magnetic structures (I do not mean the photospheric flux tubes only) and their diffusive vanishing must be common phenomena for different regions where the appropriate conditions are fulfilled. There is, of course, the coalescence effect, but it is out of the frame of this talk. For negative energy waves the region of their existence can be characterized by plasma β as follows: $\sqrt{\frac{\beta}{1+\beta}}\,a < u < \sqrt{\frac{1+2\beta}{1+\beta}}\,a$, which shows that for low β the lower threshold for critical shear velocity can be very low which makes a low β region very favourable for negative energy waves and for all accompanied phenomena. With increasing β the critical velocity threshold increases and the region of existence of negative energy waves becomes more narrow. (ii) I do not think so. The general philosophy and practice show that nonlinear effects are invaluable for efficient heating, at least when waves are involved. (iii) As I already mentioned, the nonlinear phase shift which leads to the breaking of the synchrony in time is the earliest effect which can saturate the explosive instability.

ROBERTS: You have mentioned several mechanisms for the generation of waves in tubes. Which mechanism do you think is likely to be the most important in photospheric flux tubes, and do you expect impulsive phenomena to occur?

RYUTOVA: The most important effect for the excitation of photospheric flux tube oscillations is the resonance absorption of sound waves by flux tubes which takes place if the Cerenkov condition is fulfilled: $s = v_\varphi \cos\theta$(s is sound speed, v_φ is the phase velocity of tube waves and θ is an angle between the sound propagation and the tube axis). Besides that, a very important effect is noncollinearity of flux tubes (Ryutov and Ryutova, 1976). For excitation of negative energy waves the presence of shear flows along flux tubes is necessary. Concerning impulsive phenomena, as you have shown they can occur and I agree with that.

RESONANT ABSORPTION OF P-MODES BY SUNSPOTS

S. M. CHITRE [1] and JOSEPH M. DAVILA
Laboratory for Astronomy and Solar Physics
NASA-Goddard Space Flight Center
Greenbelt, MD 20771

1. Introduction

A substantial amount ($\lesssim 50\%$) of solar acoustic (p-mode) absorption in sunspot regions has been reported by Braun, Duvall and LaBonte (1988). A number of suggestions have been advanced to account for the observed p-mode absorption in sunspots (see review by Davila in these proceedings). These include processes involving resonance absorption, radiation damping, and non-dissipative mechanisms. The main purpose of the present work is to demonstrate that any dissipative process like radiative, viscous, or resistive dissipation leads to the resonant absorption of acoustic waves incident on the sunspot tube and that the resultant heating rate can be shown to be consistent with the observed absorption of the p-mode power impinging on an isolated inhomogeneously structured sunspot.

2. Model Assumptions and Governing Equations

Sunspot is assumed to be a localized cylindrical (r, θ, z) flux-tube bounded by $r = R_{spot}$, which is in transverse pressure equilibrium with a field-free homogeneous medium. The spot magnetic field is taken to be vertical (z-direction) with a dependence on the radial coordinate, r, only, i.e. $\mathbf{B_0} = (0, 0, B_0(r))$. The equilibrium physical quantities; pressure, p_0, density, ρ_0, and temperature, T_0, are assumed to be functions of r, but gravity and vertical stratification are neglected. No specific assumption is made about the magnitude of $\beta = 8\pi p/B^2$. For the sake of illustration, the fluid in the flux tube is taken to be fully compressible and inviscid, with infinite electrical conductivity but with finite thermal (radiative) conductivity. However the results discussed do not depend on the specific dissipation mechanism. The basic equations are the usual conservation equations which can be linearized and then combined to get a fourth-order differential equation for the perturbed variables.

The heating rate may be calculated by integrating the incident Poynting and acoustic flux over the surface of the resonance layer (Davila, 1987). Given the existence of the resonance layer in the spot tube (resulting from some dissipation process in the plasma) it suffices to compute the heating rate by invoking the ideal MHD solution that is valid on the surface and outside the resonance layer. The resultant heating rate per unit area contributed by the Poynting and acoustic flux comes out to be

$$H = \frac{\rho_0}{2\omega}[v_A^2 \frac{k_z^2 r^2}{m^2}(-\frac{1}{v_A^2}\frac{dv_A^2}{dr})]\pi | V |^2 \tag{1}$$

All the physical quantities here are measured at the resonance layer. With a view to determine

[1] NAS-NRC Resident Research Associate on leave from Tata Institute of Fundamental Research, Bombay

E. R. Priest and V. Krishan (eds.), Basic Plasma Processes on the Sun, 187–188.

whether this heating rate is consistent with the observed absorption of p-mode power, we equate H to $\alpha_{max} = < \frac{1}{2}\rho_0'(V_{in}^2 - V_{out}^2)c_s >$, the maximum observed difference between the incoming and out-going acoustic flux. Here V_{in} is the amplitude of the incoming p-wave, V_{out} of the reflected wave, and ρ_0' is the density at the surface of the spot tube.

Choosing typical values for the sunspot tube; $c_s = 10^6\ cm - s^{-1}$, $v_A = 3 \times 10^6\ cm - s^{-1}$, $k_z^2 r^2/m^2 = 0.5$, $R = 0.5$, $\nu = 3.3\ mHz$, $\mid (1/v_A^2)(dv_A^2/dr) \mid = 1/R_{spot} = 2 \times 10^{-9}\ cm^{-1}$, and $\rho_0'/\rho_0 = 2$, we get the wave amplitude,

$$\mid V \mid \simeq \frac{1}{5} V_{in} \simeq 100 \quad m - s^{-1} \tag{2}$$

assuming a typical RMS value of 400 $m - s^{-1}$ for the photospheric velocity amplitude of p-modes.

3. Discussion and Conclusions

The required heating rate yields the wave amplitude of the order of 100 $m - s^{-1}$ at the resonance layer, which should be observable within the spot-tube at the surface. The heating rate scales like $1/m^2$ and this is also borne out by the detailed numerical calculation of the absorption coefficient by Lou (1989) . Lou has examined the influence of viscous dissipation on the Alfven resonance; we expect that the absorption coefficient should be independent of the precise dissipation mechanism.

The solution to the problem of the absorption of axisymmetric modes may be sought by considering more realistic sunspot models. An isolated single spot is not expected to be a perfectly circular absorber or scatterer. The cylindrical waves converging onto a non-axisymmetric spot will inevitably couple the $m = 0$ mode to $m > 0$ modes. It is also unreasonable to imagine that a sunspot would remain as a vertical monolithic structure. The sunspot may be made up of a number of individual flux tubes which could also couple the modes. Finally, it can be demonstrated that the Alfven resonance, responsible for resonant heating, survives even when m=0 when the flux tube is twisted (see also Hollweg, 1988).

References

Braun, D. C., T. L. Duvall, and B. J. LaBonte, 1988, *Ap. J.*, **335**, 1015.

Davila, J. M., 1987, *Ap. J.*, **317**, 514.

Hollweg, J. V., 1988, *Ap. J.*, **335**, 1005.

Lou, Y., 1989, Viscous Magnetohydrodynamic Modes and p-Mode Absorption by Sunspots, in press.

WAVE PROPAGATION IN SUNSPOTS

S.S.HASAN
Indian Institute of Astrophysics
Bangalore 560034, India

ABSTRACT. In this paper, wave propagation in sunspot umbrae is analysed. The stratification in a typical umbra is approximated by a model atmosphere, extending vertically from a depth of a few thousand kilometres below the photosphere, to the transition region. No boundaries are imposed in the horizontal direction. It is assumed that this atmosphere is embedded in a uniform vertical field. Using a Rayleigh-Ritz variational technique, the normal mode frequencies of the umbra are calculated for different values of the horizontal wave number k. A theoretical diagnostic diagram is generated. An interesting feature of the solutions is the existence of 'avoided crossings', similar to those found in the study of global oscillations. The nature of the wave modes is examined by decomposing the eigenvectors into longitudinal and transverse components. In general, it is found that the character of a mode changes with height. For large k, modes with periods in the 2-3 min range, correspond to low order modes in the present calculation. Above the photosphere, they resemble slow waves.

1. Introduction

Oscillations in the umbrae of sunpots have been widely reported (e.g. Beckers and Schulz,1972; Balthasar and Wiehr,1984; Lites and Thomas,1985; Lites,1986; Abdelatif *et al.*,1986; Thomas et al.,1987 and Gurman,1987). The periods of these oscillations typically lie in the range 2-3 min, although larger periods of some 300-400 s (Bhatnagar et al.,1972; Soltau et al.,1976 and Balthasar et al.,1987) or smaller ones around 100 s (Schröter and Soltau,1976) cannot be ruled out.

A number of theoretical investigations have been carried out to examine wave propagation in sunpot umbrae (e.g., Uchida and Sakurai,1975; Scheuer and Thomas,1981; Thomas and Scheuer, 1982; Leroy and Schwartz,1982 Zhugzhda et al.,1983). Yet a number of uncertainties remain. These are related to the precise values of the oscillation frequencies, their dependence upon the horizontal wave number and the nature of the modes. Scheuer and Thomas (1981) suggested that umbral oscillations are essentially fast waves trapped in a photopheric cavity. On the other hand, Zhugzhda et al. (1983) and Gurman and Leibacher (1984) have argued in favour of a slow mode, trapped in a chromospheric cavity above the temperature minimum. The aim of the present analysis is to: compute the normal mode spectrum for the umbra of a sunspot, generate a diagnostic diagram and classify the modes.

2. Equilibrium Model

In order to compute the wave modes in an umbra, we require an equilibrium model which mimics a real atmosphere in a sunspot. We selected the core umbral model M of Maltby et al. (1985) (kindly provided by T. Abdelatif). Beneath the photosphere, the atmosphere is matched to a convection zone model.

3. Wave Equation

We assume that a uniform vertical magnetic field is embedded in the model atmosphere. The normal

E. R. Priest and V. Krishan (eds.), Basic Plasma Processes on the Sun, 189–191.

modes can be determined by linearising the MHD equations, which results in the following wave equation

$$\rho\frac{\partial^2\xi}{\partial t^2} = -\nabla\delta p + \mathbf{g}\delta\rho + \frac{1}{4\pi}(\nabla\times\delta\mathbf{B})\times\mathbf{B} \quad (1)$$

where $\delta\rho, \delta p, \delta\mathbf{B}$ denote perturbations in density, pressure and magnetic field respectively and ξ is a small displacement of a fluid element. Equation (1) can be recast into an energy integral, which lends itself to a variational treatment (for details see Hasan and Sobouti 1987). Assuming a time dependence of the form $e^{i\omega t}$, the normal frequencies can be calculated using a Rayleigh Ritz variational method.

4. Results

Equation (1) was solved using a Rayleigh Ritz variational method. Rigid boundary conditions were assumed i.e., $\xi_z = \xi_x = 0$ at z=-500 km and $z = 2000$ km. A constant vertical field $B = 2000$ G, corresponding to $v_a/c_s = 0.84$ at $z = 0$ was used, where v_a and c_s denote the Alfvén and sound speeds respectively.

4.1 DIAGNOSTIC DIAGRAM

Figure 1 shows the variation of the wave frequency ν with k (diagnostic) diagram for umbral oscillations, where k is the horizontal wave number. For a fixed value of k, a number of solutions exist satisfying the boundary conditions. These correspond to the normal mode frequencies or harmonics, which form a discrete spectrum. The curves depict the variation of ω with k. The various numbers above each curve denote the order n of the solution. An interesting feature is the absence of accidental degeneracy in the solutions of different orders. It is found that when curves of adjacent orders draw close to one another, an 'avoided crossing 'occurs. The frequency varies from some 2.2 mHz to about 7.5 mHz (corresponding to periods of 450 s and 133 s respectively) for n between 1 and 5.

5. Mode Classification

In order to classify the modes, we decompose the linear displacements, using Helmholtz' theorem, into longitudinal and transverse components and examine the height dependence of the components to see the nature of a mode. It turns out that in the photosphere, the longitudinal and transverse components have comparable magnitude. However, with increase in height, the modes tend to become dominantly longitudinal. In the chromosphere, the character of the mode can be rougly described in terms of a slow wave.

6. Discussion and Conclusions

We now compare the theoretical oscillation spectrum with observations. Assuming an umbral radius of 4200 km we find k= 0.9 Mm^{-1} (i.e., for the lowest order mode which has a vanishing radial displacement at the umbral boundary). Modes with periods around 180 s correspond to n=4 and 5 in the present calculation. The lower order modes with longer periods (around 300 s) probably correspond to the modes observed for example by Balthasar et al. (1987).

The main conclusions of the study are that firstly, the theoretically computed spectrum of umbral oscillations shows the presence of modes with periods in the 2-3 min range. Wave modes with larger periods (around 300s) are also present. Secondly, the diagnostic diagram reveals the presence of 'avoided crossings'. Lastly, the physical nature of the modes is height dependent –

however, in the chromosphere the oscillations resemble slow waves.

References

Abdelatif,T.E.,Lites,B.W.,and Thomas,J.H.(1986), Astrophys. J. **311**,1015.
Antia,H. and Chitre,S.M. (1979),Solar Phys. **63**,67.
Balthasar,H. and Wiehr,E. (1984),Solar Phys. **94**,99.
Balthasar,H.,Küveler,G.,and Wiehr,E. (1987), Solar Phys. **112**, 39-48.
Beckers,J.M. and Schulz,R.B. (1972),Solar Phys. **27**,61.
Bhatnagar,A.,Livingston,W.C.,and Harvey,J.W.(1972), Solar Phys. **27**,80.
Gurman,J.B. and Leibacher,J.W. (1984), Astrophys. J. **283**, 859.
Gurman,J.B. (1987),Solar Phys. **108**,61.
Hasan,S.S. and Sobouti,Y. (1987),Mon. Not. R. astr. Soc. **228**,427.
Leroy,B. and Schwartz,S.J. (1982),Astron. Astrophys. **112**,84.
Lites,B.W. and Thomas,J.H. (1985),Astrophys. J. **294**,682.
Lites,B.W. (1986),Astrophys. J. **301**,922.
Maltby,P.,Avrett,E.H.,Carlsson,M.,Kjeldseth-Moe,O.,Kurucz,R.L. and Loeser,R. (1986), Astrophys. J. **306**,284.
Scheuer,M.A. and Thomas,J.H. (1981),Solar Phys. **71**,21.
Schröter,E.H. and Soltau,D. (1976),Astron. Astrophys. **449**,463.
Soltau,D.,Schröter,E.H.,and Wöhl,H. (1976),Astron. Astrophys. **50**,367.
Thomas,J.H.,Lites,B.W.,and Gurman,J.B. (1987), Astrophys. J. **312**,457.
Uchida,Y. and Sakurai,T. (1975),Publ. Astron. Soc. Japan **27**,259.
Zhugzhda,Y.D. (1979),Sov. Astron. **23**,42.
Zhugzhda,Y.D., Locan, V., and Staude, J. (1983) Solar Phys. **82**, 369.

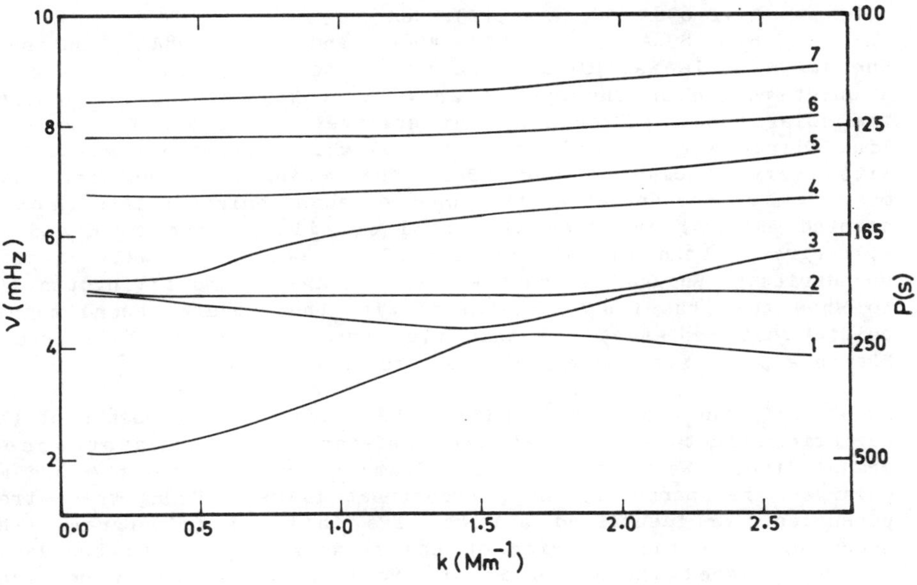

Figure 1. Variation of frequency ν with k in a sunspot umbra, assuming a uniform vertical magnetic field of 2000G.

ON THE LOCATION OF THE FOOT-POINTS OF SUB-ARC-SECOND MAGNETIC STRUCTURES IN THE QUIET SUN PHOTOSPHERE

K.R.Sivaraman and S.P.Bagare
Indian Institute of Astrophysics
Bangalore 560034, INDIA

and

Laurence J.November
National Solar Observatory
Sunspot, New Mexico 88349, USA

OBSERVATIONS: We present the results of our study of a long sequence of high spatial resolution filtergrams obtained with the Universal Birefringent filter (UBF) at the Vacuum Tower Telescope of the National Solar Observatory, Sacramento Peak on July 14, 1987. The UBF has a band pass of 0.25A and was set sequentially in the following wavelengths, and filtergrams were obtained at each of the settings.

5172.698A 5173.098A

5172.898A 5173.498A and 5174.698A (continuum).

The image scale was 10 arc sec mm^{-1} and the region of study was a quiet region on the sun of area 145" x 210" around the centre of the solar disc. Through a beam splitter, light was fed to a Ca+K line filter with a band pass of 0.6A which operated simultaneously with every exposure of the UBF. The seeing was excellent and in most part of the frames the sub-arc second bright points (henceforth denoted as BPs) are seen with good clarity in the Ca^{+} K and Mg b filtergrams, both in the interior of the network as well as at the boundaries. Enough evidence exists (Sivaraman and Livingston 1982) to show that the BPs are cospatial with the sub-arc second magnetic points that represent the thin flux tubes and so we have used the BPs as a proxy for the magnetic points in this study.

ANALYSIS: Our aim is to find out where are the foot-points of these magnetic structures located with reference to the photospheric granulation. We chose the best frames from the sequence and made enlarged transparencies on low contrast films. Using these transparencies, we identified all the BPs in the K filtergrams (which occur both in the interior of the network and on the boundaries) and established these BPs in the Mg b core picture by coaligning the K and Mg b pictures with an accuracy better than 0.5 arc sec. We then coaligned the Mg b core and the continuum pictures via the Mg b wing pictures that served as a reliable intermediate stage to

E. R. Priest and V. Krishan (eds.), Basic Plasma Processes on the Sun, 192–193.

confirm the correctness of the identification of the BPs and the network boundaries. Whereas the network boundaries served as a good overall reference, the actual alignment was done using the finer features on the boundaries and the BPs.

RESULTS: We did the visual correlation for nine frames covering a total of 1800 BPs. Of these, 77.8% are located in the dark intergranular lanes of the photospheric granulation and 14.4% are on the very periphery of the bright granules and the rest in the granules themselves.

Mehltretter (1974) and Muller (1983) used a 16A pass band K filter and showed that the network BPs (called facular points by them) occur in the dark intergranular lanes. Such a filter could have shown only the network and the most intense bright points sitting on the network boundaries but not those in the cell interiors. The latter in general are not as bright as the ones in the cell boundaries and so tend to be missed when a large band pass is used for observing them. Title et al (1987) observed an emerging flux region with their tunable filter and found that the sub-arc sec magnetic features seen in the map of Fe I 5250A line correlated with regions of downdrafts. Comparison with the continum pictures confirmed that these coincide with the intergranular lanes.

Our results illustrate the state of affairs in the quiet regions of the sun and the use of the UBF filter enabled us to have a large sample of the BPs both in the cell interior as well as in the cell boundaries.

We most gratefully acknowledge the extraordinary hospitality of Dr.Raymond Smartt and his entire staff at SPO and the very able help provided to us by the staff at the solar Vacuum Tower Telescope.

REFERENCES:

Mehltretter, J.P. 1974, Solar Phys. 38, 43.
Muller, R. 1983, Solar Phys. 85, 113.
Sivaraman, K.R. and Livingston, W.C. 1982, Solar Phys. 80, 227.
Title, A.M., Tarbell, T.D. and Topka, K.P.. 1987, Astrophys. J. 317, 892.

VI. CHROMOSPHERIC and CORONAL HEATING

THE HEATING OF THE QUIET SOLAR CHROMOSPHERE

Wolfgang Kalkofen
Harvard-Smithsonian Center for Astrophysics
60 Garden Street, Cambridge, Massachussetts 02138, USA

ABSTRACT: The quiet solar chromosphere shows three distinct regions. Ordered according to the strength of the emission from the low and middle chromosphere they are (1) the magnetic elements on the boundary of supergranulation cells, (2) the bright points in the cell interior, and (3) the truly quiet chromosphere, also in the cell interior. The magnetic elements on the cell boundary are associated with intense magnetic fields and are heated by waves with very long periods, ranging from six to twelve minutes; the bright points are associated with magnetic elements of low field strength and are heated by (long-period) waves with periods near the acoustic cutoff period of three minutes; and the quiet cell interior, which is free of magnetic field, may be heated by short-period acoustic waves, with periods below one minute. This paper reviews mainly the heating of the bright points and concludes that the large-amplitude, long-period waves heating the bright points dissipate enough energy to account for their chromospheric temperature structure.

1. Introduction

Chromospheric heating remains one of the fundamental problems in solar physics. Biermann (1946, 1948) and Schwarzschild (1948) suggested that the solar chromosphere was heated by acoustic waves that are generated by the turbulence of the hydrogen convection zone and travel outward through the photosphere, where the decreasing density of the gravitationally stratified gas causes the velocity amplitude of the waves to increase and the profile to steepen; shocks form at the base of the chromosphere and beyond, thereby heating the chromosphere. Ulmschneider (1970, 1974) proposed that the waves had to meet two conditions. They had to have the periods and initial velocity amplitudes that would enable them to produce the first shock at the observed base of the chromosphere, approximately 500 km above $\tau_{5000} = 1$ in model C of the average quiet atmosphere by Vernazza, Avrett & Loeser (1981, henceforth VAL81), and the resultant time-averaged temperature structure had to agree with the temperature structure of the empirical models, the HSRA (Gingerich *et al.* 1971) or that of VAL81. Ulmschneider and collaborators (*cf.* Ulmschneider 1986; also Ulmschneider & Kalkofen 1977) were able to construct such plane-parallel models with acoustic waves having periods near 1/2 min, but both ground-based and space-based observations failed to support the claim of shocks (*cf.* Cram 1987). From this, Cram concluded that the chromosphere was not heated by shocks and suggested that mechanisms other than wave heating were responsible for the chromospheric temperature rise.

In spite of this conclusion against acoustic heating, evidence has been accumulating that suggests wave heating of chromospheric regions that are associated with magnetic fields. Thus, Jensen & Orrall (1963) observed velocity oscillations in the K line of Ca II, and Orrall (1966) observed similar oscillations in the Hα line and noted, furthermore, that the wave

E. R. Priest and V. Krishan (eds.), Basic Plasma Processes on the Sun, 197–206.

period was positively correlated with the strength of the magnetic field. Liu (1974) found that the bright points (also called cell points, K grains, or grains), which are associated with magnetic elements in the interior of supergranulation cells, were heated by waves with periods near the acoustic cutoff period of about 3 min, where no propagating waves were expected (Schatzman & Souffrin 1967), nor predicted (Stein 1967, 1968; Leibacher & Stein 1981). And Damé (1983, 1984) studied waves with periods longer than the acoustic cutoff period in the magnetic elements on the boundary of supergranulation cells. Propagating waves in the bright points were described by Liu (1974) and by Cram (1974) in the K line, by Cram & Damé (1983) in the H line, by Damé (1983, 1984) in the H and K lines, and by Lindsey & Roellig (1987) at 350 μm and 800 μm in the H^- free-free continuum.

The mechanically heated layers that are of primary interest for the energy balance of the outer layers of the sun are the densest and, hence, lowest layers, namely, those of the low and middle chromosphere. They extend from the temperature minimum, where the temperature in the semiempirical model of VAL81 is 4200°K and the height is 500 km above $\tau_{5000} = 1$, to the end of the temperature plateau of about 7000°K at a height of 2000 km. For these layers, the VAL81 model C implies that the gas is largely neutral. Most of the radiative losses due to mechanical heating occur in the lines of ionized calcium (which in model C account for more than half the radiative cooling), in the analogous lines of ionized magnesium, in the core of the Hα line, in free-free transitions of the H^- ion, in bound-free transitions of some metals such as Si, Fe, and C, and in the far wings of the Lyman-α line and the Lyman continuum. Propagating waves heating the layers of the low and middle chromosphere have been observed in all the above line transitions in the visible part of the spectrum and in the infrared continuum, but not in transitions in the ultraviolet.

The paper is organized as follows: Section 2 discusses the association between magnetic elements, *i.e.*, photospheric areas with a typical size well below 1″ of intense vertical magnetic field, and regions of strong chromospheric emission. The paper then deals mainly with the heating of the chromosphere in the internetwork bright points: Section 3 presents evidence for outward propagation of disturbances with a typical repetition time of 3 min, discusses the expectations for vertical wave propagation from the linearized gas dynamic equations, and speculates on the nature of the "waves" that heat the bright points, and Section 4 gives estimates of the dissipated mechanical energy; the final Section presents some conclusions on chromospheric heating.

2. Association of Chromospheric Heating and Magnetic Fields

It is well known that there is a close relation between enhanced chromospheric emission and magnetic elements on the boundary of supergranulation cells (*cf.* Zirin 1988). And there is a similar association between enhanced emission from the bright points in the cell interior and internetwork magnetic elements (Sivaraman & Livingston 1982). The chromospheric emission from magnetic regions is strongly enhanced. By contrast, the emission from the field-free regions of the cell interior is much lower (Cram 1987).

The magnetic field on the cell boundary is very strong (*cf.* Stenflo 1989); its contribution to the pressure is higher than that of the gas. The field extends from the photosphere at least into the transition region between chromosphere and corona (*cf.* Athay 1976), and it may reach as high as the 2×10^{6}°K level in the corona (*cf.* Piddington 1974). The heating in these flux tubes (*cf.* Spruit 1981) occurs by means of waves with periods of 300 s, 375 s,

500 s, and 750 s (Damé 1983); the waves are regarded as internal gravity waves (*cf.* Damé; see also the discussion below) or perhaps as downward-propagating, magnetically-guided magneto-acoustic waves (Deubner 1988, Deubner & Fleck 1989). Whatever their nature, these waves carry the highest energy flux of any heating mechanism in the quiet solar chromosphere.

The bright points in the interior of the supergranulation cells also show a one-to-one correspondence with magnetic elements (Sivaraman & Livingston 1982). Thus, when a magnetic element moves, the associated bright point moves with it; a bright point is never observed without its associated magnetic element; and a magnetic element without a bright point can be understood in terms of the life time of bright points (Sivaraman & Livingston). While a magnetic field thus appears to be essential for the generation of the waves, it is probably too weak to influence their propagation. The magnetic elements appear to be located in intergranular lanes (Cram 1974, Sivaraman, Bagare & November 1990), suggesting that they arise because of the action of granular flows sweeping the magnetic field into the granular boundary regions. The heating in the bright points occurs by means of large-amplitude disturbances with a typical repeat time of 3 min, which is identical to the acoustic cutoff period of the atmosphere at the temperature minimum.

The magnetic field-free region in the cell interior, sometimes labelled the truly quiet chromosphere, is heated at a lower level. In pictures of the chromosphere in the resonance lines of Ca II, or in free-bound continua formed in the upper photosphere (*cf.* Foing & Bonnet 1984), this region appears dark. It is surmised that the atmosphere there is heated by means of short-period waves, with periods between 1/4 min and 1 min (*cf.* Ulmschneider 1990), but significant heating cannot extend higher than about 800 km since beyond that level all waves with periods shortward of the acoustic cutoff period are predominantly standing waves (Deubner 1988). However, some field-free regions may not receive a significant amount of mechanical heating at any height. This is implied by the low temperature observed in lines of the CO molecule in the infrared, which indicate temperatures well below 4000°K (Ayres 1981). However, the filling factor of these very cool, CO-emitting regions may be as small as 10% (*cf.* Athay and Ayres in the proceedings of the 1989 Cool Star Conference). It is a plausible inference from the observations of the chromospheric emission from cool stars, which imply a lower limit on the Ca II emission for very slowly rotating, old stars, that the field-free medium contributes some enhanced emission, a limit that is referred to as a basal chromosphere (*cf.* Schrijver 1987). But the heating there is likely to be significantly below the heating predicted by the acoustic heating theory (Cram 1987).

3. Waves in Bright Points

The "waves" that propagate outward into the chromosphere to heat the bright points have repetition times, or periods, that fall mainly between 2 min and 4 min, with the peak occurring near 3 min (*cf.* Liu 1974, fig. 7), *i.e.,* the cutoff period of of the gas at the temperature minimum. The waves are first observed as intensity enhancements when they are still deep in the photosphere, near $\tau_{5000} = 1$. They travel outward, gaining strength as they propagate into the chromosphere, where in most cases only the blue emission peak, K_{2v}, gets enhanced. The typical increase of the intensity of the peak is by a factor of three, and in exceptional cases a factor of five (Liu, fig. 2). Enhancement of the red peak, K_{2r}, occurs also, but much less often (*cf.* Cram & Damé 1983, fig. 2). The ratio of frequencies of occurrence of blue peak and red peak enhancements is given as 7:1 by Grossmann-Doerth, Kneer & v. Uexküll (1974) for bright points, and as 3:1 for the flux tubes of the cell boundary. At the same time, two thirds of the shifts of the K_3 absorption core are redward, implying downward velocities of the K_3-forming layer (Grossmann-Doerth *et al.*). Note that these profile statistics have never been reproduced in the modelling of bright point heating.

The importance of the 3 min waves in the heating of bright points is unexpected. Unlike the situation in a homogeneous medium, where acoustic waves may have any period and where they propagate at the sound speed, waves in a gravitationally stratified atmosphere are restricted to values of the wave period below the cutoff period, $P_{ac} = 4\pi H/a$, where H is the (projected) pressure scale height and a is the sound speed; for vertical propagation in a magnetic field-free, isothermal atmosphere, no other wave type is allowed. In particular, there is no propagation of internal gravity waves in the purely vertical direction.

When the direction of propagation is not restricted to the vertical, the wave equation admits solutions for both acoustic and internal gravity waves (*cf.* Bray & Loughhead 1974, Stein & Leibacher 1974). But for directions only moderately inclined to the vertical, the permitted gravity waves have very long period. Therefore, the inference drawn by Damé (1983) that the long-period waves he observed in the network on the boundary of supergranulation cells are gravity waves may be in error, but the conditions under which the solutions of the linearized equations are valid must be kept in mind as well. These are, above all, that the waves be plain, have small amplitude, and that the magnetic field strength be negligible. None of these conditions is satisfied.

The heating of the field-free medium inside supergranulation cells by means of short-period waves presents no difficulties to our understanding, provided such waves are present with sufficiently large energy flux. Measurements of phase delays by Deubner (*cf.* 1988) and by Lites, Chipman & White (1982) are consistent with the assumption that such waves are present, but others question heating by such waves (*cf.* Cram 1987).

The heating of the bright points by acoustic waves with a period equal to the acoustic cutoff period presents problems of a different sort. The analysis for small-amplitude waves implies that such waves cannot heat the bright points, because, if the wave period is longer than the acoustic cutoff period, the waves are evanescent, and if it is only slightly shorter, their group velocity is very small, which would allow only little energy to penetrate through the temperature minimum region into the chromosphere.

Numerical calculations of the generation of sound and internal gravity waves by the turbulence of the hydrogen convection zone are consistent with the above results on the propagation of acoustic waves. Thus, the power obtained by Stein (1967, 1968) peaks below 1 min, *i.e.,* for short-period waves, and the flux is zero at the acoustic cutoff period. This is also in agreement with the analytical estimates of Stein & Leibacher (1981), which show the power growing from zero at the cutoff period, with the rate of increase towards shorter periods depending on the assumed spectrum of the turbulence. On the other hand, we note that the cutoff period is also an eigenperiod of the atmosphere (*cf.* Lamb 1908) so that many processes should be able to excite the atmosphere to oscillations at the cutoff period.

The observational data on the waves might shed some light on the nature and the generation of the waves heating the bright points, but the data are ambiguous. Damé (1983, 1984) describes the waves as occurring typically in groups of three pulses, of which the first one is the strongest and subsequent ones are weaker (*cf.* Cram & Damé 1983, fig. 5, frame 114). This suggests perhaps that the disturbance is generated as a pulse followed by wakes oscillating at the cutoff period with diminishing amplitude, as has been investigated by Rae & Roberts (1982). This view is also supported by the narrow width evident in the disturbance (*cf.* Liu 1974, fig. 3b). However, the observational data do not exclude a periodic excitation at the cutoff period (*cf.* Cram & Damé 1983, figs. 4, 5).

Since the observational evidence so strongly suggests that acoustic waves with periods near the cutoff period of the atmosphere at the temperature minimum heat the bright points, and that the energy dissipated by these waves may be sufficient to provide all the energy support needed for the layers of the low and middle chromosphere in the bright points, we must confront the problem of the propagation of these waves. How should we imagine this? As detailed in the subsequent section, the energy dissipated by the waves is large enough to raise the degree of ionization from the low value it has in the VAL model to about 10%. This has a major effect on the value of the cutoff period, which might be doubled in the layers of the temperature plateau of 7000°K in the middle chromosphere, both because of the higher temperature there and because of the higher degree of ionization, which lowers the ratio of specific heats and thereby reduces the sound speed (*cf.* the definition of P_{ac} above). Thus, the problem of the waves, assuming that they have been generated, is only to tunnel through the layers of the temperature minimum. The damping length of waves with periods near the cutoff period is much longer than the scale height (Leibacher, Gouttebroze & Stein 1982). A significant energy flux can therefore pass through the evanescent region. In the layers of the middle chromosphere, the waves would then be again propagating.

The high degree of ionization might also solve another mystery, namely, the small size of the bright points. If the waves were confined by flux tubes, they would be expected to have spread to fill most of the available area; similarly, if the magnetic field were described by a potential field, we would expect it to have formed the canopy. In either case, the heated region should extend well beyond the observed size of 1″ to 2″. Here the high ionization and the increased temperatures caused by the waves could be imagined to create channels allowing the waves to propagate in the chromosphere. These propagation channels would be surrounded by gas that is less ionized, and hence has a higher ratio of specific heats, in which the waves would remain evanescent, *i.e.*, non-propagating. Perhaps the approximate equality of the horizontal size of the channels and the wavelength of the acoustic waves supports the picture of channels of propagation.

4. Wave Energy Flux

There are several ways in which the energy carried by the waves can be estimated: the outward propagation of the intensity enhancement observed in a spectral line yields the energy flux, the excess emission in a line and the cooling rate or the temperature function of an empirical model give the dissipation rate.

Liu (1974) estimated the wave energy flux from the rms velocity amplitude and the phase speed of the waves, *i.e.*, the speed with which the intensity enhancement in the K line propagates through the atmosphere. For this he used an empirical model with which he related the position of an intensity increase in the line profile to the position of the wave front in the atmosphere. Assuming that the waves were acoustic he then calculated the group velocity and thus the energy flux. This flux, which may be uncertain by a large margin, falls short of the flux estimated by VAL81 (*cf.* their Table 29) by only a factor of about 2. This suggests that the waves observed in the K line may carry sufficient energy for the heating of the low and middle chromosphere in the bright points.

Cram & Damé (1983) compared the average H line emission in their observations with the H line emission calculated from the empirical model of the average chromosphere, model C of VAL81. They determined the observed emitted energy from the difference between the average profile of all their observations and the average profile of the lowest decile, which was assumed to reflect only very little mechanical heating. Their estimate fell short of the emission of the empirical model by only 20%. Thus, one may conclude again that the heating of the layers in which the H line arises is consistent with heating by only these long-period waves.

Anderson & Athay (1989) determined the dissipation rate of the chromospheric heating mechanism from the empirical temperature structure of the average chromosphere, VAL model C, using a model atmosphere code. They matched the temperature curve of the empirical model in the layers of the temperature plateau below about 7000°K with a constant flux divergence per unit mass of $dF/\rho dx = 4.5 \times 10^9$ erg g^{-1} s^{-1}. In order to deduce the nature of the heating mechanism they assumed dissipation by acoustic waves, which implied a velocity amplitude of half the sound speed, a result that is consistent with heating by shock waves. As further support for their hypothesis of acoustic wave heating they pointed to the value of the microturbulent velocity in the same layers, which is approximately equal to the sound speed in these layers. Thus, the energy input into the chromosphere in the region of the temperature plateau below about 7000°K is consistent with the assumption of heating by compressional waves alone.

If we assume that VAL81 model C describes the atmosphere in the bright points, a plausible assumption for a model of the average chromosphere, and if we assume that the heating occurs by 3 min waves, the dissipation rate in the temperature plateau near 7000°K amounts to 1 eV per particle per wave pulse. Note that this occurs in a medium in which the average energy per particle is 1/2 eV per particle. These waves thus have very large amplitude (*cf.* Kalkofen 1989), and they provide enough energy to raise the degree of ionization of the gas to about 10%. The cooling of this time-dependent medium then would occur by means of radiative recombination, as suggested by Lindsey (1981).

Weak-shock theory (*cf.* Ulmschneider 1970, Bray & Loughhead 1974) permits an estimate of the wave period and the Mach number of the waves if the shocks have reached their limiting strength, where the growth of the wave amplitude due to the outward propagation in a gravitationally stratified gas is balanced by the decay of the amplitude due to dissipation. The theory is not directly applicable to the waves that heat the chromosphere in the bright points since they are not plane and the shocks are not weak. But using the theory for a rough estimate of wave properties, the dissipation rate found by Anderson & Athay implies that the shocks are fairly strong, with a Mach number of nearly 2, and that the wave period is about 1 min, not much different from the observed periods. These results are thus broadly consistent with heating of the chromosphere in the layers of the temperature plateau by compressional waves with periods near the acoustic cutoff period.

5. Conclusions

The properties of the internetwork bright points are intermediate between those of the network elements and those of the truly quiet chromosphere: The bright points are associated with relatively weak magnetic fields, whereas the network fields are very strong and the truly quiet chromosphere is probably field-free. The bright points are heated by waves with periods near the acoustic cutoff period of the atmosphere at the temperature minimum; the wave periods observed in the network elements are longer, and the quiet regions are presumed to be heated by waves of shorter period. The bright points are intermediate also in the flux of the waves heating the gas. The equality of the emission widths in the Wilson-Bappu effect of both the bright points and of the sun as star (Bappu & Sivaraman 1971) suggests that the bright points are an important element in the chromospheric emission from the sun.

The magnetic field associated with the bright points is critical for wave generation since it determines the location of the heating, but the field is probably too weak to have a major influence on wave propagation. The confinement of the waves may occur by means of propagation channels created by the waves, which heat and ionize the gas and thereby increase the cutoff period locally. In the channels, the waves would be running waves. This medium would be surrounded by gas in which the waves are evanescent.

The radiative cooling times implied by the empirical models of the chromosphere are too long to agree with the observed properties of the 3 min waves. Instead of cooling by free-free radiation of H^-, the gas may cool by radiative recombination.

I have benefitted from discussions with Dr. Venkatakrishnan. The support of this research by NASA is much appreciated.

References

Anderson, L. S. & Athay, R. G. 1989. Chromospheric and coronal heating. Astrophys. J. 336: 1089-1091.

Athay, R. G. 1976. The Solar Chromosphere and Corona, Reidel Publ. Co., Dordrecht.

Ayres, T. R. 1981. Thermal bifurcation in the solar outer atmosphere. Astrophys. J. 244: 1064-1071.

Bappu, M. K. V. & Sivaraman, K. R. 1971. K Emission-Line Widths and the Solar chromosphere. Solar Phys. 17: 316-330.

Biermann, L. 1946. Zur Deutung der chromosphärischen Turbulenz und des Exzesses der UV-Strahlung der Sonne. Naturwissenschaften, 33: 118-119.

——— 1948. Über die Ursache der chromosphärischen Turbulenz und des UV-Exzesses der Sonnenstrahlung. Z. Astrophysik 25: 161-177.

Bray, R. J. & Loughhead, R. E. 1974. The Solar Chromosphere (London: Chapman and Hall).

Cram, L. 1974. High resolution spectroscopy of the disk chromosphere. Solar Phys. 37: 75-83.

——— 1987. In: Cool Stars, Stellar Systems, and the Sun, eds. J. L. Linsky and R. E. Stencel (Berlin: Springer Verlag), pp 123-134.

Cram, L. E. & Damé, L. 1983. High spatial and temporal resolution observations of the solar Ca II H line. Astroph. J. 272: 355-361.

Damé, L. 1983. Eruptions solaires en lumière blanche et structures fines, oscillations et chauffage de la chromosphère solaire calme. Thèse, Université de Paris VII.

——— 1984. Small-scale dynamical processes in the solar atmosphere. In: Small-Scale Dynamical Processes in Quiet Stellar Atmospheres, ed. S. L. Keil (Sacramento Peak), pp 54-64.

Deubner, F.-L. 1988. Observations of solar oscillations. In: Pulsation and Mass Loss in Stars, eds. R. Stalio and L. A. Willson (Dordrecht: Kluwer Acad. Publ.), pp 163-179.

Deubner, F.-L. & Fleck, B. 1989. Dynamics of the solar atmosphere III. Cell-network distinctions of chromospheric oscillations. preprint.

Foing, B. & Bonnet, R. M. 1984. On the origin of the discrete character of the solar disk brightness in the 160 nanometer continuum. Astrophys. J. 279: 848-856.

Gingerich, O., Noyes, R. W., Kalkofen, W. & Cuny, Y. 1971. The Harvard-Smithsonian Reference Atmosphere. Solar Phys. 18: 347-365.

Grossmann-Doerth, U., Kneer, F. & v. Uexküll, M. 1974. Properties of the Solar Ca II K Line in High Spatial Resolution. Sol. Phys. 37: 85-97.

Jensen, E. & Orrall, F. Q. 1963. Observational study of macroscopic inhomogeneities in the solar atmosphere. IV. Velocity and intensity fluctuations observed in the K line. Astrophys. J. 138: 252-270.

Kalkofen, W. 1989. Chromospheric heating. Astrophys. J. (Letters) 346: L37-40.

Lamb, H. 1908. On the theory of waves propagated vertically in the atmosphere. Proc. London Math. Soc. (2) 7: 122-141.

Leibacher, J. & Stein, R. F. 1981. Oscillations and pulsations. In: The Sun as a Star, ed. S. D. Jordan (Nasa: Monograph Series on Nonthermal Phenomena in Stellar Atmospheres), pp 263-287.

Leibacher, J., Gouttebroze, P. & Stein, R. F. 1982. Solar atmospheric dynamics. II. Nonlinear models of the photospheric and chromospheric oscillations. Astrophys. J. 258: 393-403.

Lindsey, C. A. 1981. Heating of the solar chromosphere by ionization pumping. Astrophys. J. 244: 659-677.

Lindsey, C. & Roellig, T. 1987. Submillimeter diagnostics of the response of the solar chromosphere to compressional waves. Astroph. J. 313: 877-892.

Lites, B. W., Chipman, E. G. & White, O. R. 1982. The vertical propagation of waves in the solar atmosphere. II. Phase delays in the quiet chromosphere and cell-network distinctions. Astrophys. J. 253: 367-385.

Liu, S.-Y. 1974. Direct observational evidence for the propagation and dissipation of energy in the chromosphere. Astroph. J. 189: 359-365.

Orrall, F. Q., 1966. Observational study of macroscopic inhomogeneities in the solar atmosphere VIII. Vertical chromospheric oscillations measured in K_3. Astroph. J. 143: 917-927.

Piddington, J. H. 1974. The Chromospheric Energy Balance. In: I.A.U. Symp. No. 56, Chromospheric Fine Structure, ed. R. G. Athay (Dordrecht: Reidel Publ. Co.), p 269.

Rae, I. C. & Roberts, B. 1982. Pulse Propagation in a Magnetic Flux Tube. Astroph. J. 256: 761-767.

Schatzman, E. & Souffrin, P. 1967. Waves in the solar atmosphere. Ann. Rev. Astron. Astroph. 5:67-84.

Schrijver, C. J. 1987. Heating of stellar chromospheres and coronae: evidence for non-magnetic heating. In: Cool Stars, Stellar Systems, and the Sun, eds. J. L. Linsky and R. E. Stencel (Berlin: Springer Verlag), pp 135-144.

Schwarzschild, M. 1948. On noise arising from solar granulation. Astrophys. J. 107: 1-5.

Sivaraman, K. R. & Livingston, W. C. 1982. Ca II K_{2v} spectral features and their relation to small-scale photospheric magnetic fields. Sol. Phys. 80: 227-231.

Sivaraman, K. R., Bagare, S., P. & November, L., J. 1989. *cf.* paper in proceedings of this I.A.U. symposium.

Spruit, H. C. 1981. Magnetic flux tubes. In: the Sun as a Star, ed. S. Jordan (Nasa: Monograph Series on Nonthermal Phenomena in Stellar Atmospheres), pp 385-412.

Stein, R. F. 1967. Generation of Acoustic and Gravity Waves by Turbulence in an Isothermal Stratified Atmosphere. Solar Phys. 2: 385-432.

—— 1968. Waves in the solar atmosphere. I. The acoustic energy flux. Astrophys. J. 154: 297-306.

Stein, R. F. & Leibacher, J. 1974. Waves in the solar atmosphere. In: Ann. Rev. Astron. Astroph. 12: 407-435.

—— 1981. Wave generation. In: The Sun as a Star, ed. S. D. Jordan (Nasa: Monograph Series on Nonthermal Phenomena in Stellar Atmospheres), pp 289-300.

Stenflo, J. O. 1989. Small-Scale Magnetic Structures on the Sun. The Astron. and Astroph. Rev.,. subm.

Ulmschneider, P. 1970. On Frequency and Strength of Shock Waves in the Solar Atmosphere. Solar Phys. 12: 403-415.

—— 1974. Radiative Loss and Mechanical Heating in the Solar Chromosphere. Solar Phys. 39: 327-336.

—— 1986. The present state of wave heating theories of stellar chromospheres. Adv. Space Res. Vol. 6, No. 8, pp 39-46.

—— 1990. *cf.* . proceedings of the 1989 Cool Star Conference, Seattle, WA.

Ulmschneider, P. & Kalkofen, W. 1977. Acoustic waves in the solar atmosphere III. A theoretical temperature minimum. Astron. & Astroph. 57: 199-209.

Vernazza, J. E., Avrett, E. H. & Loeser, R. 1981. Structure of the solar chromosphere III. Models of the EUV brightness components of the quiet sun. Astroph. J. Suppl. 45: 635-725 (**VAL81**).

Zirin, H. 1988. Astrophysics of the Sun (Cambridge: Cambridge University Press).

DISCUSSION:

HOLLWEG: Your results sound a lot like the spicule model by myself and Sterling . Also, I believe the NRL group found UV features resembling spicules over cell interiors. Is there a relation to the bright point waves?

KALKOFEN: It does indeed seem plausible to assume that the 3-minute oscillations, with several unequal pulses of which the first one is the strongest, are generated by an impulse followed by eigenoscillations of the atmosphere at the cutoff period, as in your spicule model and as described in work by B. Roberts. But I do not know whether there is a connection to spicules in the interior of super-granulation cells.

KNEER: I could not get the point about how the 3-min. oscillation is generated and supplied with enough energy to heat chromospheric grains.

KALKOFEN: It should not be surprising that waves of 3-min. period are found in the chromosphere, since they are an eigenperiod of the atmosphere at the base of the chromosphere. But how these wave are generated in not known. The theory by Stein (1967,1968) and by others gives zero energy flux at the acoustic cutoff. That result clearly is not in accord with the bright point observations.

UBEROI: The magnetic fields do not play any role in your work. Please comment especially about the frequency ranges you are dealing with.

KALKOFEN: The magnetic field is probably of no importance for the propagation of waves in the bright points since the value of the plasma-β is likely to be much larger than unity, perhaps as large as 100. But the observations (Sivaraman & Livingston) suggest that the magnetic field is of critical importance for the generation of the waves. The wave periods are near the eigenperiod of the atmosphere at the temperature minimum, *i.e.*, the acoustic cutoff period. The range of periods, according to Liu, is 2 min. to 4 min.

KUNDU: (i) Several years ago we tried to look for 5-min. oscillations at centimeter wavelengths using either single dish telescopes or interferometers. We did not find any 5-min. oscillations at 3, 11 or 1.3 cm wavelengths using interferometers. However, we found one period at 3-min. at 1.3 cm. This seems to be relevant to the 3-min. waves that you discussed.

(ii) Can you explain why there should be 3-min. waves and not 5-min. waves in the upper chromosphere?

KALKOFEN: (i) I am not surprised that you saw 3-min. waves at 3 cm or 1.3 cm. This is consistent with observations by Yudin (1968) at 3 cm and by Simon and Shiwabukuro (1971), who saw them at 3.3 min. and 3.5 min. The latter concluded that the 3 min. waves are unimportant for chromospheric heating based on the long cooling times (much longer than the waveperiod) given by Spiegel (1957) and by Ulmschneider for free-free transitions of H^-. The actual thermal relaxation of the medium is probably controlled instead by ionization and recombination processes, as suggested by Lindsey, with cooling times significantly shorter than the waveperiod. Hence the 3-min. waves probably are very important for chromospheric heating.

The 3-min. period is an eigenperiod of the system (Lamb 1908), and the waves are likely to be allowed to propagate in the channels created by the wave in the bright points, whereas the 5-min. oscillations may still be evanescent.

(ii) The 5-min. waves are evanescent in the upper photosphere, and so they are damped in the chromosphere. The radiation at cm wavelengths is formed in the middle chromosphere, at about 2000 km above $\tau_{5000} = 1$. At that height the wave amplitude of 5-min. waves may be too small to be detectable.

ULMSCHNEIDER: Your energy estimates depend on accurate time-dependent calculations of waves with CaII-NLTE emission. How accurate are the calculations you quoted?

KALKOFEN: The estimate by Cram & Dame (1983) of the average emission in the H line is purely observational and depends on the determination by Pasachoff of the absolute intensity in H. The VAL81 model C, used for calculating the H line emission of the average chromosphere, is a spatially and temporally averaged representation of the average chromosphere, based on UV continua. The uncertainty of using such a model for computing line emission is evident. But for the purpose of estimating the time-averaged chromospheric H-line emission, the model should be adequate.

KNEER: Besides the 3-min. oscillation the 5-min. oscillation shows up clearly in chromospheric data analysis (CaK, Hα,CaIR). In fact, it is not settled among observers that the 3-min. "chromospheric" mode is so much dominant in the chromosphere since it is not seen as a modal structure in the power spectra in the k-ω-plane. Rather what I see in the power spectra is a very broad distribution from 6-min. periods to 1-min, or shorter. The 5-min. oscillations show up as modes and the low frequency power can be identified as the signature of stochastic processes.

KALKOFEN: The k-ω diagram may be too blunt a tool for studying waves in bright points, which have small size ($\approx$ 2") and filling factor ($\lesssim$ 10%). In addition, the 3-min. "waves" are not sinusoidal oscillations but consist of (unequal) pulses separated by 3 minutes.

CORONAL HEATING BY D.C. CURRENTS

J. HEYVAERTS
Observatoire de Paris and Université Paris VII
DAEC
92195 Meudon Principal Cédex
FRANCE

ABSTRACT. Present views on DC current coronal heating are presented. The relation to AC mechanisms, the importance of MHD turbulence in both processes, and the convergence of presently proposed ideas is outlined.

1. Introdution

In the past few years, the problem of coronal heating and evolution has been expressed as follows : a low β plasma fills the half space z>0. Line tying is assumed to hold at the "boundary" (ill defined). which has given, random, motions. Calculate the dissipative electric response of the corona. It is assumed that there is no slip of concentrated photospheric flux tubes with respect of the motion of the less magnetized photosphere. This motion, also, is assumed to be free of any back-reaction, via Lorentz forces, from the coronal part. Most, if not all, of these assumptions can be challenged. Nevertheless, we shall discuss this simplified model because it is (hopefully) devoid of non-essential complications.

Wave and D.C. current heating mechanisms have often been regarded as terms of an alternative. But in fact they are different regimes of a common mechanisms : the dissipative response of the corona to the excitation introduced by the driver. This response depends on the space and time properties of the driver and on their matching to the space-time characteristics of coronal loops (and open field lines). Incidentally, the former actually determine, the latter via the statistical effect of reconnection and the statistical effect of evaporation and coronal rain.

Loops have a resonance Alfvèn frequency spectrum. According to whether these frequencies and their harmonics cover a large part of the frequency spectrum of the driver or not, the wave mechanism will take more, or less, importance compared to the D.C. one. Then, long loops, which have lower fundamental frequencies, are more likely to tap the driver spectrum in the A.C. (wave) mode than short loops which mostly react to it as D.C. loads. The matching of the space properties of the driver and of the loops (driver's correlation length as compared to the size of the footprints of loops, for example) is also a significant issue.

For DC currents, and this is largely shared by A.C.'s, the main problem is to understand the process by which the scales of the electric currents are so drastically reduced in the corona as compared to the characteristic lengthes of the driver. The latter cannot be much less than 100m, since this is the photospheric dissipation scale. Some colleagues like Parker (1983) or Van Ballegooijen (1986) developed the ("extrinsic") view that this is all the result of increasing coronal complexity, driven by boundary motions, and the idea that this evolution, by itself, leads to singular states is sometimes developed (Parker 1983, Syrovatski, 1978). Others think in terms of internally developping instabilities, as a response to boundary stressing ("intrinsic" view) (Chiuderi, 1980, Heyvaerts and Priest, 1984, Chiueh and Zweibel, 1987, Vekstein, 1987). This distinction should however not be overemphasized since the effects, in terms of scale reduction, of both are not so different.

E. R. Priest and V. Krishan (eds.), Basic Plasma Processes on the Sun, 207–214.

2. About the conept of direct singularity formation

The fact that boundary motions may generate current singularities is known from studies of the evolution of X points under boundary changes in 2-D MHD (Priest and Raadu, 1975). 3-D analogs obviously exist. Such singular layers, judging from the 2D problem, should be exceptional features. However, Parker (1983), develops the more radical view that boundary motions of small scales and amplitudes drive an initially potential coronal field into a state of non equilibrium involving a great many singular layers, reflecting the complexity of the driver. He solved (Parker 1972) the equilibrium equation around an initially uniform field in infinite space, this being justified to him by the smallness of the driver scale-length as compared to loop length. He found that the only possible solution should have translational invariance. Since boundary conditions, being random, cannot spontaneously lead to such a state while keeping the flux frozen, he concluded that the symmetric state was only accessible through reconnection driven by unbalanced Lorentz forces.

Parker's views have been challenged by a number of authors, for example by Rosner and Knobloch (1982) or Van Ballegooijen (1985). Sakurai and Levine (1981) and more recently Zweibel and Li (1987) found the exact solution for arbitrary, small, perturbations with boundaries at finite distance which turned out to be perfectly regular. Then, we cannot expect direct buildup of singularities in the linear approximation. But certainly finite thickness current sheets do form, and may well become unstable or be driven to a dissipative state. So, ultimately, the end result may be quite the same as if non equilibrium and singularities indeed developed.

3. Routes to turbulence

Actually, a number of works have identified plausible primary instabilities, generally of the reconnection type (tearing modes) (Arion, 1984, Chiuderi and Van Hoven, 1979, Bodo et al., 1987). These develop non linearly, driving coalescence and secondary reconnection (Finn and Kaw, 1977, Biskamp, 1985). The effect of such large-scale forces on current sheets present in the system has sometimes been modeled by an externally imposed force (Chiueh and Zweibel, 1987, Heyvaerts and Kuperus (1978)). In this latter paper, originally concerned with flares, the 2-D time dependant response of a thin sheet to a boundary pressure pulse has been calculated analytically, and shown to lead to a growth of current density as exp. $(t/\tau)^2$. Secondary instabilities have been identified in reconnection flows, of the Kelvin-Kelmholtz type (Chiueh and Zweibel, 1987) or of the tearing type (Biskamp, 1985).

The non linear development of complex reconnection is one route to turbulence, but there are others as well, which rely on the evolution of waves in the corona (Heyvaerts and Priest 1983, Hollweg 1983). Phase mixing of shear Alfvèn waves leads to secondary instabilities of the Kelvin. Helmholtz or tearing type (Heyvaerts and Priest 1983, Browning Priest 1984) and more generally their coupling (even in absence of phase mixing) leads in the weakly non-linear regime, to cahotic behaviour and cascades towards smaller scales (Nocera et al. 1986, Pettini et al. 1985). There may be, then, less fundamental difference between DC and AC phenomena than generally appreciated, since they may merely differ by the mode of injection of the turbulent energy, cascade process and eventual dissipation being otherwise quite similar. Incidentally, this means that diagnostics of the motions associated with heating, by line profile measurements for example, may fail to unambiguously determine the wave or DC nature of the basic mechanism.

In brief, there are many potential routes from an initial equilibrium to instability and then to turbulence.

4. About the role of field line stochasticity in coronal structures

In this context, Tsynganos et al. (1984) discussed the possibility that intrinsic stochasticity of field lines appear in coronal loops. Let us remind the reader that only a few symetric MHD states have for sure "magnetic surfaces" (Edenstrasser, 1980). In such cases an analogy with an integrable hamiltonian system may be derived, an "invariance coordinate" playing the role of time, and the projection of a point following a field line onto a plane perpendicular to that direction behaving as a 1-D hamiltonian system in its phase space. Tearing modes, however, are likely to introduce "symmetry-breaking" current perturbations which could be compared to the addtion of a time dependent hamiltonian perturbation in the equivalent hamiltonian system. The result is that some trajectories (field lines)become stochastic (i.e. Finn, 1985). This can be understood by analogy in terms of the KAM theorem of mechanics. See e.g. Whiteman (1977). Tsynganos et al. (1984) argue that, in such an intrinsically stochastic state, no equilibrium is generally possible since if the pressure is to be constant, but different, on the various field lines, it should become a random function in the stochastic zone ! Does this conflict with Sakurai and Levine's and Zweibel and Li's work ? I believe not. On the one hand, for vanishingly weak perturbations, of order ε, the width of the stochastic zone is known to vary as $\exp(-1/\varepsilon)$ (negligible and non-analytic in ε). On the other hand, the stochastic character of the field lines becomes only explicit for loop lengths much larger than the length it needs for an undisturbed field line to make one turn around the loop (the pitch-length, say). Though this automatically is the case in toroïdal systems, which close onto themselves, boundaries at the photosphere constitute, by contrast, a strong-limitation in the solar case, where the length of the loop is expected to be much less than the pitch-length. We should certainly not expect stochasticity in finite length loops to result in any field line ergodically filling (even in projection onto a cross-section plane) all the stochastic zone ! I would even expect, but this certainly is conjoncture, that most of the unbalanced forces which would develop in the infinite system will actually find their balance near the boundary. Something could still remain of the exponential behaviour of the separation between two neighbouring field lines which is a characteristic feature of stochasticity. Not much is known about this, as by now, however.

It is interesting to mention, in relation to that, that Similon and Sudan (1989) have recently studied the propagation of shear Alfvèn waves in a stochastic field, which proves very effective.

5. A tentative theory of "extrinsic" type.

In an effort to deal with the field-braiding idea, Van Ballegooijen (1986) took an ambitious attack to the problem. Assuming flux freezing and knowledge of the statistical properties of the driver in two parallel horizontal boundary plates separated by l_B, he went on to derive the spatial power spectrum of the magnetic field between the plates as a function of time. Weak perturbations, perpendicular to an initial uniform field $B_o e_z$, at all time in force free equilibrium, are assumed. These equilibria are non-linear, though. Their explicit evaluation being impossible, only statistical properties can be derived, under some assumptions. Attention can be transfered to the vertical velocity gradient, which is related to the field perturbation **b** by $D\mathbf{b}/Dt = B_o \partial v/\partial z$, where D/Dt is the lagrangean derivative. It has been assumed that $\partial v/\partial z$ has the same correlation time τ_v and horizontal correlation length, l_v, as boundary velocities. Its vertical correlation length, $l(t)$, is assumed to equal l_B / $n_{(braids)}$ where n(braids) is the average number of braids introduced in the system (a time-dependant quantity). It is considered that one new braid is introduced in the system when the time has elapsed, which is necessary for the fluid elements on the boundary to suffer an average displacement equal to the correlation length l_v. A random walk theory gives the probability that points separated by D_1 at t_1 become separated by D_2 at t_2. This is enough to evaluate the field and current spectrum as a function of time. It is found that these spectra suffer a boundary-driven cascade towards smaller scales at an exponentiating rate. A situation where dissipation becomes important is soon reached after a time :

$$t_1 = t_{braid} \ln R_m \qquad \text{where } R_m = (\mu_o l_v^2 / t_{braid})$$

The Poynting flux at that time should not differ very much from the stationnary heating rate, then it is found that :

$$F_H = \frac{Bo^2}{\mu_o} \frac{v^2 \tau_v}{l_B} \ln (R_m) \qquad (1)$$

6. Tentative theories of the "intrinsic" type

Let us now consider the case when the coronal system gets rid of the energy injected in it by some internally developed turbulence. A formulation which would allow the calculation of the heating without resorting to the details of the non-linear phase which leads to dissipation is what is seeked for here. The existence, in turbulent systems, of quasiinvariants, which, for undriven turbulence, decay at a rate much slower than other global quantities is of some interest, because we may expect that rapidly decaying quantities reach a minimum value subject to the constraint that quasiinvariants keep their initial value. Quasiinvariants are associated with quantities suffering inverse cascades towards large scales, virtually subject to no decay. In 3D-MHD magnetic helicity is a quasiinvariant, while energy decays. The idea of selective decay towards a minimum energy state seems to be supported for undriven systems, by laboratory experiments in "reverse field pinches" (see i.e. the review by Taylor, 1986).

It is interesting to hear that a theory of driven turbulent systems is in progress (Montgomery, 1989). Previously lacking such a tool, some effort has been put into devising a description of coronal turbulent heating based on what was known of undriven systems. The true evolution is idealized as a succession of elementary steps during which a complete cycle of perfect MHD evolution followed by a turbulent decay episode is assumed to take place. At the end of each step, the system is idealized as having returned to the minimum energy compatible with its acquired magnetic helicity. The heating in one step is the difference between the magnetic energy hypothetically acquired by perfect MHD evolution and that of the relaxed (linear force free) configuration. In reality stressing and relaxation act simultaneously, and the system would never be found exactly in the lowest energy state. Nevertheless, this schematic approach should be adequate when departure from this state is always small, that is, when the rate of strain is small ($v\tau_D / l_B << 1$ in eq. (2) below). This idealized evolution has been calculated for a few simple field and velocity geometries (Heyvaerts and Priest 1984, Browning et al.1986, Dixon, 1987). Assuming a certain value of the time.needed to complete the stress/relaxation cycle, τ_D, the average heating flux, and is dependance on geometry, can be calculated, sometimes analytically. The result is of the form :

$$F_H = f \frac{B_v^2}{\mu_o} \left(\frac{v\tau_D}{l_B} \right) G \qquad (2)$$

where f is a numerical factor of order 1, pertinent to the geometry considered, v the characteristic boundary velocity, l_B the loop length, and G a factor less than unity, which may exceptionnally become much less if the boundary motions happen to spontaneously generate by perfect MHD evolution an almost exactly linear force free field, leaving very little energy available for dissipation in the relaxation.

The duration of the stress-relaxation episodes cannot be determined from considerations which involve only initial and final states, but no dynamics of the developing turbulence. This indeterminacy is a major shortcoming of such theories.

In a more recent paper, Vekstein (1987) tried to take simultaneous account of stressing and relaxation, by writing for the field evolution a phenomonological equation which incorporate the fact that in the absence of relaxation the field should follow a perfect MHD evolution, $B_M(t)$ say, while in the opposite limit, it should remain very close to the lowest energy state corresponding to its present helicity, $B_R(t)$. He proposes the equation

$$\frac{\partial}{\partial t}(B - B_M) = -(B-B_R)\ /\ \tau_D \tag{3}$$

which correctly takes these two limits. The heating is the Poynting flux, minus that part of the energy that will remain in magnetic form after relaxation. Vekstein (1987) evaluates B_M, B_R and B by a linearization procedure for each frequency component of the driver spectrum. He then obtains :

$$F_H = \frac{B^2}{2\mu_o}\frac{\tau_D}{l_B}\int d\omega \frac{1}{1+\omega^2\tau_D^2}\left(<v^2(\omega)> - \Gamma^2(\omega)\right) \tag{4}$$

where $<v^2(\omega)>$ is a space average of the power spectrum. The negative contribution Γ^2 represents the non-decaying part of the energy flux and plays the role of the factor G in (2). Neglecting it for simplicity, eq. (4) can be written as

$$F_H = \frac{B^2}{2\mu_o}\frac{\tau_D}{l_B} <<v^2>> \qquad \text{where } <<v^2>> = \int d\omega \frac{<v^2(\omega)>}{1+\omega^2\tau_D^2} \tag{5}$$

$<<w^2>>$ is a weighted mean of the driver's velocity power spectrum. Mainly those motions which are coherent on a time scale τ_D contribute to the heating. Otherwise their effect is reduced by the "incoherence penalty factor" $(1+\omega^2\tau^2D)^{-1}$

7. Convergences and differences in D.C. theories

Only simple configurations have been calculated for heating by the constrained minimum energy hypothesis described above. However, the same idea (with the same provisos) should be applicable to randomly braided systems, since they too seek for a minimum energy state compatible with their acquired quasiinvariants. This actually is Vekstein's idea. Note that Parker and Van Ballegooijen implicitly assume no net helicity injection, but this makes no difference of principle.

It is then no surprise that Parker's heating formula (Parker, 1983), $F_H = (B^2/\mu_0)\,(v^2/u)\,(h/l_B)$, where u is the "reconnection velocity" and h the "typical flux tube thickness" is just the same as eq. (2) if we adopt Parker's favourite value of τ_D, namely $\tau_D = (h/u)$, and ignore that f and G are not exactly unity. It is also interesting to note that, as expectable, van Ballegooijen's result (eq. 1) reduces to a form very similar, though slightly different, to Vekstein's result, if his time t_1 is adopted as τ_D and the velocity correlation time t_v is taken as $(1/\omega)$ in Vekstein's terms. We actually get eq. (1) in the form :

$$F_H = \frac{B^2}{\mu_o}\frac{\tau_D}{l_B}\left[v^2\left(\frac{v}{\omega l}\right)^2\right] \tag{6}$$

which contains an incoherence penalty factor $(v/\omega l)^2$ similar to Vekstein's $(1+\omega^2 t^2)^{-1}$. In the limit of motions which remain coherent on the time scale τ_D or (l/v), both (5) and (6) reduce to the form (2). However this relative convergence of the results should not hide the fact that the effective dissipation time τ_D remains a parameter, or, if not, is very ill-defined in these theories. It is in incorporating the turbulence dynamics in the calculation of dissipative response that progress should be made. Van Ballegooijen's work is a first step in that direction.

Reference

Arion, A., 1984, PHD Thesis, Maryland University

Biskamp, D., 1985, in Magnetic Reconnection and Turbulence, MA Dubois et al. ed., Editions de Physique (Orsay), p. 19.

Bodo, G., Ferrari, A. Massalia, S., Rosner, R., 1987, APJ. 313, 432

Browning, P. and Priest, E.R. : 1984, Astron. Astrophys. 131, 283

Chiuderi, C., 1980 in Solar Phenomena in stars and stellar systems NATO adv. stud. inst. series, RM Bonnet and A.K. Dupree ed., Reidel Pub. Co., p 269

Chiuderi, C. and van Hoven, G., 1979, APJ Letters, 232, L69

Chiueh, T. and Zweibel, E.G., 1987, APJ, 317, 900

Dixon, A., 1987, PHD Thesis, University of St. Andrews

Edenstrasser, J.W., 1980, J. Plasma Physics, 24 (3), 515

Finn, J.M., 1975, Nuclear Fusion, 15, 845

Finn, J.M., and Kaw, P.K., 1977, Physics of Fluids, 20, 72

Heyvaerts, J. and Kuperus, M., 1978, Astron. Astrophys. 64, 219

Heyvaerts, J. and Priest, E.R., 1983, Astron. Astrophys. 117, 220

Heyvaerts, J. and Priest, E.R., 1984, Astron. Astrophys. 137, 63

Hollweg, J. 1983 in solar Wind V, M. Neugebauer ed., NASA conf. pub, P.5

Montgomery, D., 1989, these proceedings

Nocera, L., Priest E.R. and Hollweg, J., 1986, Geophys. Astrophys. Fluid Dynamics 35, 111

Parker, E.N. 1972, APJ 174, 499

Parker, E.N., 1983, APJ 264, 642

Pettini, M., Nocera, L., Vulpiani, A., 1985, in Chaos in Astrophysics, J.R. Butler et al. ed., Reidel Pub. Co., p. 305

Priest, E.R. and Raadu, M., 1975, Solar Physics 43, 117

Rosner, R., and Knobloch, E., 1982, APJ, 262, 349.

Sakurai, T., and Levine, R.H., 1981, APJ 248, 817

Similon, P.L., and Sudan, R.N., 1989 : Preprint

Syrovatski, S., 1978 : Solar Physics 58, 89

Taylor, J.B., 1986 : Rev. Mod. Phys. 58 n°3, 741

Tsynganos, K.C., Distler, J., and Rosner, R., 1984, APJ 278, 409

Van Ballegooijen A.A., 1985, APJ 298, 421

Van Ballegooijen A.A., 1986, APJ 311, 1001

Vekstein, 1987, Astron. Astrophys. 182, 324

Whiteman, K., 1977 : Report Progr. Phys., 40, 1033

Zweibel, E.G., and Li H.S., 1987, APJ 312, 423

DISCUSSION

RUDERMAN: As far as I understand, the proof of magnetic surface existence is correct not only in the case of translational symmetry, but also in the cases of rotational and helical symmetries as well. Is that right?

HEYVAERTS: Yes. However, the development of single island structure in toroidal symmetry does not lead to a state of any such symmetry and gives rise to stochasticity indeed.

ROBERTS: In the scenario of shuffling footpoints in the photosphere the effects of gravitational stratification are ignored. This is understandable for the corona, where the scale-height is large. But it is less satisfactory for the photosphere-chromosphere. So while we have a boundary layer in the lower atmosphere as far as the coronal loop is concerned, it is far from a boundary layer as far as the footpoint motion are concerned. In the photospheric-chromospheric region, changes at a level where tubes are isolated from one another are transmitted over several photospheric scale-heights by the time they reach the relatively unstratified coronal atmosphere. Do you think this could be important for the derivation of the heating rates from footpoint motions?

HEYVAERTS: It would be unimportant if the motion of the fluid in the part of the field lines which pierce these layers is approximately independent of height. However, since these are regions where forces other than magnetic could play a role, it is *a priori* conceivable that motions be quite different over a scale height. But if this is so, then a significant $\partial B_{\mathrm{Horizontal}}/\partial z$ will be built up, and then a large horizontal current in the photosphere/chromosphere would be expected. It is possible to show that this effect would lead to very significant Lorentz forces in the photosphere/chromosphere, and the availability of other forces strong enough to fight this tension is by no means obvious (Heyvaerts,1974). This means that even a partial bend of the field in these layers could have a feed-back effect on the flow at these altitudes, which could then not be regarded as an "independent driver". However, if we nonetheless accept that (independent or not) the relevant statistical properties of the flow at some altitude are known, the "boundary surface" theory can still be used. If one expresses what happens in the corona in terms of a really independent "driver", one maybe should then follow the field down to the convective zone. There, the problem of flux freezing could be disputed, especially in the weakly ionized part of the photosphere. The situation becomes there similar to electric currents driven by quasi-neutral atmospheric winds when they flow across field lines in the ionosphere, and the physics becomes somewhat different in this "driver".

CHOUDHURI: Can we really take the calculations of Zweibel & Li as a conclusive proof that Parker's theory is wrong? They considered small footpoint motions, whereas one would expect finite footpoint motions to entangle field lines enough to produce complicated topologies and give rise to current sheets. So perhaps it is not surprising that Zweibel & Li did not find current sheets.

HEYVAERTS: I think the work of Zweibel and Li is a valid criticism of Parker's views as they have been presented in the literature. I do not expect small finite displacements to lead to more singularity in the weakly nonlinear regime (Van Ballegooijen). This does not mean that current concentrations of finite thickness are not going to be formed, in equilibrium to begin with, then perhaps (for significant displacements) in non-equilibrium. Anyway, I believe the "singularity war" is perhaps not worth fighting, since the important idea in

Parker's as well as others' views is that the system finds some way to release by (complicated) dissipative processes part of the energy the boundary wants to pump in. Whichever way it finds to go to small scales is of secondary interest, I believe.

VAN BALLEGOOIJEN: It should be noted that Parker (1972) considered a linearization in terms of the aspect ratio of the perturbations, whereas Zweibel and Li (1987) used a linearization in terms of the displacement amplitude (ratio of displacement amplitude to velocity scale at the boundary).

RELAXED STATES OF MHD TURBULENCE: MINIMUM DISSIPATION OR MINIMUM ENERGY?

David Montgomery
Department of Physics & Astronomy
Dartmouth College
Hanover, NH 03755, USA

ABSTRACT. Driven, dissipative MHD fluids often seem to undergo relaxation processes. After a turbulent formation phase, a geometrically simpler and less disordered configuration emerges. The best known example is the laboratory reversed-field pinch (RFP); similar field topologies have been proposed for solar prominences and astrophysical "flux ropes." In a transient situation, the more rapid decay of kinetic and magnetic energy relative to magnetic helicity provides a mechanism for generating an MHD configuration with several similarities to observed RFP states. (This is the Taylor hypothesis, not unrelated to turbulent inverse magnetic cascades.) For the driven steady state, however, all quantities are supplied at the same time-averaged rate at which they are dissipated, by definition; nothing decays relative to anything else. Some other unifying principle, beyond "minimum energy" or "selective decay," seems necessary to describe the results of driven, steady-state MHD computations. We have been attempting to adapt the principle of minimum energy dissipation rate to MHD. It is a 19th century principle that achieved some success in hydrodynamics and separately in dissipative electrodynamics.

Suppose that a particle is trapped around the minimum of a potential well and at the same time experiences a velocity-dependent frictional energy loss. Eventually, the particle will come to rest at the minimum of the well with zero kinetic energy. This is the simplest example of a "minimum energy state," established subject to a constraint.

Minimum energy states, in more elaborate guises, appear all over physics. States of minimum energy, subject to constraints and boundary conditions, have played a role in ideal magnetohydrodynamic (MHD) stability theory. More recently, Taylor [1,2] has argued that the laboratory toroidal Z pinch ("screw pinch") should relax to a state of minimum energy compatible with approximate conservation of magnetic helicity. Minimum energy states have also been used as a basis for describing solar prominences.

Some physical justification for relaxation to states of minimum energy can be given for MHD in the context of the decaying initial value problem for isolated systems. The arguments involve the more rapid turbulent decay of energy, relative to magnetic helicity, or "cross" helicity, or whatever other ideal invariants might decay less rapidly because of

E. R. Priest and V. Krishan (eds.), Basic Plasma Processes on the Sun, 215–222.

tendencies to being transferred preferentially to the long-wavelength end of the spectrum [3-8].

However, the minimum energy formulation seems to be useful mainly for discussing decaying initial-value situations. Its implications come to seem doubtful in driven, steady-state situations in which the excitation of the system is balanced by dissipation. To take the example again of a particle trapped in a potential well, it will clearly not settle into its minimum energy state if it is continually acted on by some external force. Rather, it will stay on the average at some finite level above its minimum possible energy at a place where its rate of absorption of energy from the external force will be on the average balanced by its rate of frictional energy loss. It cannot go to a "minimum energy" state.

Driven, dissipative conditions are often closer to the realities of high-Reynolds-number or high-Lundquist-number MHD fluids than is the model of initially-excited, isolated decay. This is true both in the laboratory and on the sun. Driving mechanisms can be voltage drops, pressure drops, or mechanical agitation such as thermal convection. The dissipation is resistive and viscous. The question arises as to whether there is a variational principle that is as useful for the driven, steady-state problem as the minimum energy principle is for isolated mechanical systems, particularly for fluids and magnetofluids.

We are not completely sure of the answer, but we think there is. It is the principle of *minimum energy dissipation rate.* It seems to have been first formulated by Kirchoff, who asked the following question.

Given a rigid electrical conductor of fixed non-uniform electrical conductivity $\sigma(\mathbf{x})$ with the electrostatic potential $\Phi(\mathbf{x})$ fixed over a closed surface, how will the electric current density $\mathbf{j}(\mathbf{x})$ distribute itself inside the surface? The equation of continuity relates the electric charge density ρ_e to $\mathbf{j}$:

$$\frac{\partial \rho_e}{\partial t} + \nabla \cdot \mathbf{j} = 0. \tag{1}$$

For either a steady state or a time average, the $\partial \rho_e / \partial t$ drops out and we are left with $\nabla \cdot \mathbf{j} = 0$. However, $\mathbf{j} = -\sigma(\mathbf{x})\nabla\Phi(\mathbf{x})$. Making this substitution, the differential equation

$$\nabla \cdot (\sigma(\mathbf{x})\nabla\Phi(\mathbf{x})) = 0 \tag{2}$$

then determines $\Phi(\mathbf{x})$ and $\mathbf{j}(\mathbf{x})$, given $\Phi(\mathbf{x})$ over the surface. Notice that Eq. (2) is exactly the Euler-Lagrange equation that one would have gotten if one had set out to minimize the total Ohmic energy dissipation rate R_j,

$$R_j \equiv \int \sigma^{-1} \mathbf{j}^2 d^3x = \int \sigma (\nabla\Phi)^2 d^3x, \tag{3}$$

where the integral runs over the volume interior to the surface over which Φ is given.

The answer to Kirchoff's question is that $\mathbf{j}(\mathbf{x})$ arranges itself in such a way as to minimize the energy dissipation rate. Note that the given constrained value of $\Phi(\mathbf{x})$ over the boundary is the constraint that gives individual cases their content. This is essentially a rigorous proof.

Somewhat less rigorous, but still correct, is the application of the minimum dissipation rate principle to viscous fluid flows in hydrodynamics [11] made later by Helmholtz, Korteweg, and Lord Rayleigh. If one minimizes the viscous energy dissipation rate R_ω,

$$R_\omega \equiv \int \mu\omega^2 d^3x = \rho_0 \int \nu\omega^2 d^3x, \tag{4}$$

where $\boldsymbol{\omega} = \nabla\times\mathbf{v}$ is the vorticity of a uniform incompressible fluid, μ is its viscosity, and ρ_0 and ν are its density and kinematic viscosity, one is quickly led to the elementary steady flows. Plane Poiseuille flow, pipe flow, plane Couette flow, rotating Couette flow are all derivable by minimizing R_ω subject to the relevant boundary conditions. (For the case of moving boundaries, R_ω differs from the dissipation rate expressed as an integral involving the rate-of-strain tensor by an integral whose first variation vanishes for trial functions $\delta\mathbf{v}(\mathbf{x})$ which vanish on the boundary, so that minimizing R_ω is still minimizing the energy dissipation rate.) As in the electric current case, it is the boundary constraints that give individual cases their content.

It came to seem natural to us to re-examine driven, dissipative MHD from the point of view of minimum-dissipation-rate principles. In a familiar set of dimensionless variables, the relevant integral to be minimized for incompressible, resistive, viscous MHD is ($\eta = \sigma^{-1}$)

$$R = R_\omega + R_j = \int (\nu\omega^2 + \eta \mathbf{j}^2) d^3x, \tag{5}$$

subject to appropriate constraints. We have been particularly interested [12-15] in the case of voltage-driven cylinders of magnetically-supported MHD fluid with periodically-identified ends (to simulate toroidal periodicity). This situation is one for which extensive numerical computations due to Dahlburg, et al. [16-18] and Theobald, et al. [19] have for some time been accumulating evidence that only in some respects did minimum-energy predictions seem to represent what happened, and that a fundamental revision was in order.

Noteworthy among the previously unexplained features, that could not be squared with the minimum energy predictions, were: (1) a strong spatial dependence of the ratio $|\mathbf{j}|/|\mathbf{B}|$ (here, $\mathbf{B}$ is the magnetic field and $\mathbf{j} = \nabla\times\mathbf{B}$ in the relaxed, approximately force-free state); (2) a persistent helical contribution to $\mathbf{j}$ and $\mathbf{B}$, in addition to the (larger) approximately axisymmetric components; and (3) a persistent kinetic energy of flow, at the level of a few percent, also tending to be helically distributed over space. These results had been around for some time before we found any way to get at them analytically.

The principal difficulty in minimizing the dissipation rate R (or, for that matter, in minimizing the energy, if anyone still feels that that is what he should do) is the role played by the Ohm's law, which for MHD can be written in the equivalent forms

$$\frac{\partial \mathbf{B}}{\partial t} = \nabla\times[\mathbf{v}\times\mathbf{B} - \eta\mathbf{j}] \tag{6}$$

or

$$\mathbf{E} = \mathbf{E}_0 + \nabla\tilde{\Phi} = -\mathbf{v}\times\mathbf{B} + \eta\mathbf{j}. \tag{7}$$

Here, $\mathbf{E}$ is the (dimensionless) electric field, $\mathbf{E}_0$ is its spatial average, $\tilde{\Phi}$ is a scalar potential determined by

$$\nabla^2\tilde{\Phi} = -\nabla\cdot(\mathbf{v}\times\mathbf{B}) \tag{8}$$

if the resistivity η is spatially uniform.

Eq. (6), or (7), is the source of whatever constraints apply to this system beyond simple boundary conditions, including all those involving global magnetic helicity or its rate of supply. But it is much more than that. Since $\tilde{\Phi} = \tilde{\Phi}(\mathbf{v}, \mathbf{B})$ as a consequence of Eq. (8), Eq. (7) is in effect a *pointwise constraint* relating $\mathbf{v}$, $\mathbf{B}$ and $\nabla\times\mathbf{B}$ at every point in space. Its complexity precludes elimination of either $\mathbf{v}$ or $\mathbf{B}$ in favor of the other and substitution into R. Most importantly, Eq. (7) mandates a finite flow velocity $\mathbf{v}$ for all states except the relatively uninteresting one of uniform $\mathbf{j} = \mathbf{E}_0/\eta$. This $\mathbf{v}$ is not the fancy, statistically-distributed small-scale turbulence invoked in the "alpha effect," it is macroscopic and present in the simplest cases. It requires a sort of zeroth-order revision in our ideas of what an MHD equilibrium can be. It may devalue, forever, the ideal, cylindrical equilibria with zero-flow, axisymmetric current distributions $\mathbf{j} = (0,\ j_\varphi(r),\ j_z(r))$ that have been such a preoccupation in much of MHD.

If one takes the hydrodynamic point of view that ideal equilibria are only physically significant if they can be approached as a limit, as the small but finite dissipation coefficients approach zero, then Ohm's law calls into question most of the ideal axisymmetric zero-flow equilibria that have been proposed. It turns out to be rather difficult to construct *resistive* MHD equilibria which obey not only force balance according to the equation of motion, but also flux balance as required by Eqs. (6) and (7), with small but non-zero η. The reason, not at all unfamiliar in hydrodynamics, is that η multiplies the highest spatial derivative of $\mathbf{B}$ in the problem.

It has been possible [13-15] to construct some solutions to the full set of MHD equations which obey Eqs. (6) and (7) as well as the $\partial\mathbf{v}/\partial t$ equation with viscosity. The solutions so far are perturbation-theoretic. The small parameter is the ratio of helical to the axisymmetric parts of $\mathbf{j}$ and $\mathbf{B}$. The solutions go through second order in this ratio. Graphically [13], they look in many respects like the results of the MHD computations [16-19] in their principal features. The agreement is better at not too high values of the pinch ratio Θ, as might be expected. The analytical solutions, like the numerical computations, are lengthy, and it is not our purpose to reproduce them here, but to direct attention to the original papers. Further extensions, including predictions of the amplitudes of the helical components, will be reported elsewhere [20].

Discussion

What are the implications of these considerations for solar physics? Perhaps their immediate impact is not great, since the quantities being revised, such as the aforementioned helical $\mathbf{v}$ and $\mathbf{B}$ contributions, are likely out of reach observationally in the solar context in the immediate future. Perhaps of more value is a reminder of how little of what passes for magnetohydrodynamic theory is really on very solid ground. In all astrophysics, there is a tendency to concentrate on the observations and to regard MHD as a consumer regards an off-the-shelf product. Like all consumers, the observer would like to believe the product he is buying has been thoroughly tested and is reliable. Nobody wants to hear that the propositions of MHD theory, which have been repeated so many times, are shaky ones about which the buyer had better beware. But that is, probably, the case. It did not need to be this way. For a small fraction of the expense of the many observations, a little investment in internally-diagnosable laboratory MHD would have left us far ahead of where we are at present. We would at least know whether or not to expect flows in pinch situations!

Acknowledgements

This work was supported in part by NASA Grant NAG-W-710 and U.S. Department of Energy Grant DE-FG02-85ER53194.

References

1. Taylor, J.B. (1974) Relaxation of toroidal plasmas and generation of reverse magnetic fields, *Phys. Rev. Lett.*, **33**, 1139.
2. Taylor, J.B. (1986) Relaxation and magnetic reconnection in plasmas, *Revs. Mod. Phys.* **58**, 741.
3. Montgomery, D., Turner, L. and Vahala, G. (1978) Three-dimensional MHD turbulence in cylindrical geometry, *Phys. Fluids,* **21**, 757.
4. Matthaeus, W.H. and Montgomery, D. (1980) Selective decay hypothesis at high mechanical and magnetic Reynolds numbers, Proc. Int. Conf. on Nonlinear Dynamics, *Ann. NY Acad. Sci.,* **357**, 203.
5. Riyopoulos, S., Bondeson, A. and Montgomery, D. (1982) Relaxation toward states of minimum energy in a compact torus, *Phys. Fluids,* **25**, 107.
6. Ting, A.C., Matthaeus, W.H. and Montgomery, D. (1986) Turbulent relaxation processes in magnetohydrodynamics, *Phys. Fluids,* **29**, 3261.
7. Dahlburg, J.P., Montgomery, D., Doolen, G.D. and Turner, L. (1987) Turbulent relaxation of a confined magnetofluid to a force-free state, *J. Plasma Phys.,* **37**, 299.
8. Montgomery, D. and Phillips, L. (1989) MHD turbulence, relaxation processes and variational principles, *Physica,* **D37**, 215.
9. Kirchoff, G.D. (1848) in *Ann. Phys.,* **75**, 189.
10. Jaynes, E.T. (1980) The minimum entropy production principle, *Ann. Rev. Phys. Chem.,* **31**, 579.
11. Lamb, H. (1945) *Hydrodynamics*, 6th ed., Dover, New York, pp. 617-619.
12. Montgomery, D. and Phillips, L. (1988) Minimum dissipation rates in magnetohydrodynamics, *Phys. Rev.* **A38**, 2953.
13. Montgomery, D., Phillips, L. and Theobald, M.L. (1989) Helical, dissipative, magnetohydrodynamic states with flow, *Phys. Rev.,* **A40**, 1515.
14. Montgomery, D. (1989) Relaxed states in driven, dissipative magnetohydrodynamics: helical distortions and vortex pairs, to appear in *Trends in Theoretical Physics,* **Vol. I**, ed. by P.J. Ellis and Y.C. Tang, (Addison-Wesley, New York).
15. Montgomery, D. (1990) Minimum dissipation states and vortical flow in MHD, to appear in Proc. 1989 IUTAM Symposium, H.K. Moffatt, Ed. (Cambridge University press, Cambridge, UK).
16. Dahlburg, J.P., Montgomery, D., Doolen, G.D. and Matthaeus, W.H. (1986) Large-scale disruptions in a current-carrying magnetofluid, *J. Plasma Phys.,* **35**, 1.
17. Dahlburg, J.P., Montgomery, D., Doolen, G.D. and Turner, L. (1986) Turbulent relaxation to a force-free field-reversed state, *Phys. Rev. Lett.,* **57**, 428.
18. Dahlburg, J.P., Montgomery, D., Doolen, G.D. and Turner, L. (1988) Driven steady-state RFP computations, *J. Plasma Phys.,* **40**, 39.

19. **Theobald, M.L., Montgomery, D., Doolen, G.D. and Dahlburg, J.P. (1989) Sawtooth oscillations about helical current channels, *Phys. Fluids,* B1 766.**
20. **Montgomery, D., Chen, H. and Shan, X, to be published.**

DISCUSSION

VAN BALLEGOOIJEN: How does the velocity scale with magnetic diffusivity (η)? Is the scaling linear?

MONTGOMERY: We do not really know in any comprehensive sense. There is not any velocity, for example, until the applied electric field E_0 crosses a threshold, E_c. Just above that, the rms velocity seems to go linearly with $\eta^{1/2}$ but it is also multiplied by (E_0-E_c), in the expression we have been able to calculate. E_c itself depends on magnetic diffusivity, though, so the answer does not appear to be too simple.

HEYVAERTS: Has it been possible to recognize any sign of intrinsic stochasticity of field lines in your numerical solutions?

MONTGOMERY: No, not a great deal. We have made only a few attempts to track field lines the requisite large number of times around the torus and make Poincaré-type plots of their interesections with planes of constant z. The results of one such attempt appear in Figs.8 of *Phys.Fluids*B1, 766 (1989), which to me, at least, look pretty regular. I sometimes feel that the question of "where the field lines go" receives too much emphasis in MHD, mostly for historical reasons. If one is concerned with a population of hot electrons trying to escape from a tokamak, then it *is* needed information. But if MHD alone is one's concern, all the same information is available in the various fields ($\underline{b}$,($\underline{x}$,t), $\underline{j}$($\underline{x}$,t), $\underline{v}$($\underline{x}$,t), $\underline{A}$($\underline{x}$,t) *etc.* as functions of space and time. These are what the computer provides, and it is a relatively straightforward operation to ask it to make 3D perspective plots, contour plots, *etc.*, of these various fields. It is a delicate and risky matter to try to follow a turbulent field line, which is perhaps why people seldom try in hydrodynamics. Also, when resistivity is present, there is no unique way to identify a field line from one time to the next. I would like to see us declare a moratorium on anthropomorphic speculations about field line behaviour.

BUTI: The appearance of chaos in a magnetic field would depend on the strength of the driver. It is possible that the driver you have taken is not strong enough and it is possible that you would see the chaos if you take a stronger driver.

MONTGOMERY: I guess by "chaos in the magnetic field" one might mean two things: (i) chaotic Poincaré plots of magnetic field intersections with a plane of constant z at a fixed time; or (ii) chaotic, Lorenz-like, temporal behaviour in the computed expansion coefficients for the magnetic field. Let me remark on these separately.

(i) In *Phys. Fluids*,B**1**,766 (1989) we tried to follow field lines at fixed times and make plots of their successive intersections. In those plots (which refer to the "tokamak" regime), we could see relatively little indication of "chaos". Chaotic patterns may be more likely in the "RFP" regime. It is not easy to design a field-line-tracing algorithm that will follow a field line very many times around a torus reliably, in the presence of genuine 3D turbulence. When one of the components is locked away from zero, as in the Strauss approximation, it is somewhat easier. But the turbulence is also somewhat suppressed, perhaps artificially, by the approximation.

(2) In *J. Plasma Phys.* 35,1 (1986), a very low-order truncation of the dynamical equations in transform space did, when numerically integrated, yield Lorenz-like, "chaotic" behaviour in the phase space. However, as more terms in the expansion were added, and the truncation became less severe, the exotic "strange attractor" behaviour went away.

VENKATAKRISHNAN: How does the kinetic energy of final relaxed state scale with the energy input at the boundaries?

MONTGOMERY: Energy input is something that, unlike magnetic helicity input, is not fixed by the boundary conditions we are imposing. It is up to the plasma, which behaves as an anomalously resistive element due to the $\langle \underline{v} \times \underline{B} \rangle$ opposing the applied electric field, and the amount of energy it accepts is not fixed by prescribing the electric field at the wall. Roughly speaking, the two quantities go up together, but their basic dependence on a prescribable number is better found by asking how both go up as a function of applied electric field at the wall, or "toroidal loop voltage". We plan to give an answer to that for the "tokamak regime" (pinch ratio $\Theta < 1$) in an upcoming manuscript.

KRISHAN: You mentioned that the hydrodynamicists made a mistake by neglecting viscosity. What are the consequencies of including viscosity?

MONTGOMERY: There is not much similarity, in hydrodynamics, between the case of zero viscosity and very small viscosity; it is a kind of discontinuous limit. When the Reynolds' number becomes $\gtrsim$ 1000 or 2000, turbulence sets in in the presence of velocity shear. The instabilities which signal it are intrinsically viscous. Also, since viscosity (like resistivity in MHD) multiplies the highest-order spatial derivatives in the problem, its neglect changes in fundamental ways the number and kind of boundary conditions required for a well-posed problem. There is absolutely no guarantee that, if a zero-viscosity solution can be found, there is a similar solution with small but finite viscosity. More often than not, there is not.

HOLLWEG: (i) What do you think the effects of line-tying boundary conditions will be?

(ii) What is the "Alfvénicity" of the final relaxed state?

MONTGOMERY: (i) This is pretty hard to say. For the "line-tied" case, in which the magnetic field lines intersect the conducting boundary, a lot more needs to be said in order to feel confident that we have a set of boundary conditions that will determine a well-posed problem.

(ii) The ratio of kinetic energy of motion to magnetic energy per unit volume is always small for the case we have mostly considered, where there are rigid, conducting boundaries and the driving is through applied voltages. Typical numbers would be two to four per cent. In the presence of these highly "magnetic" boundary conditions, the kinetic energy gets clamped at a fairly low level, and the tendencies toward equipartition at the larger scales that might be expected for periodic boundary conditions and no imposed dc magnetic field just do not appear.

A CASE FOR ALFVEN WAVE HEATING

F. CALIFANO, C. CHIUDERI and G. EINAUDI
Department of Astronomy and Space Science, University of Florence
Largo Enrico Fermi, 5
50125 Florence, Italy

ABSTRACT:

The resistive dissipation of Alfvén waves in magnetically structured media is examined within the framework of an analytically solvable model in plane geometry. A new class of rapidly oscillations solutions is found, for which the role of resistivity extends to the whole system.

1. INTRODUCTION

Alfvén waves are the inevitable result of the response of a magnetized plasma to relatively slow perturbations. By travelling along field lines Alfvén waves transport energy away from the source region. They might therefore constitute an ideal heating mechanism, if their energy could be damped at the appropriate rate where heat is needed. The dissipation of Alfvén waves is tied to the formation of small spatial scales across the ambient field. In homogeneous fields the only transverse spatial scale is given by the wavelength in the direction normal to the field. To achieve an efficient dissipation, the ratio between perpendicular and parallel wavelengths has to be unacceptably small. Perfectly homogeneous systems, on the other hand, are not likely to be found in nature.

If the system is non-uniform, the propagation and dissipation properties of Alfvén waves may change dramatically, as a large number of studied has shown in recent years (Sedlacek 1971, Tataronis and Grossman 1973, Kappraff and Tataronis 1977, Hasegawa and Uberoi 1982, Heyvaerts and Priest 1983, Dewar and Davies 1984, Pao and Kerner 1985, Mok and Einaudi 1985, Lee and Roberts 1986, Einaudi and Mok 1987, Davila 1987).

Most of these studies model the non homogeneous nature of the problem in slab geometry by assuming that the magnetic field and/or the density depend on a single coordinate. They also adopt asymptotically flat profiles for the above quantities and assume that resistivity can be neglected altogether in the asymptotic regions. The solutions then describe the so-called resonant absorption phenomenon, consisting of an enhanced resistive dissipation localized in a narrow region around the point where the propagating frequency matches the local Alfvén frequency. These models however, do not exhaust the possible solutions of the fourth-order resistive differential equation. Another class of solutions is investigated in the present paper and their relevance to the heating problem discussed.

E. R. Priest and V. Krishan (eds.), Basic Plasma Processes on the Sun, 223–229.

2. BASIC EQUATIONS

We shall work in a resistive MHD perturbative scheme in plane geometry and assume that the equilibrium quantitites depend only on the x coordinate. The equilibrium field is supposed to have a constant direction that defines the z-axis.

We adopt as basic scales: a, the typical scalelength of the equilibrium field, $\rho(\infty)$ and $B(\infty)$, the asymptotic values of the density and magnetic field. All other scales are derived from them.

The absence of structural variations with y and z in the equilibrium allows to write the perturbed quantities as:

$$f(\mathbf{r}, t) = f(x) \exp [i (k_y y + k_z z - \omega t)]$$

The linearized resistive MHD equations can be combined into a single fourth-order differential equation for the x-component of the velocity, w, that reads (Mok and Einaudi 1985):

$$(\varepsilon w')' - k^2 \varepsilon w = - i \omega \frac{(\rho_o c^2{}_A)}{S} \left\{ \frac{1}{(\rho_o c^2{}_A)} [(\rho_o w')' - k^2 \rho_o w] \right\}'' \tag{1}$$

where

$$\varepsilon = \rho_o (k_z^2 c_a^2 - \omega^2 - i \omega k^2 / S)$$

$$k = (k_y^2 + k_z^2)^{1/2} ;$$

$$S = \tau_r / \tau_a = (4 \pi \sigma / c^2) \, a \, c_a(\infty) ,$$

σ being the electrical conductivity and τ_r the resistive timescale. As usual, a subscript "o" indicates equilibrium quantities and a prime the x-derivative.

All the quantities entering the equations must be understood as non-dimensional, each having been normalized with respect to the appropriate scale.

Since B_o and ρ_o are asymptotically constant, in the asymptotic region the coefficients or equation (1) are constant and four independent solutions can be easily found. They read,

$$w_{1,2} \sim e^{\pm k_z x} ; \quad w_{3,4} \sim e^{\pm \lambda x}$$

where

$$\lambda = \left(\frac{i S \varepsilon}{\rho_o \omega} \right)^{1/2} . \tag{2}$$

We thus see that besides the well known "ideal" solutions ($w_{1,2}$), we have a couple of "resistive" solutions. One solution in each pair is well - behaved at $x \to + \infty$, the other at $x \to - \infty$.

The solutions of the asymptotic fourth order resistive equation thus separate naturally in two distinct classes. The asymptotically ideal solutions and their extension to the whole x-axis have been extensively studied in the past years. They give rise to the well-known "resonant"solutions, where resistivity plays a role only in the vicinity of the locations where $\omega = k_z c_a$.

On the other hand, the existence of "resistive" asymptotic solutions implies the possibility of finding complete solutions of equation (1) for which the resistive terms are important everywhere. The physical distinction between the two possible choices of asymptotic solutions becomes more apparent when we move into the non-homogeneous region. In the "ideal" case we essentially insist in keeping the system non-dissipative as long as possible. As a result the dissipation is concentrated in a thin boundary layer. By choosing the "resistive" asymptotic solutions we investigate the possibility of a more widespread influence of the resistivity in the non-uniform region.

Equation (2) shows that the resistive solutions vary asymptotically on a spatial scale proportional to $S^{-1/2}$. We notice that, if the complete solution (i.e. valid for every x) also varies on the same scale, the dominant resistive term of equation (2.1) (~ to w'^{IV}) is always comparable with the dominant ideal term (~ to w''), since each derivative introduces a factor proportional to $(S^{-1/2})$.

Retaining only the dominant terms of equation (1), we finally arrive at the equation that governs the resistive solutions:

$$(k_z^2 c_a^2 - \omega^2)\, w'' = -\frac{i\,\omega}{S}\, w'^{IV} \quad . \tag{3}$$

3. THE MODEL

We assume for $c_a^2(x)$ the only quantity entering equation (3), the following form:

$$c_a^2(x) = 1 + \Delta \mathrm{sech}^2(x) \quad , \tag{4}$$

When $\Delta > 0$, such a profile may represent a configuration where the magnetic field stays essentially constant and the density shows a central depression (as in solar coronal holes) or the cut through a magnetic flux tube where the field intensity is considerably larger than outside. Introducing equation (4) into equation (3) we get:

$$(w'')'' + i\,S\,[\,(\,\omega - \frac{k_z\,\Delta}{\omega}\,) - \frac{k_z^2\,\Delta}{\omega}\,\mathrm{sech}^2(x)\,]\; w'' = 0 \tag{5}$$

Defining:

$$\xi = \mathrm{tgh}\,(x) \quad , \quad \phi(\xi) = w''(x) \quad ,$$

$$\lambda^2 = -\,i\,S(\omega - \frac{k_z^2}{\omega}) \quad , \quad \nu(\nu+1) = -\,\frac{i\,S\,k_z^2\,\Delta}{\omega} \quad . \tag{6}$$

we obtain, from equation (5),

$$(1-\xi^2)\,\frac{d}{d\,\xi}[\,(\,1-\xi^2)\frac{d\phi}{d\xi}] + [\,\nu\,(\,\nu+1) - \frac{\lambda^2}{1-\xi^2}\,]\;\;\phi = 0 \tag{7}$$

It is possible to show that in order to satisfy the boundary conditions: $\phi \to 0$ when $|\xi| \to 1$, ν and λ cannot be arbitrary but must satisfy:

$$\mathrm{Re}\,\lambda > 0 \qquad \text{and} \qquad \nu - \lambda = n \;, \quad n = 0, 1, 2 \ldots \tag{8}$$

The solution of equation (7) can then be conveniently written as:

$$\phi(\xi) = (1-\xi^2)^{\lambda/2}\, C_n^{\lambda+1/2}(\xi) \;, \tag{9}$$

where $C_n^{\lambda+1/2}(\xi)$ is the n-th Gegenbauer polynomial (Magnus et al. 1966).
Recalling the definitions (6) we see that the conditions (8) completely determine the (complex) eigenvalues ω as functions of k_z, Δ and S. General solutions of the dispersion relation, eq. (8), must be found numerically However, since S>>1, approximate solutions can be found by expanding ω in powers of $(k_zS)^{-1/2}$. Retaining only the leading terms we get:

$$\omega = k_z \,\{(1+\Delta)^{1/2} - (1+i)\,(n+1/2)\,(\Delta/2)^{1/2}\,(1+\Delta)^{-1/4}\,(k_zS)^{-1/2}\} \tag{10}$$

It can further be shown that Re $\lambda > 0$ implies the existence of a maximum value of n, n_{max}, that scales as $n_{max} \sim \Delta^{1/2}\,(k_z\,S)^{1/2}$. For a given set of parameters therefore only a finite number of solutions exist.
The results of the numerical evaluation of the dispersion relation are shown in Fig. 1 as solid lines. The open circles give the values predicted by the approximate formula, eq. (10). The agreement turns out to be remarkably good for the physically interesting range $k_zS > 10^6$.

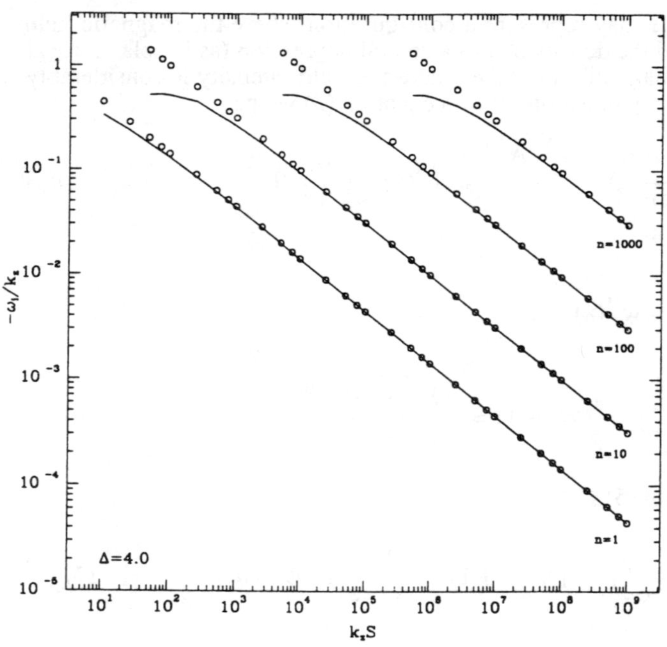

Fig. 1.

The damping rate, ω_i/k_z, as a function of k_zS for different values of n. Solid line: numerical evaluation of eq. (8). Open circles: approximate expression, eq. (10).

Eq. (9) allows us to figure out the spatial structure of the eigenfunctions. The term $(1-\xi^2)^{1/2} \equiv [\text{sech}\,(x)]^\lambda$ is a rapidly oscillating one, that quickly damps out for large values of x. The spatial scale of both the oscillation and the damping are proportional to $\lambda^{-1} \sim S^{-1/2}$, as expected. The polynomial term is a modulating factor that, for large n's becomes increasingly small in an increasingly wide region around x = 0. The combined effects of the two terms for large n's is to confine the oscillations in two symmetrical regions that move away from the origin as n increases. This behaviour is clearly born out by Fig. 2.

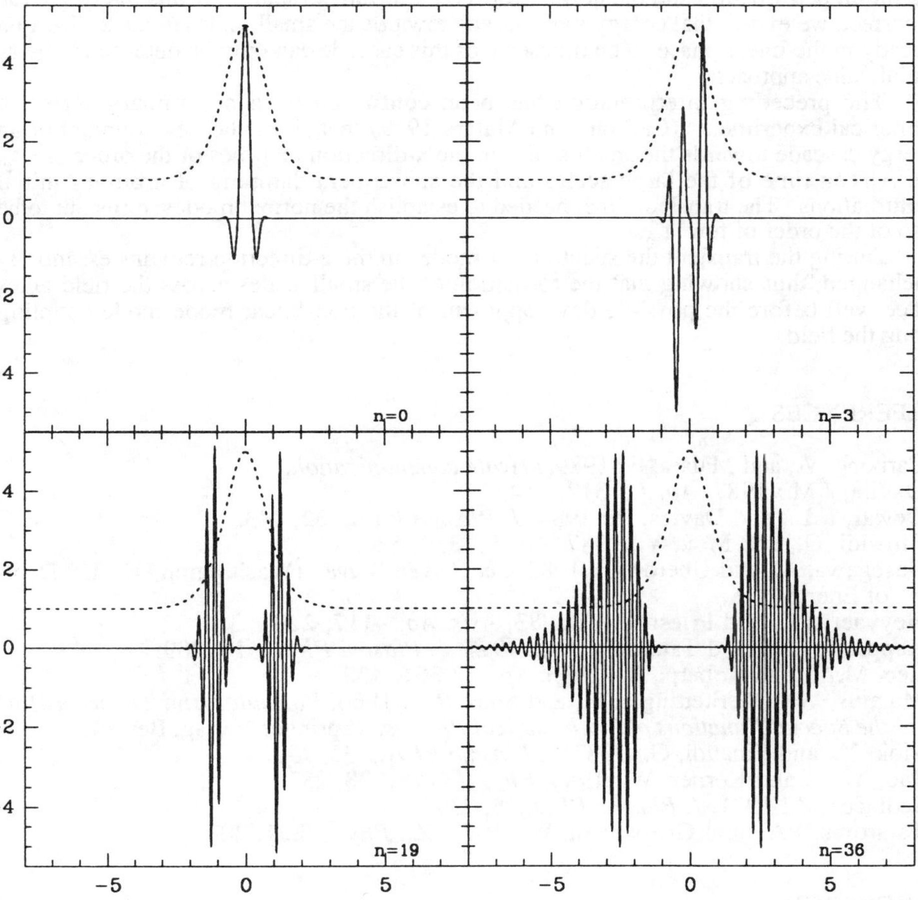

Fig. 2. - The real part of the eigenfunction, eq. (9), for $S = 10^4$, $k_z = 0.1$, $\Delta = 4$. For this case $n_{max} = 38$. The broken line is the profile of c_a^2, eq. (4).

4. CONCLUSIONS

We have shown that consideration of non-ideal asymptotic boundary conditions gives rise to a new type of solutions for the resistively damped Alfvén waves. These solutions are characterized by the appearance of very small scales in the x-direction. Their relevance to the heating problem can be appreciated by noticing that for each k_z there are now n_{max} solutions, whose superposition spans the entire non-uniform region. The spatial structure of the normal modes, previously discussed, makes likely an efficient resistive dissipation. Since the normal modes can be thought as the asymptotic state of the temporal evolution of an initial perturbation of arbitrary scale, we may interpret the appearance of the small scales as the result of the bilinear interaction of the initial perturbation and the non-uniform equilibrium structure. If this interpretation is correct, we expect that energy will cascade towards the small scales in the x-direction already in the linear phase. The timescale of this cascade can only be determined by an initial value approach.

The preceding interpretation has been confirmed by a preliminary series of numerical experiments (Carbone and Malara 1989) that show the development of an energy cascade towards the small scales in the x-direction on times of the order of $\sim\tau_a$, the Alfvén time of the large scales and the subsequent damping of arbitrary initial perturbations. The transient time, needed to establish the normal modes, turns out to be also of the order of few τ_a.

During the transient the spectrum of modes in the z-direction remains essentially unchanged, thus showing that the formation of the small scales across the field takes place well before the possible developpment of the non-linear mode-mode coupling along the field.

REFERENCES

- Carbone, V., and Malara, F.: 1989, *private communication.*
- Davila, J.M.: 1987, *Ap. J.,* **317**, 514.
- Dewar, R.L., and Davies, B.: 1984, *J. Plasma Phys.,* **32**, 443.
- Einaudi, G., and Mok, Y.: 1987, *Ap. J.,* **319**, 520.
- Hasegawa, A., and Uberoi, C.: 1982, *The Alfvén Wave,* (Washington DC: US Dept. of Energy).
- Heyvaerts, J., and Priest, E.R.: 1983, *Astr. Ap.,* **117**, 220.
- Kappraff, J.M., and Tataronis, J.A.: 1977, *J. Plasma Phys.,* **18**, 209.
- Lee, M.A., and Roberts, B.: 1986, *Ap. J.,* **301**, 430.
- Magnus, W., Oberhettinger, F., and Soni, R.J.: 1966, *Formulas and Theorems for the Special Functions of Mathematical Physics,* (Springer-Verlag, Berlin).
- Mok, Y., and Einaudi, G.: 1987, *J. Plasma Phys.,* **33**, 199.
- Pao, Y.-P., and Kerner, W.: 1985, *Phys. Fluids,* **28**, 287.
- Sedlacek, Z.: 1971, *J. Plasma Phys.,* **5**, 239.
- Tataronis, J.A., and Grossmann, W.: 1973, *Zs. Phys.,* **261**, 203.

DISCUSSION

UBEROI: The fourth-order differential equation of MHD resistive effects has been previously considered by many workes who found an absorption rate varying as $S^{1/2}$, similar to your calculations. What is new in your work?

CHIUDERI: There are essentially two new features. Firstly, the choice of "resistive" rather than ideal asymptotic boundary conditions makes the dominant resistive term balance the dominant ideal term everywhere and not simply at the resonant layer. Secondly, we have worked out an analytical solution, showing that, even with a single k_z, instead of one (or few) "resonant" region we have an extended region where large gradients of the physical quantities are present.

BUTI: I believe that your model is restricted to classical resistivity. Turbulent resistivity could be more important. Also one should consider a space-dependent resistivity consistent with the inhomogeneous magnetic field. Hasan has done this in connection with fusion plasmas and we are doing this exercise for the solar corona. Also, Alfvén waves, in the presence of a driver, can become chaotic and lead to very efficient heating.

CHIUDERI: We wanted to prove a point of principle. Therefore we used the simplest possible model. One could add other effects, but at the cost of not being able to find analytical solutions. Anyway, if dissipation works with classical resistivity it will work even better with an anomalous one.

RYUTOVA: Did you estimate the energy density of the power, released in the region of n_{max} where the scale in the x-direction is a minimum? Due to your cascade the heating power must have a very characteristic shape with a pronounced maximum. Calculations similar to yours show that the heating power grows with increasing n and at n_{max} reaches the maximum, and then rapidly decreases. This effect can lead to strong heating and brightening of the region where $n = n_{max}$.

CHIUDERI: So far we wanted simply to invesigate the possible existence of solutions different from the resonant one. A more complete study of the energetics is in progress. I was aware of your results and I am very pleased at the similarity you pointed out.

RUDERMAN: How does the damping rate of Alvén waves depend on the ratio of the magnetic field scale to wave length?

CHIUDERI: There is no "wave length" along x, since the problem is non-homogeneous in that direction. As far as the z-direction is concerned the dependences are those shown in the paper.

DAVILA: If I have understood your presentation correctly, the results you have presented demonstrate the existence of two new non-singular modes which exhibit resistive effects, and these are in addition to the better known singular modes of the ninth order dissipative wave equation.

CHIUDERI: Yes, completely correct.

RECENT ADVANCES IN ACOUSTIC HEATING

P. ULMSCHNEIDER
Institut für Theoretische Astrophysik,
Universität Heidelberg,
Im Neuenheimer Feld 561
D-6900 Heidelberg, Federal Republic of Germany

ABSTRACT. Turbulent surface convection zones of stars generate acoustic waves which contribute to the heating of chromospheres and coronae. The dissipation of limiting strength acoustic shock waves agrees well with the empirically determined chromospheric radiation loss rates. Acoustic waves with frequency and energy required for the chromospheric heating are observed in the solar atmosphere. Acoustic heating can explain the basal chromospheric emission of slowly rotating stars and constitutes a weak background in faster rotating stars; it can not explain the emission-rotation correlation and the surface variation of emission which are due to magnetic heating.

1. Acoustic heating theory and limiting shock strength

For late-type stars the acoustic heating theory works as follows: acoustic waves generated in the surface convection zone run down the steep density gradient of the outer stellar atmosphere, and, due to energy conservation, grow to large amplitude and form shocks. Shock dissipation heats the outer layers to high temperatures. From the turbulent nature of the convection zone it is clear that acoustic waves *must always* be generated. Due to the large density decrease in the outer stellar layers even very small acoustic disturbances will be amplified to sizeable amplitudes higher in the atmosphere. It is thus clear that acoustic heating is always present. The acoustic heating mechanism also works for early-type stars where surface convection zones no longer exist. Here the intense radiation field amplifies small acoustic disturbances until strong acoustic shock waves develop. For recent, more detailed reviews of acoustic heating see Narain and Ulmschneider (1989) and Ulmschneider (1990).

Bohn (1984) has computed the theoretical acoustic energy generation rates in the turbulent stellar surface convection zones. These calculations, which have been criticised recently because the buoyancy forces have not been properly taken into account in the treatment of turbulence, depend on three parameters: effective temperature T_{eff}, gravity g, and the mixing length parameter α. Bohn's results exhibit a strong increase towards higher T_{eff}. This is because the acoustic energy generation depends on a high power of the convective velocity u. In fairly efficient convection zones one has $\sigma T_{eff}^4 \approx \rho u^3$. As the convective velocity of stars is greatest near the surface where the density is lowest, the acoustic energy generation peaks near the F-stars, before the convection zones disappear towards earlier spectral type. After Bohn (1984), the acoustic energy generation also increases when going from dwarfs to giants, because the density in the atmospheres of giant stars is much smaller than in dwarfs, and thus requires larger convective velocities to transport the same total flux σT_{eff}^4.

E. R. Priest and V. Krishan (eds.), Basic Plasma Processes on the Sun, 231–235.

In his thesis Bohn (1981, Univ. Würzburg, Germany, unpublished) has computed the acoustic frequency spectra of main-sequence stars (see Fig. 2 of Ulmschneider 1990). These spectra extend roughly over a period range of $P_A/10 < P < P_A$, where $P_A = 4\pi c_S/(\gamma g)$ is the acoustic cut-off period. γ is the ratio of specific heats, c_S the sound speed. As acoustic waves of shorter periods form shocks and dissipate in the low photosphere, this spectral range should not be much affected by the above mentioned criticism. The maximum of the acoustic spectrum is found near $P_A/10$ and shifts towards $P_A/5$ and longerwards for late-type dwarf stars. Thus acoustic waves typically have wave periods of $P = P_A/5$ to $P_A/10$.

In an acoustic wave calculation in the solar atmosphere there are two effects which severely influence the behaviour of the waves. First, the acoustic wave energy flux is strongly affected by *radiation damping*, when the wave propagates through the radiation damping zone, which in the sun extends to heights of about 200 km, but for other stars can be much more extended (Ulmschneider 1988). Second, it is a persistent result of time-dependent acoustic wave calculations that acoustic shock waves, once formed, tend to quickly reach *limiting shock strength*. This behaviour, where the wave amplitude becomes essentially constant with height, and independent of the initial amplitude, results from the balance of shock dissipation which decreases the wave amplitude, and amplitude growth which is caused by the steep density gradient. In an isothermal atmosphere the limiting shock strength is given by (Ulmschneider 1970, 1989)

$$M_S^{Lim} = 1 + \frac{\gamma g}{4 c_S} P \quad . \tag{1}$$

From wave periods discussed above one finds $M_S^{Lim} = 1.3,\ 1.6$ for $P = P_A/10,\ P_A/5$, respectively. Note that this result is roughly valid for all late-type stars, independent of gravity and T_{eff}. Consequently the velocity-, temperature- and pressure amplitudes of limiting acoustic shock waves are also roughly the same for all late-type stars. With $v \approx 2c_S(M_S^{Lim} - 1)/(\gamma + 1) \approx 0.23 c_S$ one finds $F_M^{Lim} = \rho v^2 c_S/3 \approx 2.4 \cdot 10^4\ p$ for $P = P_A/10$, and $F_M^{Lim} \approx 9.2 \cdot 10^4\ p$ for $P = P_A/5$ ($P \approx 22,\ 44\ s$ for the sun). These limiting acoustic wave fluxes, which are roughly valid for all late-type stars, have been plotted in Fig. 1. In addition Fig. 1 shows the net radiative cooling rate Φ_R and empirical heating flux F_M which were recently determined by Anderson and Athay (1989) using semiempirical chromosphere models. It is seen that there is a relatively good agreement between the empirical heating fluxes and the limiting shock strength results, both showing the same pressure dependence.

2. Solar observations of acoustic waves

The recent extensive work by Endler and Deubner (1983), Deubner (1988) and Deubner et al. (1988) shows that propagating acoustic waves are present in the solar atmosphere. The success of this group is mainly due to the extensive use of Fourier analysis methods in the data handling, to the data acquisition using long time series by which spectral resolution is greatly improved, and due to the powerful method to remove from the data seeing effects caused by the earth's atmosphere. In a typical observation by Deubner et al. a spectrograph with a slit in equatorial direction is used to observe simultaneously narrow spectral bands with e.g. the lines CaII 8542Å, FeI 8496Å at the center of the solar disk. Cross correlating the observations in different lines allows to compute the cross power spectrum $CP(\omega)$ and the phase spectrum $\Delta\phi(\omega)$. By removing the influence of seeing, Endler and Deubner (1983) showed on the basis of their corrected phase spectra that there are propagating acoustic waves in the solar atmosphere with periods as low as 40 s. The authors state that this limit is a detection limit caused by the great width of the line contribution functions relative to the wavelength of the acoustic waves and does not represent a high frequency limit of solar acoustic waves. To derive numerical values of the acoustic flux expressions like

$$F_M \approx \overline{\rho v^2 c_S} \approx \bar{\rho} \int CP_{SCM}(\omega) c_G d\omega \quad , \tag{2}$$

have been used by Deubner et al. (1988). Here $\bar{\rho}$ is a mean density between the two lineforming layers, c_G is the group speed which is essentially the sound speed, and $CP_{SCM}(\omega)$ is the cross power, corrected for the three important effects of seeing, coherence and modulation transfer function (for a more detailed description of these corrections see also Ulmschneider 1990). Deubner (1988) finds for the low and middle photosphere acoustic fluxes of $F_M = 2.0 \cdot 10^7 erg\ cm^{-2}\ s^{-1}$, for the height of NaI 5896Å, $F_M = 1.2 \cdot 10^6 erg\ cm^{-2}\ s^{-1}$ and for the height of CaII 8542Å roughly $F_M = 4.5 \cdot 10^5 erg\ cm^{-2}\ s^{-1}$. He notes that these values tend to be lower bounds because the acoustic flux of short period waves, where seeing overcomes the solar signal, is not included. Taking his estimates of the heights of these lines, 300 km, 800 km and 1500 km, respectively, the acoustic fluxes have been plotted in Fig. 1. In view of the uncertainties these empirical fluxes agree relatively well with the empirical heating fluxes of Anderson and Athay (1989) and with the limiting shock strength results.

3. Acoustic heating and stellar observations

Solar observations indicate that the areas which show the largest chromospheric emission e.g. in the Ca II H+K line cores, are strongly correlated with the magnetic regions at the supergranulation boundaries and with plage regions. Stellar observations (e.g. Rutten 1987) show, that stars can have up to a factor of ten different Ca II emission fluxes which implies a corresponding difference in the magnetic flux coverage. Stars having similar T_{eff} and g are thus shown to have a considerable *chromospheric emission variability.* This variability is attributed to rotation. Stars with high rotation rates are thought to generate much magnetic flux, while stars with low rotation rates generate little magnetic flux and thus little chromospheric emission. Neither the stellar chromospheric variability nor the variation of the chromospheric emission across the solar surface can be explained by acoustic heating which predicts a unique value for the heating flux for a given T_{eff} and g. Here magnetic heating operates (see Narain and Ulmschneider 1989).

An interesting idea particularly by Schrijver (1987) was to separate the chromospheric emission into two components: a *nonmagnetic component* which is independent of rotation and only depends on T_{eff} and possibly slightly on g, and a *magnetic field related component* which depends on rotation. The separation of the two components can be achieved using the correlation of the magnetic component with the magnetically dominated X-ray emission. For late-type stars the heating by the nonmagnetic component leads to a *basal chromospheric emission flux* which may be due to acoustic heating. This basal flux constitutes a low background emission observable in stars of very low rotation rate. For faster rotating stars the acoustic component is usually greatly exceeded by the more energetic magnetic heating component.

The idea, that in the heating of chromospheres two components, a basal nonmagnetic, probably acoustic component and a magnetic, rotation dependent component are at work, can explain several other observations. The low chromospheric variability of the F-stars may be explained as the addition of a strong acoustic heating component to a given variable magnetic heating component. Red giant stars, due to their low rotation rate resulting from angular momentum conservation during the large evolutionary increase in radius and from angular momentum loss by massive stellar winds, are another class of stars, where the two component chromospheric heating theory can be tested. Middelkoop (1982, Fig. 4a-c) showed that the chromospheric emission variability decreases very much toward late spectral type and there becomes a low basal emission. Here only the nonmagnetic acoustic heating component seems to be present.

REFERENCES

Anderson, L.S., Athay, R.G..: 1989, Astrophys. J., in press

Bohn, H.U.: 1984, Astron. Astrophys. **136**, 338

Deubner, F.-L.: 1988, in Pulsation and Mass Loss in Stars, R. Stalio, L.A. Willson Eds., Kluwer Academic Publ., p. 163

Deubner, F.-L., Reichling, M., Langhanki, R.: 1988, in Advances in Helio- and Asteroseismology, IAU Symp. **123**, J. Christensen-Dalsgaard, S. Frandsen Eds., p. 439

Endler, F., Deubner, F.-L.: 1983, Astron. Astrophys. **121**, 291

Middelkoop, F.: 1982, Astron. Astrophys. **113**, 1

Narain, U., Ulmschneider, P.: 1989, Space Science Reviews, in press

Rutten, R.G.M.: 1987, Astron. Astrophys. **177**, 131

Schrijver, C.J.: 1987, in Cool Stars, Stellar Systems and the Sun, J.L. Linsky, R.E. Stencel Eds., Lecture Notes in Physics **291**, Springer, Berlin, Germany, p. 135

Ulmschneider, P.: 1970, Solar Phys. **12**, 403

Ulmschneider, P.: 1988, Astron. Astrophys. **197**, 223

Ulmschneider, P.: 1989, Astron. Astrophys. **222**, 171

Ulmschneider, P.: 1990, in Cool Stars, Stellar Systems and the Sun, G. Wallerstein Ed., to be published

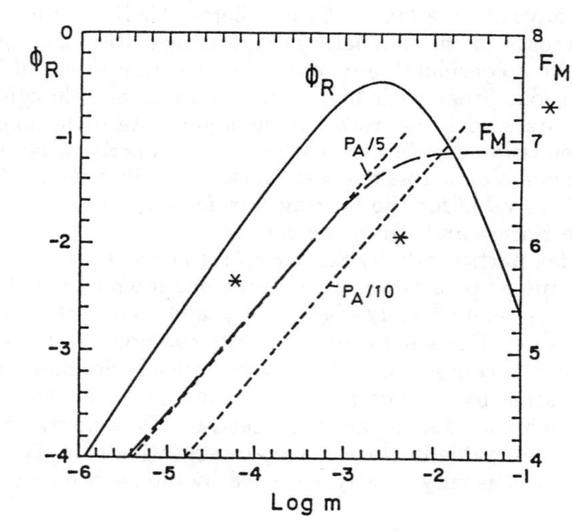

Fig. 1 Net radiative cooling rate Φ_R ($erg\ cm^{-3}\ s^{-1}$) and mechanical Flux F_M ($erg\ cm^{-2}\ s^{-1}$) versus mass column density m (g/cm^2) after Anderson and Athay (1989), together with theoretical limiting strength acoustic fluxes F_M^{LIM}, labeled $P_A/5$, $P_A/10$, and directly observed acoustic wave fluxes by Deubner (1988), labeled by $*$.

DISCUSSION

ROBERTS: Your comments about a limiting shock strength and basal acoustic heating presuppose that the star's magnetic canopy is sufficiently high that the sound wave has time to propagate unaffected by magnetism. In active stars, the magnetic canopy will be very low and sound waves are therefore more likely to reach that height before they have had time to reach limiting strength. Thereafter, the interaction of sound with the magnetic canopy has to be considered.

ULMSCHNEIDER: That is correct. But in these stars acoustic heating is insignificant. They could modify the magnetic heating rate, however.

VENKATAKRISHNAN: Uchida & Bappu have a scenario for production of magnetic fields by interaction of convection with the differential rotation caused by spinning up of the contracting core of G and K giants. Hence they show that surface rotation does not necessarily mean zero magnetic field.

ULMSCHNEIDER: Stars in the red giant region have very different internal structure due to the fact that they are in different evolutionary states. If the dredging up of magnetic fields from the core was important, then one would expect different magnetic fields on the stars in the same region of the HR Diagram. This would show itself as a variability of the chromospheric emission for these stars. Observations show that all these late-type giants have similar chromospheric emission which argues against magnetic fields and thus for an acoustic heating mechanism.

DALSGAARD: You demonstrated that the limiting strength shocks account for the acoustic flux required to heat the chromosphere. But surely a necessary condition must be that the convection zone produces sufficient acoustic energy to reach this level?

ULMSCHNEIDER: This is correct and both theoretical acoustic energy generation calculations and Deubner's observations show that there appears to be enough energy. Limiting shock strength waves have lost their memory about how they were created. If the wave starts with too much energy, dissipation dominates; with too little energy, amplitude growth prevails. Thus when the limiting shock strength is reached in the middle and high chromosphere then the acoustic flux can be predicted independently of the original source.

SIVARAM: (i) Could you elaborate on how the radiation field amplifies acoustic waves in early-type stars?

(ii) What are the relative magnitudes of acoustic and magnetic heating and at what values of the field does one exceed the other?

ULMSCHNEIDER: (i) Radiative amplification is the opposite of radiative damping. If at the point of maximum compression in the wave cycle you add energy (by *absorbing* photons due to an increased opacity, for example) the pressure amplitude will increase and the wave will be amplified. If you lose energy (by *emitting* photons) at that phase you damp the energy.

(ii) That depends on the amount of magnetic flux on the star (or the rotation rate) and on the efficiency of mhd wave energy generation in magnetic flux tubes.

NONLINEAR SURFACE ALFVÉN WAVE PROPAGATION IN SOLAR ATMOSPHERE

M.S.RUDERMAN
Institute for Problems in Mechanics
USSR Academy of Sciences
Prospect Vernadskogo 101
117526 Moscow USSR

The paper deals with the surface wave propagation in the solar atmosphere. The plasma motion is supposed to be described by magnetohydrodynamic equations. In the first part of the paper the surface wave propagation on a single magnetic interface in the solar corona is considered. The plasma is assumed to be cold. Using the reductive perturbation method we derive the equation governing the evolution of nonlinear small-amplitude surface waves as follows

$$\frac{\partial h}{\partial x} - R\left\{\frac{\partial}{\partial t}\left(H\left(h\frac{\partial h}{\partial t}\right) - hH\left(\frac{\partial h}{\partial t}\right)\right) - \frac{\partial h}{\partial t}H\left(\frac{\partial h}{\partial t}\right)\right\} - \frac{\partial^2 h}{\partial t^2} = 0 \quad (1)$$

This equation is written in a coordinate system moving with the phase velocity of linear waves. Dimensionless variables are used. The symbol H denotes the Hilbert transform. The shape of the interface is defined by the equation $z = h(t,x)$ in the Cartesian coordinates x, y, z. The Reynolds number R determines the relative contributions of nonlinearity and viscosity. We take a source radiating a sinusoidal wave. Then the wave evolution is calculated numerically. We get that at large R the wave steepening takes place. This steepening leads to a strong increase of wave damping.

In the second part we study wave propagation on the magnetic structure consisting of three magnetic interfaces parallel each other. The fluid is considered to be incompressible. The viscosity is not taken into account. In the long-wavelength approximation the dispersion equation is approximately as follows $f = ck + ak|k| + bk^3$. There are some singular directions of wave propagation for which $a = 0$. Using the reductive perturbation method we derive for waves propagating in a singular direction the governing equation similar to (1). The solutions of this equation in the form of algebraic solitary waves are found.

E. R. Priest and V. Krishan (eds.), Basic Plasma Processes on the Sun, 237–238.

DISCUSSION

ROBERTS: Have you considered applying your interface studies to the field-free magnetic interface on the penumbra of a sunspot, using appropriate speeds for that region?

RUDERMAN: Considering surface wave propagation on a single magnetic interface, I used the cold-plasma approximation that is convenient for the corona but inappropriate for the penumbra. Therefore my study cannot be applied to the field-free magnetic interface on the penumbra of a sunspot.

BUTI: (i) What is your small parameter ε ?
(ii) Are you getting solitary waves to second order in ε ?

RUDERMAN: (i) The non-dimensional wave amplitude is used as a small parameter.
(ii) No, I get solitary waves only to first order in ε .

HOLLWEG: I am pleased to see the inclusion of coronal viscosity and of course nonlinearity. But if you put in coronal numbers (viscosity coefficient = $10^{-16}T^{5/2}$ cgs and velocity amplitude $\approx$ 30-40 km/s), can you in fact get the waves to dissipate in a few loop lengths?

RUDERMAN: Estimates show that surface wave propagation can be described by the MHD equations provided their periods are larger or of the order of 5s. On the other hand the plasma can be considered to be homogeneous (outside the magnetic interface) provided wave periods are smaller than about 50s. Linear theory shows that the rate of wave damping depends very strongly on the undisturbed plasma parameters and the direction of wave propagation. In particular, there is a small region of these values for which waves with periods from 5 to 50s damp significantly in a few loop lengths. As the nonlinearity increases the rate of wave damping, it expands this region: the larger the nonlinearity, the wider this region of parameters. The observed velocity amplitudes of the order of 30-40 km/s are large enough. In some cases the inclusion of nonlinearity can increase the damping rate by up to 10 times for waves with such amplitudes.

INTERACTION OF FLUX TUBES WITH SOUND WAVES

C UBEROI
Department of Applied Mathematics
Indian Institute of Science
Bangalore 560012
INDIA

ABSTRACT. The Landau damping of sound waves in a plasma consisting of ensemble of magnetic flux tubes is discussed. It is shown that sound waves *cannot* be Landau damped in general but under certain restricted conditions on plasma parameters the possibility of absorption of these waves can exist. The possibility of radiative damping of the acoustic waves along the magnetic filaments is also discussed. It appears that the most plausible mechanism of damping of sound waves in a plasma consisting of magnetic filaments can be due to scattering of a sound wave by the filaments.

In a recent paper [1], from the point of view of looking at the possibility of confining hot plasma by a cold gas blanket Uberoi and Somasundaram [1] discussed the dispersion characteristics of an Alfvén surface wave along a cylindrical plasma column embedded in a neutral gas by discussing the phase velocity of the surface waves as a function of the parameter $\alpha = v_A/s$, where v_A is the Alfvén speed in the plasma column and s is the sound speed of the surrounding neutral medium. In this paper I would like to reinterpret the results obtained from this parametric study of the dispersion equation to discuss some aspects of sound wave interaction with flux tubes in the solar plasma. This work is an extension of the preliminary study given in an earlier paper [2].

Consider a cylindrical plasma column of radius 'a' with mass density ρ_0 and an *internal* magnetic field $\mathbf{B}_0$ surrounded by a compressible gas of mass density ρ_{g0} and magnetic field $\mathbf{B}_{g0}$. Both $\mathbf{B}_0$ and $\mathbf{B}_{g0}$ are directed along the axis of the column. The dispersion relation obtained by solving the linearized MHD equations inside the flux tube, the equations for the outer neutral gas and matching the solutions with the help of boundary conditions at r=a is given as:

$$\frac{v_{ph}^2}{v_A^2}\left[1 + \eta\,\frac{h_m(ka)g_m(\tau a)}{(1-\alpha^2\, v_{ph}^2/v_A^2)}\right] - \beta^2 F_m\,(ka) - 1 = 0, \qquad (1)$$

where

$$h_m(ka) = I_m'(ka)/I_m(ka)$$

$$g_m(\tau a) = -\,K_m(\tau a)/K_m'(\tau a)$$

$$F_m(ka) = -\,I_m'(ka)\,K_m(ka)/K_m'(ka)\,I_m(ka).$$

E. R. Priest and V. Krishan (eds.), Basic Plasma Processes on the Sun, 239–243.

$v_{ph} = \omega/k$, $\alpha = v_A/s$, $\eta = \rho_{g0}/\rho_0$, $\beta = B_{g0}/B_0$ are the interface parameters and $\tau\alpha = \kappa\alpha(1 - v_{ph}^2/s^2)^{1/2}$.

Without going into details, which are discussed in [1], I like to point out the following important results obtained in the parametric study of dispersion equation (1). The dispersion characteristics for $\alpha < 1$ do not show any change from those obtained for the plasma-plasma interface [3]. But, when $\alpha > 1$ many interesting features of the surface wave propagation are obtained. It is seen that $v_{ph} \sim s$ only for values of $\alpha > 1$. So, the acoustic surface waves are generated along the plasma column surrounded by a compressible gas only when the strength of the magnetic field is such that $v_A > s$. As η is increased, i.e. ρ_{g0} increases or the neutral gas pressure increases, the value of α for which $v_{ph} \sim s$ increases. Thus, the critical magnetic field $\mathbf{B}_0$, above which Alfvén surface waves do not exist but acoustic surface waves can be generated increases with increasing gas pressure. In a subsequent paper [4] it was found that for large ka, when curvature effects are negligible, the critical magnetic field for low η say $\eta = 0.1$ or $\rho_{g0} = 0.1\rho_0$ above which Alfvén surface waves do not exist is given by $v_A \sim s$ and for $\rho_{g0} = 0.5\rho_0$ it is given by $v_A \sim 1.6s$. For example if $\rho_0 = 10^{-7}$ gm/cm^3 and s = 15 km/s in the neutral gas it can be shown that for $\eta = 0.1$ which gives $\rho_{g0} = 0.1\rho_0$ and gas pressure $P_{g0} = 0.105 \times 10^2$ Torr, the critical value of the magnetic field above which sound waves are excited is given as B ~ 1680G when $\eta = 0.5$, $P_{g0} = 0.53 \times 10^2$ Torr, B = 2690G. Similarly for $\eta = 0.8$, $P_{g0} = 0.84 \times 10^2$ Torr, B ~ 3030G. Thus, for flux tubes with magnetic fields lying say in the range of 1000-2000 Gauss, for a given value of $\eta = 0.1$ the flux tubes which can support sound surface waves will be in the narrow range 1680-2000G. For higher values of $\eta \geq 0.5$, no resonant tube will exist. I would like to point out here that in a preliminary work where the compressibility of the medium inside the flux tube is taken into account it is found that the value of α for which $v_{ph} \sim s$ increases when v_A/c increases or s decreases.

The effect of curvature makes the Alfvén surface waves dispersive [3]. In order to understand the effect of dispersion due to curvature on the transformation of Alfvén to surface sound waves Uberoi and Somasundaram [1] study the dispersion characteristics of Alfvén surface waves for m = 0 and m = ±1 modes as function of α for different values of the wave number ka. They find that due to the curvature effects for magnetic fields above the critical magnetic field such that $\alpha > 1$, there exists a lower cut-off wave number $k_c = k_c(\alpha,\eta,m,\beta)$, such that Alfvén surface waves do not exist for $k < k_c$. Hence, the surface sound waves for a cylindrical plasma will be restricted to the wavelength region $0 < \lambda < \lambda_c$.

It was also seen that the value of k_c increases when the magnetic field above the critical value, is increased but, decreases with increasing η. Therefore, for large gas pressures k_c can be made small, i.e. sound waves will be of long wavelengths. For the axisymmetric modes, the value of k_c is large compared to that of m=0 modes for any given value of the interface parameters.

In the solar or stellar convection zone and chromospheric atmosphere one meets with the situation when an ensemble of magnetic flux tubes concentrated in the plasma are embedded in the neutral medium lying far from one another. Sound wave propagation in the system consisting of 'magnetic filaments' was first considered by Ryutov and Ryutova [5] and more recently by Bogdan and Zweibel [6] and Bogdan [7]. Ryutov and Ryutova

[5] by drawing the analogy of such a system to a liquid with gas bubbles, study on the basis of the magnetohydrodynamic equations the long wavelength sound oscillations of such a system. They note a very interesting property of this system: for long wave lengths $\lambda >> i$, where i is the average distance between the tubes, even if there are no dissipative processes, the sound oscillations are absorbed because of an effect that is in a certain sense analogous to Landau damping in a rarefied plasma and which consists of the resonant excitation of flexure waves that propagate along the magnetic filaments. The Landau damping of sound oscillations arises due to resonance interaction of sound oscillations, of the flux tubes. They find that the Landau damping rate $\nu_i \sim kfs$, where f is the total fraction of the volume occupied by flux tubes. As they consider the tubes with radius a lying far from one another $f \approx a^2/i^2 << 1$. They also calculate the radiation damping of the surface oscillations of the flux tubes and find the damping rate $\nu_{rad} \sim k^2a^2$. The damping rate due to scattering is also calculated as $\nu_{scatt} \sim ks(ka)^4/f$. Comparing the damping rates ν_i, ν_{rad} and ν_{scatt}, it is easily seen that for long wavelengths $\lambda >> i$, $\nu_i > \nu_{rad}$ and ν_{scatt} so the important source by which sound wave energy can be absorbed by the plasma system is the Landau damping of the sound waves arising due to resonant excitation of surface waves along the flux tubes. The energy of the sound waves thus gets transferred to the energy of the filament oscillations over a period of time of order ν_i^{-1}; over a considerably longer time of the order of $1/\nu_{rad}$ the filaments give off their energy through secondary sound waves.

Using the results obtained for the surface wave propagation along a single magnetic flux tube as reported above we show that the possibility of Landau damping of the acoustic waves can exist only under restricted conditions on plasma parameters. For a cylindrical plasma column surrounded by a neutral gas it is seen that acoustic waves along the magnetic filaments will exist when magnetic fields are above the critical magnetic field only for wavelengths $\lambda < \lambda_c$. Therefore only if $\lambda_c >> i$, is the Landau damping of acoustic waves possible. The value of λ_c depends on various plasma parameters. As shown by Uberoi and Somasundaram [1], the value of k_c can be made very small, i.e. sound waves will be of long wavelengths, for large values of the neutral gas pressure only but then the value of the critical magnetic field above which flux tubes will be of resonating class also increases with higher values of the gas pressure thus restricting the general generation of the sound waves. The possibility of Landau damping and absorption of sound wave energy by this mechanism thus is very much restricted and cannot exist in general as discussed by Ryutov and Ryutova (1976).

In the opposite limit, when $ka << 1 << ki$, both ν_i and ν_{rad} are very small, so damping of the sound waves is mainly determined by the scattering. The resonant scattering again will depend on the critical magnetic field. The energy transfer of acoustic waves to a plasma embedded with flux tubes thus depends very critically on the strength of the magnetic field of the flux tubes and the values of the interface parameters.

References

1. Uberoi, C and Somasundaram, K (1982), 'Alfvén Surface Waves along a cylindrical plasma column surrounded by a neutral gas', *Plasma Physics*, **24**, 465-474.
2. Uberoi, C (1984) 'On the Landau damping of sound waves in a plasma physics', **26**, 1037.

3. Uberoi, C and Somasundaram, K (1982), 'Acoustic surface wave generation along a cylindrical plasma column surrounded by a compressible gas', *Phys. Rev. A.*, **25**, 2414-2416.
4. Uberoi, C and Somasundaram, K (1980), 'Alfvén Surface Waves along cylindrical plasma columns', *Plasma Physics*, **22**, 747-752.
5. Ryutov, D D and Ryutova, M P (1976) 'Sound oscillations in a plasma with "magnetic filaments"', *Sov. Phys. JETP*. **43**, 491-497.
6. Bogdan, T J and Zweibel, E G (1987), 'Propagation of compressible waves through fibril magnetic fields', *Ap. J.*, **312**, 444-456.
7. Bogdan, T J (1989) 'On the resonance scattering of sound by slender magnetic flux tubes', *Ap. J.* **345**, 1042-1049.

DISCUSSION

FORBES: I thought you were going to propose that resonant absorption would affect reconnection by heating the plasma and so reducing the electrical conductivity. However, you seem to be proposing that the concentration of surface waves in a singular layer will drive the tearing-mode.

UBEROI: Yes, the final picture of my analysis is that the surface waves can drive magnetic reconnection across the neutral layer by inducing the tearing mode instability. But this kind of reconnection is associated with the Alfvén resonance process. The higher frequencies of the excited spectrum of surface waves can give rise to resonant coupling, whereas the low frequencies produce reconnection.

PRIEST: How does your analysis differ from that of Sakai *et al* (*Solar Phys* **91**,103,1984), who showed (in connection with solar flares) that tearing mode instability could be driven faster than normal by waves incident on a current sheet?

UBEROI: I have not seen Sakai's paper. In my work I model surface-wave-driven magnetic reconnection by using the analysis of Alfvén waves propagating in a system with resonant layers.

HOLLWEG: (i) In my description of resonance absorption, I consider driven harmonic oscillator equations. Are your "well-behaved" solutions just the homogeneous solutions to my oscillator equations?
(ii) Is your temporal increase of z-current density simply due to phase-mixing?

UBEROI: (i) I appreciate your physical description of resonance absorption, though I am not sure the harmonic oscillator equation is equivalent to the Alfvén wave equation.
(ii) The temporal increase of z-current density is not due to phase-mixing. This increase is for time intervals $0 < t < t^*$, and it is only for $t > t^*$ that phase-mixing begins.

ROBERTS: Why did you say that your solution holds only for a finite amount of time t? That limitation was not clear to me. I would have expected your solution to hold for all time, even if some quantities grow secularly, provided one took a sufficiently small amplitude for linear theory to be valid.

UBEROI: The well-behaved solutions only hold good in an interval $0 < t < t^*$, where t^* is calculated from the phase-mixing time. Even if the solutions are of small amplitude and obtained by linear theory, the effect of the continuous spectrum is to give rise to phase mixing for $t > t^*$. Once the phase mixing starts there is no meaningful well-behaved solution for the Alfvén wave equation.

RYUTOVA: The paper [5] mentioned by C. Uberoi was apparently the first to consider sound wave propagation in a plasma containing an ensemble of randomly distributed magnetic flux tubes. The problem of flexure oscillations of a single flux tube is also considered there, including radiative damping and resonance scattering. Uberoi's paper represents an important step in generalising this work. We found in agreement with C. Uberoi that for short wavelength acoustic waves their damping is determined by the resonant scattering. Moreover, the nonlinear estimates were made for the maximum energy which can be transferred by a magnetic flux tube due to the interaction (resonance absorption or scattering) with a finite-amplitude sound wave.

Resonant absorption of Alfven Waves and the Associated Phenomenon of Magnetic Reconnection.

C. Uberoi
[Department of Applied Mathematics, Indian Institute of Science, Bangalore 560 012].

STRACT.The mathematical analysis of the Alfven Wave equation in homogeneous magnetic fields which explain the resonance sorption of Alfven surface waves near a resonant layer can also used to show that magnetic reconnection process can arise near e zero frequency resonant layer driven by very low frequency fven surface waves. The associated phenomena of resonant sorption and magnetic reconnection in inhomogeneous plasmas can plain the recent obsrevations of intense magnetic activity in e long period geomagnetic micropulsations range, at gnetospheric cusp latitudes, during the time of occurence of ux transfer events.

The Alfven wave equation governing the dynamics of the eal MHD system in the presence of inhomogeneous magnetic elds with variation in the direction perpendicular to the ane of the magnetic field, shows a singular behaviour at the int where the local Alfven speed becomes equal to the phase eed of the wave. Extensive analysis of this equation shows at due to this singularity the system can absorb energy when is driven externally if there are well defined scontinuities in the Alfven speed profile then the energy sorption is due to the resonant interaction between the shear fven and surface Alfven Waves. In the latter case there is a llisonless damping of the surface waves. When Alfven speed ofile passes through a zero value, the singularity occurs at e zero-frequency and the resonant surface is now the neutral rface. The aim of this note is to show that the analysis of

. Priest and V. Krishan (eds.), Basic Plasma Processes on the Sun, 245–249.

the Alfven wave equation carried out to understand the resona[nt] phenomena of Alfven waves can be used to show that very low frequency surface waves can drive magnetic reconnection acros[s] the neutral layer.

Considering an incompressible medium with density ρ and [t]magnetic field

$$B_o(x) = [o,\ B_{o1}(x),\ B_{o2}(x)]\ , \qquad (1)$$

the linearized ideal MHD equations give the Alfven wave equati[on]

$$\nabla \cdot \left\{ \rho_o \left[\frac{\partial^2}{\partial t^2} - \frac{(B_o \cdot k)^2}{4\pi\rho_o} \right] \nabla u_{xk} \right\} = 0 \qquad (2)$$

where $u_{xk}(x,k,t)$ is the Fourier transform in y and z of t[he] perturbed velocity component $v_x(x,y,z,t)$. The wave numb[er] $k=(o,k_y,k_z)$.

Recently it was shown[1] that for a given set of initi[al] conditions Eqn.(2) has a well behaved harmonic solution in t[he] closed form valid only for short intervals of time. Th[e] solution is not valid for arbitrary large times as t[he] calculation of current densities J(x) contain terms of order and therefore become very large as $t \rightarrow \infty$. To be more specif[ic] consider a linear variation of the initial magnetic field:

$$B_o(x) = \frac{B_{o1}x}{a}\hat{y}\ , \qquad a \lesssim x \lesssim a\ ,$$

$$= \text{constant}\ , \qquad |x| > a\ . \qquad (3)$$

The current density component $J_z(x)$ is given as

$$J_z(x) = -\frac{1}{4\pi}\left[\frac{B_{o1}x}{a}\frac{\partial^2 \xi_{xk}}{\partial x^2} + \frac{2B_{o1}}{a}\frac{\partial \xi_{xk}}{\partial x} - k^2 \frac{B_{o1}x}{a}\xi_{xk} \right], \qquad (4)$$

where the displacement vector $\frac{\partial \xi_{xk}}{\partial t} = u_{xk}$, calculated from the solution of (2) given in Ref.1, is given as

$$\frac{\partial \xi_{xk}}{\partial x} = A(x,k,o)\frac{\cos[\omega_A(x)t]}{[\omega_A(x)]^2} + B(x,k,o)\frac{\sin[\omega_A(x)t]}{\omega_A(x)}\ , \qquad (5)$$

re A and B are constants determined by the initial conditions $\omega_A(x)$ is the Alfven wave frequency. From (4) and (5) we that $\partial^2\xi/\partial x^2$ will contain a term of order t which makes x) assume large values for increas-ing values of t.

sidering the values of $J_z(x)$ near theneutral layer we e $\frac{\partial}{\partial x}$ >>k and for simplicity choosing A(z,k,o)=0, Eqn.(4) es

$$J_z(o) = B(x,k,o)\,\frac{B_{o1}}{a}\,t\ . \qquad (6)$$

ch increases with time t.

The growth of solutions like (4) and (6) can be understood m the results obtained by the normal mode analysis[2] of the .(2), which predicts only singular modes and an entirely tinuum spectrum. No dispersion relation exists. Thus we can clude that current density increases at every point in space the region in which magnetic field varies continuously luding the neutral point (which forms the end point of the tinuum spectrum) as the disturbance cannot be carried away by Alfven waves due to lack of dispersion. It is interesting to nt out here that Dungey [3], in his historical paper on netic reconnection explains the accumulation of energy near a tral point but not elsewhere in the following way: "If there a small disturbance in a region of non-zero magnetic field current density will not become large in this region, but disturbance will spread in the form of Alfven waves, if, ever, there is a neutral point anywhere a small disturbance cause the current density to become large in the ghbourhood of the neutral point". Here we see that the rent density increases in the neighbhourhood of every point space not just at the neutral point as due to lack of persion the Alfven wave cannot propagate away from any point space in the region in which the magnetic field varies tinuously. The normal mode analysis also shows that the gularity of the Eqn.(2) at $\omega=\omega_A(x)$ implies a process of orption of energy by the plasma when it is driven externally steady state at frequency ω. The Laplace transform method to understand the initial value problem of the Eqn.(2), ever, shows that if magnetic field profile has finite number of discontinuities, then the energy is absorbed from the Alf surface waves propagating along the discontinuities. At neutral surface the energy is derived from the zero freque surface waves. The surface waves, which form the o collective modes of colisionless system are damped away. T can be noted from the large time asymptotic solution given Ref.2.

It is seen that the physical processes near the reson and neutral points do not show any different behaviour for ideal MHD system with magnetic fields of type (1). To furt understand the dissipation of the accumulated energy near th prints, we take finite resistivity into account. The Alf wave equation is now a fourth order differential equation w no singular point. The singularity is removed by the fin resistivity term and a resistive boundary layer forms around singular point. The time behaviour of this eqn. shows that for $t<t_h$ the current density shows a linear increase as in ideal MHD case but for $t>t_h$ the solution becomes oscillato where

$$t_h=\left(\frac{24}{\omega_A[x_o]^2\eta}\right)^{1/2}\times\left[\frac{2B'_o(x_o)}{B}-\frac{\rho'(x_o)}{\rho}\right]^{-2/3}. \qquad (7)$$

For linear variation $t_h=2^{-1/3}\tau_A^{2/3}\tau_R^{1/3}$, $\tau_A=a/v_A$ and $\tau_R=4\pi a^2/\eta$. The analysis of the fourth order differential equation ho good at the neutral point as this equation remains unchanged shifting x to x_o+x.

Considering the mechanism of dissipation of accumulated energy in the resistive layer it was shown[5] t near the Alfven resonant point the surface wave resonan interacts with the shear Alfven wave modified by the fin conductivity effects. The energy of the surface w transformed to the damped Alfven wave thus gets dissipated the collisional effects.

The surface wave energy in the neighbourhood of the neut point, however, cannot propagate away as the damped Alfven wa The nonoscillating type of surface purterbations (as the surf e frequency is nearly zero) can give rise to tearing mode stability and the reconnection of the magnetic lines across neutral layer.

We suggest that the associated phenomena of resonant sorption and magnetic reconnection can explain the recent servations [6] of intense magnetic activity in the long period magnetic micropulsations range, at cusp latitudes, during the ne of occurrence of flux transfer events. The k-H instability the magnetopause generates a spectrum of Alfven surface ves. The higher frequency surface waves are associated with excitation of micropulsation activity [7]but the low quency surface waves can give rise to the magnetic connection process, which is associated with the flux transfer nts at the magnetopause.

References

Uberoi, C. 1988, The Alfven Wave Equation for Magnetospheric Plasmas J. Geophys. Res. 93, 295.

Hasegawa, A. and Uberoi, C. 1982 The Alfven Wave, Technical Information Center, U.S. Department of Energy.

Dungey, J.W., Conditions for the Occurrence of Electrical Discharges in Astrophysical Systems, Phil. Mag., 44, 725, 1953.

Kappraff, J.M. and Tataronis, T.A. Resistive effects on Alfven Wave heating, J. Plasma Physics 18, 209.

Uberoi, C. 1977 Kiev A Note on Resonant Absorption of Alfven Waves in Nonuniform Plasma, Conference on Plasma Physics, Trieste.

Lonzerotti, L.J. and Maclennan, C.G. 1988, Hydromagnetic waves associated with possible flux transfer events, Astrophys. Space Sci. 144, 279.

Chen, L. and Hasegawa, A. 1974, A theory of long-period magnetic pulsations: 2. Impulse excitation of surface eigenmode, J. Geophys. Res. 79, 1033.

FABRY-PEROT INTERFEROGRAM PROFILES IN λ5303 IN RELATION TO CORONAL STRUCTURES : 1980 AND 1983 ECLIPSES

J.N.Desai, K.P.Raju, T.Chandrasekhar, N.M.Ashok
Physical Research Laboratory
Ahmedabad 380 009, India
and
J.M.Pasachoff
Williams College, Hopkins Observatory
Williamstown, MA 01267, USA

ABSTRACT. Fringe profiles of Fabry-Perot interferograms taken in the λ 5303 line during the total solar eclipses of 1980 and 1983 are studied in relation to coronal structures observed in white-light pictures. Evidence for doppler-shifted discrete components and large random velocity components in specific regions is presented here.

During the total solar eclipses of 1980 (INDIA) and 1983 (INDONESIA) Fabry-Perot interferograms were taken in the emission line λ 5303 using an etalon of FSR 4.69 A and a finesse ~ 15. The spatio-spectral information provided by the fringe profiles was compared with corresponding coronal structures as seen in white light, after careful orientation matching. Some interesting features that emerge are reported here.

i) Fringes observed in the 1980 eclipse show profiles in which the existence of doppler-shifted components are apparent, the largest observed being ~ 70 km/sec at a position angle of ~ 285° [Chandrasekhar et al 1981]. Such features are almost entirely missing in the 1983 eclipse [Fig. 1(a) & (b)].

ii) Interfringe spacings were carefully measured at many locations to observe dispersion velocities, the accuracies attained being ~ 6 km/sec. We find that the fringe peaks do not show velocity dispersion within this error anywhere. However, discrete doppler components with velocity shifts ranging up to 50-70 km/sec are seen in many regions. This suggests that doppler-shifted components are associated with discrete features in the coronal plasma.

iii) Interferograms for 1980 show fringes with low contrast (~ 5-8) at many locations, in comparison with the expected value of ~ 20 for a 2.5×10^6 Kelvin gaussian line for this etalon. However, in specific locations (Fig.2) the contrasts do attain high

E. R. Priest and V. Krishan (eds.), Basic Plasma Processes on the Sun, 251–252.

values close to those expected. For the 1983 eclipse, a good interferogram was obtained in a limited region (Chandrasekhar et al 1984) at a position angle $\sim$260°, where the contrast was high.

A comparison with white-light pictures show that the low values of contrast cannot be explained in terms of the contribution of the continuum; but they could be due to the existence of large random velocities ($\sim$70 km/sec) in these regions. It is significant that the profiles in these regions are usually quite broad at the base.

REFERENCES

Chandrasekhar T., Desai J.N. and Angreji P.D. (1981) Appld. Opt. 20, 2172.

Chandrasekhar T., Ashok N.M., Desai J.N., Pasachoff J.M. and Sivaraman K.R. (1984) Appld. Opt. 23, 508.

Loucif M.L. and Koutchmy S. (1989) Astron. Astroph. Suppl. 77, 45.

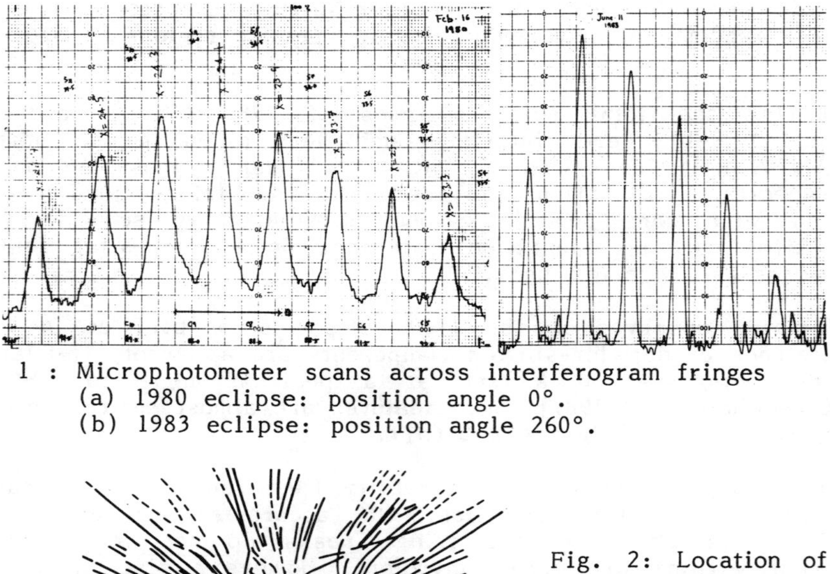

Fig. 1 : Microphotometer scans across interferogram fringes
(a) 1980 eclipse: position angle 0°.
(b) 1983 eclipse: position angle 260°.

Fig. 2: Location of some high and low fringe contrast regions on the map of coronal structures for 16 Feb. 1980 eclipse. Structure map from Loucif M.L. and Koutchmy S. (1989).

DENSITY IRREGULARITY OF THE INNER CORONA DETERMINED FROM SIMULTANEOUS MEASUREMENTS OF THE XUV AND THE K CORONAL BRIGHTNESS

M. GUHATHAKURTA
G.J. ROTTMANN
Laboratory for Atmospheric and Space Physics
U. of Colorado, Campus Box 392
Boulder, CO, 80309-0392

F.Q. ORRALL
Institute for Astronomy
University of Hawaii
2680 Woodlawn Drive
Honolulu, HI 96822

R.R. FISHER
High Altitude Observatory
National Center for Atmospheric Research
PO Box 3000
Boulder, CO 80307-3000

ABSTRACT. In this paper we report preliminary results from a study of the inner corona based on the direct comparison of XUV resonance emission line λ174.53 FeX with that of the white-light emission from the K corona. The data sets were obtained 17/18th of March, 1988, during a total solar eclipse of the sun and consists of co-spatial and co-temporal measurements of these two quantities as a function of position angle and height above the solar limb. The local emission of a coronal resonance line is proportional to the electron density squared, the chemical abundance, and the electron kinetic temperature, while the white-light emission (arising from Thomson scattering by electrons) depends directly on the electron density and the local radiation field. Taken together these measurements yield an estimate of the quantity $(\overline{n^2})/(\overline{n})^2$ where n is the electron density. This quantity, called "X the coronal irregularity factor" by C.W. Allen, is found to be >1.

Several studies of the density irregularity factor in the low solar corona have been conducted and the subject remains one of active inquiry. The most recent study of the irregularity has been carried out by Orrall et al. (1990) who used co-spatial and co-temporal observations of the intensity of λ625 MgX and pB (polarized brightness) of the K corona as a function of height above the limb to derive X for an active and a quiet region on the sun. In the present study we use co-spatial and co-temporal observations of the intensity I of λ174.53 FeX line and K-coronal brightness B as a function of height above the limb for two positions in the north polar region where the corona appeared relatively free of structures. A description of the XUV imager instrument and its photometric calibration has been given by Guhathakurta, M. (1989) and Rottman, G.J. (1989). The white-light coronal camera with a CCD detector and its calibration has been described by Fisher, R.R. (1989).

A model corona that is spherically symmetric, isothermal, and in hydrostatic equilibrium is used to estimate the XUV emission line intensity and the white-light emission brightness of the corona as a function of height above the limb. We use the same notation and approximations as Guhathakurta, M. (1989), and Orrall et al.(1990). Since XUV intensity, I, varies as $(\overline{n^2})$ and white-light brightness, B, varies as $(\overline{n})^2$ we can write the equation for X as follows:

$$X \equiv \frac{(\overline{n^2})}{(\overline{n})^2} = C\,\frac{I'(x)}{B'^2(x)}\,f(T),\quad I'(x) = I(x)x^{-3/2},\quad B'(x) = \frac{B(x)x^{-3/2}}{[A(x)+B(x)]}$$

where C is a constant, I(x) and B(x) are the intensities at a height x above the solar limb in the plane of the sky and intersecting the line of sight and f(T) is a slowly varying function of T.

In Table 1, we have summarized the results of the analysis for two north polar regions. We find a good correlation between B(x) and I(x) (Figure 1) while Figure 2 presents the theoretical fit to observed profile of B(x) for region I.

We find the irregularity to be greater than 1 and increasing with height for the two regions analyzed. In addition, the scale height temperature estimated from XUV observations is higher than the correspond-

E. R. Priest and V. Krishan (eds.), Basic Plasma Processes on the Sun, 253–254.

ing white-light scale height temperature. An increase in X with height can account for this behavior. We have specifically chosen two regions that seemed relatively free of structures and we still obtained X > 1. Thus the coronal irregularity at least in this region seems to be a feature of the inner corona. The detailed analysis of these data are in progress and results will be published elsewhere.

TABLE I. X inferred from observations of the North polar region

Position Angle	n=1.1	x=1.2	$\frac{x(1.2)}{x(1.1)}$	$T_{white\ light}$ in 10^6 K	T_{XUV} in 10^6 K
I. 356-0-5	3.5±3.7	4.3±4.5	1.23	1.23	1.42
II. 337-346	1.32±1.7	2.35±1.4	1.78	1.08	1.40

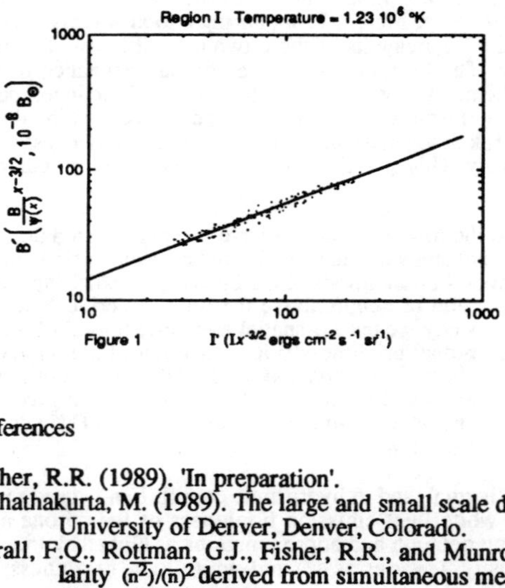

Figure 1

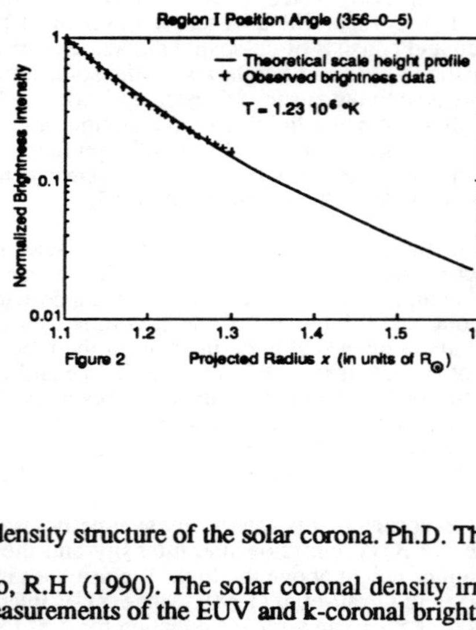

Figure 2

References

Fisher, R.R. (1989). 'In preparation'.
Guhathakurta, M. (1989). The large and small scale density structure of the solar corona. Ph.D. Thesis, University of Denver, Denver, Colorado.
Orrall, F.Q., Rottman, G.J., Fisher, R.R., and Munro, R.H. (1990). The solar coronal density irregularity $\langle n^2\rangle/\langle n\rangle^2$ derived from simultaneous measurements of the EUV and k-coronal brightness. *Ap.J.*, **349**, in press.
Rottman, G.J. (1989). 'In preparation'.

SIMULTANEOUS ORGANIZATION OF (V,B) : THE SPICULES

V.Krishan
Indian Institute of Astrophysics
Bangalore 560034, India

Formation of large structures in a nonlinear turbulent energetically open magnetohydrodynamic system has been discussed using the concepts of self-organization processes Krishan and Mogilevskij (this proceedings). The presence of magnetic field allows the existence of long lived Alfven waves which can be easily excited by any external disturbance. The Alfven waves generate a state of aligned velocity and magnetic fields introducing large cross helicity in the system. Assuming that (V,B) are related as $\vec{V} = (1/4\pi\rho)^{1/2}\vec{B}$, the equation of motion becomes $\rho\,\partial/\partial t(\vec{\nabla}\times\vec{V}) = -(\vec{B}_0\cdot\vec{\nabla})c/4\pi\,(\vec{\nabla}\times\vec{B})$ where $\vec{B}_0$ is the ambient field and ρ is the mass density. In a compressible medium, the Alfven waves are known to decay and the spectrum condensates towards lower frequencies and hence towards smaller parallel wavenumbers K_z such that $K_\perp >> K_z$ developing large anisotropy in the process, Hasegawa (1985) until $(B_\perp/B_{0z}) \approx K_z/K_\perp$. This gives rise to long (in the z direction) and narrow (in $\perp$ direction) MHD structures which could be identified with the spicules since they are believed to be associated with the transport of energy by Alfven waves as suggested by Osterbrock (1961). Using $V_\perp \sim 0.3$ km/sec, the velocity associated with supergranules, $\rho = 8 \times 10^{-11}$ gm/c.c. at a height of 1000 km above the photosphere, and $B_{0z} = 10$ Gauss one finds $\frac{K_z}{K_\perp} = \frac{L_\perp}{L_z} = \frac{\sqrt{4\pi\rho}\,V_\perp}{B_{0z}} \simeq 0.1$ which agrees with the dimension of spicules, where $L_\perp$ and L_z are the horizontal and the vertical extents of spicules. The aligned turbulence is further known to evolve to a state of dominant polarity with either $\vec{V}\cdot\vec{B} > 0$ or $\vec{V}\cdot\vec{B} < 0$, depending upon the initial conditions. Thus anisotropy seems to be the crucial characteristic of turbulent media.

References

Hasegawa,A., 1985. Advances in Physics, 34, 1.
Krishan,V. and Mogilevsij,E.I.,1989, This proceedings.
Osterbrock,D. 1961. Ap. J. 134, 347.

E. R. Priest and V. Krishan (eds.), Basic Plasma Processes on the Sun, 255.

MAGNETIC HELICITY OF OSCILLATING CORONAL LOOPS

V.Krishan[1] and E.R.Priest[2]

[1] Indian Institute of Astrophysics, Bangalore 560034, India

[2] Univ. of St. Andrews, Mathematical Institute, St.Andrews KY 169 SS, Fife, Scotland

The dynamics of the velocity and magnetic field in a coronal loop is studied using ideal MHD equations and the Chandrasekhar-Kendall representation. The complete dynamics is described by a set of infinite, coupled nonlinear ordinary differential equations which are first-order in time for the expansion coefficients of the velocity and magnetic field. Here, the coronal loop plasma is represented by a superposition of the three [n = 0 = m; n = m = 1 and n = m = 1] lowest-order C-K functions. This system, when perturbed linearly from its equilibrium state exhibits sinusoidal oscillations. The frequency S of these oscillations is given by (Krishan et al 1988):

$$S^2 = A\eta_a^2 + B\eta_b^2 + C\eta_c^2 \tag{1}$$

where A, B and C are constants and η's are the equilibrium amplitudes of velocity field which are also equal to magnetic field amplitudes. The three quadratic invariants of this system are

$$\text{the total energy } E = 2[\lambda_a^2\eta_a^2 + \lambda_b^2\eta_b^2 + \lambda_c^2\eta_c^2]$$

$$\text{the magnetic Helicity } H_m = \lambda_a\eta_a^2 + \lambda_b\eta_b^2 + \lambda_c\eta_c^2 \tag{2}$$

and the cross helicity H_c becomes equal to the total energy under the conditions of equilibrium $\vec{V} = \vec{B}$, which is an aligned Alfvenic state. λ's are the characteristic wave-vectors of the three modes. From equations (1) and (2), we found that the frequency S can be expressed in a very simple form as

$$S^2 = \frac{A H_m}{\lambda_a} \tag{3}$$

Here λ_a, where $a \equiv (0,0)$ mode, can be expressed in terms of the ratio of poloidal ψ_p to toroidal ψ_t magnetic flux; λ_b and λ_c are numerical values

E. R. Priest and V. Krishan (eds.), Basic Plasma Processes on the Sun, 256–257.

related to zeros of Bessel functions (Montgomery et al (1978)) obtained from

$$(\psi_t/\psi_p) = -\frac{R}{L}\frac{|\lambda_a|}{\lambda_a}\frac{J_0'(\lambda_a R)}{J_0(\lambda_a R)}$$

$$\frac{2\pi R}{L}\gamma_b J_1'(\gamma_b R) + \lambda_b J_1(\gamma_b R) = 0$$

and $$\frac{2\pi R}{L}\gamma_c J_{-1}'(\gamma_c R) + \lambda_c J_{-1}(\gamma_c R) = 0$$

where $\lambda = \left[\gamma^2 + (2\pi R)^2/L^2\right]^{1/2}$ and (L,R) are the length and radius of the cylindrical plasma loop. By measuring the periods of oscillating loop prominences often observed in coronograph movies, one has now a way of estimating magnetic helicity which eludes any direct measurement.

References

Krishan,V. Berger,M. and Priest,E.R., 1988 in "Solar and Stellar Coronal Structure and Dynamics" proceedings of the ninth Sacramento peak summer symposium Aug. 1987.

Montgomery,D., Turner,L. and Vahala G. : 1978 Phys. Fluids 21, 757.

NONLINEAR ALFVÉN WAVES WITH LARGE LARMOR RADIUS EFFECT

NAGENDRA KUMAR and KRISHNA M. SRIVASTAVA
Department of Mathematics
University of Roorkee
Roorkee 247667 INDIA

The nonlinear behaviour of Alfvén waves propagating parallel to a uniform magnetic field in a compressible fluid with large Larmor radius effect is investigated. We have considered an inviscid non-heat conducting gas of infinite electrical conductivity with scale-length of interest L such that $L << \rho_i$; where ρ_i is the ion gyroradius. Its motion is governed by basic MHD equations. A reductive perturbation analysis, including weak spatial variation in the transverse direction, based upon the coordinate stretchings $\xi=\epsilon(x-\lambda t), \eta=\epsilon^{3/2}y, \zeta=\epsilon^{3/2}z$, $\tau = \epsilon^2 t$ is performed. The transverse components of magnetic field and velocity, denoted by means of the complex quantities $\tilde{V}(= V_y+iV_z)$ and $\tilde{B}(= B_y+iB_z)$, have expansions of the form $\tilde{V} = \epsilon^{1/2}(\tilde{V}^{(1)} +\epsilon\tilde{V}^{(2)} +\dots)$ and $\tilde{B} = \epsilon^{1/2}(\tilde{B}^{(1)}+\epsilon\tilde{B}^{(2)}+\dots)$, while the other variables have expansions $B_x = B_o+\epsilon B_x^{(1)}+\dots$, $V_x = \epsilon V_x^{(1)}+\epsilon^2 V_x^{(2)}+\dots$, $\rho = \rho_o+\epsilon\rho^{(1)}+\dots$, $p = p_o+\epsilon p^{(1)}+\dots$. Using the above stretchings and expansions, we obtain the set of equations governing the evolution of the transverse $\tilde{B}$ and parallel B_x magnetic field perturbations. If the spatial variation in the transverse direction is negligible, then we get the "Derivative Nonlinear Schrödinger (DNLS) equation"

$$\frac{\partial\tilde{B}}{\partial\tau} + C_2 \frac{\partial}{\partial\xi}(|\tilde{B}|^2\tilde{B})+iC_1 \frac{\partial^2\tilde{B}}{\partial\xi^2} = 0; \; C_2 = v_A^3/(4B_o^2(v_A^2-\alpha^2))$$

The solution of the DNLS equation and its properties have been discussed by many researchers. When C_1 $(= Mcv_A^2/2eB_o) = 0$, the dispersive term in the DNLS equation vanishes and higher-order dispersive effects are required to contain the nonlinear growth of the wave. It means that the LLR effect provides a dispersive term to balance the nonlinear growth of the Alfvén wave. It is conjectured that the present calculations may be applicable to the study of the structure of MHD waves in both laboratory and space plasmas e.g. plasma flow near small planetary bodies such as comets, plasma dynamics near collisionless shock fronts.

Reference:

Mjølhus, E. and Wyller, J.(1988) 'Nonlinear Alfvén waves in a finite-beta plasma', J. Plasma Phys. 40, 299-318.

E. R. Priest and V. Krishan (eds.), Basic Plasma Processes on the Sun, 258.

COSMIC RAY SIGNATURES OF DIFFERENT TYPES OF SOLAR WIND STREAMS

P.K. Shrivastava* and S.P. Agrawal

Vikram Space Physics Centre (Physics Dept.)
A.P.S. University, Rewa (M.P.) 486003
*Also at Govt. New Science College,
Rewa (M.P.) 486001

INTRODUCTION The earlier concept of average solar wind speed has changed with time. Besides quiet periods of low/average solar wind speeds, two different kinds of solar sources (solar flares and coronal holes) have been identified to produce high speed solar wind streams. In an earlier investigation, it was reported that the high speed streams associated to these sources produce distinctly different effects on the cosmic ray intensity (Venkatesan, et. al., 1982).

RESULT AND CONCLUSIONS The results of the chree analysis for days –5 to +8 days have been plotted, as percent deviations of the individual stations data for the two intervals of 1979–82 and 1983–86. We have used same criteria and method of analysis as adopted earlier by Venkatesan & Co-workers. It is observed that the solar flares associated high speed solar wind streams are dominant in 1979–82, and produce much larger cosmic ray decreases at the high latitude neutron monitor stations as compared to meson monitors or low latitude neutron monitors. In contrust coronal hole associated streams are dominant during the interval of 1983–86, and produce almost equal decrease at all the stations.

REFERENCES Venkatesan, D., Shukla, A.K., and Agrawal, S.P. (1982), Cosmic ray intensity variations and two types of high speed solar streams, Solar Physics, 81, 375–81.

E. R. Priest and V. Krishan (eds.), Basic Plasma Processes on the Sun, 259–260.

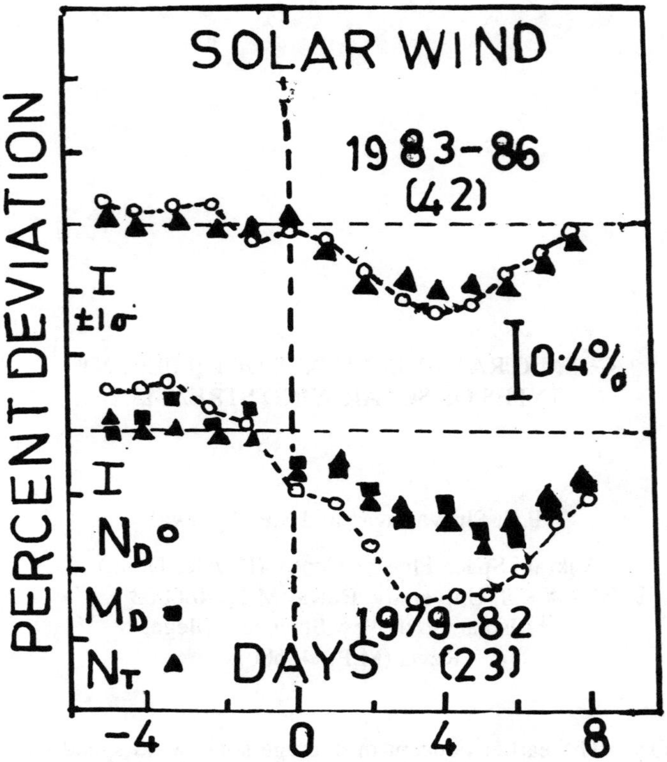

Figure 1. Shows the percent deviations of the daily mean cosmic ray intensity observed by Deep River neutron and meson monitors and those of Tokyo neutron monitor for a number of events.

Calcium K line profiles as a function of latitu de and solar cycle phase

Jagdev Singh
Indian Institute of Astrophysics
Bangalore

Summary The sun as a star has been studied by many observers by monitoring the calcium K line profile. Skumanich et al (1984) proposed a three component model of the solar cycle variability of calcium K emission using extant contrast and fractional area parameters for (1) cell (2) network and (3) plage components. The computed line profile agreed well with the observed one at the solar minimum by taking the contribution of only cell and network features and using extant limb-darkening laws. The occurrence of plages during the growth of the solar cycle was found to be insufficient to account for the increase in K emission and therefore, they introduced an additional network component, 'Active network' in excess of the quiet sun value to explain the observed excess emission during the maximum phase.

To estimate the active component during different phases of the solar cycle, a technique has been developed to monitor the calcium K line profile as a function of latitutde and integrated over the visible longitude. The data have been analysed to study the variation of calcium line parameters as a function of latitude. The sunspots and related features tell us only about middle latitude belts but the changes in the chromosphere at polar regions are not well known. These spectra will provide this vital information about the polar regions as a function of the phase of the solar cycle. Singh and Livingston (1987) have shown that a power spectral analysis of the K-index yields a reliable value of the chromospheric rotation rate. Therefore, these data will be used to study chromospheric differential rotation and the variation of rotation rate (if any) with time or phase of the solar cycle. Finally the data will provide a clue to the movement in global activity and magnetic flux on the solar surface as a function of the solar cycle. To improve the quality of the data an optical arrangement has been developed which makes use of the unidirection focusing property of a cylindrical lens.

References

Singh,J. and Livingston,W.C. 1987, Solar Phys. 109, 387.
Skumanich,A., Lean,J.L., White,O.R. and Livingston,W.C. 1984, Ap.J. 282, 776.

E. R. Priest and V. Krishan (eds.), Basic Plasma Processes on the Sun, 261.

ON THE EXISTENCE OF HYDROMAGNETIC INTERFACE WAVES AT A STRUCTURED ATMOSPHERE

K.SOMASUNDARAM , S.MANTHIRAMOORTHI
DEPARTMENT OF PHYSCIS , GANDHIGRAM RURAL INSTITUTE
GANDHIGRAM 624 302, TAMIL NADU, INDIA

AND

A.SATHYA NARAYANAN , DEPARTMENT OF APPLIED MATHEMATICS, INDIAN ISNTITUTE OF SCIENCE, BANGALORE 560 012, INDIA.

ABSTRACT: The conditions under which the hydromagnetic interface waves can exist at a magnetic interface is deduced. Using these conditons, it is shown that a slow interface wave with a phase velocity about 5Km/s and a fast interface wave with a phase velocity 6.5 to 8km/s at the photospheric level can exist.

1.INTRODUCTION: It is a well established fact that the solar atmosphere is highly structured. Such structured atmosphere supports a spectrum of Alfven waves(Uberoi,1972). When the stratification is approximated to a single discontinuity, it has been shown that the interface supports hydromagentic interfacial waves (Wentzel, 1979; Roberts,1981a; Uberoi and Somasundaram, 1980). Studies of such waves help to understand the observation of the wave motion on solar surface and the associated wave phenomena. Several studies of the interfacial waves at plane (Wentzel,1979), slab(Roberts, 1981b) and cylindrical(Uberoi and Somasundaram, 1980) geometries have been reported. Such studies, even for the simple plane interface, were done by numerical evaluation of the dispersion relation of the interfacial wave. Therefore, given an atmospheric condition, it is difficult to infer whether such surface can support an interface wave or not. In this note , the interface wave propagation along a plane interface separated by two compressible plasma embedded in a magnetic field is considered. The conditions underwhich the slow and fast magnetoacoustic interface wave can exist are deduced. These conditions are applied to test the existence of the interfacial wave at the photospheric level.

2.DISPERSION RELATION. Solving the linearized MHD equations for the perturbations of the form $f(x,z,t) \equiv \exp i(kz + \omega t)$, for the two plasma media in $x>0$ and $x < 0$, embedded in a manetic field $B_{01} = B_{01}\hat{z}$ and $B_{02} = B_{02}\hat{z}$ with equilibrium

E. R. Priest and V. Krishan (eds.), Basic Plasma Processes on the Sun, 262–263.

mass density ρ_{01} and ρ_{02} in the two media respectively and applying the boundary conditons one obtains the dispersion relation as (Roberts 1981a):

$$\rho_{01} (k^2 V_{A1}^2 - \omega^2) m_2 + \rho_{02} (k^2 V_{A2}^2 - \omega^2) m_1 = 0 \quad (1)$$

where $V_{A1,2}$ are the bulk ALfven wave velocities in the two media 1 and 2,

$$m_{1,2}^2 = \frac{(k^2 V_{A1,2}^2 - \omega^2)(C_{1,2}^2 k^2 - \omega^2)}{(C_{1,2}^2 + V_{A1,2}^2)(k^2 C_{T1,2}^2 - \omega^2)} \quad (2)$$

$$C_{T1,2}^2 = C_{1,2}^2 \, V_{A1,2}^2 / (C_{1,2}^2 + V_{A1,2}^2) \quad (3)$$

3.DISCUSSION: For an interfacial wave , real roots of equation (1) exists only when

$$\max (V_{A1}, V_{A2}) < \omega/k < \min (V_{A1}, V_{A2}) \quad (4)$$

and m_1 and m_2 are to be positive for a interface wave. It can be shown from equations(1)-(4), that a slow magnetoacoustic interface wave can exist only when

$$V_{A1} < \omega/k < \min (C_1 , C_{T2}) \quad \text{for } V_{A1} < V_{A2} \quad (6)$$
$$\text{or } V_{A2} < \omega/k < \min (C_2 , C_{T2}) \quad \text{for } V_{A1} > V_{A2}$$

and a fast moagnetoacoustic interface wave can exist only when

$$\max(V_{A1} , C_2) \omega/k < \min (C_1 , V_{A2}) \text{ for } V_{A1}<V_{A2}$$
$$\text{or } \max(V_{A2}, C_1) \omega/k < \min (C_2 , V_{A1}) \text{ for } V_{A1}>V_{A2} \quad (7)$$

Earlier results (Roberts, 1981a; Somasundaram and Uberoi, 1982; for the case l=0; Miles and Roberts,1989) can be easily obatined using the conditions in equation(6) and (7). Consider the case C_1=8 km/s , V_{A1}=5 km/s at the photospheric level and C_2=8 km/s , V_{A2}=8 km/s above the photosperic level as given by Nye and Thomas(1974). At this interface,one can expect a slow manetoacoustic interface wave with a phase velocity about 5km/s and a fast interface wave with a phase velocity 6.5 to 8km/s.

REFERENCES

Miles, A.J. and Roberts,B.:1989, Solar Phys. 119,257.

Nye, A.H. and Thomas, J.H.:1974, Solar Phys. 38, 399.

Roberts, B.:1981a, Solar Phys. 69 , 27.

Roberts, B.:1981b, Solar Phys. 69 , 39.

Somasundaram, K. and Uberoi, C.:1982, Solar Phys. 81 , 19.

Uberoi, C. and Somasundaram, K.:1980, Plasma Phys. 22 , 747.

Uberoi, C.:1972, Phys. Fluids 15 , 1673.

Wentzel, D.G.:1979, Astrophys. J. 227 , 319.

ACOUSTIC WAVE GENERATION IN VERTICAL MAGNETIC FIELDS

H.S. Yun and J.W. Lee
Astronomy Department
Seoul National University
Seoul 151-742, Korea

ABSTRACT. We extended Stein's hydrodynamic theory of acoustic wave generation to accommodate the effect of a vertically uniform magnetic field. It is shown that in the presence of a uniform vertical field with pure horizontal field fluctuations the acoustic waves are produced both from a non-magnetic monopole source of the Reynolds stress and by the magnetic field fluctuations. The computed acoustic flux is found to be confined to the topmost of the spot's convection zone with a small range of depth.

In Stein's approach (Stein, 1967) the hydrodynamic equations are combined to inhomogeneous wave equation for one variable, the perturbed pressure P_1, with a source function S_o $(\vec{x},t)$ for acoustic waves driven by the pure hydrodynamic turbulent stress f_o. In the presence of uniform and vertical magnetic field, the linearized momentum equation is given by

$$\rho_o \frac{\partial \vec{v}}{\partial t} + \vec{\nabla} p_1 - \rho_1 \vec{g} - \frac{1}{4\pi}(\vec{\nabla}\times\vec{B}_1)\times\vec{B} = \vec{f}_o$$

where B is assumed to be made up with B_o and its fluctuations B_1. The magnetic force $\vec{f}_B = \frac{1}{4\pi}(\vec{\nabla}\times\vec{B})\times\vec{B}$ comprises two terms, the linear f_L and non-linear f_N components,

$$\vec{f}_B = \vec{f}_L + \vec{f}_N = \frac{1}{4\pi}(\vec{\nabla}\times\vec{B}_1)\times\vec{B}_o + \frac{1}{4\pi}(\vec{\nabla}\times\vec{B}_1)\times\vec{B}_1 .$$

The linear force term f_L is the term which affects the wave propagation, and the non-linear term f_N is associated with the source function via $\vec{\nabla}.\vec{f}_N$ and $\vec{g}.\vec{f}_N$. In general, $\vec{\nabla}.\vec{f}_L$ and $\vec{g}.\vec{f}_L$ do not vanish. However, we can make them vanish by forcing $\vec{B}_1$ to have only the horizontal component.

Thus as long as $\vec{B}_1$ lies in the horizontal direction, the wave propagation becomes independent of the presence of uniform vertical field. Under such a condition, the non-linear force $\vec{f}_N$ simply becomes $\vec{f}_N = -\frac{\partial}{\partial z}\left(\frac{B_h^2}{8\pi}\right)\hat{z}$ which acts along the vertical direction. In this case the

E. R. Priest and V. Krishan (eds.), Basic Plasma Processes on the Sun, 264–265.

source function $S(\vec{x},t)$ can be readily evaluated, since the terms associated with the magnetic contribution $\vec{\nabla}.\vec{f}_N$ and $\vec{g}.\vec{f}_N$ are given in terms of $h=B_h/\sqrt{4\pi\rho_o}$ as $\vec{\nabla}.\vec{f}_N = -\frac{\partial^2}{\partial z^2}(\frac{1}{2}\rho_o h^2)$ and $\vec{g}.\vec{f}_N = g\frac{\partial}{\partial z}(\frac{1}{2}\rho_o h^2)$.

The turbulent stress in fluid motion can be expanded in monopole, dipole and quardrupole terms (Unno 1964). In the present work only the monopole term is considered in view of the strong tendency of fluid motion to be aligned along vertical field lines. In such a case the source function comprises two terms, one arising from pure turbulent motions and the other, associated with the magnetic fluctuations. Regarding the choice of the correlation functions needed for the calculation of the Fourier transform of $S(\vec{x},t)$, we adopted the Gaussian form given by Unno (1964) both for the turbulent and magnetic fluctuations.

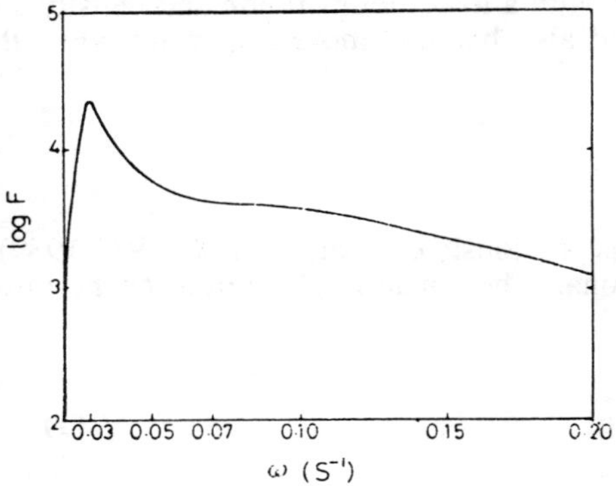

Fig.1 Computed acoustic flux spectra

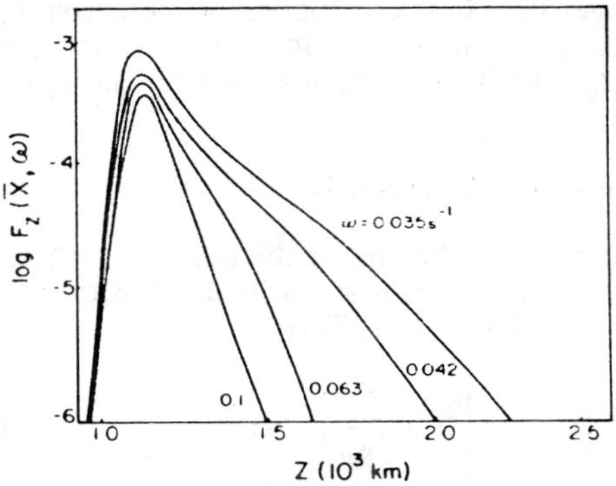

Fig.2 Computed acoustic emission rate per unit volume

We estimated the horizontal field strength from Yun's magnetostatic model No 7 (T_{eff}=4000K, B=3200 Gauss) with the use of Mullan's magnetic diffusion theory (Mullan, 1974). Figure 1 is the computed acoustic flux spectra where the flux F is in erg/(cm^2 sec H_z). Figure 2 shows the acoustic emission rate per unit volume which is confined to the topmost of the spot convection zone.

* This work is supported by Korean Science and Engineering Foundation. We wish to thank Dr. D.J. Mullan and Dr. W. Unno for their helpful comments.

References

Mullan D.J. (1974) "Sunspot models with Alfven wave emission" Ap.J. 187, 621.

Stein, Robert F. (1967) "Generation of acoustic and gravity waves by turbulence in an isothermal stratified atmosphere" Solar Physics 2, 385.

Unno, W. (1964) Transaction I.A.U. XII B Academic Press, New York, p555

Yun, H.S. (1968) "Theoretical Sunspot Models" Indiana University.

Wave Energy Dissipation in The Solar Atmosphere

Zhou Aihua
(Purple Mountain Observatory, Academia Sinica,
Nanjing, P.R.C.)

ABSTRACT. There are two regions of rapid dissipation when Alfvén waves propagate from the transition region to the corona. They occur respectively in a range of several hundred kilometers above the base of the transition region and in the corona at about 1-3R$_\odot$. The heating of the atmosphere by wave dissipation could be one order of magnitude larger than heat conduction in the coronal part with a lower temperature and density and stronger magnetic field. Wave heating could also become more important when the magnetic field divergence becomes stronger.

Basic Equations and Results

Alfvén wave heating in the transition region has previously estimated (Zhou (1988,1989)). In this paper we discuss its dissipation in corona. The equation of motion for the mean flow is (Jacques (1977))

$$\rho \frac{dU}{dt} = - \nabla \left[P + \frac{B^2}{8\pi} \right] + (B \cdot \nabla) \frac{B}{4\pi} - \nabla \cdot P + \rho g, \tag{1}$$

In the case of Alfvén waves the wave stress tensor is $P_w = WKK - WI/2$ (Li *et al* (1984)) and W is the wave energy density. The energy equation is

$$\frac{\partial E}{\partial t} + \nabla \cdot G = 0; \tag{2}$$

where

$$E = \rho U^2 + P/(r-1) + B^2/(8\pi) + W; \tag{3}$$

$$G = [\rho U^2 + rP/(r-1) - \rho GM_\odot/R]U - (U \times B) \times B/(4\pi) + VaW + U \cdot P_w + G_1 \tag{4}$$

and G_1 represents the contribution to G from processes such as radiation and heat conduction (Vanveberen and De Loore (1976); Rosner and Tucker (1978)). The equation of continuity is

$$\frac{\partial \rho}{\partial t} \nabla \cdot (\rho U) = 0, \tag{5}$$

and the equation of wave action density is (Jacques (1977))

E. R. Priest and V. Krishan (eds.), Basic Plasma Processes on the Sun, 266–267.

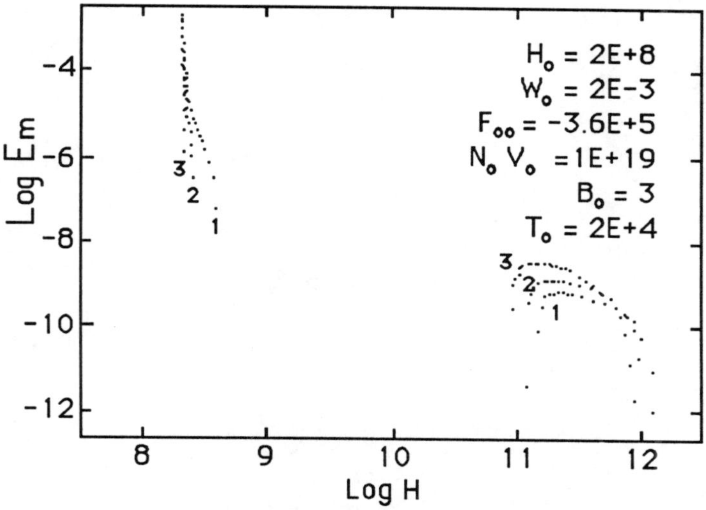

Figure 1. The distributions of Em for radial flow for $N_0 = 1.3 \times 10^{10}$ (1), 1.8×10^{10} (2) and 3×10^{10} (3).

$$\nabla \cdot (VaS) = -VaS/d, \tag{6}$$

where S is the wave action density and d is the Alfvén wave damping length (Kaplan and Tsytovich (1973)).

The results show that Alfvén waves have two dissipation regions in the transition region and in corona at about 1-3R⊙. As T_0 and N_0 decrease and B_0 increases, Alfvén wave heating by the dissipation becomes more important and the height ranges of the dissipation move to larger heights. Also, when the magnetic field becomes more divergent, the ratios of the wave heating to the heat conduction Em/Ec become larger. This nonradial effect could explain coronal-hole cycle variations of the basic heliosphere quantities (Zhou (1990)).

References

[1] Zhou, A H *et al* (1988), Chinese Journal of Space Science, 8, 81.
[2] Zhou, A H *et al* (1989), Adv. Space Res. 9pp (4) 33.
[3] Jacques, S A (1977), Ap. J. 215, 942.
[4] Li, X Q *et al* (1984), Solar Phys., 91, 289.
[5] Vanbeveren, D and De Loore, C (1976), Solar Phys., 50, 99.
[6] Rosner, R and Tucker, W H (1978) Ap. J. 220, 643.
[7] Kaplan, S A and Tsytovich, V N (1973) Plasma Astrophysics, Pergamon, Oxford, P. 25.
[8] Zhou, A H (1990), to be published.

VII. MAGNETIC RECONNECTION and CORONAL EVOLUTION

MAGNETIC RECONNECTION ON THE SUN

E R PRIEST
Mathematical Sciences Dept
The University
St Andrews
KY16 9SS
Scotland, UK

ABSTRACT. Magnetic reconnection is a fundamental process in astrophysics which plays many different roles on the Sun. The classical theory of fast steady reconnection due to Sweet, Parker, Petschek and Sonnerup has been unified in a theory which possesses many new regimes that depend on the boundary conditions at large distances. For example, the *flux pile-up regime* possesses diverging flows, a long central diffusion region and a reconnection rate that is much larger than the Petschek value. Recent numerical experiments however, often possess three features that are not present in the earlier theoretical models, namely highly curved inflow field lines, separatrix jets and reversed current spikes, and so an attempt is described to include these features in a new theoretical model.

Several solar phenomena where reconnection is believed to be operating are described. Cancelling magnetic features in photospheric magnetograms are probably evidence of reconnection submergence. Most prominences are of inverse polarity and have upflows driven by reconnection below them. Coronal heating may well be due to reconnection in many small current sheets. In solar flares reconnection may be driven below an eruptiing active-region prominence and hence power the high-temperature flare loops and chromospheric ribbons.

1. Introduction

It was realised 30 years ago that the energy for a flare can only come from the coronal magnetic field and that it is being released very fast - at a significant fraction of the Alfvén speed - and so this was the stimulus for the discovery of fast reconnection mechanisms by Sweet, Parker and Petschek. The emphasis was on steady mechanisms because the overall energy release continues for many thousands of Alfvén times, although of course it is modulated in a time-dependent manner.

In a vacuum, reconnection is a trivial process, but, in a plasma atmosphere such as the Suns, normally the plasma is attrached very effectively to the magnetic field. It is only where the magnetic gradients are, say, a million times stronger than normal that the magnetic field can slip through the plasma and reconnect (Figure 1). There are three important effects of such a process. Firstly, the global topology of the magnetic field may be changed since the connectivity of the field lines may be altered. For example, initially in Figure 1 the point A is joined to point B and finally it is connected to point C. This may affect the behaviour of particles and heat which tend to travel along the field lines. Secondly, inflowing stored magnetic energy is converted into heat, bulk kinetic energy and fast particle energy. Thirdly, reconnection creates large electric currents, electric fields, shock waves, filamentation, each of which may be involved in the acceleration of fast particles.

E. R. Priest and V. Krishan (eds.), Basic Plasma Processes on the Sun, 271–291.

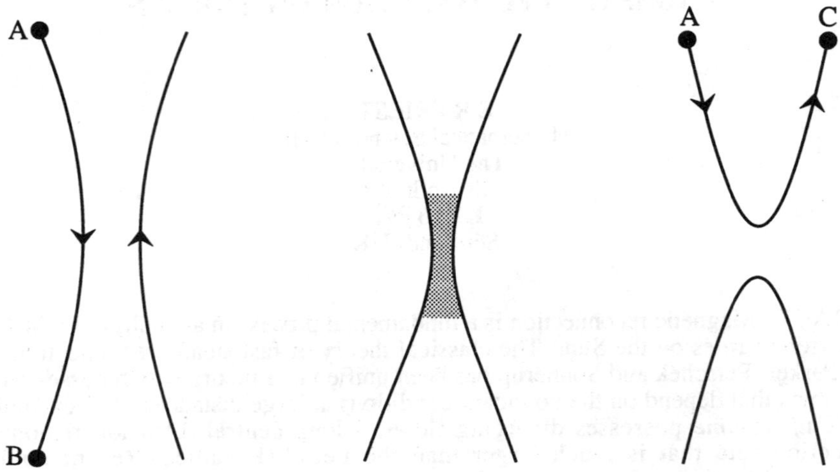

Figure 1. The breaking and reconnecting of field lines in a region (shaded) of very strong magnetic gradient.

Now reconnection is realised to be important in a wide range of solar phenomena, such as coronal heating, coronal mass ejections, magnetoconvection, dynamo generation, prominences, and bright points, as well as the many types of solar flare. We have seen in the talks by Heyvaerts, Hollweg and Van Ballegooijen how reconnection in many current sheets may provide the heating of the solar corona. Thus all

Figure 2. X-ray picture of the solar corona (courtesy L Golub)

of the beautiful 3/4 arcsec structure in the spectacular X-ray picture from Golub's Normal Incidence X-ray Telescope on the Sept 11 (1989) rocket flight may be a direct consequence of reconnection. The X-ray bright points that are often present in coronal X-ray pictures, although strangely not in Figure 2, are now thought to be produced by magnetic reconnection, though not always by emerging flux. Reconnection also takes place below most prominences and releases the energy in a solar flare, creating both the Hα ribbons and the hot flare loops after an active-region prominence has erupted. Ahead of an erupting prominence one sees a coronal mass ejection (Figure 3), a huge magnetic bubble which is sometimes observed to disconnect from the Sun.

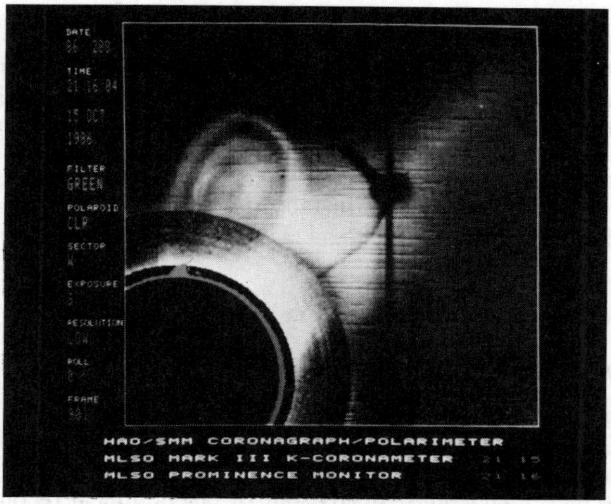

Figure 3. A coronal mass ejection (courtesy A Hundhausen)

In magnetohydrodynamics the plasma velocity (**v**) and magnetic field (**B**) are primary variables, with the electric current

$$\mathbf{j} = \nabla \times \mathbf{B}/\mu \approx B/(\mu L) \tag{1.1}$$

and electric field

$$\mathbf{E} = - \mathbf{v} \times \mathbf{B} + \mathbf{j}/\sigma \tag{1.2}$$

deducible from them if required. (This is quite different from laboratory electromagnetism, where **E** and **j** are primary.) The flow and field are determined by the equation of motion

$$\rho \frac{d\mathbf{v}}{dt} = - \nabla p + (\nabla \times \mathbf{B}) \times \mathbf{B}/\mu, \tag{1.3}$$

and the induction equation

$$\frac{\partial \mathbf{B}}{\partial t} = \nabla \times (\mathbf{v} \times \mathbf{B}) + \eta \nabla^2 \mathbf{B} \tag{1.4}$$

where $\eta = (\mu\sigma)^{-1}$ is the magnetic diffusivity. Equation (1.4) implies that the magnetic field changes in time due to two effects. The first term on the right represents the advection of magnetic field lines with the plasma. The second represents diffusion through the plasma and at the same time magnetic energy is converted to heat ohmically at a rate j^2/σ, where j is given by (1.1).

The time-scale for magnetic dissipation is given by equating the first and third terms in (1.4), namely

$$\tau_d = \frac{L^2}{\eta} \approx 10^{-9} L^2 \; T^{3/2}, \tag{1.5}$$

which is enormously long (10^{14} seconds) for a typical global coronal length-scale ($L = 10^7$m) and coronal temperature ($T = 10^6$K). Thus, in order to release magnetic energy, one needs to create extremely small length-scales in sheets or filaments and therefore very large magnetic gradients and electric currents (Eqn 1.1). This may be done in three ways: as propagating sheets in shock waves; as stationary sheets in magnetic equilibria near X-points or separatrices; and as a result of instabilities such as the tearing mode or the coalescence instability.

2. Theory of Fast Magnetic Reconnection

2.1 CLASSICAL THEORY

A region of oppositely directed or sheared magnetic field may be linearly unstable to the breaking and reconnection of field lines by the tearing mode instability. When the tearing mode develops nonlinearly or when a current sheet forms dynamically or when magnetic sources are driven together, then a quasi-steady state of fast nonlinear reconnection may be reached. The classical models are sketched in Figure 4. The Sweet-Parker (1958) model is a simple diffusion region. In the Petschek mechanism (1964) the diffusion region occupies only a small central location, while most of the energy conversion occurs at standing slow-mode shock waves that accelerate and heat the plasma to form two hot fast outflowing streams. In Sonnerup's (1970) model an extra set of discontinuities is standing in the flow ahead of the slow shocks, but Vasyliunas (1975) pointed out that these are slow-mode expansion waves which need to be generated externally at discrete points in the flow which would not be present in astrophysical applications.

Sonnerup and Priest (1975) discovered an exact solution of the nonlinear MHD equations in which an incompressible stagnation point flow

$$v_x = -\frac{V_0 x}{a}, \qquad v_y = \frac{V_0 y}{a}$$

satisfying $\nabla \cdot \mathbf{v} = 0$ carries in oppositely directed (but straight) magnetic field lines. The equation of motion (1.3) for steady flow determines the pressure

$$p = \text{const} - \frac{1}{2}\rho v^2 - B^2/(2\mu)$$

and for a field $B(x)\hat{\mathbf{y}}$ Ohm's Law becomes

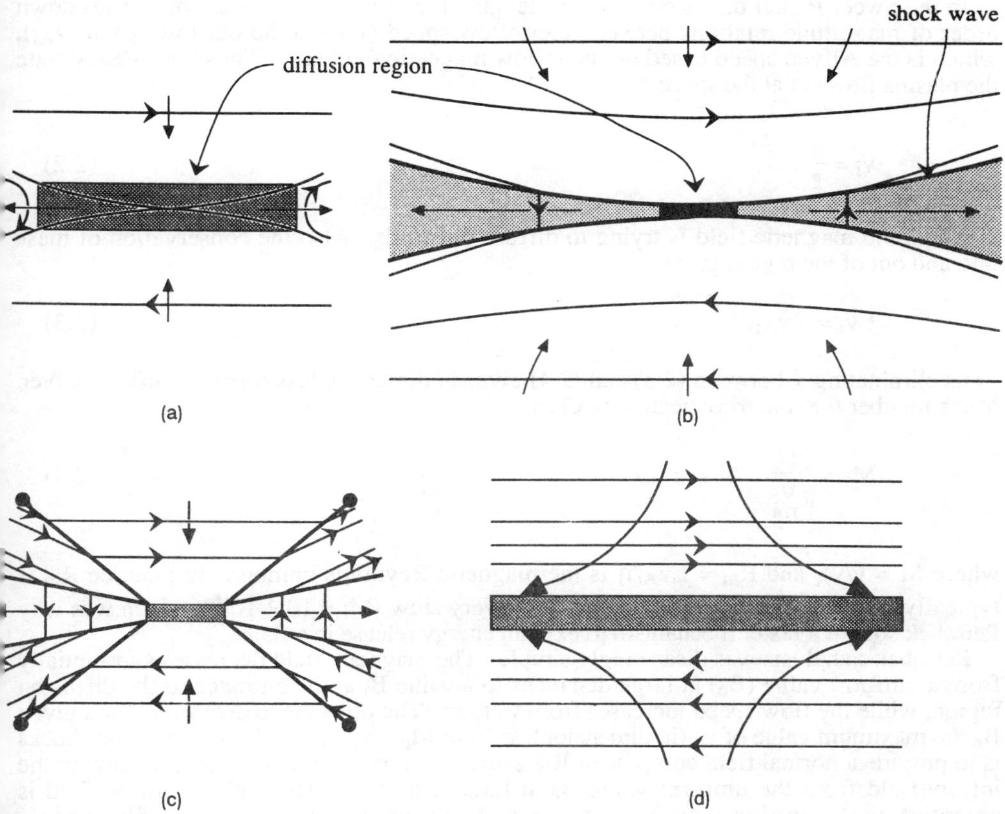

Figure 4. The classical energy conversion mechanisms of (a) Sweet-Parker, (b) Petschek, (c) Sonnerup and (d) Sonnerup and Priest.

$$E_0 - \frac{V_0 x}{a} B = \eta \frac{dB}{dx} \tag{2.1}$$

where the electric field (E_0) and flow scale (V_0) are constant. The solution of (2.1) at large distances has the field (B) increasing like x^{-1} as it is carried in, but eventually at a distance of order $(\eta a/v_0)^{1/2}$ the diffusion term on the right becomes important and causes B to decrease to zero at the origin. This solution has been generalised to three-dimensions and also, by Gratton *et al* (1988) to include viscous flows.

In the Sweet-Parker diffusion region of length 2L and width 2ℓ one simply writes down order of magnitude relations between the inflow speed (v_i) and the outflow speed (v_{Ai}), which is the Alfvén speed based on the inflow magnetic field (B_i). Thus for a steady state the plasma flows in at the speed

$$v_i = \frac{\eta}{\ell} \tag{2.2}$$

at which the magnetic field is trying to diffuse outwards. Also the conservation of mass into and out of the region gives

$$L v_i = \ell v_{Ai}, \tag{2.3}$$

Thus eliminating ℓ between (2.3) and (2.4) gives in dimensionless form the inflow Alfvén Mach number (i.e. the reconnection rate) as

$$M_i = \frac{1}{R_{mi}^{1/2}} \tag{2.4}$$

where $M = v/v_A$ and $R_m = Lv_A/\eta$ is the magnetic Reynolds number. In practice R_m is typically 10^6-10^{12} and so the reconnection is very slow ($M_i \sim 10^{-3}$-10^{-6}), which was why Petschek sought a faster mechanism to explain energy release in a flare.

Petschek's analysis was disarmingly simple. The magnetic field decreases substantially from a uniform value (B_e) at large distances to a value B_i at the entrance to the diffusion region, while the flow speed increases from v_e to v_i. The object is to determine for a given B_e the maximum value of v_e (in dimensionless form $M_e = v_e/v_{Ae}$). The effect of the shocks is to provide a normal field component B_N which essentially produces the distortion in the inflow field from the uniform value B_e at large distances. Thus, if the inflow field is potential, the distortion may be regarded as being produced by a series of monopole sources along the x-axis between $|x| = L$ and $|x| = L_e$, say. The result is that, as the diffusion region is approached the field strength falls to

$$B_i = B_e - \frac{2}{\pi} \int_L^{L_e} \frac{2B_N}{x} dx$$

or

$$B_i = B_e - \frac{4B_N}{\pi} \log \frac{L_e}{L} . \tag{2.5}$$

At the shock waves the condition that they be standing is

$$\frac{B_N}{\sqrt{(\mu\rho)}} = v_e,$$

and Petschek estimates the maximum reconnection rate ($M_e{}^*$) by putting $B_i = \frac{1}{2} B_e$ in (2.5) to be

$$M_e^* = \frac{\pi}{8 \log R_{me}} .$$

In practice this is typically 0.01, much greater than the Sweet-Parker rate.

2.2 UNIFIED THEORY OF FAST STEADY ALMOST-UNIFORM RECONNECTION

Vasyliunas (1975) clarified the physics of Petschek's mechanism by pointing out that the inflow region has the character of a fast-mode expansion with the pressure and field strength decreasing and the flow converging as the magnetic field is carried in. A fast-mode disturbance has the plasma and magnetic pressure increasing or decreasing together, while a slow-mode disturbance has the plasma pressure changing in the opposite sense to the magnetic pressure. An expansion makes the pressure decrease while a compression makes it increases, even in the incompressible limit. Sonnerup's model possesses slow-mode expansions that are unlikely because they are discrete. Vasyliunas suggested that a Sonnerup-like solution may be possible with a diffuse slow-mode expansion spread throughout the inflow region, making the field strength increase, the pressure decrease and the flow diverge as the field lines are carried in, although he was unable to find such a solution.

I wanted to understand Vasyliunas's distinction mathematically and was also puzzled at many strange features of some of the numerical reconnection experiments such as much longer diffusion regions than Petschek, diverging flows and large pressure gradients. Also what is the relation to the stagnation point flow solution? Can a Sonnerup-like solution be found without the extra discontinuities? And can a model in a finite region be produced, like the numerical experiments?

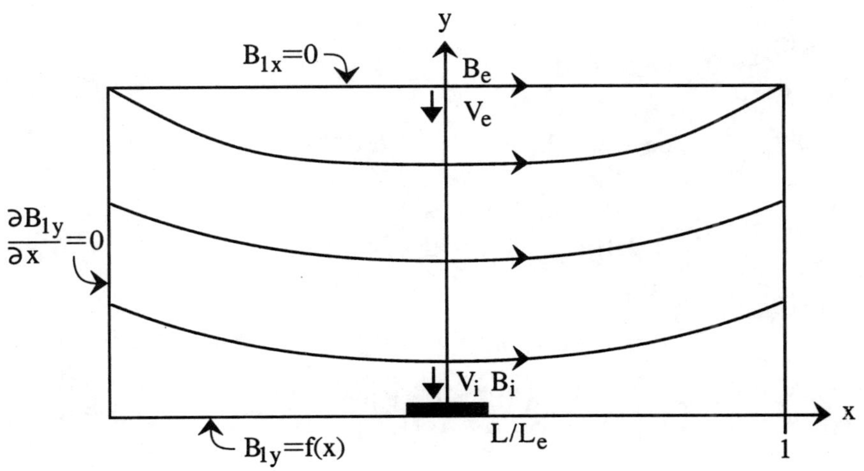

Figure 5. Notation and boundary conditions.

During one summer Terry Forbes and I (Priest and Forbes, 1986) tried to answer such questions by seeking fast, steady almost-uniform reconnection solutions to the equations for two- dimensional, incompressible flow, namely

$$\rho(\mathbf{v}\cdot\nabla)\mathbf{v} = -\nabla p + (\nabla \times \mathbf{B}) \times \mathbf{B}/\mu \tag{2.6}$$

$$\mathbf{E} + \mathbf{v} \times \mathbf{B} = \mathbf{0} \tag{2.7}$$

$$\nabla\cdot\mathbf{v} = \nabla\cdot\mathbf{B} = 0. \tag{2.8}$$

The solutions are almost-uniform in the same sense as Petschek's, namely that one performs a linear expansion about a uniform field

$$\mathbf{B} = B_0\hat{\mathbf{x}} + \mathbf{B}_1 + \ldots$$

$$\mathbf{v} = \mathbf{v}_1 + \ldots .$$

Neglecting the pressure gradient (2.6) then reduces to a potential field

$$\nabla^2 A_1 = 0 \tag{2.9}$$

where

$$B_{1x} = \frac{\partial A_1}{\partial y}, \qquad B_{1y} = -\frac{\partial A_1}{\partial x}.$$

The solution to (2.9) subject to the boundary conditions that

$$B_{1x} = 0 \quad \text{on} \quad y = L$$

$$\frac{\partial B_{1y}}{\partial x} = 0 \quad \text{on} \quad x = L$$

$$B_{1y} = 0 \quad \text{on} \quad x = 0$$

$$B_{1y} = f(x) = \begin{cases} 2B_N\, x/L & 0 \le x \le L \\ 2B_N & L \le x \le L_e \end{cases}$$

is

$$A_1 = -\sum_0^{\infty} \frac{a_n}{\left(n+\frac{1}{2}\right)\pi} \cos\left[\left(n + \frac{1}{2}\right)\pi \frac{x}{L}\right] \cosh\left[\left(n + \frac{1}{2}\right)\pi\left(1 - \frac{y}{L}\right)\right] \tag{2.10}$$

where

$$a_n = \frac{4B_N \sin\left[\left(n+\frac{1}{2}\right)\pi L/L_e\right]}{L/L_e\left(n+\frac{1}{2}\right)^2\pi^2\cosh\left[\left(n+\frac{1}{2}\right)\pi\right]}.$$

This represents a Petschek-type solution with a weak fast-mode expansion. From (2.7) the first-order flow ($\mathbf{v}_1 = (E/B_0)\hat{\mathbf{y}}$ is uniform but the second-order flow is converging. By calculating B_i one can deduce a relation between M_e and M_i, which shows that, as Petschek had expected, M_e does indeed possess a maximum value which is close to Petschek's estimate.

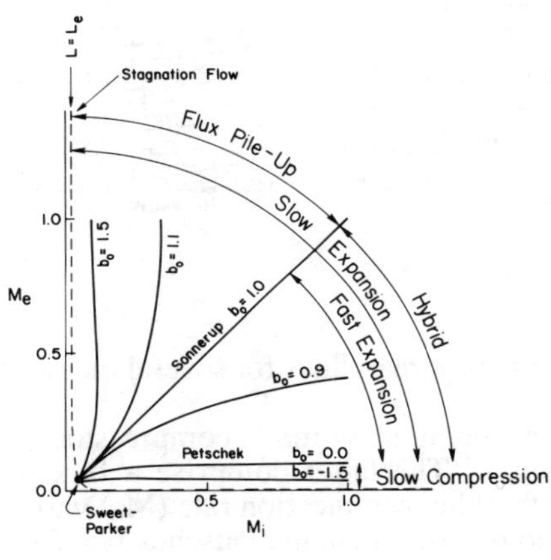

Figure 6. Inflow Alfvén Mach number M_e as a function of M_i for several values of b.

As well as putting Petschek's analysis on a firmer basis, one can, however, generalise the theory by including a first-order pressure gradient, which simply changes (2.9) to

$$\nabla^2 A_1 = -\frac{\mu}{B_0}\frac{dp_1}{dy}$$

and the solution (2.10) to

$$A_1 = -\sum_0^\infty \frac{a_n}{\left(n+\frac{1}{2}\right)\pi}\left(by - \cos\left[\left(n+\frac{1}{2}\right)\pi\frac{x}{L}\right]\cosh\left[\left(n+\frac{1}{2}\right)\pi\left(1-\frac{y}{L}\right)\right]\right).$$

When b = 1, the inflow field on the y-axis is uniform and we have found the Sonnerup-like solution with a slow-mode expansion across the whole inflow region. However, there are many other solutions for the other values of the parameter b, which is determined by the nature of the flow on the inflow boundary, since the horizontal flow speed at the corner $(x,y) = (L_e,L_e)$ is proportional to $(b - 2/\pi)$. As b increases so the inflow turns from being

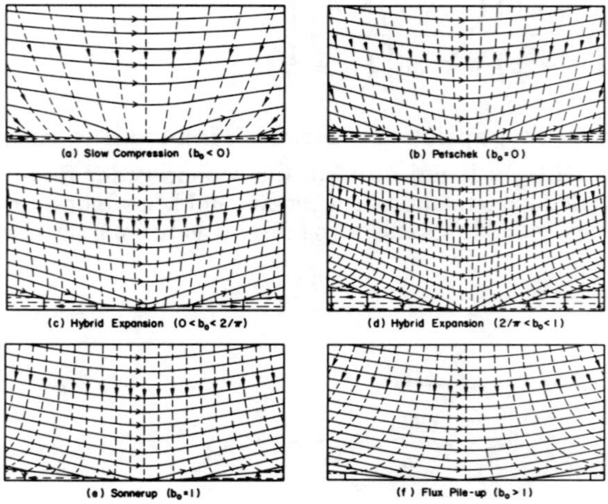

Figure 7. Magnetic field lines and streamlines for several members of the unified theory.

converging (and therefore producing slow-mode compressions) to being diverging (with strong slow-mode expansions). The latter comprise a flux pile-up regime with long diffusion regions. The way that the reconnection rate (M_e) varies with M_i and b is shown in Figure 6, where it is seen to be faster than the Petschek rate for regimes with $b > 0$.

The main results from the above analysis are that the type of reconnection regime and the rate of reconnection depend sensitively on the inflow boundary conditions, with the Petschek (b=0) and Sonnerup-like (b=1) solutions being particular members of a much wider class. Jardine and Priest (1988) have recently extended the theory to include higher orders, compressibility and energetics. Also, it has been compared with a variety of numerical experiments (Forbes and Priest, 1987).

2.3 NONUNIFORM RECONNECTION WITH SEPARATRIX JETS

Numerical experiments such as that shown in Figure 7 reveal four puzzling features that are not present in the classical models of reconnection:

(i) different types of inflow;
(ii) separatrix jets;
(iii) reversed current spikes;
(iv) highly curved field lines in the inflow region.

The first feature is the one which Priest and Forbes (1986) attempted to model with their unified almost-uniform theory. The other three features have been tackled in a new nonuniform theory by Priest and Lee (1990).

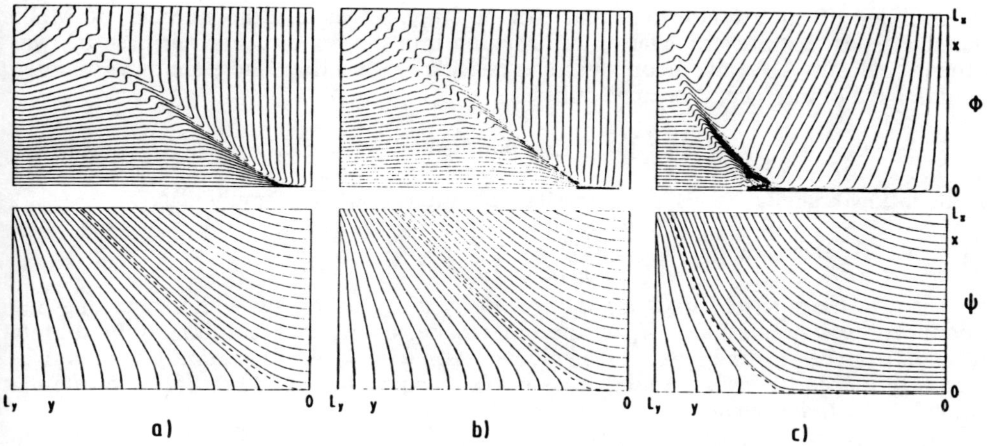

Figure 7. Numerical experiment on reconnection (Biskamp, 1986)

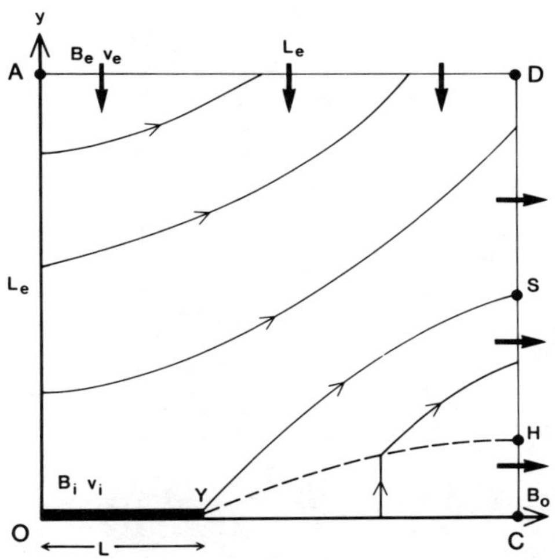

Figure 8. The notation for reconnection with a highly curved inflow.

Feature (iii), the reversed current spikes at the ends of the diffusion region, slow down the streams of plasma that are emerging from the diffusion region and partially divert them along the separatrix jets. This is a consequence of imposing boundary conditions at the outflow boundary that give a mismatch with the outflow from the diffusion region.

By contrast, feature (iv), the highly curved inflow field lines and the associated wide shock angle, are a direct result of the form of the inflow boundary conditions. In general the number of such conditions that can be imposed equals the number of MHD characteristics that are propagating information into the region. For our case of two-dimensional, sub-Alfvénic, incompressible, essentially ideal flow, there are three imposed conditions. For instance, if one prescribes

$$v_x = 0, \qquad v_y = \text{constant}, \qquad p = \text{constant} \tag{2.11}$$

on the inflow boundary (AD), then the MHD equations (2.6)-(2.8) imply that

$$B_x = \text{constant}, \qquad B_y = 0,$$

so that the straight field lines are carried in by a uniform flow without curving and reconnection is impossible. If, on the other hand, boundary conditions only slightly different from (2.11) are imposed, then reconnection with weakly curved inflow field lines may be produced. Thus it is entirely reasonable to expect that conditions greatly different from (2.11) could produce a highly curved inflow.

Now equations (2.8) may be satisfied identically by writing **v** and **B** in terms of a stream function (Ψ) and flux function (A), namely

$$v_x = \frac{\partial \Psi}{\partial y}, \quad v_y = -\frac{\partial \Psi}{\partial x}, \quad B_x = \frac{\partial A}{\partial y}, \quad B_y = -\frac{\partial A}{\partial x}.$$

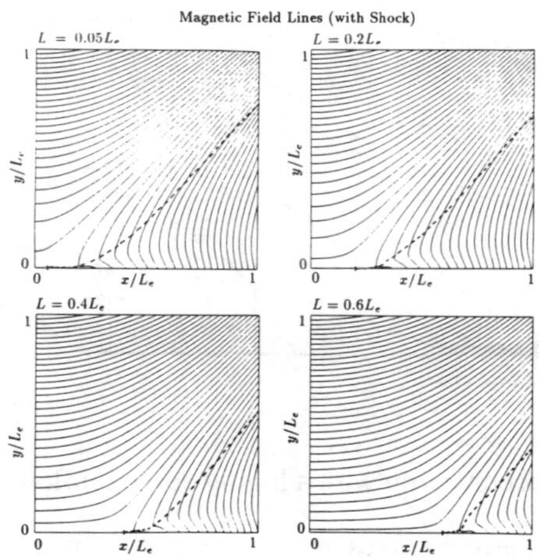

Figure 8. Field lines for one quadrant of nonuniform reconnection.

Then the problem of nonuniform reconnection with an imposed inflow (v_e) and field strength (B_e) on the inflow boundary producing a diffusion region (OY) of half-length L with a separatrix YS and slow shock YH may be tackled in three steps.

First of all, in the upstream region ahead of YH suppose for simplicity that both the plasma speed (v) and sound speed are much smaller than the Alfvén speed (v_A). Then (2.6) implies that $j = 0$ and so for a potential field with a current sheet we may use complex variable theory to pick

$$B_y + iB_x = B_i\left(\frac{z^2}{L^2} - 1\right)^{1/2} \tag{2.12}$$

where $z = x+iy$ and there is a cut (a current sheet) from $z = -L$ to $z = L$. Then (2.7) implies that the flow velocity may be deduced from

$$\Psi = v_e B_e \int \frac{ds}{B} \tag{2.13}$$

where the integral is along a field line.

The second step is to calculate the position of the shock from the characteristic curve

$$\Psi + A = \text{constant}$$

that passes through the end point (Y) of the diffusion region. Then the shock relations are applied to deduce the conditions just downstream of YH. Finally, one needs to solve the MHD equations in the downstream region subject to the appropriate boundary conditions at the shock and at the outflow boundary CH. In general these equations may be written

$$\mathbf{v}\cdot\nabla A = -v_e B_e \tag{2.14}$$

$$\mathbf{v}\cdot\nabla\,\omega = \mathbf{B}\cdot\nabla j \tag{2.15}$$

where $\omega = -\nabla^2\Psi$ is the vorticity.

For example, the results of assuming $v >> v_A$ and so taking $\omega = 0$ or

$$\nabla^2\Psi = 0$$

with Ψ imposed along the boundary YHC and Ψ = constant on YC are shown in Figures 8 and 9. The shock (strictly speaking an Alfvénic discontinuity of slow-mode compressional type in this incompressible model) is shown dashed but is rather weak and has little effect on the magnetic field. The effect of the reversed current spike downstream of the diffusion region shows up in the field lines of abnormal curvature and in the spreading of the streamlines. Also the separatrix jet is prominent and makes streamlines follow the separatrix as they pass through it. As the current sheet decreases in length, so the inflow speed increases up to a value that depends on the inflow Alfvén Mach number. Results have also been obtained by solving the full equations (2.14) and (2.15).

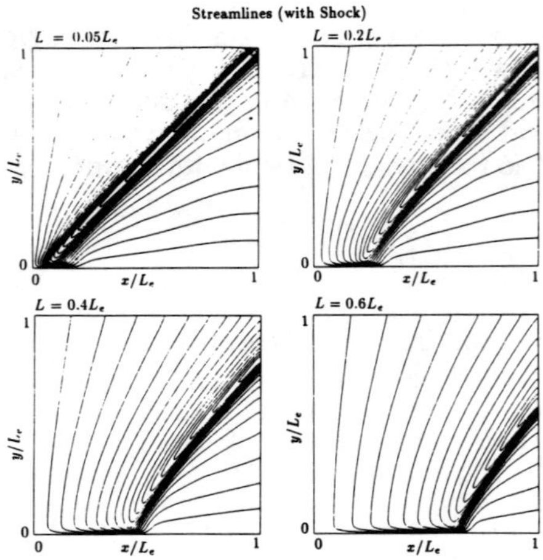

Figure 9. Streamlines corresponding to Figure 8.

Figure 10. Cancelling magnetic features shown by rectangles in a set of photospheric videomagnetograms (Martin et al, 1985)

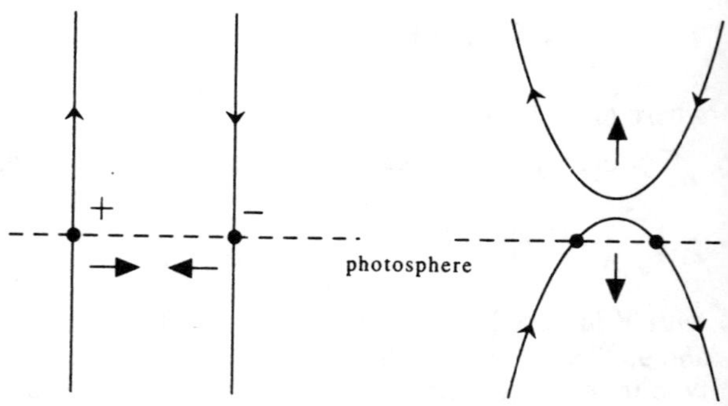

Figure 11. Reconnection submergence as an explanation for the cancellation of magnetic fragments.

3. The Role of Reconnection on the Sun

3.1 CORONAL HEATING IN CURRENT SHEETS

There may well be several mechanisms at work in heating different parts of the corona. Almost certainly X-ray bright points, for example, are caused by reconnection, though not always, as was believed until recently, by emerging flux. They certainly occur above pairs of oppositely directed magnetic fragments, but Martin (1986) has demonstrated with a videomagnetograph that such fragments are often cancelling rather than emerging (Figure 10). It is my view that, as such fragments come together, their magnetic field lines reconnect in the atmosphere above, creating one loop which moves out while another submerges - such a process may be called *reconnection submergence* (Priest, 1987). Such an explanation has natural theoretical grounds and would in a natural manner create the X-ray bright points, subflares, macrospicules and minifilaments that are often observed to be associated with cancelling magnetic features.

Reconnection could also be responsible for energy release in current sheets created if smooth magnetic equilibria do not always exist when photospheric footpoint motions are complex (Parker, 1972). Mikic *et al* (1988) have recently performed a 3D ideal MHD numerical experiment to show how such current sheets could be created. They assume the pressure is zero, the flows are much slower than the Alfvén speed, and they adopt a $64 \times 64 \times 64$ point mesh. The footpoints of an initially straight uniform magnetic field are

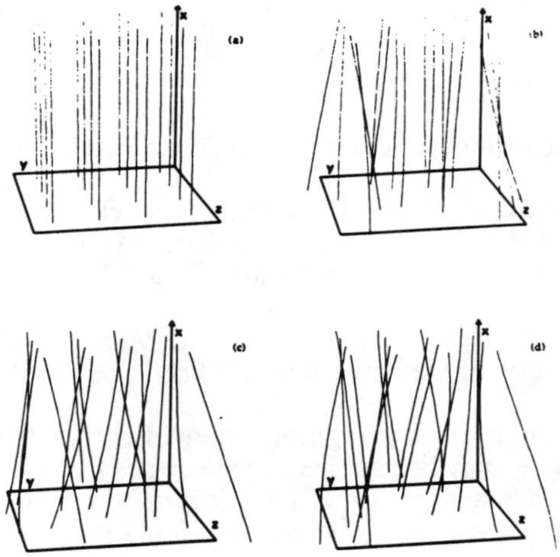

Figure 12. Field lines at intervals of $40\tau_A$ (Mikic *et al*, 1988)

braided with random footpoint motions. A series of smooth equilibria are produced but there is a transfer to small scales with filamentary electric currents that grow like $\exp(0.01t/\tau_A)$.

3.2 PROMINENCES AND GLOBAL FLUX BALANCE

Quiescent prominences may well lie at the boundaries of giant cells and most have Inverse Polarity, so that the magnetic field threads through the prominence in the opposite direction from what one would expect on the basis of the observed photospheric line of sight components. The observed upflows of plasma in prominences on the disc have therefore

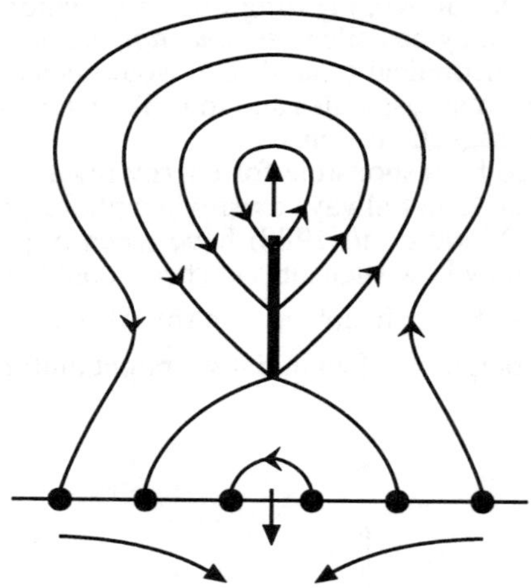

Figure 13. A prominence of Inverse Polarity with flux cancellation

been suggested to be created by magnetic reconnection below the prominence in response to converging motions of the photospheric footpoints (Malherbe and Priest, 1983; Priest, 1987). In addition such a process of flux cancellation below a prominence would be consistent with Sara Martin's observations and would build up helical structure in the prominence magnetic field (Van Ballegooijen and Martens, 1989) and so create the structure that forms the basis of the new Twisted Flux Tube Model of prominences (Priest *et al*, 1989)

An important question is how the global magnetic flux balance in the corona is maintained. Magnetic buoyancy brings flux to the surface over a wide range of scales, but then what happens to it? On a large scale flux is lost in coronal mass ejections at a rate of about 0.5 per day carrying a flux of 10^{21}Mx. When the field in a coronal mass ejection becomes stretched out, a disconnection is needed eventually to prevent an indefinite build-up of flux in the interplanetary medium although it is rarely observed, perhaps because it usually occurs after the mass ejection fades from view. On a small scale we have seen that cancelling magnetic features often create X-ray bright points and perhaps small plasmoids (and HRTS jets) are associated with such events. The plumes that overlie bright points in coronal holes (Ahmad and Webb, 1978; Holt and Mullan, 1986) may have an outflow of 100 km s^{-1}, which would be enough to supply the high-speed solar wind. Furthermore, a flux escape of 10^{21}Mx per day would provide the heat required for the active-region corona, while 10^{20}Mx per day would be enough for a quiet region.

3.3 RECONNECTION IN SOLAR FLARES

Reconnection plays several roles in solar flares. Small flares may be created by the interaction of neighbouring flux systems, either when new flux emerges or when existing bipolar flux moves laterally or when separate magnetic elements cancel by reconnection submergence. Large flares involve three phases: the slow preflare rise of a prominence; the flare onset and impulsive phase when reconnection begins at many points below the prominence; the main phase when quasi-steady reconnection continues and creates hot coronal loops and bright chromospheric ribbons as the opened-out field closes back down. The role of reconnection in a large (two-ribbon) flare is therefore to release the energy in the impulsive and main phases and, in some cases, to trigger the initial eruption by emerging flux.

Forbes (1989) has conducted a numerical experiment on emerging flux in which new bipolar flux is injected from below into a uniform horizontal field. Flux pile-up reconnection occurs with two streams of plasma produced by the reconnection and slowed down by termination shocks. The new flux rises and pinches off, forming a plasmoid and eventually approaching a potential field of lowest energy. The power output shows a slow increase in a driven phase, followed by an impulsive bursty release of energy and a slow decline, similar to what is observed in flares.

Some numerical experiments of the creation of hot flare loops have focussed on the close-down process by beginning with a line-tied open vertical field and watching it reconnect (Forbes and Priest, 1982; Forbes *et al*, 1989). Recent results show: an impulsive bursty regime which explains sudden jumps in loop height; a fast shock which slows the downflowing stream, increases the density and triggers a condensation; a reversed deflection current which deflects the downflow around a stagnation region; the splitting of the slow shock into a conduction front and an isothermal subshock; and the presence of evaporation in the hot loops to produce the observed density and temperature. Other numerical experiments (Mikic *et al*, 1988; Biskamp and Welter, 1989; Forbes, these proceedings) have modelled the global eruption followed by reconnection.

4. Conclusion

We have seen that reconnection may play many different roles on the Sun in a wide variety of phenomena from dynamo action to solar flares. At the present time new theoretical advances are being stimulated by surprises in numerical experiments, whose understanding is in turn being deepened by such theories. In future a better understanding is hoped both for the coupling to microscopic processes such as particle acceleration, for three-dimensional and time-dependent features and also for a closer comparison with observations.

References

Ahmad, I and Webb, D (1978) *Solar Phys.* **58**, 323.
Biskamp, D (1986) *Phys. Fluids* **29**, 1520.
Biskamp, D and Welter, H (1989) *Solar Phys.* **120** 49.
Forbes, T G (1989), submitted.
Forbes, T G and Priest, E R (1982) *Solar Phys.* **81**, 303.
Forbes, T G and Priest, E R (1987) *Rev. Geophys.* **25**, 1583.
Forbes, T G, Malherbe, J M and Priest, E R (1989) *Solar Phys.* **120**, 285.
Gratton, F, Heyn, F, Biernat, H, Rijnbeek, R and Gnavi, G (1988) *J. Geophys. Res.* **93**, 7318.

Holt, R and Mullan, D (1986) *Solar Phys.* **107**, 63.
Jardine, M and Priest, E R (1988) *J. Plasma Phys.* **40**, 143.
Martin, S F, Livi, S H B and Wang, J (1985) *Aust. J. Phys.* **38**, 929.
Martin, S F (1986) in *Coronal and Prominence Plasmas* (ed A Poland), NASA CP 2442, p.73.
Malherbe, J M and Priest, E R (1983) *Astron. Astrophys.* **123**, 80.
Mikic, Z, Barnes, D C and Schnack, D D (1988) *Astrophys. J.* **328**, 830.
Parker, E N (1958) *Astrophys. J.* **128**, 664.
Petschek, H E (1964) AAS-NASA Symp. on Phys. of Solar Flares, NASA SP-50, p.425.
Priest, E R (1987) in *The Role of Fine-Scale Magnetic Fields on the Structure of the Solar Atmosphere* (ed E Schroter, M Vazquez, A Wyller), Cambridge Univ. Press, p.297.
Priest, E R and Forbes, T G (1986) *J. Geophys. Res.* **91**, 5579.
Priest, E R, Hood, A W and Anzer, U (1989) *Astrophys. J.* **344**, 1010.
Priest, E R and Lee, L C (1990) *J. Plasma Phys.*, submitted.
Sonnerup, B U O (1970) *J. Plasma Phys.* **4**, 161.
Sonnerup, B U O and Priest, E R (1975) *J. Plasma Phys.* **14**, 283.
Sweet, P A (1958) *Nuovo Cimento* **8**, 188.
Van Ballegooijen, A and Martens, P (1989) *Astrophys. J.* **343**, 971.
Vasyliunas, V M (1975) *Rev. Geophys.* **13**, 303.

DISCUSSION

MOGILEVSKIJ: (i) The time of reconnection is usually short but in some flares the proton acceleration continues for longer than 5000 sec.
(ii) Can you give experimental arguments that the energy for the solar flare will be *only* by the reconnection of magnetic fields in the corona?

PRIEST: (i) Impulsive bursty reconnection in the impulsive phase may accelerate particles rapidly over short times, but then the quasi-steady reconnection in the main phase may continue to accelerate the particles over longer periods.
(ii) One cannot *prove* the existence of reconnection in a flare, since we shall not this century be able to resolve the reconnection region or map the field lines observationally with high precision. However, there are many indirect observations which are consistent with our theoretical understanding and with numerical experiments, and so reconnection is highly likely, especially since there is no other way of releasing the required energy from the magnetic field.

WEISS: As you know, the latest numerical experiments by Biskamp show complicated dynamical behaviour with self-similar structure associated with resistive instabilities. Biskamp asserts that reconnection has little resemblance to your models. Do you have any comments on this?

PRIEST: Biskamp's experiments are excellent but their interpretation is a matter of lively debate. Numerical and theoretical approaches complement one another and each has advantages and limitations. For example, the numerical experiments apply to a small range of magnetic Reynolds number and for specific boundary conditions. My view is that the theory does indeed enable us to understand the experiments better and, in particular, isolates the effect of a wide range of different boundary conditions. In particular, Forbes and Priest (1987, *Rev. Geophys* **25**,1583) have indeed been able to interpret the results of Biskamp in terms of their models. I would expect new reconnection experiments in future with a more comprehensive treatment of the boundary conditions to show Biskamp's results to be part of a wider class of solutions. In particular, we have shown how the type of reconnection depends crucially on the details of both inflow and outflow boundary conditions.

SMITH: (i) D. Biskamp is of the opinion that for 2-D steady reconnection you cannot exceed the Sweet-Parker rate.
(ii) Also your classification depends on a small parameter which the numerical experiments show is not small.

PRIEST: (i) What we have shown conclusively is that you can indeed exceed the Sweet-Parker rate provided you have the appropriate boundary conditions. Indeed you can obtain regimes similar to the Petschek regime or much faster regimes. It is not surprising that Biskamp did not find Petschek-like reconnection when you see what boundary conditions he has imposed. However, what I have presented here are, in a way, generalisations to the Petschek analysis to include extra effects that one finds in the experiments of Biskamp and others.
(ii) I agree that the Priest-Forbes theory does rely on assuming a small deviation from a uniform magnetic field. However, the resulting physical understanding in terms of slow- and fast-mode expansions is more general. The new Priest-Lee models are an attempt to drop the assumption of a nearly uniform field and we plan to try and generalise them even further.

FORBES: I have myself looked at Biskamp's numerical experiments in detail, and I believe they do show that you can get reconnection rates faster than the Sweet-Parker rate and the Petschek rate. Biskamp's definition of the reconnection rate is not the same as the standard one we use. When I re-evaluated his reconnection rate using the standard definition (Forbes and Priest, *Rev. Geophys*,**25**,1583 1987) I found that he does indeed have reconnection rates in excess of both the Sweet-Parker and Petschek rates in some of his cases.

LANG: (i) What is the driving force that makes magnetic fields come together?
(ii) Why does reconnection occur rarely?
(iii) Does reconnection happen in the photosphere, or in the corona?
(iv) Is there any observational evidence for magnetic fields actually coming together?

PRIEST: (i) It depends on the application. For example, in X-ray bright points the driver is the motion of photospheric magnetic footpoints, whereas in a two-ribbon flare I suggest that the eruption of a prominence drives reconnection below it.
(ii) I suspect that it actually occurs quite often, continually relaxing the field, reducing its complexity and energy and heating the corona.
(iii) In both. Photospheric magnetic fields tend to be dominated by plasma motions, which may drive the fields together, as in cancelling magnetic fields. Coronal magnetic fields tend to dominate the plasma and so reconnection can occur there either spontaneously by, say, the tearing mode instability or in response to footpoint motions.
(iv) Yes, for example in cancelling magnetic features and also there are cases where opposite polarity regions approach one another in active regions before flares.

KUNDU: I can cite several examples of evidence for magnetic field reconnection in solar flares, from VLA observations. The preflare region is generally heated (up to 10^7K) prior to flare onset; however, unless there is a newly emerging flux with appropriate polarity, the region does not flare. Here the new flux interacts with a pre-existing flux to produce a current sheet, which reconnects and produces a flare. Similarly, impulsive release of energy takes place in a gradual burst source when new flux emerges, produces a current sheet and leads to an impulsive burst via reconnection. These evidences are primarily dependent upon obtaining a spatial resolution of 1 arcsec.

UBEROI: There is a considerable literature on time-dependent magnetic reconnection in magnetospheric physics. Have these theories been applied to solar physics?

PRIEST: Yes, certainly. I have limited myself to steady reconnection but nonsteady effects have been considered analytically and numerically. The numerical studies of reconnection resulting from tearing or from an eruptive instability are strongly time-dependent. For regimes in which the central current sheet is small enough one tends to produce a quasi-steady reconnection, but when the central sheet is too large it goes unstable to secondary tearing (Forbes and Priest,1987), in which case one has an impulsive bursty modulation of the basic quasi-steady reconnection.

GOKHALE: (i) To what extent is the reconnection rate enhanced in your model?
(ii) If there are radiative losses from the reconnection region, will it not enhance the reconnection rate further?

PRIEST: (i) In a driven situation the reconnection may be slow or fast, depending on the speed of the driver, but for the almost-uniform models when $b < 0$ the maximum rate exceeds the Petschek value. For the nonuniform Priest-Lee models, for a given outflow speed the reconnection rate (*i.e.* inflow speed at large idistances) is a maximum at the smallest values of the central sheet length and it is typically a half of the outflow speed.
(ii) The energy balance will determine the plasma temperature in the central current sheet and so the magnetic diffusivity, but in a driven or freely reconnecting situation the main effect is to determine the size of the sheet rather than the rate of reconnection.

MONTGOMERY: We did a few preliminary 3D computations (incompressible MHD, periodic boundary conditions, initial-value problems) in order to watch the development of 3D perturbations of 2D current sheets. We saw very little that we could recognize from the well-studied 2D problem. (This involved no stabilizing, externally-applied, dc magnetic field.) The time-scales were fast, and the current sheets and field-surface perspective plots soon began to look like Swiss cheeses. What features of the 2D "X-point problem" do you expect to be able to recognize in 3D?

PRIEST: I agree very much that we need to go to 3D with numerical experiments and these will certainly give some surprises, just as the 2D experiments have done. Also, it is important to have a good interplay between numerical experiments and analytical models, using the former as a guide to constructing the latter if possible. However, the 2D calculations are still of great value since one can go to much higher magnetic Reynolds number and higher resolution and so they complement the 3D experiments. I would expect that locally near current sheets the 2D results may be valid in many cases. In some applications, such as coronal heating, the 3D aspects should be crucial, but in others such as two-ribbon solar flares with straight Hα ribbons and arcades the process is essentially 2D with a slow 3D modulation.

DRYER: Will reconnection occur only when an external forcing function or configuration is present?

PRIEST: No. In a sheared magnetic field or a current sheet, reconnection can be driven locally by a resistive instability. Indeed, the special feature of the Petschek regime is that it is, in a sense, "free" reconnection because it is the only regime where all the MHD characteristics are propagating information from the reconnection region.

THE ROLE OF MAGNETIC RECONNECTION IN FLARES AND PROMINENCE ERUPTIONS

T.G. FORBES
Institute for the Study of Earth, Oceans, and Space
University of New Hampshire
Durham, NH 03824, USA

ABSTRACT. Magnetic Reconnection is often invoked as the primary mechanism for driving a flare or a prominence eruption. Here I argue that a catastrophic loss of mechanical equilibrium, rather than reconnection, is probably the primary mechanism for driving these phenomena. However, reconnection is still essential in order for any significant amount of energy to be released. To illustrate this idea we present some recent results from an MHD simulation based on a catastrophe mechanism first proposed by Van Tend and Kuperus. In order for this mechanism to be effective a substantial amount of reconnection must occur within a few Alfvén-scale times. Such rapid reconnection is plausible since the loss of mechanical equilibrium can generate flows which drive the reconnection at a rapid rate.

1. Introduction

It is often stated that magnetic reconnection causes flares and prominence eruptions. However, I would like to suggest that reconnection is not the basic mechanism behind these phenomena. This does not necessarily mean that reconnection is unimportant, since in the scenario that I will discuss in this talk, reconnection is still an essential process without which two-ribbon flares and prominence eruptions would be impossible.

My present view of how reconnection works in these eruptive phenomena is based on recent analytical and numerical MHD models developed by myself and E.R. Priest. Our models were inspired by the early circuit models of Van Tend and Kuperus (1978) and Kaastra (1985) which treat the prominence, or pre-flare filament, as a line current suspended above the photosphere (see also Martens and Kuin, 1989). The basic magnetic mechanism which drives the current filament upwards in these models is illustrated in Figure 1. In Figure 1a, a current filament (shaded region) is stably suspended above the photosphere (hatched region) because the field lines surrounding the current filament are line-tied to the photosphere. If the filament is perturbed, it simply oscillates up and down around its equilibrium location as indicated in the energy diagram in Figure 1b. However, this stable equilibrium vanishes if the field lines surrounding the filament are disconnected from the photosphere so that the magnetic configuration approaches that shown in Figure 1c. As the energy diagram in Figure 1d indicates, the only stable equilibrium location for the configuration shown in Figure 1c occurs at infinite height.

Van Tend (1979) showed that a transition like that shown in Figure 1 can occur either continuously or discontinuously as the strength of the background field relative to the current in the

E. R. Priest and V. Krishan (eds.), Basic Plasma Processes on the Sun, 293–302.

filament is decreased. In the purely two-dimensional system considered by Van Tend, the transition is discontinuous only if the photospheric background field falls off with height faster than $1/y$, where y is the vertical coordinate. In order for the field to decrease with height in this fashion, the photospheric background field must have a two-dimensional dipole, or higher order

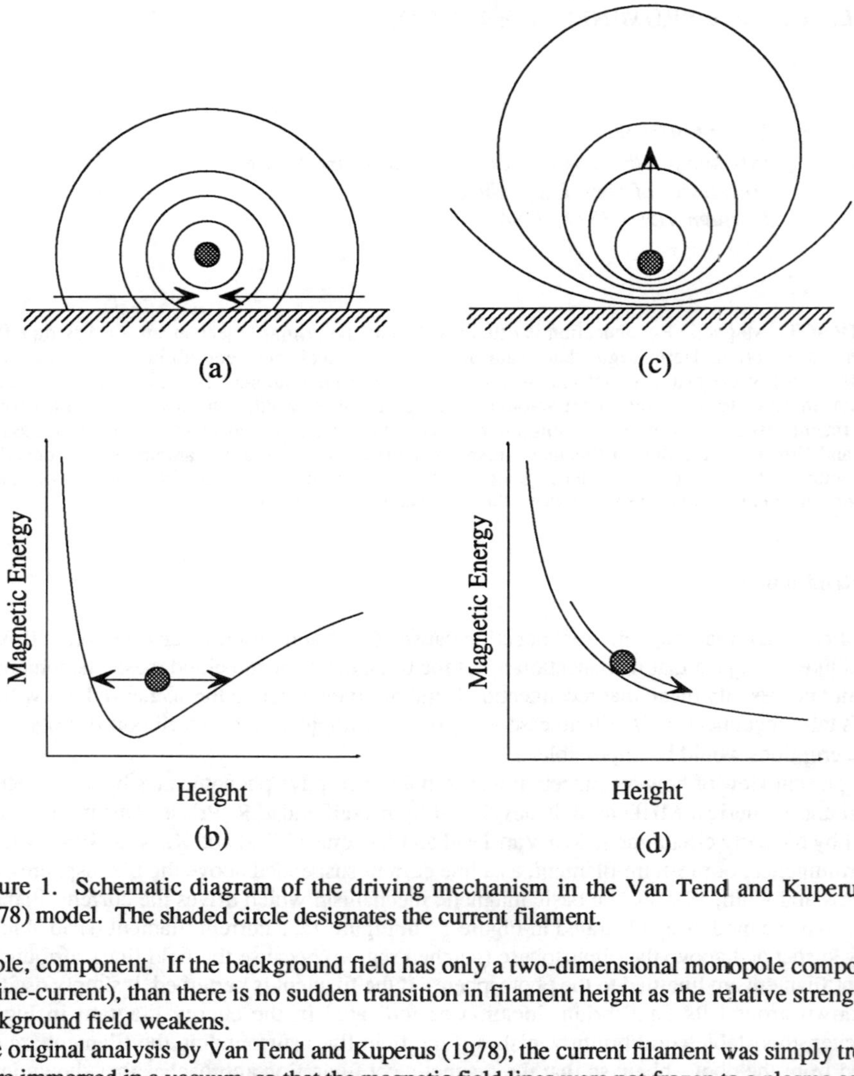

Figure 1. Schematic diagram of the driving mechanism in the Van Tend and Kuperus (1978) model. The shaded circle designates the current filament.

multipole, component. If the background field has only a two-dimensional monopole component (*i.e.* a line-current), than there is no sudden transition in filament height as the relative strength of the background field weakens.

In the original analysis by Van Tend and Kuperus (1978), the current filament was simply treated as a wire immersed in a vacuum, so that the magnetic field lines were not frozen to a plasma as they are in ideal MHD. Consequently, in their model, reconnection occurs freely. However, in a realistic plasma environment any attempt to change rapidly a configuration with an x-line in it leads to the formation of a current sheet at the x-line since reconnection is inhibited by the conductivity of the plasma. Kaastra (1985) and Martens and Kuin (1989) attempted to address this problem by incorporating current sheets within the circuit framework of Van Tend and Kuperus, but from an MHD point of view neither Kaastra's nor Martens and Kuin's model is self-consistent. Both models use Syrovatskii's (1971) solution for a current sheet to prescribe the properties of the current sheet. Although Syrovatskii's solution is a valid solution of the ideal MHD equations in the limit of low plasma β (*i.e.* strong magnetic fields), it has boundary conditions at infinity which are not consistent with the Van Tend and Kuperus model.

2. An Analytical MHD Model

Recently, E.R. Priest and myself developed a relatively simple Van-Tend-and-Kuperus-type model which is a self-consistent solution of the ideal-MHD equations in the limit of low plasma β (Priest and Forbes, 1990). In complex notation the magnetic field of this model is

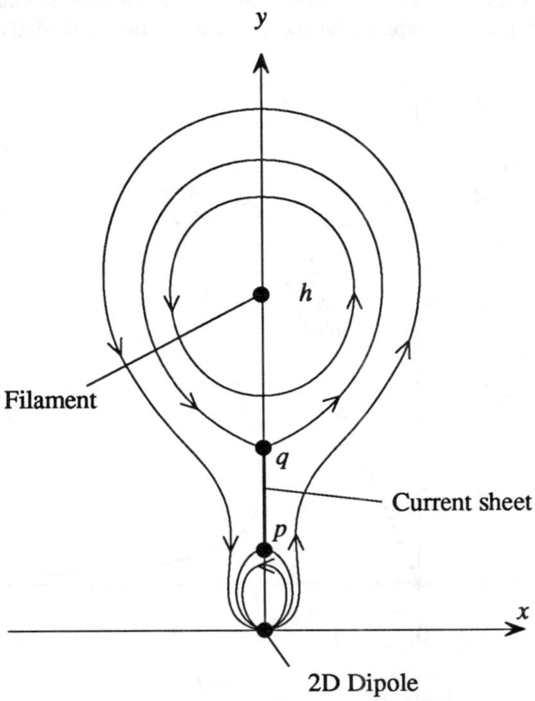

Figure 2. Magnetic field of the analytical MHD model .

$$B_y + iB_x = i\,\frac{mh^2\sqrt{(z^2+p^2)(z^2+q^2)}}{pq\,z^2\,(z^2+h^2)} \tag{1}$$

where $z = x + iy$, h is the height of the filament, and p and q are the upper and lower end points of an infinitely thin current sheet located on the imaginary axis as shown in Figure 2. The quantity m is the strength of a dipole at $z = 0$ and this dipole represents the photospheric background field. Application of the frozen-flux condition gives the relation:

$$2hpq\sqrt{(h^2-p^2)(h^2-q^2)} = m/I\,(h) \tag{2}$$

where $I(h)$ is the filament current. In the limit that the radius of the filament current tends to zero, $I(h)$ becomes constant.

When the background dipole is placed at a depth, h_b, below the photosphere, equilibria occur at

$$h = m/I - h_b \pm [(m/I)^2 - 2h_b(m/I)]^{1/2} \tag{3}$$

as shown in Figure 3 for $h_b = 1$. This set of equilibria has both stable and unstable branches, but there are no equilibria when m is less than the critical value of $2Ih_b$. As h_b tends to zero, the unstable branch approaches $h = 2m/I$, while the stable branch approaches $h = 0$.

Figure 4 compares the energy curve of this ideal-MHD model with no reconnection) to the energy curve of a comparable Van Tend and Kuperus model with uninhibited reconnection. In both models the magnetic energy of the system decreases with increasing height. Thus despite the formation of a current sheet in the MHD case, the filament still moves upwards. However, in the absence of reconnection the energy release is only 12.3% of that occurring when the reconnection is uninhibited. The remaining 87.7% of the available magnetic energy is stored in the field produced by the current sheet, and this energy can only be released by reconnection.

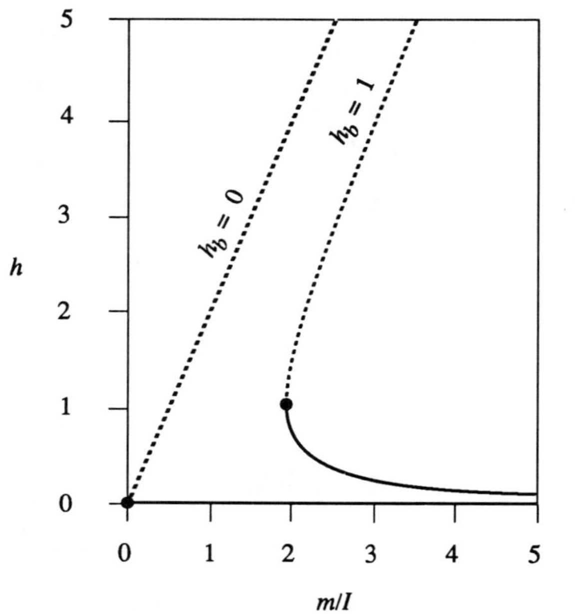

Figure 3. Equilibrium filament heights as a function of the ratio of the relative dipole strength. The dashed curves denote unstable equilibria, and h_b is the dipole depth below the surface (arbitrary units).

3. A Numerical MHD Model

Although Van Tend and Kuperus's model was originally discussed within the framework of circuit theory, it is possible to formulate the model as an initial-boundary-value problem for the time-dependent MHD equations. To illustrate how this can be done, consider the following system of equations:

$$\partial \mathbf{B}/\partial t = -\nabla \times (\mathbf{v} \times \mathbf{B}) + \eta \nabla^2 \mathbf{B} \tag{4}$$

$$\partial \rho/\partial t = -\nabla \cdot (\rho \mathbf{v}) \tag{5}$$

$$\rho[\partial \mathbf{v}/\partial t + (\mathbf{v}\cdot\nabla)\mathbf{v}] = -\nabla p + \mathbf{j} \times \mathbf{B} + \rho \nu \nabla^2 \mathbf{v} \tag{6}$$

$$\rho c_v T(\partial/\partial t + \mathbf{v}\cdot\nabla)\log(p/\rho^\gamma) = \eta j^2 + \rho\nu(\nabla\cdot\mathbf{v})^2 \tag{7}$$

$$p = R \rho T \tag{8}$$

in the domain $-\infty < x < \infty$, $0 < y\ \infty$. For this system the dependent variables are the magnetic field, $\mathbf{B}$, the velocity, $\mathbf{v}$, the density, ρ, and the pressure, p. The constants η, ν, c_v, γ, and R are the electrical resistivity, the viscosity, the specific heat at constant volume, the ratio of specific heats (= 5/3), and the universal gas constant (= $2c_v/3$), respectively. For the numerical simulation ν and η are set to the smallest possible values which, with a grid mesh of 100 x 100 points, gives viscous and magnetic Reynolds numbers of about 200 each.

The initial magnetic field is the sum of a dipole background at depth h_b, a current filament with the field

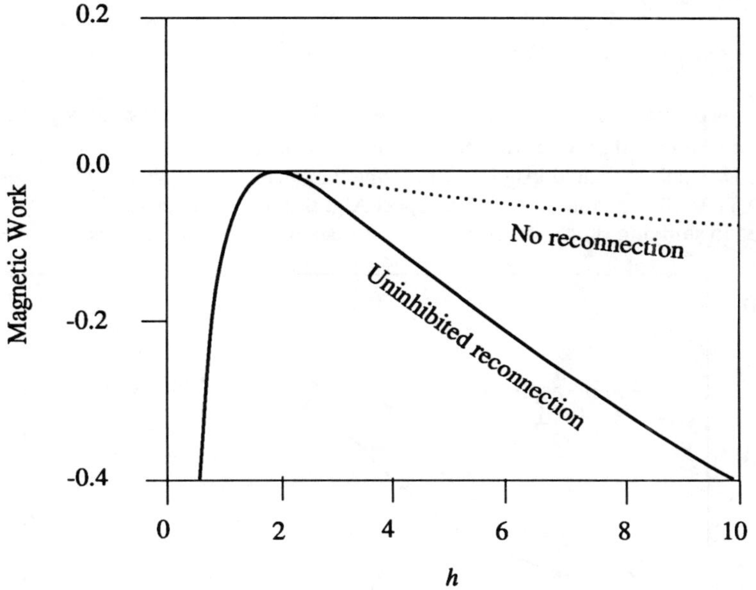

Figure 4. Magnetic work as a function of filament height, h, for the MHD (dashed curve) and non-MHD (solid curve) models. In both models $h_b = 0$.

$$B_\phi = \begin{cases} B_o(r/r_o) \ , & r \leq r_o \\ B_o(r_o/r) \ , & r \geq r_o \ , \end{cases} \tag{9}$$

centered at $y = h$, and a corresponding image filament at $y = -h$. Here r_o, the radius of the filament, is set at 0.05 times the half-with of the numerical domain, and initially the plasma is at rest and has a uniform entropy. The initial pressure distribution inside the filament is determined from the requirement that $\nabla p = \mathbf{j} \times \mathbf{B}$, while the pressure outside the filament is uniform (see Forbes 1990 for details).

The boundary conditions at the photosphere ($y = 0$) are:

$$j = 0 \qquad \text{(Line-Tying)} \tag{10}$$

$$v_x(x,0,t) = 0 \qquad \text{(No - Slip)} \tag{11}$$

$$v_y(x,0,t) = 0 \qquad \text{(Wall)} \tag{12}$$

The line-tying condition, $j = 0$, follows from the requirement that field lines be anchored to the base. In two-dimensional MHD, field lines correspond to contours of constant A, where A is the magnitude of the vector potential. Thus Faraday's equation combined with Ohm's law gives

$$E = -\partial A/\partial t = -(\mathbf{v}\times\mathbf{B})_z + \eta j = 0. \tag{13}$$

Since the wall and no-slip boundary conditions give $\mathbf{v} = 0$ at the base, j is necessarily zero.

In general this definition of line-tying is not equivalent to the frozen-flux condition. In MHD the field line velocity, $\mathbf{v_b}$, is

$$\mathbf{v_b} = \mathbf{v} + \eta\,(\mathbf{J}\times\mathbf{B})/B^2. \tag{14}$$

and line tying occurs at the base when $\mathbf{v_b} = 0$, while frozen flux occurs when $\mathbf{v_b} = \mathbf{v}$. Only for the special case when $\mathbf{v} = 0$ are line tying and frozen flux equivalent.

The MHD system of equations (4) - (8) is solved by using the explicit, flux-corrected transport code SHASTA (SHarp And Smooth Transport Algorithm) of Weber *et al.* (1979) . SHASTA is well-suited to studying shock waves since it can sustain a sharp shock transition over only two or

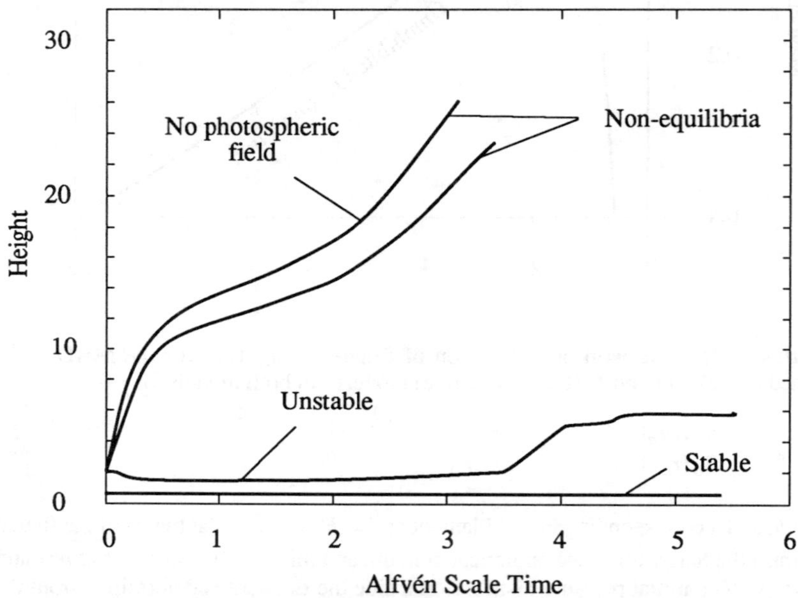

Figure 5. Filament height as a function of time with the height in units of h_b

three mesh points.

One can test the equilibrium properties of the system by varying the filament's height and current relative to the background dipole field. Figure 5 shows the filament height as a function of time for 4 cases all of which start at $h = 2h_b$ except for the case labeled 'stable' which starts at $h = 0.25\ h_b$. The nonequilibrium case labeled 'no photospheric field' is for $m = 0$, while the other nonequilibrium case is for $m = 0.5\ h_b/I$. The case labeled 'unstable' has $m = 2.25\ h_b/I$, and corresponds to a case which is initially on the unstable branch of the equilibrium curve shown in Figure 3.

As m increases, the background field becomes stronger, and the filament's upward motion becomes increasingly constrained by the tension that develops in the line-tied field lines as they are stretched upwards. When $m = 2.25\ h_b/I$ is reached, there is no initial force on the current filament, and the filament is in equilibrium. However, this equilibrium is unstable, and the filament moves upwards until a new equilibrium is reached at about $h = 8.1\ h_b$.

The final height which the filament reaches depends strongly upon the initial radius, r_o, of the filament. As r_o tends to zero the final equilibrium height tends to infinity because the magnetic energy available to drive the filament upwards increases without limit. The final equilibrium height in the analytical model increases only as $1/\log(r_o)$ (*cf.* Figure 4), and decreasing the radius by even several orders of magnitude causes only a modest increase in the final equilibrium height. Thus in the absence of significant reconnection, the numerical and analytical solutions together suggest that

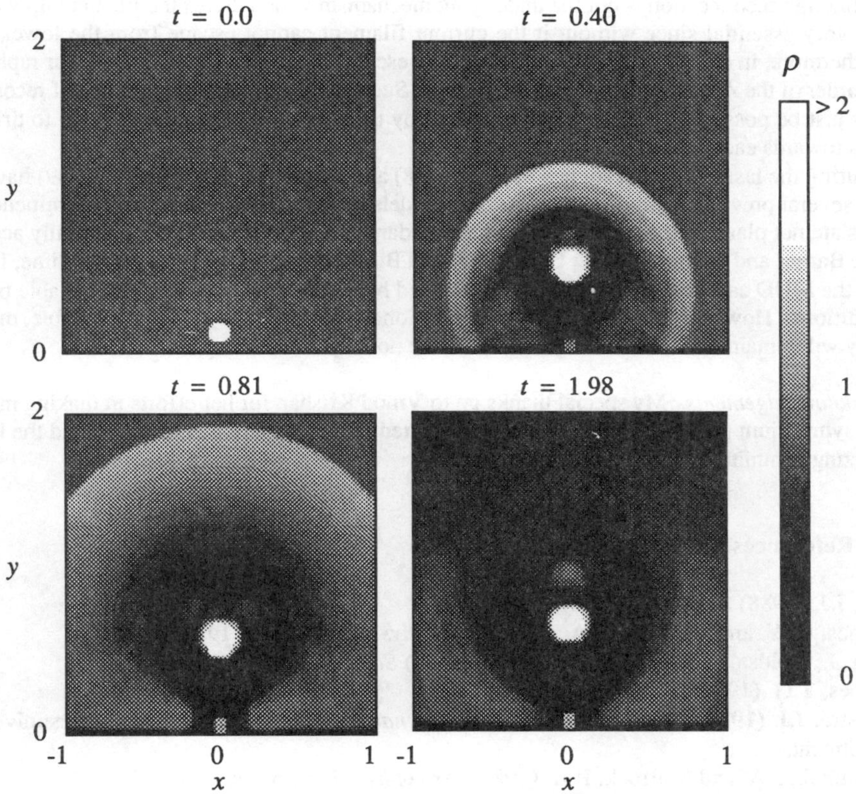

Figure 6. Grey-scale contours of the mass density at 6 different times for the nonequilibrium case with $m = 0.5\ h_b/I$. The filament moves upwards at about 0.3 times the Alfvén speed and is preceded by a fast-mode shock. Times are shown in units of the Alfvén scale-time, and $\rho = 1$ corresponds to the ambient coronal density.

the filament is not likely to rise more than roughly 10 times its initial height for any realistic value of the filament radius, r_o.

4. Conclusion

Using analytical and numerical MHD equivalents of Van Tend and Kuperus's circuit model, I have argued that a catastrophic loss of ideal MHD equilibrium, rather than reconnection, is the underlying mechanism which drives two-ribbon flares, and prominence eruptions. All of the models are based on current filaments suspended in the corona by a balance between magnetic compression and magnetic tension. The compression of field lines trapped between the photosphere and the filament pushes the filament upwards, while the tension of field lines tied to the photosphere and passing over the filament pulls the filament downwards (*cf.* Figure 1). MHD equilibrium is lost when the strength of the photospheric field relative to the filament field decreases below a critical value.

Although reconnection is not the underlying mechanism which drives the filament upwards, it is still very essential since without it the current filament cannot escape from the lower corona. Furthermore, in order for the current filament to escape, the reconnection must occur rapidly – on the order of the Alfvén scale-time of the system. Such an extraordinarily rapid rate of reconnection may just be possible since the flows produced by the loss of ideal equilibrium act to drive field lines towards each other (*cf.* Forbes 1990).

During the last few years, studies by Aly (1988) and Klimchuk and Sturrock (1989) have found that several previously proposed catastrophe models for two-ribbons flares and prominence eruptions are not plausible because they require boundary conditions that are not physically acceptable (*e.g.* Barnes and Sturrock, 1972; Low, 1977; and Birn *et al.*, 1978). At the present time, I believe that the MHD catastrophe model I have presented here does have physically acceptable boundary conditions. However, until a fully three-dimensional version of the model is available, the possibility will remain that the boundary conditions are not physically acceptable.

Acknowledgements. My special thanks go to Vinod Krishan for her efforts in making my trip to this symposium possible. This work was supported by NASA Grant NAGW-76 and the local organizing committee of IAU Symposium 142.

5. References

Aly, J.J. (1988) *Astron. Astrophys.*, **203**, 183.
Barnes, C.W. and Sturrock, P.A. (1972) *Astrophys. J.*, **174**, 659, 1972.
Birn, J., Goldstein, H., and Schindler, K. (1978) *Solar Phys.* **57**, 81.
Forbes, T.G. (1990) *J. Geophys. Res.*, in press.
Kaastra, J.J. (1985) *Solar Flares: An Electrodynamics Model*, PhD. Thesis, Rijksuniversiteit, Utrecht.
Klimchuk, J.A. and Sturrock, P.A. (1989), *Astrophys. J.*, in press.
Low, B.C. (1977) *Astrophys. J.*, **212**, 234.
Martens, P.C.H. and Kuin, N.P.M. (1989) *Solar Phys.*, **22**, 263.
Priest, E.R. and Forbes, T.G. (1990) *Solar Phys.*, in press.
Syrovatskii, S.I. (1971) *Soviet Phys. JETP*, **33**, 933.
Van Tend, W. (1979) *Solar Phys.*, **61**, 89.
Van Tend, W., and Kuperus, M. (1978) *Solar Phys.*, **59**, 115.
Weber, W.J., Boris, J.P., and Gardener, J.H. (1979) *Computer Phys. Comm.*, **16**, 243, 1979.

DISCUSSION

KUIJPERS: In Kaastra's work the position of the curret sheet is not arbitrary. In fact his self-consistent electrodynamical calculation gives an energy dissipation rate into particle acceleration in agreement with the observations. However he used as an input to the model the observed rise-speed of the filament. Of course in your work it is an important step forward that you can derive the filament speed.

FORBES: I cannot agree that Kaastra's current was calculated self-consistently since he used a current sheet solution of Syrovatsky and just added this current to the other currents. Since Syrovatsky's solution is for much different boundary conditions, Kaastra's solution is not a self-consistent solution of the MHD equations. However, I would still like to emphasize that Kaastra's solution was very worthwhile as a first attempt to construct a quantitative model based on the Van Tend-Kuperus mechanism.

KUNDU: You have discussed the question of particle acceleration in fast CME's. As you know Hundhausen *et al* at HAO have observed many CME's with slow speeds (50-200km/s). Everytime we have overlapping radio observations at meter-decameter wavelengths, we find radio emission in the form of type IV and/or type II bursts. Obviously particle acceleration is involved. Have you given any thinking as to how slow-mode shocks can accelerate particles responsible for the observed radio emission?

FORBES: I myself have not looked at the possibility of energetic particle acceleration at slow-mode shocks, but Marty Lee and Phil Isenberg in our group at the University of New Hampshire have done so. In their opinion there is, as of yet, no clearly established mechanism which will accelerate particles at slow shocks. However, they have also told me that it is still an open question.

DRYER: The slow-mode waves that are associated with the slow CMEs observed by the NRL and HAO coronagraphs, must be preceded by fast-mode waves. These fast-mode waves can easily steepen in the context of a decreasing density profile in the corona. If such steepening occurs and results in shock developments (as for the not-so-slow CMEs), then particle acceleration can be expected due to gradient-**B** drift and/or diffusion processes.

FORBES: Yes, the compressive wave generated by a slow CME will always steepen into a fast-mode shock, even if the CME is moving at a speed less than the fast-mode wave speed. Such a fast-mode shock can produce energetic particles by the well-known diffusion process, but I am not sure if these particles would appear in time to explain Kundu's Type IV and Type II radio bursts in the lower corona.

DAVILA: Would you expect spherical geometry, *i.e.* additional radial gradients in **B**, to affect your conclusions regarding the expulsion of the filament from the solar atmosphere?

FORBES: Yes. Probably, the decrease of the magnetic field with height should help the current filament to escape, since the concentration of field near the base of the photosphere would produce a "melon-seed" effect.

UBEROI: Considering the acceleration of protons, it appears at least for auroral phenomena that Alfvén's resonance may be a better mechanism.

FORBES: Again, it is difficult for me to say anything definite about kinetic processes within the MHD framework of my calculations. Resonance absorption could very well be important in the process.

MOGILEVSKIJ: Can such a mechanism of solar flares accelerate protons to 10^8-10^{10} eV?

FORBES: With an MHD model it is not possible to say much about the acceleration of energetic particles. However, I believe that any reconnection process which produces supermagneto-sonic outflow jets has the possibility to produce energetic protons by a Fermi-type-A-diffusion process operating at the fast-mode, termination shocks of the jets. Whether this process is quantitatively feasible, I do not yet know.

BUTI: Extending your argument about proton acceleration, could we accelerate heavier ions by this process?

FORBES: I assume so, but really I just do not know.

VAHIA: There is a lot of information that can be derived about such mechanisms, and, in particular, solar cosmic ray studies allow us to put limits on several parameters in the processes discussed here.

STRUCTURE AND EQUILIBRIUM OF CORONAL MAGNETIC FIELDS

A. A. VAN BALLEGOOIJEN
Harvard-Smithsonian Center for Astrophysics
60 Garden Street
Cambridge, MA 02138, U.S.A.

ABSTRACT. In "closed" magnetic structures (i.e., coronal loops) the random shuffling of magnetic footpoints in the photosphere causes twisting and braiding of field lines in the corona. If the motions are sufficiently slow, the coronal field evolves through a sequence of force-free equilibrium states. Numerical simulations are presented for a simplified model in which the overall curvature of the coronal loop is neglected. It is shown that magnetic fine structures develop on spatial scales significantly smaller than those of the imposed "photospheric" velocity field.

1. Introduction

It has been suggested that coronal loops are heated by dissipation of field-aligned electric currents (e.g., Gold 1964; Parker 1972; Tucker 1973; Rosner *et al.* 1978). These currents arise as a result of footpoint motions driven by subsurface convective flows. Since the magnetic field is nearly "frozen" into the plasma, the random walk of the footpoints across the solar surface causes twisting and braiding of field lines in the corona. If the footpoint motions are slow compared to the Alfvén time L/v_A and the plasma pressure is negligible compared to the magnetic pressure, then the force balance requires that the Lorentz force nearly vanishes, $\mathbf{j}\times\mathbf{B} = 0$; hence, electric currents associated with such twisted fields flow nearly parallel or anti-parallel to the magnetic field lines.

In this paper I present results from a numerical model in which the overall curvature of the coronal loop is neglected: the initial field is assumed to be uniform, extending between two parallel plates which represent the solar photosphere at the two "ends" of the loop. The field is assumed to evolve according to ideal MHD. The formation of current sheets in this model was discussed by Parker (1972, 1983, 1986a,b) and van Ballegooijen (1985, 1986, 1988a). The latter showed that spatially continuous flows at the boundary plates produce spatially continuous force-free fields inside the volume. Numerical results were presented earlier by van Ballegooijen (1988b).

2. Numerical Method and Results

A Cartesian reference frame (x,y,z) is adopted, with the boundary plates located at $z = 0$ and $z = L$. The initial field (time $t = 0$) is given by $\mathbf{B}(x,y,z,0) = B_0\hat{\mathbf{z}}$. For simplicity I assume

E. R. Priest and V. Krishan (eds.), Basic Plasma Processes on the Sun, 303–308.

that the field lines are fixed at $z = 0$, and that the velocity at $z = L$ consists of a series of sinusoidal flow patterns with "wavelength" $\lambda \ll L$. Specifically, the velocity $\mathbf{v}_n(x,y,L)$ for times t in the range $(n-1)\Delta t < t < n\,\Delta t$ (with n = 1, 2, ...) is given by:

$$v_{n,x} = 0, \qquad v_{n,y} = A \sin(kx + \phi_n), \qquad \text{for } n \text{ odd}, \tag{1a}$$

$$v_{n,x} = A \sin(ky + \phi_n), \qquad v_{n,y} = 0, \qquad \text{for } n \text{ even}, \tag{1b}$$

where A is the velocity amplitude, k (= $2\pi/\lambda$) is the wavenumber, and the ϕ_n are arbitrary phase angles ($v_z = 0$). The magnetic field is described in terms of the paths of the field lines, $x = X(x_0,y_0,z,t)$ and $y = Y(x_0,y_0,z,t)$, where x_0 and y_0 are the initial coordinates of a field line at time $t = 0$. Then the magnetic field is:

$$\mathbf{B}(x_0,y_0,z,t) \approx \frac{B_0}{J}\left[\frac{\partial X}{\partial z},\frac{\partial Y}{\partial z},1\right], \tag{2}$$

where $J(x_0,y_0,z,t)$ is the Jacobian of the transformation $(x_0,y_0) \rightarrow (X,Y)$. The functions X and Y satisfy the following boundary conditions:

$$X(x_0,y_0,0,t) = x_0, \qquad Y(x_0,y_0,0,t) = y_0, \tag{3a}$$

$$X(x_0,y_0,L,t) = X_{top}(x_0,y_0,t), \qquad Y(x_0,y_0,L,t) = Y_{top}(x_0,y_0,t), \tag{3b}$$

where X_{top} and Y_{top} are the positions of the footpoints at $z = L$, as determined from equations (1a) and (1b). In the limit $\lambda \ll L$ the magnetic *free* energy ΔW of the system is given by

$$\Delta W \approx \frac{B_0^2}{8\pi}\int_0^L\int_0^\lambda\int_0^\lambda \left[\left(\frac{\partial X}{\partial z}\right)^2 + \left(\frac{\partial Y}{\partial z}\right)^2 + (J-1)^2\right] dx_0\,dy_0\,dz. \tag{4}$$

By minimizing ΔW subject to the constraints (3a,b), a force-free magnetic field is obtained (Sakurai 1979). I introduce a Lagrangian grid tied to the field lines. Let X_{ijk} and Y_{ijk} denote the values of $X(x_0,y_0,z)$ and $Y(x_0,y_0,z)$ at a grid point. Replacing the partial derivatives in equation (4) by their finite-difference analogs and the integrations by sums, the magnetic energy ΔW may be written as a function of the variables X_{ijk} and Y_{ijk} at all interior grid points. I determine the minimum of this function using the conjugate-gradient method (Press *et al.* 1986). This is a general-purpose method for finding the minimum of a function of a large number of variables.

Computations were made for nine sets of randomly chosen phase angles. A velocity amplitude $Ak\Delta t = 1$ was assumed. Calculations for time steps $n \leq 4$ were made with a grid of 64×64×8 points (including the points on the boundaries), while calculations for $n = 5$ were made with a grid of 128×128×11 points. For $n > 5$ not all nine cases could be adequately resolved, even with the larger grid. Plots of the projections of field lines onto a plane z = constant show that the field lines are more or less straight lines connecting the footpoints. This indicates a tendency of the field lines to be as short as possible. Contour plots of the electric current density at $z = 0$ were also produced. As time progresses, the electric currents become stronger and elongated regions with high current density develop ("current sheets"). Examples of such plots are given by van Ballegooijen (1988b) and by Mikic *et al.* (1989). Figure 1 shows contour plots of the average power spectrum of the current density. This clearly shows the progression of power towards higher wavenumbers, indicative of the development of small-scale structure in the electric-current distribution.

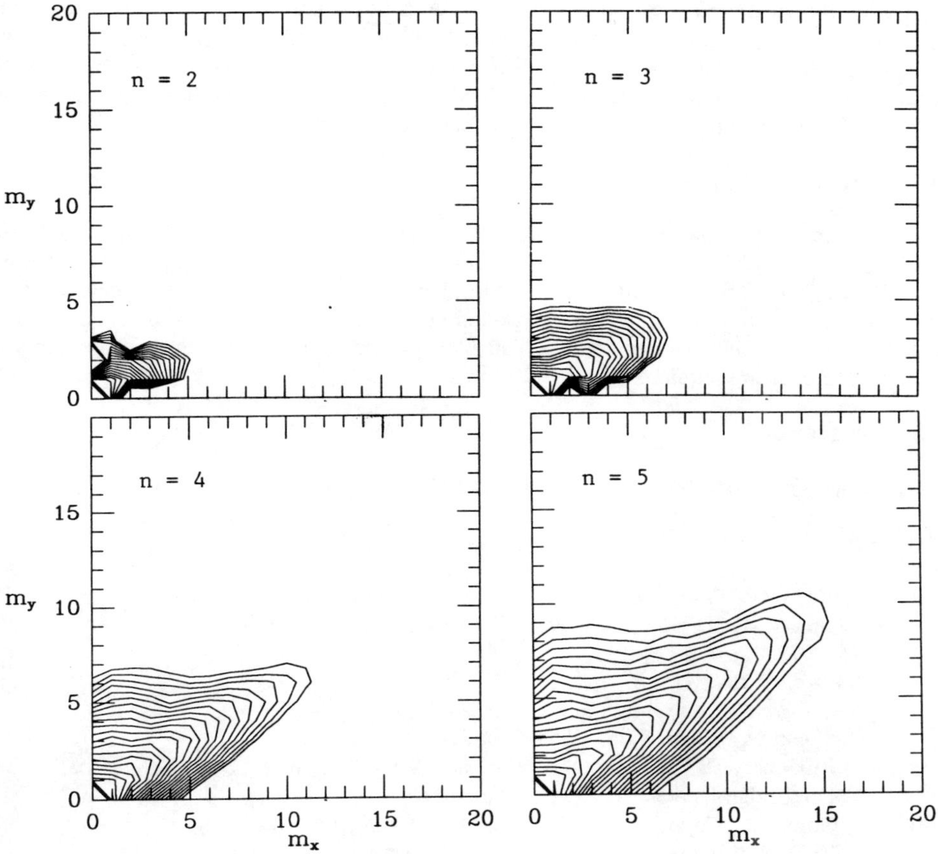

Figure 1. Contour plots of the spatial power spectrum of current-density fluctuations as measured at $z = 0$, for different times n. The average over nine realizations of random phase angles is shown. Neighboring contours are spaced by a factor 2 in power.

The fact that field lines do not deviate strongly from straight lines suggests the following linear approximation:

$$X(x_0, y_0, z, t) = x_0 + [X_{top}(x_0, y_0, t) - x_0]\, z/L, \tag{5a}$$

$$Y(x_0, y_0, z, t) = y_0 + [Y_{top}(x_0, y_0, t) - y_0]\, z/L. \tag{5b}$$

Although this does not generally represent a physically possible magnetic field (J may be negative), it does provide a *lower bound* on the magnetic free energy of the force-free field. Also, it provides the correct solution in the limit of small displacements, i.e., when $|X_{top} - x_0|$ and $|Y_{top} - y_0|$ are small compared to λ (see Zweibel and Li 1987). Inserting (5a,b) into the first two terms of equation (4), and neglecting the third term, I obtain the following expression for the lower bound:

$$\Delta W_{\min} = \frac{B_0^2}{8\pi L} \int_0^\lambda \int_0^\lambda \left\{ [X_{top}(x_0, y_0, t) - x_0]^2 + [Y_{top}(x_0, y_0, t) - y_0]^2 \right\} dx_0\, dy_0. \tag{6}$$

I find that with $Ak\Delta t = 1$ the actual energy ΔW does not exceed ΔW_{min} by more than about 15%. Also, the distributions of electric current at $z = 0$ derived from expressions (5a,b) are very similar to the computed, force-free distributions. Thus, we have not yet reached the regime where the field lines are strongly braided. Apparently, strong braiding implies the existence of even smaller spatial scales that cannot yet be resolved.

3. Conclusion

The simulations show that random motions of footpoints can produce fine structures in the electric-current distribution on scales significantly smaller than λ, the scale of the imposed velocity field. However, the field lines are still relatively simple; they can be crudely approximated as straight lines. The numerical resolution is not large enough to simulate strongly braided magnetic fields, with circuitous paths of the field lines. Further improvements in numerical methods are required.

Acknowledgement

This work was supported by NASA under grants NAGW-249 and NAGW-1568.

References

Gold, T. 1964, in *The Physics of Solar Flares*, ed. W. Hess (NASA SP-50), p. 389
Mikic, Z., Schnack, D.D., and Van Hoven, G. 1989, *Ap. J.* **338**, 1148
Parker, E.N. 1972, *Ap. J.* **174**, 499
-------- 1983, *Ap. J.* **264**, 642
-------- 1986a, *Geophys. Astrophys. Fluid Dyn.* **34**, 243
-------- 1986b, *Geophys. Astrophys. Fluid Dyn.* **35**, 277
Press, W.H., Flannery, B.P., Teukolsky, S.A., and Vetterling, W.T. 1986, *Numerical Recipes* (Cambridge University Press), Chap. 10.6
Rosner, R., Golub, L., Coppi, B., and Vaiana, G.S. 1978, *Ap. J.* **222**, 317
Sakurai, T. 1979, *Pub. Astron. Soc. Japan* **31**, 209
Tucker, W. 1973, *Ap. J.* **186**, 285
van Ballegooijen, A.A. 1985, *Ap. J.* **298**, 421
-------- 1986, *Ap. J.* **311**, 1001
-------- 1988a, *Geophys. Astrophys. Fluid Dyn.* **41**, 181
-------- 1988b, in *Solar and Stellar Coronal Structure and Dynamics*, ed. R.C. Altrock (Sunspot, NM: National Solar Observatory), p. 115
Zweibel, E.G., and Li, H. 1987, *Ap. J.* **312**, 423

DISCUSSION

MONTGOMERY: Do you have a velocity field away from the end plate? If so, how is it advanced in z and t? Physically, one would think that forcing the velocity field to have a definite temporal behaviour at the end plate would launch Alfvén waves, and that is where much of the energy might go.

VAN BALLEGOOIJEN: Yes, there is a velocity field away from the end plates, but it is not explicitly considered in the code. The force-free problem is treated as a boundary-value problem; the advance in t of (x,y) positions is done only at the boundary plates. If the transverse velocity is small compared to the Alfvén speed and the velocity varies on a timescale (Δt) long compared to L/V_A, significant Alfvén waves are probably not launched; the field evolves quasi-statically.

MOGILEVSKIJ: What is the energy density of the force-free magnetic fields?

VAN BALLEGOOIJEN: The magnetic free energy is at most a few percent of the potential-field energy; it scales with $(\lambda/L)^2$.

DRYER: (i) In order to achieve the large amount of braiding, do you think it may be necessary to go to an initial-boundary value problem with the presence of pressure, velocity field, *etc.*?
(ii) Would you comment on the conclusion of Klimchuk and Sturrock (*Ap.J.*, Oct 1989) that loss-of-equilibria, quasi-static, force-free calculations have no physical basis unless the plasma pressure and gravity are included?

VAN BALLEGOOIJEN: (i) Gas pressure is not essential, but it might be numerically advantageous to use an initial-value approach as Mikic *et al.* have done.
(ii) Loss-of-equilibrium may occur in force-free fields if 3D effects are taken into account. They may occur in the context of the Parker problem, but I have not found any evidence for non-equilibrium in my calculation, perhaps because the free energy is much smaller than the potential-field energy.

PRIEST: (i) What does $t = 5$ in your nondimensionalisation mean. How many time steps would you need to produce braiding?
(ii) Is it a problem of numerical stability or computer storage which stops you going beyond $t = 5$?
(iii) How do your results differ from those of Mikic?
(iv) How long would it take for the current densities to build up to values where magnetic dissipation is important?

VAN BALLEGOOIJEN: (i) Each velocity pattern persists for a time Δt; "$t = 5$" refers to the magnetic field at time $t = 5\Delta t$. Braiding probably requires several tens of Δt.
(ii) Numerical errors develop for $t = 6\Delta t$ for some sets of phase angles. I would not call it a problem of "numerical stability" because the energy-minimization scheme always finds a lowest energy state.
(iii) The results are nearly identical to those of Mikic *et al.* when I use the same phase angles. However, they did not do the averaging over different sets of phase angles.
(iv) Several tens of Δt.

HOLLWEG: A key issue in the field-stressing models is how much energy flux you can in fact launch into the corona. For realistic numbers, do you have a feel yet for how much flux you can launch into a loop? I still have the impression that one has to stretch the numbers to get the required energy fluxes.

VAN BALLEGOOIJEN: Earlier analytical estimates (Van Ballegooijen 1986) suggested a heating rate per unit volume $\varepsilon \simeq 0.2B^2D/L^2$, where D is the "diffusion constant" of photospheric motions. Taking $D \sim 400$ km^{2}s^{-1} (De Vore *et al.* 1985), the heating rate is a factor 40 too small, but there are many uncertainties in these estimates, and so we should not rule out DC currents as a mechanism for coronal heating.

DAVILA: In your present numerical calculation are you near the statistical steady state where the energy input at the boundary is balanced by current dissipation within the interior of the model?

VAN BALLEGOOIJEN: No, on the contrary, these calculations show only the initial development of the cascade. Magnetic diffusion (which is neglected here) becomes important only much later, when very thin current layers have developed.

KUIJPERS: The coronal field is anchored in discrete subphotospheric elements. I would expect the velocity field not to vary much over an individual magnetic element. Therefore, are your calculations applicable primarily to the boundaries between individual flux tubes rooted in separate magnetic elements?

VAN BALLEGOOIJEN: The sinusoidal velocity patterns used in these calculations are indeed not very realistic and do not take into account the flux-tube nature of the photospheric field. Indeed, if there are discrete flux tubes, magnetic discontinuities develop at the interfaces between these flux tubes in the corona. Hence, current sheets probably develop even more rapidly than suggested here.

NARAIN: How do you differ from Parker? What is your estimate for heating solar corona?

VAN BALLEGOOIJEN: Parker proposes that, in ideal MHD, current sheets form "spontaneously" (*i.e.*, sheet thickness vanishes at finite time), whereas I propose that current sheets develop "gradually" (sheet thickness decreases exponentially with time). I agree, however, with Parker that current dissipation is a promising mechanism for coronal heating.

THE EVOLUTION OF A SHEARED POTENTIAL MAGNETIC FIELD IN THE SOLAR CORONA*

J. T. Karpen, S. K. Antiochos, and C. R. DeVore
Naval Research Laboratory, Washington, DC 20375-5000

1. Introduction

Several theoretical studies have proposed that, in response to photospheric footpoint motions, current sheets can be generated in the solar corona without the presence of a null point in the initial potential magnetic field. In these analytic models, current sheets form wherever the coronal field dips down and is parallel to the photosphere. A fundamental assumption in these analyses — commonly referred to as the line-tying assumption — is that all coronal field lines are anchored to a boundary surface representing the top of the dense, gas-pressure-dominated photosphere. In theoretical arguments presented elsewhere (Karpen, Antiochos, and DeVore 1989), however, we show that line-tying is not valid for "dipped" coronal fields, and hence that the conclusions of the line-tied models are incorrect. We contend that current sheets will not form if the photosphere–corona interface is represented by a physically valid model. Here we summarize a numerical investigation of the response of a "dipped" potential magnetic field in a hydrostatic-equilibrium atmosphere to shearing motions of the footpoints. Our results show that, in the absence of artificial line-tying conditions, a current sheet indeed does not form at the location of the dip. Rather, the dipped magnetic field rises, causing upflows of photospheric and chromospheric plasma.

2. The Numerical Model

Our calculations were performed with a new, 2.5-dimensional, Cartesian finite-difference code which solves the compressible, time-dependent conservation equations of ideal Eulerian magnetohydrodynamics plus solar gravity. We assume that the plasma is fully ionized with a mean ionic charge of 1.059. The effects of momentum and magnetic field normal to the computational ($x-y$) plane are included self-consistently, assuming no gradients in this direction (z). The atmosphere begins in hydrostatic equilibrium, with temperature and density profiles typical of the photosphere–chromosphere interface. The initial magnetic configuration is a 2D potential field whose source is a pair of line quadrupoles separated by 5000 km. The initial field strength is in the range $1.3 <| \mathbf{B} |< 101$ G. We take advantage of the inherent symmetry in this field and model only the right half. Appropriate symmetry conditions were imposed at the left boundary. The bottom boundary was assumed to be sufficiently deep in the photosphere that the normal component of velocity is negligible there. Elsewhere, we applied open conditions to allow maximum freedom for the magnetic field and plasma to adjust to the shear.

E. R. Priest and V. Krishan (eds.), Basic Plasma Processes on the Sun, 309–312.

3. Results

The initial 2D potential field was sheared by imposing a localized body force for the entire run duration of 500 s. By this time, the effects of the shear clearly extend to the left axis yet no vertical current structures are visible there. Contours of the shear-induced B_z are aligned with the planar field lines in the low-β region, indicating that this region is approximately force free. Where $\beta \gtrsim 1$, significant mass flows exist along the field. Upflows of a few km s^{-1} predominate in the upper left quadrant, where $\beta \sim 0.1$, in response to the magnetic-field uplift. The magnetic flux through the left axis, $\int_0^x B_y dx$, decreases with time at a given height for $\beta \lesssim 5$, and varies smoothly with height at each time. This verifies that the upward field-line motion increases continuously from zero at the base to finite values above, and that a vertical current sheet never develops.

4. Conclusions

We find that no discontinuity or current sheet forms when the footpoints of an initial potential magnetic field are sheared at the photosphere–chromosphere interface in a gravity-stratified atmosphere. This result contradicts the conclusions of Aly (1985), Moffatt (1987), and Low and Wolfson (1988), for the following reason. Contrary to a basic assumption of these models, the transition between the photosphere and chromosphere is continuous. The field lines are not line-tied at this level, but rather can adjust to the imposed stress as long as $\beta \lesssim 5$. Hence, the effects of solar gravity cannot be neglected in determining the coronal magnetic-field reponse to photospheric footpoint motions. In addition, the dynamic response of the photospheric and chromospheric plasma lifted by the stressed field might have interesting ramifications for solar activity such as surges and spicules (*e.g.*, Blake and Sturrock 1985). On a larger scale, this process might drive flux eruption in complex magnetic topologies.

* This work was supported by ONR, the NASA STTP Program, and the NASA Solar and Heliospheric Branch. The calculations were performed on the NRL Cray X-MP with a generous grant of computer time under the 6.1 Production Run program.

References

Aly, J.J. 1985, *Astron. Astrophys.*, **143**, 19.
Blake, M.L., and Sturrock, P.A. 1985, *Astrophys. J.*, **290**, 359.
Karpen, J.T., Antiochos, S.K., and DeVore, C.R. 1989, *Astrophys. J.Letters*, submitted.
Low, B.C., and Wolfson, R. 1988, *Astrophys. J.*, **324**, 574.
Moffatt, H.K. 1987, in *Advances in Turbulence*, eds. G. Comte-Bellot and J. Matieu (Berlin: Springer-Verlag), p. 228.

DISCUSSION

VARMA: What role has gravity played, physically speaking, in your calculations?

KARPEN: Gravity is responsible for the stratification of the atmosphere, and hence the invalidation of treating the photosphere as an infinitely thin discontinuity (between high and low β plasmas) at which contacting field lines are line-tied.

FORBES: Are you saying that the Low & Moffatt results are not valid because a slight violation of the line-tying condition completely prevents the formation of the current sheets that occur in their model?

KARPEN: Essentially, yes.

TAKTAKISHVILI: Is your calculation relevant for spicule formation?

KARPEN: Our results - which show upward velocities of a few km s^{-1} in the vicinity of the dipped magnetic field - may have interesting ramifications for dynamic chromospheric events of this type. We plan to investigate this issue further in future numerical simulations.

PRIEST: A time of 400 secs is how many Alfvén times? Does the experiment approach an equilibrium.

KARPEN: 400s is approximately 10 Alfvén transit times in the low-β portion of the system. We have performed relaxation calculations of the asymmetric version of the Low-Wolfson model and do observe current sheets forming along the separatrices as predicted. Thus far, none of our calculations - either with or without gravity - exhibit a loss of equilibrium.

DRYER: (i) As I recall - your calculation did not have explicit resistivity included?
(ii) In your high-β case, am I correct in thinking that a small shearing velocity was sufficient to get dynamic motion? (In our calculations [Wu,Song, & Dryer,poster], for our $\beta = 15.4$ case, a peak shearing velocity of 5 km/sec was sufficient. For another case of $\beta = 0.06$, we required a higher value of 15 km/sec.)

KARPEN: (i) That is correct - we are solving the ideal MHD equations, although the finite difference technique always introduces a small "numerical" resistivity which does not affect the present results.
(ii) Our calculations employed a shearing velocity which was smaller than your smallest value; we did not see any evidence for the instability observed in your work, under those low-velocity conditions, but we do see upflows of a few km s^{-1} in the vicinity of the magnetic-field "dip".

HOLLWEG: Did your forcing actually get applied to field lines which touch the photosphere?

KARPEN: The forcing was applied to a range of field lines - those which extended deep in the photosphere as well as up to the $\beta \approx 1$ level. It is important to realize, however, that there is no *discrete* photosphere to "touch", due to the gravitational stratification of the atmosphere.

ROBERTS: Can you say how your results will change with varying g? I am wondering whether current sheets will again form if stratification is sufficiently weak (low g), as it will be in low g stars, or whether your results will show that sheets are always absent if $g \neq 0$. My guess would be that a critical strength of stratification would arise; if so, this might be important for ideas of heating of stellar atmospheres in the case of stars that are different from the Sun.

KARPEN: Our present intuition suggests that the effective width of the current structures which form in a gravity-stratified atmosphere should be roughly equal to the photospheric gravitational scale height. What actually happens when $g \rightarrow 0$ is an open but very interesting question, however. We plan to investigate this issue numerically, by repeating our calculations with successively smaller gravitational scale heights.

STORAGE AND RELEASE OF MAGNETIC ENERGY IN A FORCE-FREE FIELD

J.J. ALY
Service d'Astrophysique
Centre d'Etudes Nucléaires de Saclay
91191 Gif-sur-Yvette Cedex
France

ABSTRACT. For a 3D force-free field occupying a half-space $D = \{z > 0\}$, we discuss : i) the storage of free magnetic energy when the field evolves quasi-statically as a consequence of motions imposed to its footpoints on the plane $\{z = 0\}$; ii) the release of this energy during a reconnection process implying a rearrangement of the lines which is either local, occuring in the neighbourhood of spontaneously formed current sheets, or global, occuring in an explosive flare-like way.

1. INTRODUCTION

In the solar corona, a large amount of free energy may be stored in the magnetic field which is maintained in a stressed, essentially force-free state by the never-ceasing motions of the subphotospheric plasma in which its footpoints are anchored. Part of this energy may be released : i) in a more or less continuous way if the quasi-static evolution of the field leads to the spontaneous recurrent formation of thin current layers through which the lines may reconnect (thus suffering a local rearrangement) ; ii) in an explosive way if the field is brought into a metastable state which thus effects a reconnecting transition to a lower energy state (this implying a global rearrangement of the lines). The first process may lead to a general heating of the corona (Parker, 1989), while the second one may be related to such well-known phenomena as flares and coronal transients. In this communication, we discuss some theoretical points which pertain to the general processes of energy storage and release. We shall use a simple model in which the corona is represented by the half-space $D = \{z > 0\}$ and the photosphere by the plane $\partial D = \{z = 0\}$.

E. R. Priest and V. Krishan (eds.), Basic Plasma Processes on the Sun, 313–318.

2. SOME GENERAL PROPERTIES OF FORCE-FREE FIELDS IN D

2.1. How much magnetic energy can be stored in D ?

Let us consider in D a finite energy force-free magnetic field $\mathbf{B}$ (thus satisfying $\nabla.\mathbf{B} = 0$ and $(\nabla \times \mathbf{B}) \times \mathbf{B} = 0$) and assume that the function $g(x,y) = B_z(x,y,o)$ is such that $\omega^{2+\varepsilon} g = 0(1)$ for some $\varepsilon > 0$ ($\omega^2 = x^2 + y^2$). Then one has (Aly, 1984)

$$C_o[g] \overset{\text{def}}{=} \int_D B_o^2 \, d\mathbf{r} \leq C \overset{\text{def}}{=} \int_D B^2 \, d\mathbf{r} \leq C_m[g] \tag{1}$$

where $\mathbf{B}_o$ is the finite energy potential field ($\nabla.\mathbf{B}_o = 0$; $\nabla \times \mathbf{B}_o = 0$) such that $B_{oz}(x,y,o) = g(x,y)$, and $C_m[g]$ is a number depending only on g and choosen as small as possible (C_m may be explicitly bounded from above). As there are only two force-free fields in D which can be canonically determined from g - i.e. $\mathbf{B}_o$ and $\mathbf{B}_{op}$, which is an open field, potential everywhere but on a current sheet in equilibrium - it is tempting to conjecture that $C_m[g] = C_{op}[g] = \int_D B^2{}_{op} \, d\mathbf{r}$. That this conjecture is true indeed is supported by the fact that one has $C \leq C_{op}[g]$ if g is choosen to belong to a particular class (Aly, 1984) or if $\mathbf{B}$ is a constant-α force-free field (Aly, 1989a) ; and by a general result reported in § 2 below.

2.2. Scaling of the free energy with the current

The force-free equation for $\mathbf{B}$ may be also written : $\nabla \times \mathbf{B} = \alpha \mathbf{B}$, where the function α is constant along any field line ($\mathbf{B}.\nabla \alpha = 0$). Clearly, the magnitude of α is a measure of the intensity of the field-aligned current. It may be shown that (Aly, 1984, 1988)

$$|\alpha|_m \; \ell_1 \leq 1 \tag{2}$$

where $|\alpha|_m = \sup_{\partial D} |\alpha| = \sup_{D} |\alpha|$ (quite generally, all the lines of $\mathbf{B}$ are connected to ∂D) and ℓ_1 is a typical length scale depending on g and h, with $h(x,y) = \alpha(x,y,o)/|\alpha|_m$.

As for the "free energy" $c = C - C_o[g]$ of $\mathbf{B}$, one has (Aly, 1988)

$$(\ell_2 \, |\alpha|_m)^2 \, C_o \leq c \leq (\ell_3 \, |\alpha|_m)^2 \, C_o \tag{3}$$

where ℓ_2 is a function of g and h, and ℓ_3 a function of g. When $\mathbf{B}$ and α have the same typical length scale ℓ, one has : $\ell_1 \sim \ell_2 \sim \ell_3 \sim \ell$.

2.3. Linear ideal MHD stability of $\mathbf{B}$

If we assume that D is filled up with a perfectly conducting plasma and that the lines of $\mathbf{B}$ have their feet firmly tied to ∂D, then this field is found to be linearly stable if (sufficient condition) (Aly, 1989b)

$$[\int_D |\alpha|^3 \, d\underset{\sim}{r}]^{1/3} + [\int_D |f|^{3/2} \, d\underset{\sim}{r}]^{1/3} < \frac{\sqrt{3}}{2} \pi^{2/3} \tag{4}$$

$$f(\underset{\sim}{r}) = \int_{\mathfrak{L}(\underset{\sim}{r})} |\nabla\alpha|^2 \frac{ds}{B} \cdot \int_{\mathfrak{L}(\underset{\sim}{r})} B \, ds \tag{5}$$

where $\mathfrak{L}(\underset{\sim}{r})$ denotes the field line passing through $\underset{\sim}{r}$ and ds the element of arc length. For fields for which $\underset{\sim}{B}$ and α have the same typical length scale ℓ, stability is obtained if $|\alpha|_m \ell < O(1)$. A field with $|\alpha|_m \ell \lesssim O(1)$ is stable and from the results of § 2.1 and 2.2, it contains an energy which is near the maximum amount which may be stored in D at g given.

3. QUASI-STATIC EVOLUTION OF $\underset{\sim}{B}$

3.1. Simple topology field

Let us now assume that D is filled up with a perfectly conducting plasma and that a potential field $\underset{\sim}{B}_o$ having a simple topology (i.e. whose lines do not form separatrix surfaces) is made evolving quasi-statically through a sequence of force-free configurations as a consequence of a smooth stationary velocity field $\underset{\sim}{v}$ (with $v_z = 0$ and $|\underset{\sim}{v}|$ vanishing fast at infinity) imposed on ∂D. The field $\underset{\sim}{B}_t$ at time t may be written as $\underset{\sim}{B}_t [\underset{\sim}{r} (\underset{\sim}{r}_o,t)] = J^{-1} (\underset{\sim}{r}_o,t) \cdot \underset{\sim}{B}_o(\underset{\sim}{r}_o) \cdot \nabla_o \underset{\sim}{r} (\underset{\sim}{r}_o,t)$, where $\underset{\sim}{r} (\underset{\sim}{r}_o,t)$ is the position at time t of a plasma element located at $\underset{\sim}{r}_o$ at t = 0, and J = det($\nabla_o \underset{\sim}{r}$). The function $\underset{\sim}{r}$ is determined at each time t as a stationary point of

$$C [\underset{\sim}{r}] = \int_D B^2 \, d\underset{\sim}{r} = \int_D [\underset{\sim}{B}_o(\underset{\sim}{r}_o) \cdot \nabla_o \underset{\sim}{r}]^2 \frac{d\underset{\sim}{r}_o}{J} \tag{6}$$

defined over the set of all the functions $\underset{\sim}{r} (\underset{\sim}{r}_o)$ which take on ∂D the values $\underset{\sim}{r} (\underset{\sim}{r}_o,t)$ associated with the given velocity field $\underset{\sim}{v}$. Note that this problem looks similar to a problem of anisotropic elasticity ; it turns out to be more complicated, however, because of the obvious possibility to construct a field $\underset{\sim}{B}$ at t from $\underset{\sim}{B}_o$ by an infinite number of admissible mappings $\underset{\sim}{r}$.

We have recently undertaken a study of the functional (6). However, so for we have only been able to develop arguments indicating that (6) must admit an absolute minimizer when $|\underset{\sim}{r} (r_o,t) - \underset{\sim}{r}_o|$ is not too large on ∂D, the associated magnetic field being <u>smooth</u> (i.e. no current sheets do form ; compare with Parker, 1989).

Assuming the existence of $\underset{\sim}{B}_t$ for $0 \leq t < \infty$, we have also shown shown that, when $t \to \infty$, $\underset{\sim}{B}_t$ approaches asymptotically an open field, all the currents concentrating into a current sheet (we suppose here to simplify the presentation that B_z is kept invariant by $\underset{\sim}{v}$). This result strongly supports our conjecture of § 2.1.

3.2. Complex topology fields

If the lines of the initial field $\mathbf{B}_o$ have a complex topology (i.e. if separatrices do exist), then the problem of quasi-static evolution driven by $\mathbf{v}$ on ∂D may still be formulated as above as a problem of minimizing (or more generally of finding the stationary points) of the functional (6) over an appropriate set of functions at each time t. In that case, however, we are sure a priori that the solutions are not going to be smooth : current-sheets have to develop indeed along the separatrices because the "connectivity" of the lines is discontinuous accross these surfaces (Aly, 1987 ; Antiochos, 1989).

Physically, such current sheets are unsatisfactory features, as any tiny amount of resistivity would imply their dissipation, and then, to get a meaningful formulation of the problem of evolution in that case, one needs to add a prescription describing the ineluctable reconnection of the field accross them (this is still to be done). Anyway, we may conclude that deformation of a complex topology configuration leads necessarily to plasma heating and thus is a potentially interesting mechanism for heating the corona.

4. LARGE-SCALE RECONNECTION

The asymptotic result reported in § 3.1 strongly suggests that, if we allow the plasma to have a small resistivity, the field must become unstable with respect to reconnection when the shear becomes large enough. Up to now, we have been able to gain some insight into this conjecture only in the particular situation where $\mathbf{B}$ is an arcade which is axisymmetric around $\hat{z}$. In that case, we may reasonably admit that 2D reconnection, acting on a time scale much smaller than the diffusion time scale of the field, can change the topology of the lines, but not the "distribution of the magnetic fluxes" : i.e. the range of values of the "flux function" u, as well as the amount of toroidal flux between any two surfaces Σ_{u1} and Σ_{u2}, are conserved during a reconnecting transition. Then reconnection is possible in an arcade configuration only if there exists a field related to it by these flux constraints, but having a more complex topology and a lower energy. We have shown that such fields start existing at some critical time t_c when the shear exceeds a critical value. For $t > t_c$ the arcade is metastable with respect to reconnection and is able to effect an explosive transition (flare) towards a lower energy state under the action of a finite perturbation.

5. CONCLUSION

We have been able to extend to a 3D situation some of the analytical results previously obtained in a study of the quasi-static evolution of a 2D force-free field (Aly, 1989c). In particular, we have shown that a 3D field may approach asymptotically an open field. But much is still to be done, especially on the problem of the existence and smoothness of the solutions.

REFERENCES

Aly, J.J. (1984). "On some properties of force-free magnetic fields in infinite regions of space", Ap. J. 283, 349.
Aly, J.J. (1987). "On the existence of a topological heating mechanism for a plasma in a sheared magnetic field", Astron. Ap. (submitted).
Aly, J.J. (1988). "Some properties of the solutions of a non-linear boundary value problem for a force-free field in an infinite region of space. I. Energy estimates", Astron. Ap. 203, 183.
Aly, J.J. (1989a). "Some properties of finite energy constant-α force-free fields", Preprint.
Aly, J.J. (1989b). "The stability of a line-tied force-free magnetic field in an unbounded region of space", Phys. of Fluids (in press).
Aly, J.J. (1989c). "Quasi-static evolution of a force-free magnetic field", Comp. Phys. Comm. (in press).
Antiochos, S. (1989). "Magnetic topology and current sheet formation", in B.M. Haisch and M. Rodono (eds.), Solar and Stellar Flares (Poster Papers, IAU Coll. n° 104), Catania Astrophysical Observatory, p. 277.
Parker, E. (1989). "Solar and stellar magnetic fields and atmospheric structures : theory", in B.M. Haisch and M. Rodono (eds.), Solar and Stellar Flares (IAU Coll. n° 104), Kluwer Academic Publishers, Dordrecht, p. 271.

DISCUSSION

PRIEST: (i) Have you proved the existence of your open fields?
(ii) As you mentioned, the field may lose equilibrium or stability before reaching the maximum energy. Hood and Priest (1980) considered the stability of simple arcades and flux tubes. They could not find an instability for simple arcades, but flux tubes or arcades with flux tubes (which would have $\alpha\ell > 1$) become unstable when the twist is too great.

ALY: (i) For a given flux distribution $B_z(x,y,0)$, it is simple to prove that you can construct one and only one totally open field of finite energy. What is difficult to prove (and this has not yet been done) is the existence of a sequence of (closed) force-free fields approaching this open field when they are indefinitely sheared and twisted.
(ii) In 3D, a flux tube (isolated or embedded in an arcade) has to be tied to the boundary, no region magnetically disconnected from $\{z = 0\}$ generally exists. For a tube for which the current and the field have the same typical length-scale ℓ, my calculations imply existence and stability for $|\alpha|_m \ell \lesssim 1$, non-existence for $|\alpha|_m \ell \gtrsim 1$, and there seems to be not much room for an instability to exist. But they allow (but not imply!) the possibility of a tube existing and being unstable if the current is concentrated in a region of transverse scale $\ell_\alpha << \ell_B$.

FORBES: (i) Is the result you have for the maximum energy storage possible for a force-free field being twice the potential energy completely general?
(ii) Does it still hold if there are x-lines or x-points in the configuration?

ALY: (i) The maximum energy you can store in a force-free field is equal to the energy of the open field having the same B_z on $\{z = 0\}$. That is the exact result. The fact that this energy is about twice the energy of the potential field is just an estimate, which may be checked to be correct in most cases of interest.

(ii) To get my result, you have just to assume that the field is force-free, no other assumptions being required. The field may have neutral points or current sheets. The upper bound on the energy still holds true.

MAGNETIC SHEAR AND FLARES

P. VENKATAKRISHNAN
Indian Institute of Astrophysics
Bangalore 560034
India

ABSTRACT. Magnetic tension posseses low values at regions of large magnetic "shear" on the polarity inversion line. The low tension allows for a non-force-free field that depends on the density of the medium. Instabilities which can modify the density may then possibly evolve into solar flares. Heuristic arguments are given to favour instability driven flare scenarios.

1. Introduction

With the advent of vector magnetographs, the non- potential nature of pre-flare magnetic configurations could be clearly discerned from the angular deviation of the observed transverse component of the magnetic field from the transverse potential field calculated from the distribution of the line- of-sight component of the magnetic field. This angular deviation, called magnetic "shear" was seen to be large at locations of four major flares (Hagyard, *et al.*, 1984; Hagyard, Venkatakrishnan, and Smith 1989). A theoretical basis for such a correlation was suggested in terms of the loss in magnetic tension caused by a shear in the field (Venkatakrishnan, 1990). In this paper, we advance arguments in support of instability driven scenarios for flares.

2. Loss of Magnetic Tension in Sheared Fields

The magnetic pressure gradient balances the magnetic tension in a force free field. The vertical component of the tension force is given by $\mathbf{B}_T.\nabla_T B_z$ where $\mathbf{B}_T$ is the transverse component of the magnetic field and ∇_T is the lateral gradient operator, $(\mathbf{x}\partial/\partial x,\ \mathbf{y}\partial/\partial y)$. The maximum tension is available in a potential configuration, where the gradient of B_z is along $\mathbf{B}_T$. Near a polarity inversion line, this is manifest by $\mathbf{B}_T$ pointing across the line. For a sheared field, $\mathbf{B}_T$ is almost aligned with the inversion line, thereby providing a smaller tension force as compared to a potential field. A force free field must respond to this loss of tension by a decrease in the vertical gradient of magnetic pressure. This conclusion is also borne out in detailed computations (Wu *et al.*, 1990). If the gradient in magnetic shear is small, then this extention of B_T^2 must persist to greater heights. The interaction of this vertically extended transverse field with the over-lying global field through reconnection would then be able to "open" out the

E. R. Priest and V. Krishan (eds.), Basic Plasma Processes on the Sun, 319–322.

configuration and release the excess energy stored in the closed field in the form of a solar flare. There is another possibility where the gradient of B_T^2 does not have to match the low magnetic tension of the sheared field, but is balanced by the weight of the material ρg. This could happen, for instance, in an active region filament .where the density is larger than the ambient value. Such a field with magnetic pressure gradient balancing ρg will depend on the plasma density. Any decrease in ρ caused, e.g. by extra heating could possibly become a runaway process via a thermal instability. The "lifting" of the field lines in response to the decrease of ρg would drain material away from the region leading to a further decrease in ρg. The rearrangement of the field lines caused by this instability could lead to centri- fugal acceleration of plasma along the field lines (Venkatakrishnan, 1984a,b) resulting in shocks and eventually flares. Alternatively, the electric field induced by the rapidly changing magnetic field could be trapped for particle acceleration (Colgate, 1978). The details of this scenario are given in an earlier paper (Venkatakrishnan,1990).

3. Instability vs Non-Equilibrium Driven Scenarios

Onset of non-equilibrium of a slowly evolving field as it cro- sses a critical parameter was suggested as a possible way of converting the magnetic energy stored in a non-potential field on a very short time scale (Low,B.,C.,1982). In all cases where a critical parameter was identified, it was seen that a change in the parameter involved specifying unphysical boundary conditions. In cases were evolution of the field was modelled on the basis of physical boundary conditions, e.g. by moving the footpoints of the field lines, no loss of equilibrium was encountered (Zwingmann, 1989). Observational identification of such a critical parameter has also met with only limited success. A promising parameter seemed to be magnetic shear. However, there are examples where the shear was large on the neutral line for several days without any flare in the active region (Hagyard, Venkatakrishnan, and Smith, 1989). On the other hand, there is growing evidence for pre-flare activation of filaments several minutes before the impulsive phase (Kahler *et al.*, 1988). Soft X-ray enhancement is also seen several minutes before the flare. The buoyancy time scale, given by $(\Lambda/g)^{1/2}$, is $\approx$ 5 minutes for a scale height Λ of 10^9 cm. This could be construed as a support for the instability driven scenario for the flare outlined in the previous section. The observed pre-flare heating could then be interpreted as the progress of the thermal instability. The Alfven travel time, which is the relevant time scale for non-equilibrium, is tens of seconds for a similar length scale and a field strength of a few hundred Gauss. If this short time scale is equated with the timescale for the impulsive phase of the flare, then the pre-flare activation requires an additional explanation in the non-equilibrium scenario. On the other hand, an instability driven scenario would require a pre-flare stage on the buoyancy timescale, while the onset of the flare proper can be interpreted as the the onset of reconnection caused by large magnetic field gradients or electric fields created from the rapidly changing magnetic field.

4. Discussion and Conclusions

Choosing between instability driven and non-equilibrium driven flare scenarios is not a pedagogical exercise. Observers, in particular, are interested in clues that would help in predic- ting the location and time of flare onset. A non-equilibrium scenario implies the search for a critical parameter of the equilibria, while the thermal instability scenario implies the search for triggers. Although the Skylab data showed some evidence for triggers, the SMM results did not prove to be eq- ally convincing. The modelling of the 3-D magnetic configuration based on boundary conditions specified at the photosphere will also help in discriminating between potentially flaring and non-flaring regions with equal extent of magnetic shear along the polarity inversion line. A promising criterion is the vertical gradient of shear reflected, perhaps, in the lateral gradient of the shear across the neutral line. Searching for similar criteria in observed vector magnetograms will be certainly useful. The expectation is that configurations with a smaller vertical gradient of magnetic shear exert more pressure on the overlying field and thus should be more prone to erupt in a flare. In conclusion, the signature of large shear at flare sites allows us to consider instability driven scenarios for flares. The failure to identify a critical parameter for loss of equilibrium either from specifying physical boundary conditions or from observations should urge us to look beyond the non-equilibrium driven flare scenarios.

References

Colgate,S.,A. : 1978, *Astrophys. J.*, **221**, 1068.
Hagyard,M.,J., Smith,J.,B.,Jr., Teuber,D., and West,E.A. : 1984, *Solar Phys.*, **91**, 115.
Hagyard,M.J., Venkatakrishnan,P. and Smith,J.B., Jr. : 1989, *Astrophys. J. Supp. Series* (in press).
Kahler,S.W., Moore,R.L., Kane,S.R., and Zirin,H. : 1988, *Astrophys. J.*, **328**, 824.
Low,B.C. : 1982, *Solar Phys.*,**77**, 43.
Venkatakrishnan,P. : 1984a Ph.D. Thesis, Bangalore University.
Venkatakrishnan,P. : 1984b, *Kodaikanal Obs. Bull.* **4**, 19.
Venkatakrishnan,P. : 1990, *Solar Phys.*,accepted
Wu,S.,T., Song,M.,T., and Dryer,M.: 1990, *IAU Symp.* 142.
Zwingmann,W. : 1987, *Solar Phys.*, **111**, 309.

DISCUSSION

VARMA: What instability are you envisaging?

VENKATAKRISHNAN: I am envisaging a thermal instability for this process, but have not done any stability analysis to identify it precisely.

VERMA: (i) Is there any relation between magnetic shearing in the active region and the type of solar flares?
(ii) Can we know on the basis of magnetic shearing in the active region whether the flares will be impulsive or gradual?

VENKATAKRISHNAN: (i) The observations of vector magnetic fields in active regions so far have concentrated on two-ribbon flares. We do not know yet whether compact flares are associated with locations of large magnetic shear or whether these can be produced even in less sheared regions.
(ii) Since the association of flares with magnetic shear has been established for very few cases, it is difficult to give any prediction on the absence or presence of the impulsive phase based purely on the shear.

IMPLICATIONS OF TENSION FREE EQUILIBRIA FOR PRE-FLARE ENERGY BUILDUP

P. VENKATAKRISHNAN
Indian Institute of Astrophysics
Bangalore 560034
India

ABSTRACT: The downward tension force is known to be extremely small at the locations of highly sheared magnetic fields. The free energy in such tension free fields is shown to be dependent on the plasma density. The excess of this free energy over that available in a force free configuration is estimated and shown to be sufficient to drive instabilities which could possibly initiate a solar flare.

1. Introduction

Magnetic shear, (as measured by the angular deviation of the observed transverse field from a potential transverse field), was seen to be large at sites of major flares (Hagyard, *et.al.*, 1984; Hagyard and Rabin, 1986; Hagyard, Venkatakrishnan, and Smith, 1989). Large magnetic shear was shown to be equivalent to low magnetic tension, which implied that the magnetic fields were dependent on the plasma density, thereby allowing for plasma instabilities to destabilise the magnetic configurations (Venkatakrishnan, 1990). In this paper, we will estimate the excess energy stored in a tension free field over and above that stored in a force free field.

2. Estimation of Excess Energy in Tension Free Fields

A lower limit to the free energy in a sheared magnetic field can be obtained by assuming that the shear is produced by only a rotation of the transverse magnetic field from its potential configuration through the shear angle without any increase in field strength (Moore, R., L.: 1988, personal communication). This gives the free energy as

$$E \approx 4.(B_T^2/8\pi).\sin^2\vartheta/2 \quad , \tag{1}$$

where B_T is the transverse field and ϑ is the shear angle.

The shear angle for a force free field can be obtained by equating the magnetic pressure and tension forces. This gives

$$\partial B_T^2/\partial z \approx 2.\ B_T.\nabla B_z \ , \quad \text{or} \quad B_T^2/\Lambda_B \approx 2.B_T.B_z.\cos\vartheta\ /\Delta \quad , \tag{2}$$

where Λ_B is the magnetic scale height and Δ signifies the lateral scale of variation of the vertical component of the magnetic field. When the plasma density is sufficiently high, then a further reduction in the tension force is possible which would be replaced by the downward force of gravity. This reduction can thus be expressed in terms of the weight

E. R. Priest and V. Krishan (eds.), Basic Plasma Processes on the Sun, 323–324.

of the material. By equating the variation of the tension force (R.H.S. of equation 2) caused by a change $\delta\vartheta$ in the shear angle with the weight of the material ρg, we have

$$\sin\vartheta . \delta\vartheta \approx 4\pi\rho g\Delta / B_T B_z \quad (3)$$

The increase in the free energy given by (1) caused by the change in angle $\delta\vartheta$ can be similarly obtained from

$$\delta E \approx B_T^2 . \sin\vartheta . \delta\vartheta . / 4\pi \ . \text{ or } \quad \delta E \approx B_T . \rho g \Delta / B_z , \quad (4)$$

after substituting for $\sin\vartheta . \delta\vartheta$ from equation 3.

Equation 4 gives an estimate for the free energy density of the tension free field. For purposes of order of magnitude estimation, we will assume that the lateral gradient of B_z equals the vertical gradient of B_T (this is strictly true for a potential field and is a good approximation for a force free field near the polarity inversion line). Using this assumption in equation 4, we have

$$\delta E \approx \rho g \Lambda_B \ . \quad (5)$$

For order of magnitude estimates one can also assume that Λ_B equals the typical size of a bipolar active region which is $\approx 10^9$ cm. The total energy can now be written as

$$\delta\mathcal{E} \approx A . (h_{chr} n_{chr} + h_{cor} n_{cor}) m_H g \Lambda_B \ . \quad (6)$$

Assuming $A \approx 10^{19}$ cm^2, $m_H = 1.6\times10^{-24}$, $g \approx 2\times10^4$ cms^{-2}, we have $\delta\mathcal{E} \approx 10^{30}$ ergs, for $n_{chr} = 10^{14}$, and $h_{chr} \approx 10^8$ cm. For the sun, this energy is lower than the energy released in major flares, but is very similar to the energy released in the impulsive phase of typical flares. Thus the storage of energy in tension free fields can provide sufficient destabilisation via thermal convective instabilities to initiate the flare. The size of the flaring region in stellar atmospheres does not seem to be much different from that on the Sun (de Jager, *et.al.*, 1989). Thus, one cannot explain the frequent occurrence of white light flares in stars by invoking larger starspots. One clue is that the chromospheres of flare stars (dMe) are few orders of magnitude denser than normal chromospheres (Cram and Mullan, 1982). By assuming $n_{chr} \approx 10^{15}$ in equation 6, we obtain $\delta\mathcal{E} \approx 10^{31}$ ergs. In this case, there is much more energy available for destabilising the tension free fields and thus, a greater probability for producing major flares.

References

Cram, L., E. and Mullan, D., J.: 1982, *Astrophys. J.*, **234**, 579.
De Jager, *et. al.*: 1989, *Astron. Astrophys.* (preprint)
Hagyard, M., J. and Rabin, D., M.: 1986, *Adv. Space Res.*, **6**, 7.
Hagyard, M., J., Smith, J., B., Jr., Teuber, D., and West, E., A.: 1984, *Solar Phys.*, **91**, 115.
Hagyard, M., J., Venkatakrishnan, P., and Smith, J., B., Jr.: 1989, *Astrophys J. Supp. Ser.*, in press.
Venkatakrishnan, P.: 1990, *Solar Phys.*, accepted.

ENERGY BALANCE IN PROMINENCE-CORONA TRANSITION REGIONS

C. CHIUDERI AND F. CHIUDERI DRAGO
Department of Astronomy and Space Science, University of Florence
Largo Enrico Fermi 5
50125 Florence, Italy

1. INTRODUCTION

Prominences are sheets of cold ($< 10^4$ K) plasma totally imbedded in the hot ($> 10^6$ K) coronal plasma. A Prominence-Corona Transition Region (PCTR) is therefore present all around it.

The PCTR can be analyzed using observed UV line intensities and brightness temperatures at μ-waves. Radio observations are usually performed on the disk (above filaments) while UV line intensities are measured at the limb. They therefore refer to the top and to the side PCTR, respectively.

It appears that PCTR models derived from UV and radio observations strongly disagree: in fact if we take the Differential Emission Measure (DEM=p^2dz/dT) derived from UV line intensities and compute the radio brightness temperature using the following relationship between T_b and DEM:

$$T_b(\nu) = \int_{T_o}^{T_c} T\, e^{-\tau}\, d\tau + T_c(1 - e^{-\tau_c})$$

$$d\tau\,(\nu) = \frac{\xi}{4\,k^2\nu^2}\,\frac{DEM}{T^{7/2}}\,dT \qquad \tau_c = \frac{\xi\; p^2}{4\,k^2\nu^2 T_c^{\;7/2}}\,\frac{H}{2}$$

with H = coronal scale height and $\xi = 0.2$, we get systematically much larger values than the observed one.

On the other hand, radio models of PCTR (Kundu et al., 1986, and references therein) require a gas pressure much lower than that consistent with optical and UV obvservations (Vial, 1989).

In all PCTR models, derived from both UV and radio data, the effect of the magnetic field on heat conduction has been so far neglected. Its inclusion can drastically change the results, especially for the top PCTR observed at radio wavelengths. In fact following Leroy (1989) the magnetic field in prominences lies in a plane parallel to the solar surface and in this plane forms an angle α with the prominence axis. α can vary from $0°$ to $\sim 60°$, and the most commonly observed value is $30° \pm 10°$. Therefore when we observe the prominences from the top (filaments on the disk) at radio frequencies the angle θ between B and ∇T is

E. R. Priest and V. Krishan (eds.), Basic Plasma Processes on the Sun, 325–329.

$\theta \cong 90^\circ$, while when the prominence is observed at the limb (UV lines) the angle can be $30^\circ < \theta < 90^\circ$ with the most probable value given by: $\theta = 60^\circ \pm 10^\circ$.

2. PCTR MODELS

We will now show that when the correct expression of the heat conduction is taken into account in deriving PCTR models, no discrepancy between UV and radio data exists.

Let us assume the z axis parallel to the temperature gradient ∇T, the energy balance in the static case is given by (Priest, 1982).

$$\frac{d}{dz}\left(k_{//} T^{5/2} + k_{\perp} T^{-5/2} \frac{dT}{dz}\right) = E_R - E_H \qquad (1)$$

where

$$k_{//} = c \cos^2\theta \quad ; \quad k_{\perp} = 2 \cdot 10^{-11} c \operatorname{sen}^2\theta \frac{p^2}{4 k^2 B^2} \quad ; \quad c = 1 \cdot 10^{-6} \quad ;$$

p is the pressure in dyn/cm^2 and B the magnetic field in G .

The specific radiated power, E_R has been approximated with a composite power law given by Rosner et al. (1978) for $T > 5 \cdot 10^4$ and by Poland (private communication) for $T < 5 \cdot 10^4$

$$E_R = p^2/ 4 k^2 A_i \, T^{\alpha_i - 2}$$

The unknown heating function E_H has been parametrized as follows:

$$E_H = p^2/4 k^2 \, f_H(T)$$

If $\theta \neq 90^\circ$ and $d\theta/dz = 0$, eq. (1) becomes

$$k_{//} \frac{d}{dz} [\, T^{5/2}(1+ \varepsilon T^{-5}) \, dT/dz] = E_R - E_H \qquad (2)$$

where

$$\varepsilon = k_{\perp}/k_{//} = 2.6 \; 10^{20} \; p^2/B^2 \; tg^2\theta \qquad \varepsilon T^{-5} > 1 \left\{ \begin{array}{l} \theta > 85^\circ \\ T < 10^5 \end{array} \right.$$

assuming p = const. and defining:

$$\eta(T) = T^{5/2} (1 + \varepsilon T^{-5}) \; dT/dz$$

a first integral of eq. (1) is given by:

$$\eta^2(T) = \eta^2(T_o) + \frac{p^2}{4k^2 k_{//}} \int_{T_o}^{T} T^{5/2} (1 + \varepsilon T^{-5}) (A_i T_i^{\alpha_i - 2} - f_H(T)) \; dT \qquad (3)$$

It is easily seen that if $\eta(T_o) = 0$, $\eta(T) \propto \dfrac{p}{\cos\theta}$ and $p^2 \dfrac{dz}{dT} \propto p \cos\theta$.

This means that an increase of the angle θ is equivalent to a decrease of the pressure.

We have first assumed that the heating function is given by

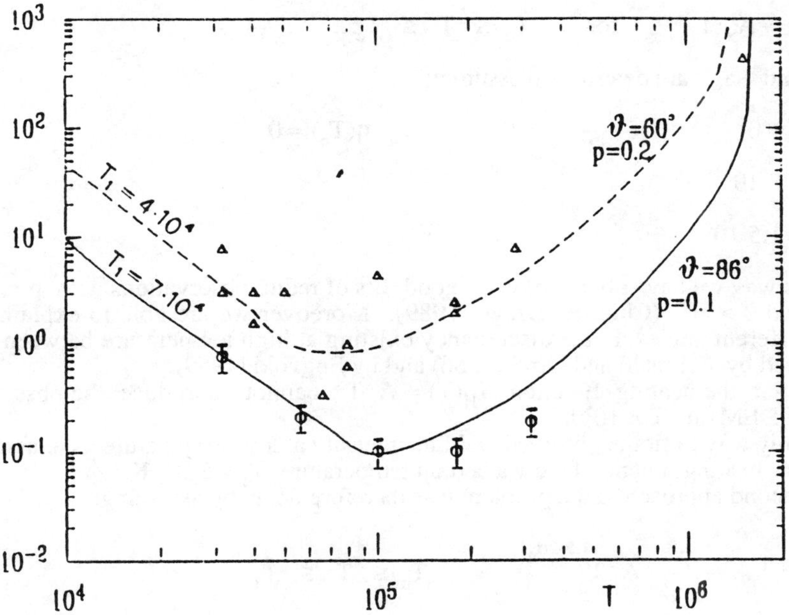
P²·dz/dT
T
T₁ = 4·10⁴
T₁ = 7·10⁴
ϑ=60°
p=0.2
ϑ=86°
p=0.1
10⁴
10⁵
10⁶
10³
10²
10¹
10⁰
10⁻¹
10⁻²

Fig. 1

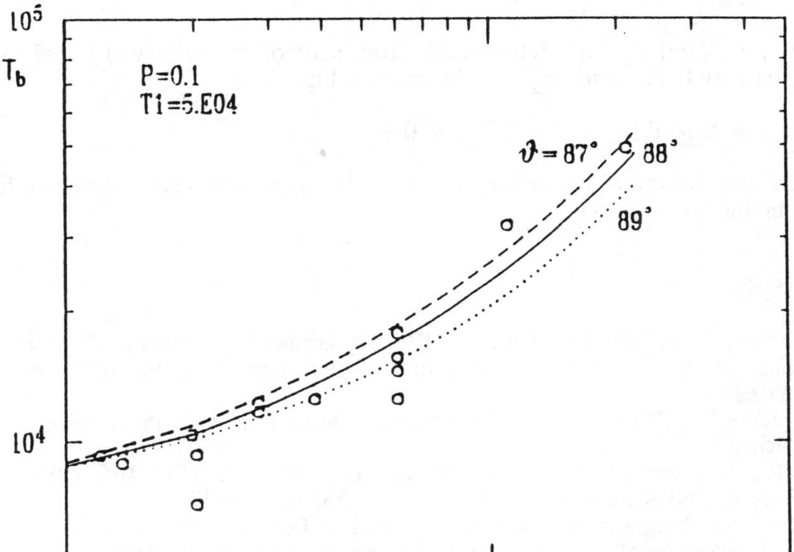
Tb
P=0.1
T1=5.E04
ϑ = 87°
88°
89°
λ(cm)
10⁰
10¹
10⁴
10⁵

Fig. 2

$$f_H(T) = A_x\, T^{a_x} \quad \text{for} \quad T_o \leq T \leq T_c \,.$$

A_x and a_x are determined assuming:

$$\eta(T_o) = 0 \qquad \left(\frac{d\eta}{dT}\right)_{T=T_o} = 0 \qquad \eta(T_c) = 0$$

$$T_o = 8\ 10^3$$

$$T_c = 1.5\ 10^6$$

In this way we have obtained very good fits of radio observations with $p \sim 0.2$ dyn/cm^2 and $\theta > 80°$ (Chiuderi Drago, 1989). Moreover we are able to explain in terms of different angles θ the discrepancy existing at high temperature between the DEM derived by Schmahl and Orral (1986) and by Engvold (1989).

However the heating function $f_H(T) = A_x T^a$ cannot reproduce the observed form of the DEM at $T < 10^5$ k.

The only way to fit the observed UV line intensity at low temperature is to assume an "ad hoc" heating function below a certain temperature $T_1 \cong 5\ 10^4$ K.

The second approach to the problem was therefore done by assuming:

$$E_{H_1} = E_R - \frac{p^2}{4\,k^2} A_1 T^{1.5-2\alpha_1} \qquad T_o \leq T \leq T_1$$

$$E_{H_2} = \frac{p^2}{4\,k^2} A_2\, T^{\alpha_2} \qquad T_1 \leq T \leq T_c$$

where A_1, α_1 and T_1 are determined from a fit of the observed DEM in its decreasing part and A_2 and α_2 by the relationships:

$$E_{H_1}(T_1) = E_{H_2}(T_1) \quad ; \quad \eta(T_c) = 0$$

Assuming this form of the heating functions in eq. 1, the results shown in fig. 1 and 2 are obtained.

REFERENCES

Chiuderi Drago, F.: 1989, IAU Coll. n. 117, "Dynamic of Prominences" (in press).
Engvold, O.: 1989, "Dynamic and Structure of Quiescent Solar Prominences", Ed. E.R. Priest.
Leroy, J.L.: 1989, "Dynamic and Structure of Quiescent Solar Promineces", Ed. E.R. Priest.
Kundu, M.R., Melozzi, M. and Shevgaonkar, R.K.: A & A 167, 166, 1986.
Pramesh Rao, A. and Kundu, M.R.: Solar Phys. 55, 161, 1977.
Priest, E.R.: Solar Magneto-hydrodynamics, Reidel Publ. Co. 1982.
Rosner, R., Tucker, W.H. and Vaiana, G.S.: Ap. J. 220, 643, 1978.
Schmahl, E. and Orral, F.: 1986, "Coronal and Prominences Plasmas", NASA Conf. Pub. 2442, 127.
Vial, J.C.: 1989, IAU Coll. n. 117, "Dynamic of Prominences", (in press).

DISCUSSION

KUNDU: After all these model computations that you have done using a sophisticated data base, can you tell us what is the thickness of the filament-corona transient sheath?

CHIUDERI-DRAGO: I cannot tell the thickness because we performed only a first integral (analytically) which gives dT/dz. However, since $dT/dz \sim pf(T)/\cos\theta$, I expect a much thinner transition region for large values of θ (radio) than for small values of θ(UV).

NUMERICAL SIMULATIONS OF SOLAR DISTURBANCES AND THEIR INTERPLANETARY CONSEQUENCES

M. DRYER[1], S. T. WU[2], and T. R. DETMAN[1]
[1]Space Environment Laboratory
National Oceanic and Atmospheric Administration
R/E/SE, 325 Broadway
Boulder, Colorado 80303-3328
USA
[2]Center for Space Plasma and Aeronomic Research, and
Department of Mechanical Engineering
University of Alabama in Huntsville
Huntsville, Alabama 35899
USA

ABSTRACT. Responses of the solar atmosphere and interplanetary medium to simulated solar disturbances were studied by time-dependent, MHD numerical simulations. This deterministic initial-boundary value problem was attacked in the classical way: a representative steady state is first established, then input parameters at the lower near-Sun boundary are perturbed. We discuss a number of 2- and 3-dimensional examples of coronal mass ejection (CME) simulations and some current controversies concerning the basic process of CME initiation. Footpoint shearing motion is tested to see whether it can provide a reasonable mechanism for CME development from arch filament configurations.

We also demonstrate possible interplanetary consequences to CME-like disturbances by using 3-D simulations to determine the dynamic response of the solar wind to a plasmoid injection from an eruptive filament or prominence. We also discuss the separate possibility whereby a plasmoid may be generated in the interplanetary medium by a solar-generated shock that propagates through a heliospheric current sheet. Application of the 3-D model for the interpretation of interplanetary scintillation observations is also discussed.

1. INTRODUCTION

1.1 Near-Sun Activity

The origin of coronal mass ejections (CMEs) is one of the major topics currently under active debate. Observations by white-light coronagraphs led to the first ideas and models for CMEs. Coronagraph images are produced by Thomson scattering of photospheric photons by coronal electrons. In addition to the problem of CME origin, the problems of CME propagation and evolution in interplanetary space are also important topics which provide the backdrop for this paper.

A variety of phenomenological descriptions have been applied to the transient white-light images detected by coronagraphs. First OSO-7 and then Skylab, P78-1, and SMM have contributed to the observations. As observed in the solar-occulted plane of sky (Howard et al., 1985), these traveling images have been called curved fronts, spikes, bubbles, loops, blobs, etc. Some workers considered them to be more-or-less planar struc-

E. R. Priest and V. Krishan (eds.), Basic Plasma Processes on the Sun, 331–340.

tures or helical, magnetically bound loops that escaped the Sun's gravitational attraction; other workers considered them to be compressions (followed by rarefactions) in the corona, produced by near-surface energy conversion that expanded quasi-spherically. Their rate of occurrence and solar-cycle dependence are in statistical dispute, with differences (one per day vis-a-vis two per day) most likely caused by variations in coronagraph design, resolution, and duty cycle. About half of the CMEs are associated with filament eruptions (easily detected at the solar limbs); some are associated with solar flares (not easily detected near the limbs because of the awkward remote-sensing line of sight from Earth); some are associated with both of the above; and sometimes there are no optical, radio, or x-ray observations temporally and spatially associated with CMEs (Munro and Sime, 1985; Webb and Hundhausen, 1987).

Three theoretical descriptions (reviewed by Dryer, 1982) have been offered: (a) White-light "loops" are magnetically driven by stresses in the curved, moving plasma column; (b) White-light "loops," followed by depleted brightness, are quasi-spherical shells of compressed coronal plasma followed by rarefactions; these "loops" are produced by a localized, near-surface change of properties in or near active regions; (c) Very-large-scale coronal magnetic topologies become unstable and trigger CMEs in some way.

Klimchuk (1989) has discussed theoretical ideas for physical mechanisms of CME initiation. He first identifies three basic questions:

"1) What causes the disruption of the large-scale magnetic field/plasma configuration?
2) How does the system evolve once the disruption begins?
3) How does the disruption trigger solar flares?"

Klimchuk addresses the first question within the framework of quasi-static evolutionary models. The second, he suggests, "will require a fully time-dependent MHD treatment." As noted by Dryer and Wu (1985),this point has been studied extensively. The third question is "likely [he noted further] to involve non-MHD plasma processes." Neither Klimchuk nor we discuss this third question.In SECTION 2 below, we discuss a numerically demonstrated MHD treatment that, in our opinion, is relevant to both the first and second questions.

1.2 Interplanetary Activity

Several radio astronomers (Hewish and Duffett-Smith, 1987; and Hewish and Bravo, 1986) have interpreted their observations of interplanetary scintillation (IPS) to be associated with geomagnetic activity. Scintillations of distant radio galaxies' radiation are caused by density fluctuations in the intervening solar wind. These fluctuations can be used to generate maps of enhanced and depleted solar wind density. These workers (see, also, Tappin et al., 1988) introduced an ability to generate "interplanetary images" of compressed and rarified solar wind plasmas once each day.

A controversy stems from the radio interpretation of these maps when the density-enhanced regions are back-projected to the Sun. The point of ejection is (according to Hewish, 1988) within (or within a 45° circle surrounding) a coronal hole. Hewish (1988) therefore inferred that an erupting stream within a coronal hole emits very-high-momentum flux that expands into a large (~ 90°) heliolongitudinal expanse and persists for several days. This high efflux of energy, he claims, is the source of geomagnetic storms. He asserts that solar flares are peripheral events.

The alternative view, as expressed by most of the solar physics community, is that the energy influx to the interplanetary medium is due to magnetic eruptions which produce a complicated interaction of shocks, compressions, and rarefactions. The net result [suggested by the 2-D and 3-D numerical MHD simulations by Dryer, Smith, and Wu (1988)] is the high dynamic pressure and IMF amplitude, negative B_z that are required for geomagnetic activity.

A number of transient interplanetary events (often preceded by shocks) have been described by spacecraft investigators as "magnetic clouds," or "plasmoids" (see the review by Burlaga, 1989). These magnetic clouds are characterized by: (1) a rotation of the IMF polarity through a large angle during a temporal interval of about a day, (2) an IMF magnitude which is higher than average, and (3) a solar wind temperature which is lower than average. It is not known if the global topology is disconnected from the Sun (i.e., a plasmoid); if the IMF is still connected to the Sun at both ends (i.e., extension of a solar loop arcade as suggested by Gold, 1959); or if the propagating shocks introduce large-amplitude MHD waves in their wake that cause the IMF to twist, then unwind, with one end rooted in the Sun and the other in interstellar space (Dryer, Wu, and Gislason, 1983). The plasmoid and extended loop are currently attracting much attention together with the notion of twisted, nearly-force-free, IMF "flux ropes."

Another interesting observational inference (based on in situ observations) is concerned with the IMF external to the magnetic cloud. Gosling (1989) has reviewed work that suggests that IMF draping around the object occurs in the sheath region between a bow shock and the presumed boundary of the "CME." Although there is no objective criterion for identifying the boundary of a "magnetic cloud" (Burlaga, 1989), this inference is reasonable, particularly if the object (CME, magnetic cloud, etc.) moves relative to the background solar wind with a velocity greater than the local magnetosonic speed.

2. RESULTS

2.1 Shear-Induced Instability

Figure 1 shows the schematic representation of a dipole magnetic field in an initial state of equilibrium in a stratified atmosphere. A 2 ½-D (i.e., non-planar) MHD model is used to simulate the response of the exponentially stratified atmosphere to a photospheric shearing motion as indicated by the sinusoidal velocity profile in Figure 1. It was found that upward plasma flow velocities are generated in the vertical direction. The velocities grow exponentially at first, with a growth rate equal to $\sqrt{8}\,(\overline{V}_A\, a)$, where $\overline{V}_A$ is the average Alfvén speed and a^{-1} is the characteristic length scale. The growth rate is saturated by the Lorentz force, but growth continues until it reaches the same order of magnitude as the Alfvén speed. MHD instability, which we suggest may be called "shearing-induced instability" (SII), occurs shortly thereafter. Physically, the simulation suggests that the central magnetic field lines are pinched, and the outer loops stretch upward with a tendency to open. This process may be considered as one of the fundamental mechanisms for CME initiation (Wu, Song, Martens, and Dryer, 1990).

The SII was studied for three values of plasma beta, β = 15.4, 1.54, and 0.06. The characteristic Alfvén velocities for these three cases are, respectively: 4.67, 46.7, and 232 km s^{-1} . Figure 2 shows the maximum upward velocity within the computational domain as a function of time. The peak shearing velocity (Figure 1) was 5 km s^{-1} for the two high values of β and 15 km s^{-1} for the (more realistic) lowest value. The growth rate for

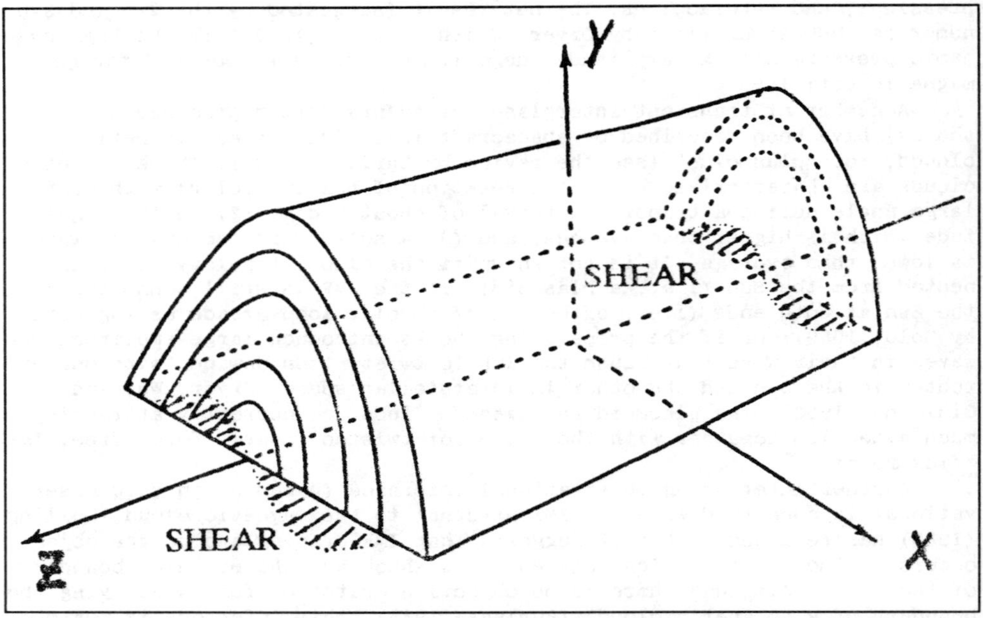

Figure 1. Schematic representation of an initial magnetic field (dipole) arcade which is subsequently sheared at the photosphere by the indicated velocity profile. The computational domain is: $x = \pm 8.4 \times 10^3$ km, and $y = 8 \times 10^3$ km. (Wu, Song, Martens, and Dryer, 1990.)

these upward velocities became unstable when the maximum deviation of the field at the coronal base reached shear angles of 63°, 48°, and 21° for β = 15.4, 1.54, and 0.06, respectively. Thus, instability is indicated for moderate shearing angles when the plasma betas are low, as expected in the lower corona.

It is important to note that the forcing function is a finite-amplitude perturbation upon a stable configuration that eventually becomes unstable. Reduction of the peak shearing velocity of 15 km s^{-1} to a more gentle value, say 0.15 km s^{-1}, could be accomplished via the principle of dynamic similitude (c.f., Wu et al., 1988). The computational run time must then be longer. In the present case of β = 0.06 (the "prototype"), the same realistic beta could be maintained for the "model," together with the same Struhal, Euler, and Froude numbers as well as the same ratio of magnetic to kinetic energy for a dissipationless fluid.

As suggested above, however, there is a problem in this particular case. The prototype ran for 7 Alfvén periods, where the Alfvén time was 35 seconds. Because of the desired hundred-fold decrease of shearing velocity, the model's rather excessive temporal requirement, τ_M, would be:

$$\tau_M = 7 \times 35 \times 10^2 = 24{,}500 \text{ s}.$$

2.2 Solar-Injected Plasmoid into the Solar Wind

Using the 3-D code of Han, Wu, and Dryer (1988), Detman et al. (1990) have simulated the injection of an initially spherical plasmoid into the solar

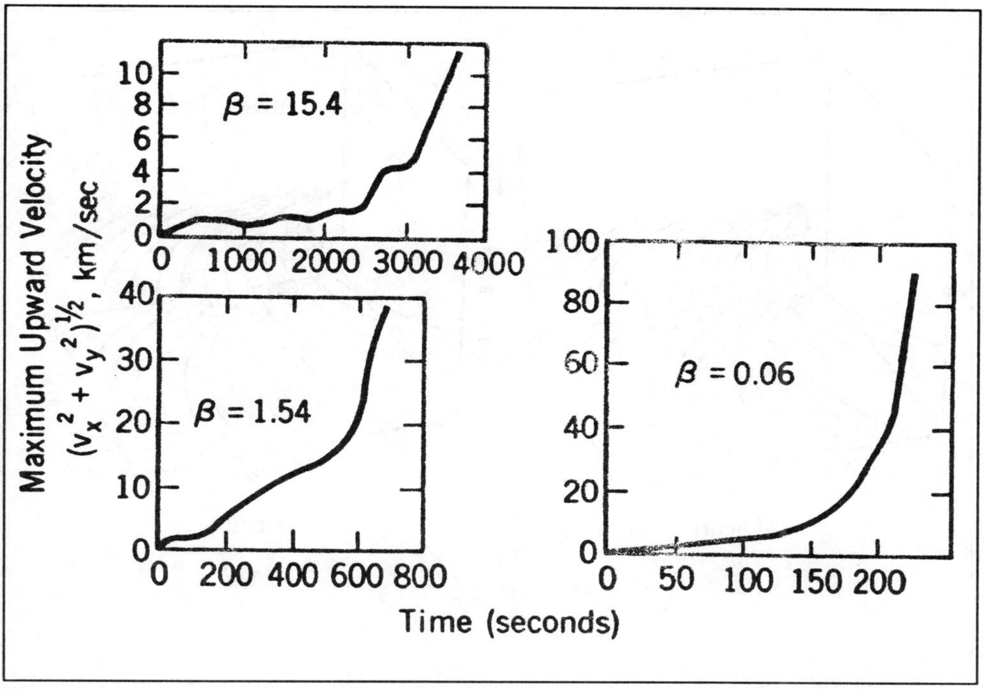

Figure 2. Maximum upward velocity in the computational domain (Figure 1) when photospheric shearing of a dipole magnetic field takes place. Note that "shearing-induced instability" takes place at t ≈ 200 s, for β = 0.06, after approximately seven Alfvén times. (From Wu, Song, Martens, and Dryer, 1990.)

wind. The plasmoid possessed both toroidal and poloidal magnetic field components, like a set of concentric "slinky toys" placed end to end. The plasmoid survived the injection and continued to propagate through the solar wind, even producing a substantial shock wave when injected at a speed greater (relative to the background solar wind velocity) than the magnetosonic speed. The approximate positions of the plasmoid and its shock wave, and the draping of the IMF around the plasmoid, were determined. Figure 3 shows a 3-D view of some representative IMF lines and their draping around the plasmoid. A representative magnetic field line within the plasmoid is also shown.

It is interesting to note that some reconnection (due to numerical diffusion) takes place between some of the plasmoid field lines and IMF lines that come into close proximity to the neutral points on the front and rear positions of the plasmoid.

2.3 Plasmoid Created at Heliospheric Current Sheet

In a separate numerical experiment, Dryer et al. (1989) showed how a cigar-shaped plasmoid might be generated by a shock wave that propagates through a flat heliospheric current sheet. The high total pressure, formed by the 3-D shock wave just within its outermost envelope, decreases to low values within the central portion, i.e., near the IMF reversal zone. The high pressure gradient, generated by the outward-moving, large-scale

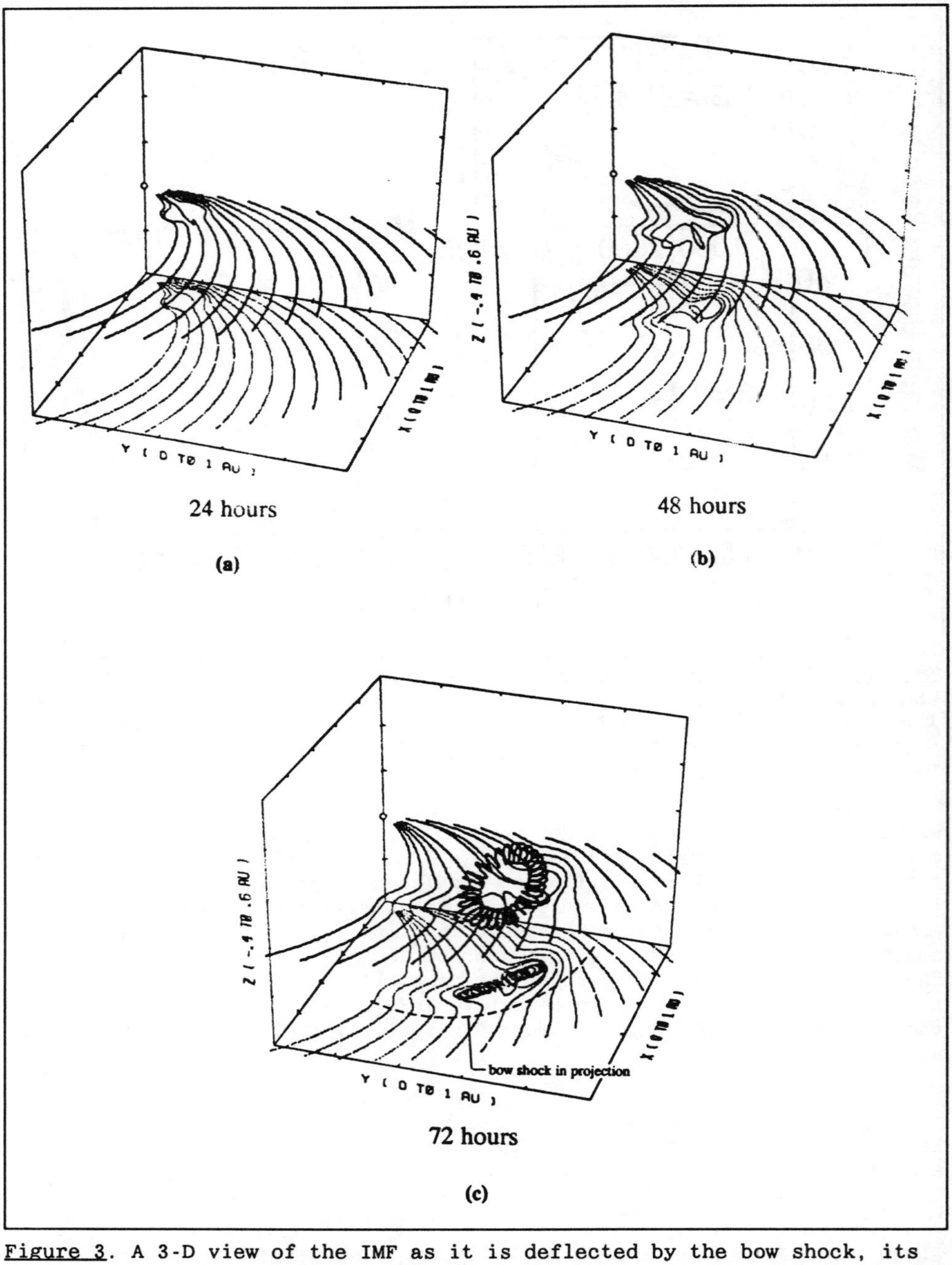

<u>Figure 3</u>. A 3-D view of the IMF as it is deflected by the bow shock, its draping around the solar-generated plasmoid, and a single helical magnetic field line within the plasmoid. Initially <u>equatorial</u> IMF lines are shown at t = 24, 48, and 72 hours in panels (a), (b), and (c), respectively. The viewing perspective is from 8 AU, $\theta = 60°$, $\phi = 20°$, where θ is the heliocolatitude and ϕ is the heliolongitudinal angle measured from the lower left of the 1 AU-sized box. (Detman et al., 1990.)

heliospheric shock wave, forces the opposite-directed IMF field lines together and causes them to reconnect.

Figure 4 shows the initial stage of reconnection at what will be the leading edge of the cigar-shaped plasmoid. Reconnection also takes place at the rear, pinching off the opposite-directed IMF as the entire structure propagates through the solar wind. The "cigar" would be oriented in a direction transverse to the outward motion of the large-scale global disturbance.

4. CONCLUDING REMARKS

We have briefly summarized some of our ongoing work in the field of non-planar and 3-D numerical simulations of solar disturbances and their possible interplanetary consequences. The classical initial boundary-value approach is scrupulously followed to ensure a deterministic response whenever a stable initial state is perturbed by a set of observationally in-

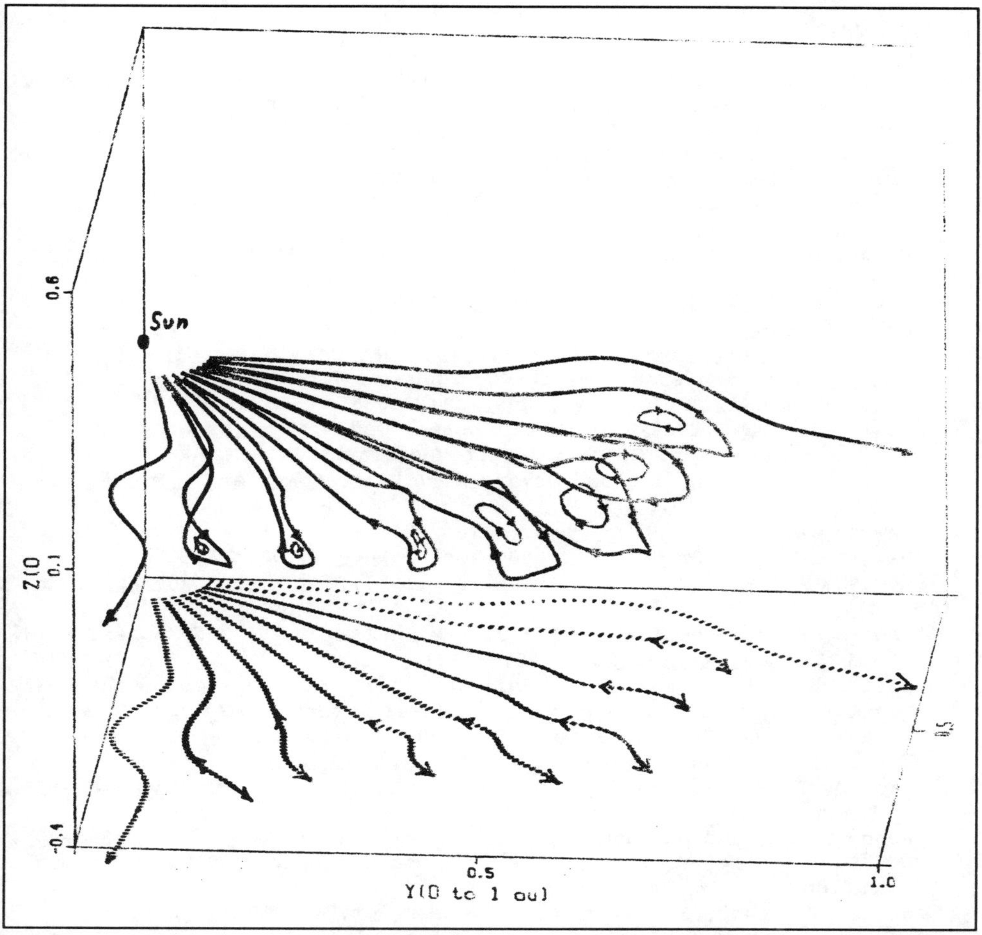

Figure 4. Initial stage of a cigar-shaped plasmoid that is formed in the interplanetary medium by the propagation of a shock wave through a flat heliospheric current sheet. (Dryer et al., 1989; S.M. Han, private comm., 1989.)

ferred parameter changes. Numerical experiments of this kind are a necessary step beyond the "cartoon" stage, and must be undertaken with the solution of the mathematical expressions for well-known physical laws together with reasonably chosen assumptions. The insight derived from simulations such as the three described here are essential for understanding large-scale global processes. Only investigation by multiple, in situ, spacecraft missions can confirm or refute the global predictions of such 3-D numerical experiments. Such missions have yet to be undertaken.

5. ACKNOWLEDGMENT

This work was supported in part by NASA Order No. 17,015 to NOAA Space Environment Laboratory (MD, TRD), and by NOAA Contract 50RANR-700099, NAGW-9, and AFOSR Grant 88-0013 (STW). We wish to thank Z. Smith, R. Zwickl, S. M. Han, S. W. Kahler, T. G. Forbes, J. A. Klimchuk, and D.J. McComas, for their help and advice.

6. REFERENCES

Burlaga, L. F. (1989), in E. Marsch and R. Schwenn (eds.), Physics of the Inner Heliosphere, D. Reidel Publ. Co., Dordrecht, in press.

Detman, T. R., Dryer, M., Yeh, T., Han, S. M., Wu, S. T., and McComas, D. J. (1990), in preparation.

Dryer, M. (1982), Space Sci. Rev., 33, 233.

Dryer, M., Detman, T. R., Wu, S. T., and Han, S. M. (1989), Adv. Space Res., 9(4), 75.

Dryer, M., Smith, Z. K., and Wu, S. T. (1988), Astrophys. Space Sci., 144, 407-425.

Dryer, M., and Wu, S. T. (1985), J. Geophys. Res., 90, 559.

Dryer, M., Wu, S. T., and Gislason, G. (1983), in M. Neugebauer (ed.), Proceedings of the Fifth International Solar Wind Conference, NASA Conference Publ. No. 2280, p. 735.

Gold, T. (1959), J. Geophys. Res., 64, 1665.

Gosling, J. T. (1989), in C. T. Russell (ed.), Physics of Magnetic Flux Ropes, Amer. Geophys. Union Monograph, in press.

Han, S. M., Wu, S. T., and Dryer, M. (1988), Computers & Fluids, 16, 82.

Hewish, A. (1988), Solar Phys., 116, 195.

Hewish, A., and Bravo, S. (1986), Solar Phys., 106, 185.

Hewish, A., and Duffett-Smith, P. J. (1987), Planet. Space Sci., 35, 487.

Howard, R. A., Sheeley, N. R., Jr., Michels, D. J., and Koomen, M. J. (1985), J. Geophys. Res., 90, 8173.

Klimchuk, J.A. (1989), in K. Phillips (ed.), Proceedings of the Thermal-Nonthermal Flares Workshop II, Rutherford-Appleton Laboratory Report, RAL-89-102, Didcot, U.K., pp. 1-6 to 1-8.

Munro, R. H., and Sime, D. G. (1985), Solar Phys., 97, 191.

Tappin, S. M., Dryer, M., Han, S. M., and Wu, S. T. (1988), Planet. Space Sci., 36, 1155.

Webb, D. F., and Hundhausen, A. J. (1987), Solar Phys., 108, 383-401.

Wu, S. T., Song, M. T., Martens, P., and Dryer, M. (1990), in preparation.

Wu, S. T., Wang, S., Wang, A. H., and Dryer, M. (1988), Adv. Space Res., 8(11), 221-226.

DISCUSSION

FORBES: Was the initial state in your sheared arcade example potential or force-free? If so, it seems to me that your result completely contradicts the work of J. J. Aly which shows that such a disruption, which opens the field, by shearing should be impossible.

DRYER: The initial state is, indeed, a potential force-free magnetic arcade. When the footpoints are moved, they are moved rather rapidly. For example, the lowest beta case ($\beta = 0.06$) had a peak shearing velocity of 15 km/sec. Consequently, the system quickly evolves into a non-force-free system with pressure gradients. Thus, the force-free results of Aly do not apply. Also, the instability only results in a rapid expansion of loops and locally fast mass flows after the mean Alfvén speed is exceeded. This instability, moreover, does not necessarily open the magnetic field. You will recall that there is no resistivity in this model, nor are there any anti-directed fields where numerical reconnection could, in principle, take place. Thus, this model does not address the question of field-line-opening.

KUNDU: I am a little confused by your referring to flares as the cause of IPS-producing shocks rather than high-speed streams from coronal holes, which Tony Hewish believes. Since you showed Hewish's data, when you talked about IP shocks, I would like to know what the present status is with regard to flares versus coronal holes as the cause of IP shocks.

DRYER: Our use of Hewish's IPS data is decoupled from his interpretation that high speed streams from coronal holes are responsible for geomagnetic storms. If a transiently-developing coronal hole suddenly (say, on a few-hour time scale) develops, a shock could certainly develop. I have a constructive and friendly disagreement with Tony who believes that flares are peripheral events *vis-à-vis* geomagnetic storms. I believe otherwise. You will recall that IPS data contains no information about the IMF (which, if southerly-directed, is important for storm triggering); hence my comment above about decoupling. Of course, even a steady-state hole could develop a shock that develops in the corotating frame. My point is that *any* temporal and/or spatial solar inhomogeneity (*c.f.*, flare, eruptive prominence, or hole) could produce a shock. Hewish's point, however, about a transient event, followed by a *long-lasting* energy output (be it a flare or whatever) is an important point that is worth investigation. To this purpose, Zdenka Smith and I have recently completed a 2D MHD parametric study that is relevant to this point. A final point is worth making: there are no observables of erupting streams from coronal holes. Transient coronal hole area changes are not sufficient, in my opinion, to claim that a shock will propagate from such an event. The case for flares is well-established.

SWARUP: How does the intensity of shocks vary with solar distance in your models?

DRYER: When the temporal duration of an input pulse is short, say less than a few hours, the strongest part of the shock will decay similarly to a classical blast wave with shock speed $\sim R^{-1/2}$ where R is the heliocentric radius. If the energy input is long-lasting, say some 5-15 hours (as suggested by long duration X-ray flares) the shock could move out at a constant velocity (*i.e.*, as a piston-driven shock) for some tenths of an AU before decelerating as noted above in the frame of the background, moving solar wind.

PRIEST: (i) Is the plasma beta much smaller than unity in magnetic clouds and in your magnetic bubble?

(ii) If so, why should the plasma density changes be directly proportional to the initial density?

DRYER: (i) Your first question relates to our "magnetic bubble" numerical experiment. We were interested in examining the dynamics of a particular configuration and the response (*c.f.*, field draping) of the ambient solar wind and its interplanetary magnetic field to its projectile-like motion. Although we were not interested at this exploratory stage to make any comparisons with spacecraft-observed "magnetic clouds" the particular choice of the parameters (n,T,B,) within our input bubble produced plasma betas greater than unity. We would expect that other, judiciously-chosen, parameter combinations could produce betas less than one - as found in the observations. It is not clear, incidentally that the latter are bubbles - or whether they are gigantic loops with both ends rooted in the Sun.
(ii) The density fluctuations that give rise to IPS are experimentally correlated with *in situ* density measurements by Tappin (1986) and more rigorously, recently, by Zwickl *et al* (AGU abstract,1988).

UBEROI: In your analogy of magnetic bubble to Hill's vortex did you take care of the fact that some conservation theorems valid for vortices do not hold good for MHD theory?

DRYER: Thank you for bringing this possibility to my attention. No, we did not take this point into consideration.

STABILITY OF A MASSIVE CURRENT SHEET SUPPORTED BY A TWO-DIMENSIONAL POTENTIAL MAGNETIC FIELD.

J.J. ALY and S. COLOMBI
Service d'Astrophysique
Centre d'Etudes Nucléaires de Saclay
91191 Gif-sur-Yvette Cedex
France

ABSTRACT. Necessary and sufficient criteria of linear ideal magnetohydrodynamic stability are derived for a massive current sheet - idealizing a solar prominence - supported against gravity by a two-dimensional potential magnetic field.

1. INTRODUCTION

A solar prominence is often idealized by a vertical infinitely thin massive current sheet extending in the "coronal half-space" $\{z > 0\}$ parallel to the x-axis and in equilibrium under the opposite actions of a uniform vertical gravitational field and of an x-invariant potential magnetic field (see e.g. the review by Anzer, 1989). In this communication, we analyze the linear ideal magnetohydrodynamic stability of this simple model, taking into account the stabilizing influence (neglected in the earlier work by Anzer, 1969), of the low-beta coronal plasma.

2. MODEL

We consider a simple model of solar prominence in which : i) the corona is represented by the half-space $D = \{z > 0\}$ (in a cartesian frame (Oxyz)) ; D is filled up with a pressureless perfectly conducting plasma ; ii) the prominence is represented by the strip : $\Sigma = \{-\infty < x < +\infty \; ; \; y = 0 \; ; \; 0 < a \leq z \leq b\}$; Σ is made of perfectly conducting matter having a mass density $\sigma(z)$ (satisfying $\sigma(a) = \sigma(b) = 0$) ; iii) Σ is in equilibrium under the opposite actions of a uniform vertical gravitational field $\underset{\sim}{g} = - g\,\hat{z}$ $(g > 0)$ and of an x-invariant magnetic field $\underset{\sim}{B}(y,z) = \nabla A(y,z) \times \hat{x}$ which is y-symmetric $(A(-y,z) = A(y,z))$, has finite energy per unit of x-length, and is potential outside Σ ($\Delta A = 0$ in D/Σ) ; thus $2\pi\,\sigma g = B_y\, B_z^+$ on Σ ($B_z^+ = B_z\,(0^+,z)$).

E. R. Priest and V. Krishan (eds.), Basic Plasma Processes on the Sun, 341–342.

3. IDEAL LINEAR STABILITY OF THE MODEL

Let us apply to the configuration we have just described a displacement ξ (x,y,z) vanishing on $\partial D = \{z = 0\}$ (line-tying assumption) and periodic in x. By computing the resulting average variation of the energy of the system per unit of x-length and using the classical "energy principle", it may be shown that :

i) the equilibrium is stable with respect to any perturbation which does not depend on x (this result is still true if D/Σ is a vacuum) ;

ii) the equilibrium is stable with respect to any arbitrary perturbation <u>if and only if</u> one has at all the points of Σ

$$K \, \partial \, |B_y| \, / \, \partial z \;\leq\; 2\pi \, \sigma \, g \tag{1}$$

$$-K \, \partial \, |B_z{}^+| \, / \, \partial z \leq B_y{}^2 , \tag{2}$$

where $K(z) = \int_{\mathcal{C}(z)} B \, ds$ ($\mathcal{C}(z)$) is a field line of B connecting a point of Σ at height z to the plane ∂D ; it may be shown that all the lines of B necessarily cut ∂D). The positive RHS of Eq. (1) and (2) are related to the stabilizing effect resulting from the lines of B being frozen in the plasma in D (thus any displacement of Σ in the x-direction necessarily leads to a bending of the lines). Would D/Σ be a vacuum, both terms would be replaced by zero (Anzer, 1969).

4. CONCLUSION

For a massive current sheet supported by a 2D potential magnetic field, we have derived stability criteria which are less severe than Anzer's. When Anzer's criteria predicted that all the models with a > 0 considered here are unstable, ours indicated indeed that stability holds at least in some cases, e.g. when σ is not too large and the lines of the "unperturbed" field B_o (potential in the whole D and satisfying $B_{oz}(y,0) = B_z(y,0)$) have a "dip" at the points where they cut Σ (such a situation may be encountered above a quadrupolar region).

5. REFERENCES

Anzer, U. (1969). "Stability analysis of the Kippenhahn - Schlüter model of solar filaments", Solar Phys. 108, 26.

Anzer, U. (1989). "Structure and equilibrium of prominences", in E.R. Priest (ed.), Dynamics and Structure of Quiescent Solar Prominences, Kluwer Academic Publishers, Dordrecht, p. 143.

NATURE OF LARGE-SCALE SOLAR MAGNETIC FIELD AND COMPLEXITY OF HCS AS OBSERVED IN INTERPLANETARY PLASMA

T E GIRISH and S R PRABHAKARAN NAYAR
Department of Physics, University of Kerala
Kariavattom, Trivandrum. INDIA 695 581

1. INTRODUCTION

The properties of the interplanetary plasma and magnetic field near 1 AU is determined by the nature of large-scale solar magnetic field and the associated structure of the heliospheric current sheet (HCS). Magnetic multipoles often present near the solar equator affect the solar wind plasma and magnetic field (IMF) near earth's orbit. The observation of four or more IMF sectors per solar rotation and the north-south asymmetry in the HCS are observational manifestations of the influence of solar magnetic multipoles, especially the quadrupole on the interplanetary medium (Schultz, 1973; Girish and Nayar, 1988). The solar wind plasma is known to be organised around the HCS. In this work, we have investigated the possibility of inferring i) the relative dipolar and quadrupolar heliomagnetic contributions to the HCS geometry from the observation of four sector IMF structure near earth and ii) the properties of the north-south asymmetry in HCS geometry about the heliographic equator from IMF and solar wind observations near 1 AU.

2. DATA ANALYSIS

a) Annual number of Bartel's rotations where four sector structure are observed near earth is determined from the IMF polarity data for the period 1947-1985. It is assumed that the HCS geometry near 1 AU can be approximated by

$$\theta = R_1 \operatorname{Sin} \phi + R_2 \operatorname{Sin} (2\phi-\delta) \quad (1)$$

Where R_1 and R_2 are the solar magnetic dipolar and quadrupolar contributions to the HCS, θ and ϕ are the heliographic latitude and longitude of the HCS and δ is the phase lag between the maxima of the heliomagnetic dipolar and quadrupolar moments. Fourier analysing the HCS data obtained from K-corona observations or potential field modelling (Korzov, 1982; Hoeksema and Scherrer, 1986), the yearly average R_2/R_1 defined by equation (1) is determined for the period 1971-1985.

b) The asymmetry parameters of the HCS, θ_T (transition lattitude

E. R. Priest and V. Krishan (eds.), Basic Plasma Processes on the Sun, 343–344.

of reversal of dominant polarity of IMF near 1 AU) and Δ (the asymmetry in the heliolatitudinal extension of HCS in opposite heliohemispheres) are obtained from the observations of IMF and high speed streams near 1 AU between 1971 and 1985. If L_n and L_s correspond to yearly average width of IMF sectors observed near earth with magnetic polarity identical to solar north pole and south pole respectively, we have for $L_n > L_s$, $\theta_T < 0$ and for $L_n < L_s$, $\theta_T > 0$. Using this concept, yearly mean sign of θ_T is inferred from IMF data for the period 1971-1985. Similarly if V_n and V_s correspond to the yearly average amplitude (Velocity maxima) of corotating high speed streams observed near 1 AU with a magnetic polarity identical to solar north pole and south pole respectively, then we have for $V_n > V_s$, Δ is negative and for $V_n < V_s$, Δ is positive. We have determined V_n and V_s and hence the sign of Δ for the years 1974-1976, 1978 and 1982 where there is a good coverage of solar wind plasma data and reported heliomagneic latitudinal organisation of solar wind velocity (Mavromichalaki et al, 1988, Newkirk and Fisk, 1985).

3. RESULTS AND DISCUSSION

i) The variation of the annual occurrences of four sector IMF near earth correlates well with the change in average (R_2/R_1) determined from HCS data. This indicates that one can infer the relative solar magnetic dipolar and quadrupolar contributions to the HCS geometry from IMF data.

ii) The yearly average sign of asymmetry parameter θ_T of HCS determined from IMF data agree well with the same determined using HCS data (Girish and Nayar, 1989) during 1971-1985. The yearly average sign of the asymmetry parameter Δ determined from solar wind velocity data agree well with the same determined from HCS data. These results indicates that the interplanetary plasma and magnetic field data can be used to infer the shape of heliospheric current sheet geometry and the related multipolar solar magnetic field. Since the conditions of interplanetary magnetic field and high speed streams can be inferred using geomagnetic data, one can infer the long term changes in the nature of the HCS and solar magnetic field from geomagnetic activity.

REFERENCES

Girish, T.E. and S.R.P.Nayar, 1988, Solar Phys. **116**, 369.
Girish, T.E. and S.R.P.Nayar, 1989, Solar Phys. (in press)
Hoeksema, J.T. and Scherrer, P.H. 1986, Rep. UAG-94, U.S. Dept. of Commerce, Colarado.
Korzhov, N.P., 1982, Sib. Izmiran Preprint 2-82, Irakutsk.
Mavromichalki, H. et al. 1988 Solar Phys. **115**, 345.
Newkirk, G, Jr., and L.A. Fisk, 1985,J. Geophys. Res. **90**, 3391.
Schultz, M., 1973, Astrophys. Space Sci. **24**, 371.

THE QUASI-STATIC EVOLUTION OF MAGNETIC STRUCTURES ON THE SUN AND THEIR TOPOLOGICAL RECONSTRUCTION

Yu.G.MATYUKHIN and V.M.TOMOZOV
664033, Irkutsk 33, P.O.Box 4026,
SibIZMIR, USSR

ABSTRACT. We have considered the problem of quasi-static evolution of a magnetic configuration as it is affected by shear motions at field line footpoints. It is found that, with a certain character of the shear motions at the field line footpoints, topological reconstruction will, by necessity, occur in the magnetic configuration. A pattern of magnetic field lines during the course of configuration restructuring is constructed.

We consider the problem of quasi-static evolution of a magnetic structure on the Sun such as a loop arcade under the action of shear motions of a certain kind of field line footpoints in the photosphere. Equations describing the quasi-static evolution of a magnetic structure are derivable from a system of MHD equations by the method of expansion in small parameter $\alpha^2 = (v/v_A)^2 \ll 1$ (v being the velocity of motions, and v_A, the Alfven velocity, respectively). A 2D-configuration of the magnetic field $\vec{B} = \{\vec{e}_z \times \nabla u, \vec{e}_z B_z(u)\}$ ($\vec{e}_z u = \vec{A}$ being a vector-potential) is considered.

The boundary-value problem (BVP) has the form:

$$\left.\begin{array}{ll} \Delta u = -\lambda f(u) & \partial u/\partial y(y = b) = 0 \\ u/s = g(x) & \partial u/\partial x(|x| = a) = 0 \end{array}\right\} (*)$$

where λ is a parameter, and $f(u)$ is a function related in some way to $B_z(u)$. If $B_z(u)$ is a unique function at a given u, then all solutions of the system (*) for a given λ, possessing a different field topology, will correspond to the same displacement. We specify the function $f(u)$ such that the BVP (*) had in a certain finite range of λ-values three solutions at least. To do this, it suffices to postulate that $f(u) \sim u^s$ and $s > 3$ when $u \to 0$ and, besides, $f(u) = 0$ when $u > u_0$ (Heyvaerts et al., 1980). The BVP (*) was solved by the method of upper and lower solutions (Matyukhin and Tomozov, 1989) using the particular values of the parameters $a = 3$ and $b = 6$ in units of the scale $\ell_0 = 10^9$ cm. Let us now clarify the character of the magnetic

E. R. Priest and V. Krishan (eds.), Basic Plasma Processes on the Sun, 345–346.

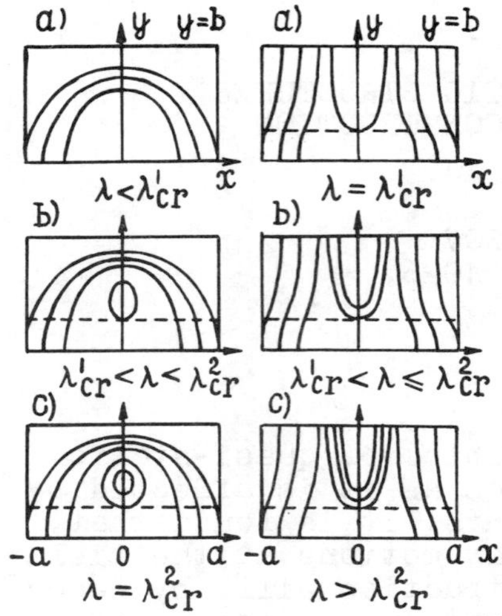

Fig. 1 Fig. 2

field topology for different solutions of the BVP (*). Fig. 1 (a,b, c) shows a typical topology of magnetic field lines for the lower solution of the BVP. From Fig. 1 it is easy to see that, for the entire range of values of the parameter λ in which the lower solution exists, the field lines have a closed topology. In the process of evolution from $\lambda = 0$ to $\lambda = \lambda^2_{cr}$, a "magnetic island" is generated inside the magnetic configuration, i.e., a region of closed field lines unembedded in the photosphere.

The magnetic field line topology for the upper solution as a function of the parameter λ is shown in Fig. 2 (a,b,c). Note that in the above range of variation of the parameter λ all magnetic configurations are open, and inside the field lines embedded in the photosphere there are open field lines with $B_z(u) = 0$ (with a zero shear of the magnetic field). Note that all solutions (upper and lower) describing the closed and open field topologies, are stable to 2D-disturbances which do not alter the problem symmetry along the Z-axis.

In order to describe the evolution of a magnetic configuration, it is necessary to investigate the dependence of the full magnetic energy of the structure on the value of shear (the parameter λ) for a closed and open magnetic structure. This suggests the following scenario of a quasi-static evolution of an arcade (Matyukhin and Tomozov, 1989). Under the action of a shear motion the originally closed equilibrium magnetic structure necessarily evolves to such an equilibrium state that it becomes linearly unstable, so that if the value of magnetic field shear reaches $\lambda = \lambda^2_{cr} = 4$, then the magnetic structure ought to reach a new stable equilibrium state with a change in magnetic field topology (from closed to open), with the release of the accumulated free energy of the field.

REFERENCES

Heyvaerts J., Lasry J.M., Schatzman M., and Witomsky G. (1980), Lecture Notes in Mathematics, 782, pp. 160-191.

Matyukhin Yu.G. and Tomozov V.M. (1989), In: AGU Chapman Conference on: Physics of Magnetic Flux Ropes, Extended Abstracts, Hamilton, Bermuda, March 27-31, pp. 146-157.

SLIGHT DISAPPEARANCE OF PROMINENCE PLASMA TO THE SOLAR CORONA

V. Rusin, V. Dermendjiev, M. Rybansky, G. Buyukliev

1 Astronomical Institute
Slovak Academy of Sciences
059 60 Tatranska Lomnica
Czechoslovakia

2 Department of Astronomy and NAO
72 Lenin Blvd.
1784 Sofia
Bulgaria

1. INTRODUCTION

The problem of prominences-corona relationship is relativelly old. Already in 1931 Lockyer [1] showed that there is a close relation between prominences distribution and the form of white-light corona. However, this problem is still debatable and poses a number of controversial questions. One of them is the question of the energy and mass exchange between prominences and the ambient corona. It is generally assumed that the mass balance exists between the corona and prominences, but unambiguous observational proofs for prominences-corona plasma exchange are very rare. There are little data [2-4], as well, that could be used to address the problem of slight plasma flows from prominences to the corona.

In this contribution we give an observational evidence for slight disappearance of prominence plasma as it was found during the H-alpha observations with Lomnicky Stit coronograph and discuss possible explanations of this phenomenon.

2. OBSERVATIONS AND DATA PROCESSING

An interesting prominence was observed with the 20-cm lens coronograph at Lomnicky Stit coronal station in August 28, 1989 (for more details see [5]). It was relatively stable but not so calm and has shown rapid changes of the shape of its upper parts. This prominence was observed in H-alpha (FWHM = 0.8 nm) during 17 minutes with one minute interval. The whole series of observations was analysed in detail using MDM 6 microdensitometer Joyce Loebl, computers and program facilities at Rozhen Observatory in Bulgaria.

3. RESULTS AND DISCUSSION

After the processing of H-alpha images of August 28 prominence we obtained 17 consecutive maps. Four of them are shown on the Figure. They ilustrate the fact that the prominence loses a noticeable, but exceedingly small quantity of plasma. The plasma at the top of two curved columns

E. R. Priest and V. Krishan (eds.), Basic Plasma Processes on the Sun, 347–349.

(dotted area) continuosly disappeared. This process can be traced in detail on the whole series of maps. In the last frames it is not already seen in H-alpha.

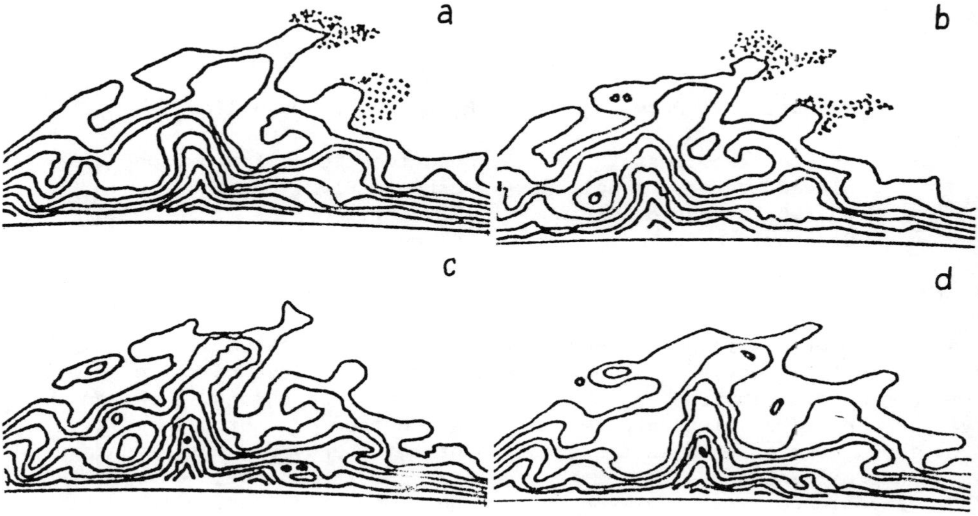

Figure. Isodensite maps of August 28, 1988 prominence (a/05:05, b/05:09 , c/05:15, d/05:20 UT). The isoline levels are 0.05; 0.15; 0.25; 0.30; 0.40; 0.50; 0.70; 0.80; 0.85; 0.95; 1.00.

The structure of this prominence was complicated but relatively stable and retained its shape. We did not observe a noticeable changes of the prominence height during the observation as for the cases of eruptive prominences.

Assuming prominence mass density of 10^{-14} g cm^{-3} [2] for the dissappeared area which has a volume of 10^{25} cm^3, we compute the total amount of prominence plasma, contained in both features, of about 8×10^{11} g.

One may suppose that this cool prominence plasma was continuosly heated to a coronal temperature. As was suggested in [6] from observations in H-alpha, 530.3 nm and 637.4 nm, fast-mode waves energy can enter the prominence and to produce an enhanced temperature region in its vicinity. A similar process could work in the case of August 28 prominence as well. However, another explanation of this observational effect is possible. In a plasma medium with a highly structured magnetic field the non-linear effects lead to generation of plasma flows, as it was shown in [7]. Along the magnetic lines of the "channels" or "edges", on the upper part of August 28 prominence, propagation of oscilations is possible. Due to the absorbtion of the momentum of such oscilations a faint prominence plasma flow could be driven. Such a flow is possible due to the action of ponderomotive force on the prominence plasma, as well.

We would like to point out that the boundary between the prominence and the corona for the most parts of prominence surface was stable during the observed sequence. It means that a flow of prominence plasma, or plasma evaporated from some parts of the prominence to the corona, could not be free, but occurs in a special magnetic field configuration. Disappearance of prominence material has also been observed during disparition brusques or from slow ascending material connected with flares [8]. However, in both these cases only several isolated knots dissipate to the solar corona.

4. CONCLUSIONS

Prominence images obtained in H-alpha showed that some of prominence plasma in the edges with special configuration can slowly enter the corona. In our opinion not only a process of heating, but some dynamic processes also are possible. We propose to consider the action of the ponderomotive force and absorbtion of the momentum of oscilation propagating along the prominence magnetic field in the places with special configuration. This observational result supports the old idea that material from underlying layers could also come into the corona through prominences.

REFERENCES

1. Lockyer, W. S. J. (1931) 'On the relationship between solar prominences and the forms of the corona', Monthly Notices Roy. Astron. Soc. 91, 707-809.
2. Engvold, O. and Malville Mc. K. (1977) 'The fine structure of prominences', Solar Phys. 52, 369-377.
3. Noens, J. C. (1988) 'Observation of coronal material associated with prominences' in R. C. Altrock, Solar and Stellar Coronal Structure and Dynamics, NSO/Sacramento Peak, 182-190.
4. Lebecq, Ch. (1988) 'High-resolution observations of prominence-corona interactions' in R. C. Altrock, Solar and Stellar Coronal Structure and Dynamics, NSO/Sacramento Peak, 170-181.
5. Rusin, V., Rybansky, M., Dermendjiev, V. and Buyukliev, G. (1989) 'Corona-prominence interface as seen in H-alpha'. Bull. of Hvar Observatory (in press).
6. Smart, R. N. and Zhang Zhenda (1984) 'Visible coronal emission associated with a quiescent prominence', Solar Phys. 90, 315-324.
7. Ryutova, M. P. (1986) 'Generation of plasma flows by oscilations of magnetic flux tubes', in Proc. of the Joint Varenna-Abastumani Inter. School & Workship on Plasma Astrophys., USSR, Sukhumi, May 1986, 71-75.
8. Svestka, Z. (1962) 'Motion in chromosphere flares', Bull. Astron. Inst. Czechosl. 13, 190-198.

CORONAL LOOP INTERACTION

Raymond N. Smartt and Zhenda Zhang[1]
National Solar Observatory/Sacramento Peak
Sunspot, New Mexico 88349 U.S.A.

Abstract: High–resolution images of post–flare loop systems in Fe XIV (5303Å) and FeX (6374Å) display transient enhancements at the projected intersections of some loops. The brightness of such enhancements gradually increases to a marked maximum and then fades with a typical lifetime of about thirty minutes. The maximum in the red line lags that of the green line by about ten minutes, while Hα reaches a maximum about ten minutes later, which gives a measure of the cooling time, and hence electron density. The observed phenomenon is interpreted as localized loop coalescence and partial magnetic reconnection, with the possibility of an increase in the current density, and heating of the plasma in the immediate vicinity of the X–point.

The coalescence of current loops leading to magnetic reconnection has been widely studied in relation to flares (see, for example, Tajima *et al*, 1987 and Tajima *et al*, 1982). In such treatments, internally–and externally–driven magnetic reconnection of the plasma is considered over different time scales and energies. Here we discuss the case of externally–driven reconnection in the low–energy case of post–flare coronal loops.

The data are from the 20–cm aperture NSO/SP emission–line coronagraph that records photographically sequential images in Fe XIV (5303Å), FeX (6374Å) and Hα, with a normal cadence of one minute. High–resolution post–flare loop images in the green line show occasional marked enhancements in the vicinity of the projected intersection of two loops. From the data it is evident that such enhancements occur as one loop develops or moves to intersect a neighbor, producing an X–point of coalescence. The morphology of such events as seen at their maximum, especially clear in cases where the plane of the loops are not strongly inclined to the line–of–sight, is a brightness at the point of intersection far greater than the simple sum of the brightness of the individual loops, and some enhancement extending partially along the loops away from the intersection point. Further, there is some partial "filling in" of the space between the loops, marked by a sharp boundary, evidently a process of redistribution of magnetic flux between non–aligned loops, with partial magnetic reconnection. Observations of over 80 events have been observed. It is found that typical enhancements increase to a

[1]On leave from Department of Astronomy, University of Nanjing, Nanjing, PRC.

E. R. Priest and V. Krishan (eds.), Basic Plasma Processes on the Sun, 350–351.

maximum over a period on average of about 15 minutes and then gradually fade with a lifetime of the order of 30 minutes. Usually at least several such events are observed in a post–flare loop development. The maximum brightness varies between different events, but all those identified as such show a brightness obviously significantly greater than the loops in gemera;. This leaves the possibility that there is a regime of weaker events that remain unidentified in data of this quality. Further, events with small spatial scale would probably not be detected except in exceptionally high–resolution data. It appears therefore that this is a relatively common phenomenon as detected in the visible emission corona, especially around solar maximum with an average of at least several events per day. Corresponding enhancements at the same location appear in the FeX (6374Å) images. The overall red–line emission image is generally much fainter than that of the green line, and the enhancements in the red–line are correspondingly weaker, but nevertheless are usually well–defined in the form of a highly localized bright patch. Each maximum lags the corresponding maximum in the green line by a few minutes, with a mean of about 10 minutes, a measure of the cooling time. About 10 minutes later again, Hα reaches a maximum at the site of the reconnection. In some cases it is completely absent just minutes before. It should be pointed out that a post–flare loop event is itself a highly dynamical system in which the loops, typically in a complex configuration, gradually increase in height. During this time period of several hours, new loops can appear, while others disappear, with a systematic trend to simpler configurations, a decreasing number of loops and fainter emission. Apparently it is this dynamical property of post–flare loops that strongly increases the probability of loop coalescence, as compared with coronal loops in general. But the dynamical property of post–flare loops also results in some uncertainty in tracing the morphology of a loop coalescence event through its life. Nevertheless, the characteristics of the enhancements as described above represent a very well–defined phenomenon.

From the temperature dependence of the thermal conductivity, we assume radiative cooling dominates for $T < 10^7$K. From the radiative cooling time, $t_c = 3KT/\Lambda_\gamma(T)n_e$, where K is the Boltzman constant, T the plasma temperature before cooling ($\sim 2 \times 10^6$K), $\Lambda_\gamma(T)$, the radiative cooling coefficient (10^{-22}erg cm^3 s^{-1}) and n_e the electron density, we find $n_e \lesssim 10^{11}$ cm^{-3}. The processes involved in heating the plasma at the X–point following coalescence remain uncertain without spectral information and data from other wavelength regimes, especially since the current density might be enhanced also in this region. For example, Dungey (1953) has pointed out that in the vicinity of a magnetic neutral point, a small disturbance can cause the current density to become large in the neighborhood of the neutral point. We have made a first estimate of the energy, E, involved in the reconnection via the coalescence instability (Tajima et al, 1982), in which $E = (LB^2a^2/2) \ln (L/a)$, where L is the length of the reconnecting region ($\sim 7 \times 10^8$cm), B is the loop magnetic field ($\sim$ 100G), and a is the current channel width ($\sim 10^8$cm), which gives $E = 7 \times 10^{28}$ ergs. The numerical values applied here are somewhat uncertain, especially the value of B; nevertheless, the value for E is thought to be a reasonable order–of–magnitude estimate, corresponding to that of a very small flare.

References:

Dungey, J.W.: 1953, Phil. Mag. **44**, 725.

Tajima, T., Sakai, J., Nakajima, H., Kosugi, T., Brunel, F. and Kundu, M.R., 1987, Ap. J. **321**, 1031.

Tajima, T., Brunel, F. and Sakai, J. 1982, Ap. J., **258**, L45.

VIII. SOLAR FLARES

PLASMA PROCESSES IN SOLAR FLARES

V.M.TOMOZOV
Siberian Institute of Terrestrial Magnetism,
Ionosphere and Radio Wave Propagation (SibIZMIR)
664033 Irkutsk, P.O. Box 4026
USSR

ABSTRACT. A rationale is presented for a conception that appearance of flares in active regions is due to the interaction of large-scale convective elements. Such an interaction gives rise to shear motions in the vicinity of the inverse polarity line of the photospheric magnetic field which generate vortical motions leading to non-equilibrium state of the magnetic configuration. Modern concepts of manifestations of turbulent plasma processes are described in terms of theoretical models for solar flares. Plasma effects arising at propagation of electron beams and thermal fluxes in the solar atmosphere are considered. Their role in the interpretation of hard X-ray and type III radio bursts is pointed out. The role of the turbulent Stark effect for diagnostics of collective plasma processes in solar flares is emphasized.

An important role in modern flare theory is played by plasma processes and, in particular, by physical effects connected with a turbulent state of plasma. In some phenomena taking place in the plasma of the solar atmosphere, collective plasma characteristics are especially clearly manifested - the interaction of plasma particles with external and self-consistent electric and magnetic fields. Not only can these fields control the plasma motion but also can, in turn, be changed and be induced themselves in the case of different plasma motions (Kaplan et al., 1977; Altyntsev et al., 1981).

One of the aspects of collective plasma features is the turbulence, a state of plasma connected with the presence of oscillations or noise whose interaction with plasma particles influences essentially its macroscopic characteristics. Such noise and oscillations arise because of some plasma instabilities (Kadomtsev, 1976). In the solar plasma such instabilities may be caused by electric

E. R. Priest and V. Krishan (eds.), Basic Plasma Processes on the Sun, 355–364.

currents flowing in the active region (AR), shock waves propagating out of the flare region, beams of energetic particles, by thermal fluxes, etc. All presently available flare models based on magnetic field dissipation mechanisms in AR take advantage of the concepts of turbulent processes in plasma.

Observational data accumulated over the past decades revealed a close relationship between flares and AR magnetic fields. It is known that most frequently flares arise in complicated multipolar sunspot groups during their fast evolution (see a review by Altyntsev et al., 1981). In the upper chromosphere and lower corona of the Sun where flares are produced, observational data do not show the existence of some other energy source capable to provide the flare energetics, except for the magnetic field energy. Therefore, at present the most well-grounded idea is that the flare energy source is provided by the magnetic field. In this connection the question of the energy storage mechanism in AR magnetic configuration is of great interest. We report here a conception proposed by Golovko et al. (1986, 1988) which, from our point of view, answers the relevant question of magnetic energy storage.

It is well known that the dynamics of the magnetic field and matter in photospheric layers of the solar atmosphere are determined by a complicated interaction between structural features of various scales. There exist a three-stage hierarchy of convection, namely granulation, supergranulation and large-scale structures consisting of elements with a characteristic scale of sizes, lifetimes and mass velocities in them (Kaplan et al., 1977). Besides the whole of this complex picture of motions is affected by the differential rotation of the Sun. For example, see a paper by Monin (1987) on the interconnection of differential rotation with the large-scale velocity field pattern at the solar surface. It is evident that, as the observed photospheric velocity field is formed, these structures must determine the evolution of the magnetic configurations - complexes of activity, AR and separate sunspots - and ultimately they must influence flare activity.

Indeed, Stepanyan (1983) has shown that although the places of AR formation do not show an obvious dependence of large-scale structures (giant cells),the activity level, however, is significantly higher near the borders of the above-mentioned structures. This conclusion may be extended to structures of smaller scales typical of supergranulation. It has been soundly confirmed by an analysis of flare development in the May 1981 complex of activity carried out by Banin (1984) on the basis of H-alpha filtergrams. Moreover, in many papers the correlation of flare location with horizontal shear motions near an inverse polarity line (IPL) was found (Zirin and Tanaka, 1973; Harvey and Harvey, 1976). There are broad lines of observational evidence now for shear motions of plasma elements with frozen-in magnetic field along the IPL. Of course, such mass motions must lead to the appearance of a shear structure of magnetic configuration. The large value of magnetic field shear serves as one of indications of flare production (Gaizauskas and Harvey, 1986; Machado et al., 1986; Moore et al., 1988). The orientation of dark filaments, emergence of new magnetic flux, and local deformation of the IPL are often also recognized as flare activity predictors. There are observational indications on vortical motions of matter close to IPL connected with flares also (Martres et al., 1973; 1983; Mein et al., 1984).

Analysis of the observational facts presented above leads to the following physical picture which permits us to establish causal relationships responsible for producing a pre-flare situation (build-up) and, ultimately, for the flare process. The interaction of structural features of different scales (supergranular and larger cells, for example) must lead to the generation of shear motions along the IPL in complexes of activity and separate AR with the production of vortices. Vortical motions of matter give rise to local enhancements of the magnetic field near the IPL and, as a consequence, to accumulation of an excess (non-potential) energy in a magnetic configuration, to current sheet formation and therefore to flare generation. Within the frames of this scheme, the excess energy in magnetic field configurations is derived from the energy of hydrodynamical motions of structural features, the scale of which may be well above the flare region size while the flare itself as a relaxation process of excess energy release in the magnetic field configuration. In other words, by changing and complicating the magnetic field the shear motions of the matter, together with a system of the vortices generated, ultimately brings the magnetic configuration to a non-equilibrium state. Once its stability threshold is attained, the accumulated energy starts to be rapidly released, thus indicating the flare onset. This scheme is illustrated in Fig. 1.

The first detailed description of this scheme was given in papers by Golovko et al. (1986, 1988) which also presented computation results on the velocity field in AR McM 11976 that produced a series of large flares in August 1972. They confirm the proposed mechanism.

In connection with the above-said, the problem of the evolution and stability of equilibrium magnetic configurations (loop structures of the magnetic field on the Sun) which are influenced by slow motions of field line footpoints at the photosphere may be of great interest for astrophysical applications. It is possible that, when a displacement of footpoints of a loop deformed by photospheric mass motions reaches some critical value, the equi-

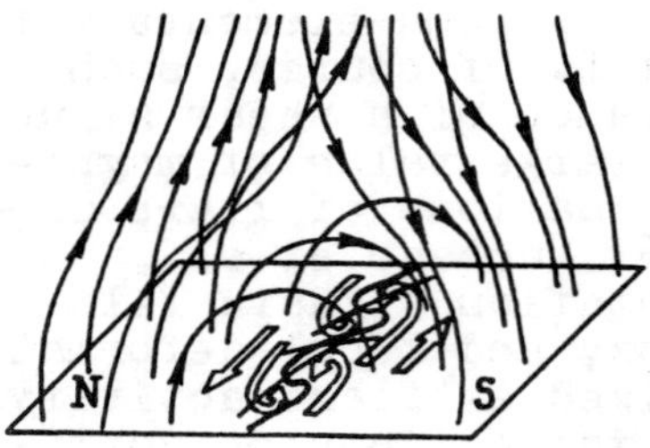

Fig. 1. Magnetic field configuration above the photospheric plane and the horizontal velocity field in the vicinity of the inversion polarity line (IPL).

Legend:
- —— inversion polarity line;
- —→— magnetic field lines of force;
- ⇒ direction of mass motions in the photospheric plane.

librium state of the configuration may, then, be upset and become unstable and a forced transition into a new equilibrium state will take place. If the magnetic energy of the new equilibrium state is less than that of the unstable state, then a free magnetic energy release will occur during the transition process. Besides, magnetic configuration topology in the new equilibrium state may be of quite a different kind, which means a topological reconstruction of the configuration will be taking place. Particularly, the closed loop structure may become partially open. Such problems are of interest with regard to understanding the solar flare nature (for example, Priest (1982)).

Recently Matyukhin and Tomozov (1986, 1989a,b) examined the problem of the quasi-static evolution of loop magnetic structures on the Sun (one magnetic loop and a loop arcade) with given boundary conditions. Slow shear (and vortical) motions of magnetic loop footpoints were considered to be boundary conditions, i.e., the kind of velocity field was analogous to that computed in papers by Golovko et al. (1986, 1988). It was shown that for a loop arcade a shear motion such as a viscous boundary layer leads to its topological reconstruction, with the energy released comparable with that released in large solar flares, i.e., 10^{32} ergs (for one magnetic loop, the value of free energy is 10^{28} - 10^{29} ergs). The topology of the original arcade configuration, when a critical value of magnetic shear is reached, becomes an open one. Note that such a scenario of the magnetic arcade evolution on the Sun explains quite well the main energy characteristics of large solar flares and predicts the origination of open magnetic structures in ARs of the Sun after flares.

We now examine the role of collective processes in different stages of a flare event. We shall restrict our attention only to some aspects which, we believe, are of particular interest. First of all, it is related to magnetic field dissipation in a solar flare energy release region in the explosive phase. This is characterized by energy release in the form of thermal fluxes and energetic particles. According to the electromagnetic nature of flares, for rapid dissipation of current energy in an electric circuit formed in AR plasma it is required that the plasma resistance should be sufficiently large. As is known, classical resistance of solar plasma is too small. Therefore, it is necessary to include collective effects which sharply enhance the plasma resistance. Collective effects connected with the generation of plasma waves and their interaction with particles are natural to arise within thin current sheets with typically high current densities. In some flare models the kinetic instabilities are able to cause a plasma sheet to change to a turbulent state. This change is possible when the current sheet rises into a lower corona where the energy balance of plasma in the sheet is upset due to thermal instability (Heyvaerts et al., 1977). As the result, the current density increases sharply in the sheet, a chain of kinetic instabilities develop, giving rise to anomalous resistance of sheet's plasma and, accordingly, the energy release rate increases rapidly (Tomozov, 1971, 1972).

Electric fields, MHD and plasma turbulence, and shock waves have customarily been treated as the main mechanisms of particle acceleration in flares. Strong electric fields are produced and plasma turbulence is generated in the flare energy release region itself where the magnetic field reconnection process (in the current sheet, for example) occurs. Theoretical investigations showed that the Langmuir and ion-sound turbulence modes may be important ones in interpreting the heating and electron acceleration in a current sheet (Tomozov, 1972; Kaplan et al., 1977; Smith, 1977a,b;1980; Brown and Smith, 1980). Laboratory simulation of current sheets confirmed the existence of these turbulence modes (Altyntsev et al., 1982; 1987). The ion-sound turbulence leads to intensive heating of the bulk of electrons. At the same time, the Langmuir turbulence effectively accelerates only fast electrons. The high efficiency of heating and electron acceleration processes in the current sheet is confirmed by laboratory experiments (Altyntsev et al., 1977). Recently in course of laboratory experiments it was succeeded in finding that instability of merging of magnetic islands plays a very important role in maintaining the high rate of magnetic field reconnection in the current sheet (Altyntsev et al., 1987). Very effective mechanisms of ion acceleration in flares are provided by shock waves and MHD turbulence (Smith and Brecht, 1988; Melrose, 1983) propagating from the region of primary energy release. A potential-jump induced acceleration mechanism along the front of a magnetosonic shock wave of large amplitude has recently been extensively discussed for purposes of interpreting a prompt acceleration of particles (protons and electrons) in the impulsive phase of flares (Ohsawa and Sakai, 1988). The existence of such an acceleration mechanism in the current sheet was recently confirmed through laboratory experiments (Altyntsev et al., 1988). The present status of the problem of particle acceleration in flares is reviewed by Vlahos et al. (1985).

As argued by a large number of authors (for example, Švestka, 1976), particle acceleration in flares consists of two stages, each of which differs by its specific mechanisms, energy characteristics of fast particles, and by time scales. As a result of the first-phase production an acceleration of electrons of up to an energy of 100 keV and of protons of up to an energy $\sim$ 20 MeV takes place directly in the energy release region and the electrons are accelerated more effectively as compared to protons. During the second phase of acceleration (the typical time of a few minutes) the most probable mechanism of which is an acceleration at shock fronts, relativistic electrons and ions with an energy over 200 MeV appear. However, Chupp's measurements (1983) revealed that there was virtually no time delay between bursts of hard X- and gamma-ray emissions, which led Chupp to the conclusion that electron and proton acceleration to maximum energies takes place directly in the energy release region by a single mechanism without involving the second phase. Of course, the above conception of two-stage acceleration of particles in flares (Švestka, 1976) disagrees with presently available observational data on hard X- and gamma-ray emissions of flares. As suggested by Bai (1986), it is possible to avoid this discrepancy, by revising the classification of flares. A very attractive model in which these results can be explained, is a topologically non-equilibrium MHD system arising during the course of merging of two magnetic loops with compensated longitudinal currents (Bardakov, 1987). In this model the magnetic field energy is transformed into fast hydrodynamical motions with formation of shocks. It may be noticed that in the frames of Golovko et al's (1986) scheme such an MHD system may well be the case.

Thus, it may be argued that the role of collective plasma processes is important during the stage of initial flare energy release. But these processes should be taken into account also when studying secondary flare processes in the solar atmosphere; we, now, shall discuss them.

Heat and energetic particle fluxes propagating from the region of initial energy release where magnetic field "annihilation" takes place, give rise to a disturbance of a large volume of the solar atmosphere, thereby causing the flare event proper. While penetrating into dense chromospheric layers, electron fluxes generate X-ray radiation in both the soft and hard spectral regions. Whereas the soft (0.5-10) keV X-ray radiation involves thermal emission of a hot plasma "evaporating" from the chromosphere and its interpretation is easily done, the origin of hard X-ray bursts (the energy of quanta is more than 15 keV) is not clear yet. The power-law character of the hard X-ray spectrum led to the construction of non-thermal models in which X-ray quanta are the bremsstrahlung radiation of fast electron beams decelerating in a dense chromospheric plasma. Naturally, this radiation must be polarized. In the beam's interpretation of the hard X-ray radiation it was necessary to understand how a directed electron beam penetrates into the chromosphere without scattering by plasma waves generated by the beam's particles. It was found that the mechanism for nonlinear stabilization similar to that suggested to interpret type III radio bursts, is able to diminish considerably the role of turbulent scattering for the electrons with an energy in excess of 10 keV (Livshitz and Tomozov, 1974). Allowance for strong Langmuir turbulence effects made it possible to extend the conclusion about the possibility of nonlinear stabilization to dense electron beams with $n_b/n_e = 10^{-2}$ as well (Vlahos and Papadopoulos, 1979). Thus, nonlinear stabilization of the beams is a striking example of collective interactions of plasma waves with particles in solar flares.

Later on, some difficulties of nonthermal models of X-ray bursts were encountered, namely uncertainties in the interpretation of power-law spectra; the need for an exceptionally large number of energetic electrons; etc. Some of the just-mentioned weaknesses of the nonthermal model of hard X-ray radiation have been overcome in a thermal model in which the existence of high-temperature plasma ($T_e \gg 10^8$ K) is supposed. In this model, the number of energetic electrons is decreased and the requirements to source energetics are reduced.

Different types of sporadic solar radio emission are related to a wide range of secondary phenomena associated with propagation of hot plasma and energetic particles into upper coronal layers and out into the interplanetary medium. Most types of solar radio bursts, intimately connected with flares, are explained by plasma mechanisms, including turbulent processes (Kaplan et al., 1977). Apparently, type II and III radio bursts are the most striking manifestations of collective plasma effects in the flare radio emission. Type II bursts are generated due to a shock wave front moving in the corona and radio emission of type III arises as a result of plasma wave generation by accelerated electron fluxes. The interpretation of type III radio emission is based upon the plasma hypothesis, according to which an electron flux excites high-frequency Langmuir waves due to a beam instability and plasma waves are, then, converted to electromagnetic ones. The generation of radio emission is by no means the only manifestation of plasma effects in type III radio bursts. It is known from observational data that the source velocity of these bursts (0.3-0.4 of the speed of light) was in fact constant from lower levels of the solar corona to a distance of 1 AU where only deceleration and expansion of the beams are observed. The mysteriosness of this result is that the well-known mechanism of quasi-linear relaxation must lead to destruction of the beam even in coronal regions of the solar atmosphere. In order to overcome this difficulty, it is necessary to identify the mechanism for beam's stabilization which is also connected with collective plasma processes (Kaplan and Tsytovich, 1972). This mechanism implies that, due to processes of nonlinear scattering off ions, Langmuir plasmons, which are excited by a beam, change their direction and phase velocity in such a way that they become off resonance with beam's particles and, therefore, beam's relaxation is proceeding much more slowly. Note that solar type III radio bursts represent a sort of a natural laboratory for studying the interaction of energetic beams with plasma. Of course, a complete theory of type III bursts consistent with observations has not yet been developed, and investigations in this area are being underway on a large scale (see, for example, Smith, 1974; Smith and Sime, 1979).

To sum up, we will remark that the turbulent Stark effect can be used, on the basis of spectral observations in the optical range, as a direct diagnostic tool for gaining insights into collective processes in solar flares. Revealing this effect is a challenging problem due to difficulties encountered when obtaining pertinent observational data and taking into account instrumental errors while processing such data as well as by virtue of incomplete theory. But it is highly important to resolve this problem because this promises to provide new insights into the nature of solar flares.

Of course, the range of problems under consideration does not include all the diversity of manifestations of collective plasma effects in solar flares. Many unsolved problems regarding our understanding of the origin of solar activity are left (in particular, with regard to solar flares). Nevertheless, a comprehensive study of these problems carried on by plasma physicists and astrophysicists promises to help towards answer to some of them in the near future.

REFERENCES

Altyntsev A.T., Krasov V.I. and Tomozov V.M. (1977), Solar Phys., 55, 69.

Altyntsev A.T., Banin V.G., Gopasyuk S.I., Maksimov V.P., Somov B.V., Stepanov V.E., and Tomozov V.M. (1981), Issled. po geomagn., aeron. i fizike Solntsa, Moscow: Nauka, vyp. 56, p. 39.

Altyntsev A.T., Banin V.G., Kuklin G.V., and Tomozov V.M.

(1982), Solar Flares, Moscow: Nauka (in Russian).
Altyntsev A.T., Krasov V.I., and Tomozov V.M. (1987), In: Solar Maximum Analysis, VNU Science Press, p. 223.
Altyntsev A.T., Lebedev N.V., and Strokin N.A. (1988), Phys. Letters A, 129, No. 5-6, 326.
Bai T. (1986), Adv. Space Res., 6, No. 6, 203.
Banin V.G. (1984), Issled. po geomagn., aeron. i fizike Solntsa, Moscow: Nauka, vyp. 68, p. 73.
Bardakov V.M. (1987), In: Solar Maximum Analysis, VNU Science Press, p. 217.
Brown J.C. and Smith D.F. (1980), Report on Progr. Phys., 43, 125.
Chupp E.L. (1983), Solar Phys., 86, 383.
Gaizauskas V. and Harvey K.L. (1986), Adv. Space Res., 6, No.6, 17.
Golovko A.A., Kuklin G.V., Mordvinov A.V., and Tomozov V.M. (1986), Contrib. of the Astron. Observatory Skalnate Pleso, vol. 15, p. 243.
Golovko A.A., Kuklin G.V., Mordvinov A.V., and Tomozov V.M. (1988), In: Solar Maximum Analysis. Additional Issue, Novosibirsk: Nauka, p. 278.
Harvey K.L. and Harvey J.W. (1976), Solar Phys., 47, 233.
Heyvaerts J., Priest E.R., and Rust D.M. (1977), Astrophys. J., 216, 123.
Kadomtsev B.B. (1976), Collective Phenomena in Plasma, Moscow: Nauka (in Russian).
Kaplan S.A. and Tsytovich V.N. (1972), Plasma Astrophysics, Moscow: Nauka (in Russian).
Kaplan S.A., Pikelner S.B., and Tsytovich V.N. (1977), Plasma Physics of the Solar Atmosphere, Moscow: Nauka (in Russian).
Livshitz M.A. and Tomozov V.M. (1974), Astron. Zh., 51,560.
Machado M.E., Gary G.A., Hagyard M.J., Hernandez A.M., Rovira M.G., Schmieder B., and Smith J.B., Jr. (1986), Adv. Space Res., 6, No. 6, 33.
Martres M.-J., Soru-Escaut I., and Rayrole J. (1973), Solar Phys., 32, 365.
Martres M.-J., Mein N., Mein P., Rayrole J., Simon G., and Soru-Escaut I. (1983), In: Active Phenomena in the Outer Atmosphere of the Sun and Stars, Pecker J.C. and Uchida Y. (eds), p. 261.
Matyukhin Yu.G. and Tomozov V.M. (1986), Report at V Annual Seminar of the Working Group on "Solar Plasma Investigations", Ashkhabad, U.S.S.R.
Matyukhin Yu.G. and Tomozov V.M. (1989a), Issled. po geomagn., aeron. i fizike Solntsa, Moscow: Nauka, vyp.87,36.
Matyukhin Yu.G. and Tomozov V.M. (1989b), AGU Chapman Conf. on: "Physics of Magnetic Flux Ropes", Extended Abstracts, Hamilton, Bermuda, p. 146.
Mein N., Martres M.-J., Simon G., Mein P., and Soru-Escaut I. (1984), Adv. Space Res., 4, No. 7, 33.

Melrose D.B. (1983), Solar Phys., 89, 149.
Monin A.S. (1987), Dokl. AN SSSR, 293, No. 1, 70.
Moore R.L., Hagyard M.J., and Davis J.M. (1988), Solar Phys., 113, No. 1-2, 347.
Ohsawa Y. and Sakai J.-I. (1988), Solar Phys., 116, 157.
Priest E.R. (1982), Solar Magnetohydrodynamics, Reidel, Dordrecht, Holland.
Smith D.F. (1974), Space Sci. Rev., 16, 91.
Smith D.F. (1977a), Astrophys. J., 212, 891.
Smith D.F. (1977b), Astrophys. J., 217, 644.
Smith D.F. (1980), Solar Phys., 66, 135.
Smith D.F. and Sime D. (1979), Astrophys. J., 233, 998.
Smith D.F. and Brecht S.H. (1988), Solar Phys., 115, 133.
Stepanyan N.N. (1983), Publ. of Debrecen Heliophys. Observ., vol. 5, 225.
Švestka Z. (1976), Solar Flares, Dordrecht-Reidel.
Tomozov V.M. (1971), Astron. Zh., 48, 556.
Tomozov V.M. (1972), Astron. Zh., 49, 502.
Vlahos L. and Papadopoulos K. (1979), Astrophys. J., 233, 717.
Vlahos L., Machado M.E., Ramaty R. et al. (1985), In: Energetic Phenomena in the Sun, Ch. 2, Proc. of the SMM Workshop, University of Maryland.
Zirin H. and Tanaka K. (1973), Solar Phys., 32, 173.

COHERENT RADIATION FROM ELECTROSTATIC DOUBLE LAYERS[1]

Jan Kuijpers
Sterrekundig Instituut Utrecht
P.O.Box 80 000
3508 TA Utrecht
The Netherlands

Abstract

The radiation properties of electrostatic double layers (DLs) are of potential significance for cosmic magnetic explosions. If the emission is sufficiently intense it can serve as a diagnostic tool for particle acceleration by localized strong potential drops in current-carrying plasmas. Moreover such intense emission may form the explanation of some of the observed intense and narrow-band bursts of radiation from stellar and planetary magnetospheres. Here we study the efficiency of two coherent radiation processes: antenna radiation and a maser process. It is found that both processes can operate in the DL to produce intense and narrow-band emission. Antenna radiation occurs if the dimensions of the double layer are smaller than the wavelength of the emitted radiation. This process is therefore relevant to laboratory rather than to astrophysical plasmas. The maser on the other hand requires an amplification length inside the double layer much larger than the emitted wavelength, and can lead to observable emission in astrophysical circumstances. The growth is exponential and the rate depends only on the electric field energy density of the DL. Since the latter is externally controlled by the electric circuit it is a constant for the emission process so as to constitute a true maser. The maximum brightness temperature is of order $10^{25}\,K$. Masing radiation from electrostatic DLs is therefore a candidate for some of the observed intense narrow-band cosmic radio emission.

1 Introduction

Electrostatic double layers (DLs) have first been proposed in astrophysical context by Alfvén (1958). Since then they have been advocated in a variety of cosmic high-energy phenomena (Alfvén, 1981). Recent reviews on the subject are given by Raadu and Rasmussen (1988) and Raadu (1989). DLs are of particular interest since the energy of a global electric circuit can be released locally in DLs (Jacobsen and Carlqvist, 1964) which have large voltage drops. DLs are therefore of importance for the explosive release of stored magnetic energy as is observed in *solar and stellar flares* (Kuijpers, 1989a). The important dynamic properties of DLs are the *acceleration of both electrons and ions and the triggering of mass motions*, effects which are indeed characteristic for many cosmic explosive phenomena. The particle acceleration is caused by the electric potential drop parallel to the magnetic field; the mass motion in the ambient ideal (that is non-resistive) mhd plasma is associated with a perpendicular potential difference, which again is connected to the parallel potential drop in a realistic finite 3-D structure (Raadu, 1984).

On the theoretical side much effort has been put into the construction of stationary DLs. A stationary DL can be considered as a coherent large-amplitude plasma wave: The electric field derives from a charge separation which is maintained self-consistently by the particle motion under the influence of the electric force. Laboratory experiments show that in practice stationary DLs are rather difficult to maintain and that "*explosive*" DLs are the rule.

Of particular interest for astrophysics are the *strong DLs*, in which the energy of an accelerated particle is much larger than the thermal energy per particle outside the DL. If indeed strong DLs are the locations of energy release in cosmic magnetic explosions the question arises how they can

[1]The calculations of the results presented in this paper can be found in Kuijpers 1989b.

E. R. Priest and V. Krishan (eds.), Basic Plasma Processes on the Sun, 365–373.

be observed. In this paper we calculate the coherent emission from DLs. As a first approach we shall restrict ourselves to ***stationary and strong*** DLs. For our model calculations we use the simple configuration of Fig. 1 and write the potential jump ϕ as

$$e\phi = uKT_0, \tag{1}$$

with $u \simeq 10 - 10^4, T_0$ the temperature outside the DL, e the electron charge and K is Boltzmann's constant. For the characteristic thickness of the DL along the electric field direction, taken to coincide with the magnetic field direction, we write

$$L = \ell\lambda_{Do}, \tag{2}$$

with $\ell \simeq 10 - 10^3$ and λ_{Do}, the Debye wavelength outside the DL. The shape of the DL is modelled as a cylinder with radius R. By assumption the electric field is confined to the DL, it is uniform within the DL and directed parallel to the cylinder axis.

In Section 2 we calculate the coherent emission from a ***non-relativistic*** DL. In Section 3 we apply the results of the calculation to laboratory experiments and cosmic plasmas. The conclusions are given in Section 4.

2 Coherent Emission

As a charged particle inside a DL is accelerated it emits electromagnetic radiation. For a single particle the power radiated is

$$P = \frac{2q^2\gamma^2}{3m^2c^3}\left\{ \mid \frac{d\vec{p}}{dt} \mid^2 - \frac{1}{c^2}\left(\frac{dU}{dt}\right)^2 \right\}, \tag{3}$$

with q the charge, m the mass, γ the Lorentz factor, $\vec{p}$ the momentum and U the energy of the particle. Inside the DL the force on the particle and its rate of change of energy are given by ($\vec{E}$ is the electric field)

$$\begin{aligned} \frac{d\vec{p}}{dt} &\simeq q\vec{E}, \\ \frac{dU}{dt} &\simeq \frac{dp}{dt}v, \end{aligned} \tag{4}$$

since for a strong DL the particle speed v is largely parallel to the electric and magnetic fields. With these expressions Eq. (3) reduces to

$$P = \frac{2q^2}{3m^2c^3} \mid \frac{d\vec{p}}{dt} \mid^2 = \frac{2q^4E^2}{3m^2c^3}, \tag{5}$$

independent of the particle energy γ. Clearly this "*linear accelerator emission*" (a term introduced by Melrose (1980) in the context of radio pulsars) is a relatively inefficient process when the (single) particle is moving at relativistic speed. For comparison: the power emitted in gyrosynchrotron radiation is given by a relation similar to Eq. (5) with E replaced by $\gamma\beta B_\perp(\beta \equiv v/c$ and $B_\perp$ is the magnetic field component perpendicular to the particle velocity v). Further Eq. (5) shows that electrons radiate much more efficiently in a DL than ions, as often happens in radiation processes. In principle the efficiency can be greatly enhanced if the particles radiate in phase *(Antenna radiation)* or if the radiation is amplified *(Maser)*. We shall investigate both possibilities in turn for non-relativistic DLs.

2.1 Antenna radiation

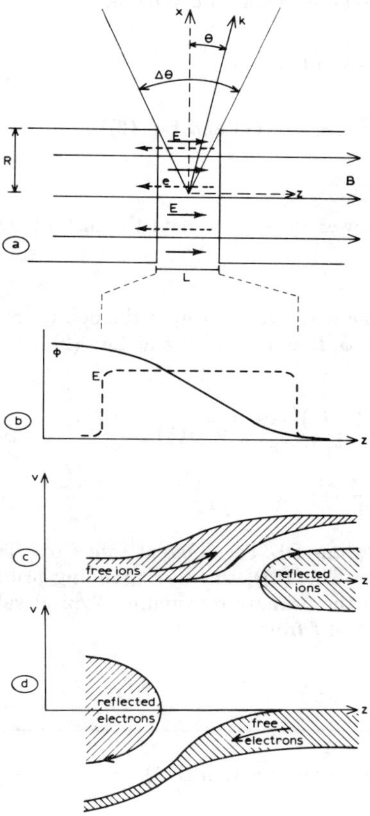

When the wavelength of the emitted radiation is larger than the dimensions of the emitting ensemble of electrons, an enhancement of the emission occurs in comparison with the case of an ensemble of independent radiators: in the former case the amplitudes of the emitted waves add coherently. The increase of the emission with respect to the incoherent case is expected to be a factor N, the number of particles in the bunch. Indeed such a result obtains from a rigorous calculation of the emission (Kuijpers, 1989b).

Let us consider the radiation from an ensemble of electrons (charge e and mass m). As the electric field is assumed to be confined to the DL (see Fig. 1) also the radiation originates from within the DL.

Figure 1: (a) shows the geometry of the DL with dimensions L and R parallel and respectively perpendicular to the magnetic field $\vec{B}$, the electric field $\vec{E}//\vec{B}$, the electron beam, and the wave vector $\vec{k}$ of escaping radiation. The lower figures show from top to bottom: the electric potential and field in the DL (b), level curves of the ion distribution function (c) and of the electron distribution function (d) with v the velocity in the z-direction.

For a strong DL the electron transit time t is practically independent of the initial electron speed:

$$t \simeq \left(\frac{2mL}{eE}\right)^{1/2} = \frac{\ell}{\omega_{po}}\left(\frac{2}{u}\right)^{1/2} \equiv \frac{\epsilon^{1/2}}{\omega_{po}}. \tag{6}$$

ω_{po} is the electron plasmafrequency outside the DL, m is the electron mass and the last equality defines ϵ. We choose the $\hat{z}$-direction to coincide with the direction of the background magnetic and electric field, and the $(\hat{x},\hat{z})$- plane to contain the wave vector $\vec{k}$. Since the acceleration occurs in the $\hat{z}$-direction, only the component j_z contributes to the emission. Then only the wave mode with polarization vector in the $(\vec{k},\vec{B})$-plane, the so- called ordinary mode, is radiated. Further we use the approximations

$$\begin{aligned} 2kR\cos\theta &\leq 1, \\ kL\sin\theta &\leq 1, \end{aligned} \tag{7}$$

where θ is the angle between the $\vec{k}$ and $\hat{x}$- direction. Condition (7b) guarantees that the contributions to the Fourier transform of the current density arising from one electron at the different z-positions along its orbit, add constructively, while condition (7a) ensures that the contributions from different x-coordinates add constructively. Since the DL is assumed to be strong, the transit time is independent of the initial particle speed, and therefore also of the initial spread in velocities. We approximate $\omega^t(\vec{k}) = (\omega_{po}^2 + k^2c^2)^{1/2}$ for $\omega_{co} << \omega_{po}(\omega_{co}$ is the electron cyclotron frequency,

ω_{po} the electron plasmafrequency, both taken outside the DL). Then the energy radiated per unit frequency, solid angle and time $p^t(\omega)$ (the differential power) for a stationary DL is

$$p^t(\omega) = \frac{N^2 e^4 E^2 \epsilon^{1/2} \cos^2\theta}{\pi^2 m^2 c^3 \omega_{po}} \frac{(1-\cos x)^2}{x^4} \left(1 - \frac{\epsilon}{x^2}\right)^{1/2}. \tag{8}$$

Here we use the abbreviations $x \equiv \omega t, \epsilon \equiv 2\ell^2/u = (\omega_{po} t)^2$ (see Eq. (1) and Eq. (2)).

2.1.1 Frequency, Bandwidth and Polarization

To determine the peak and bandwidth of the radiated power we approximate the instantaneous number of electrons in the DL with

$$N = A n_o v_{to} t, \tag{9}$$

where A is the cross-section of the DL perpendicular to the magnetic field n_o is the density of the ambient plasma and $v_{to} \equiv (KT_o/m)^{1/2}$, and we use $E \simeq \phi/L$ and Eq. (1) and Eq. (2), so that Eq. (8) can be rewritten as

$$p^t(\omega)\omega_{po} = \frac{A^2}{(\ell\lambda_{Do})^4} \frac{(KT_o)^4 \cos^2\theta}{2^5 \pi^4 m^2 c^3 e^2} u^3 G(x),$$

$$\text{with the definition } G(x) \equiv \frac{\epsilon^{5/2}(1-\cos x)^2}{x^4}\left(1-\frac{\epsilon}{x^2}\right)^{1/2}. \tag{10}$$

The function $u^3G(x)$ is a measure of the power radiated by one DL. For different values of ϵ Table 1 gives the relative peak frequency $\omega/\omega_{po} = x\epsilon^{-1/2}$ (see Eq. (6)), the corresponding primary maximum of $G(x)$ and the relative half width $\Delta\omega/\omega_{po}$ of the primary maximum. Typical values for ϵ are in the range $2 - 2\ 10^4$ as u varies from $10 - 10^4$ and ℓ from $10 - 10^3$.

TABLE 1

Peak and halfwidth of the dimensionless power function $G(x)$
$x \equiv \epsilon^{1/2}\omega/\omega_{po}$

ϵ	2000	200	20	2	0.2	0.02
ω/ω_{po}	1.19$^{(1)}$	1.11	1.04$^{(2)}$	1.34	2.09	3.57
$\Delta\omega/\omega_{po}$	0.043	0.13	0.12	0.80	1.92	3.84
max$G(x)$	48	16	1.2	0.51	3.4 10^{-3}	1.3 10^{-5}

(1) Secondary maxima above half the primary maximum exist both at lower and higher frequencies.
(2) A secondary maximum above half the primary maximum exists at a higher frequency.

From the table and Eq. (10) it follows that for a fixed temperature of the ambient plasma and a fixed shape of the DL the maximum radiation is emitted for strong (large u) and extended (large ℓ) DLs above the plasmafrequency of the ambient plasma ω_{po} and has a characteristic bandwidth of 10 % .

The emission at the source is completely linearly polarized and in the ordinary mode (electric polarization in the plane $(\vec{k}, \vec{B})$).

2.1.2 Brightness Temperature

The radiated power must always be smaller than the rate at which the electric current energy is transformed into particle acceleration:

$$\int\int p^t(\omega)\mathrm{d}\omega\mathrm{d}\Omega < I\phi \simeq An_o v_{to} e\phi = An_o v_{to} uKT_o$$

or using Eq. (10),

$$\frac{2}{3\pi^2}\frac{A}{(\ell\lambda_{Do})^2}\left(\frac{v_{to}}{c}\right)^3\frac{u}{\epsilon^{3/2}}\int_o^\infty G(x)\mathrm{d}x < 1. \tag{11}$$

Condition (11) puts an upper limit to the lateral cross section A of the DL.

An estimate of the brightness temperature T_b of the power calculated in Eq. (10) is obtained from the definition of $T_b(f)$ at a frequency $f = \omega/2\pi$:

$$\frac{f^2}{c^2}KT_b(f) \equiv \frac{p^t(\omega)\mathrm{d}\omega}{A(\theta)\mathrm{d}f}, \tag{12}$$

where $A(\theta)$ is the projected area of the DL perpendicular to the direction θ. Substituting Eq. (10) into Eq. (12) one finds

$$T_b(f) = T_o N_{Do}\frac{u^3}{\ell^2}\frac{v_{to}}{c}\frac{A^2}{(\ell\lambda_{Do})^4}\frac{(\ell\lambda_{Do})^2\cos^2\theta}{A(\theta)}\frac{G(x)}{(\omega/\omega_{po})^2} \tag{13}$$

with $N_{Do} \equiv n_o\lambda_{Do}^3$ the Debye number of the ambient plasma. At the intensity maximum ($2\pi f = \omega \simeq \omega_{po}$ and $x \simeq \epsilon^{1/2}$) the brightness temperature is enhanced with respect to the ambient temperature by a factor $N_{Do}u^3\ell^{-2}v_{to}c^{-1}$ (e.g with a factor 4 10^8 for $T = 10^5 K, n = 10^{10} cm^{-3}, u = 10^4, \ell = 10^3$).

Finally the brightness temperature (13) can never exceed the optically thick limit, given by the "effective temperature" of the coherent bunch of particles:

$$T_b(f) \leq NuT_o = T_o N_{Do} u^{1/2}\ell^3\frac{A}{(\ell\lambda_{Do})^2} \tag{14}$$

2.2 Maser

In the previous section we have considered the possibility of coherent emission from a DL in the form of antenna radiation. A necessary condition for the antenna process is that the DL should be small in comparison to the wavelength of the emission. In this section we consider the possibility of amplification of radiation in the DL in the opposite case when the lateral dimensions (perpendicular to $\vec{E}$ and $\vec{B}$) of the DL are large in comparison with the wavelength. A priori it is clear that the extension of the DL along $\vec{E}$ is always smaller than the wavelength of the radiation, since

$$k^t < \omega^t/c = O(2\pi/t)/c \simeq O\left(\frac{\pi v_f}{Lc}\right) < \frac{\pi}{L}, \tag{15}$$

where we have used Eq. (6) for the transit time of an electron in the DL and v_f is the fast electron speed obtained in a strong DL.

The process which we are looking for produces electromagnetic waves from the interaction between the electron beam created in the DL and the electric field inside the DL. Now *the electric field inside a strong DL is strong in several respects*: Firstly in a strong DL it is strong with respect to the critical runaway field ($E_c = 0.16E_D, E_D$ is the Dreicer field) in the ambient plasma,

$$\frac{E_{DL}}{E_c} \simeq \frac{\phi\lambda_{Do}^2}{Le0.16\ell n\Lambda} = \frac{u}{\ell}\frac{4\pi N_{Do}}{0.16\ell n\Lambda} >> 1. \tag{16}$$

Secondly the energy density in the electric field inside a strong DL is at least comparable to the thermal energy density of the ambient plasma

$$\frac{E^2/8\pi}{3n_oKT_o} = 4\pi\left(\frac{u}{\ell}\right)^2. \tag{17}$$

Here we shall consider only weakly nonlinear effects as a first approach; It is expected however that the efficiency will be comparable or higher in a realistic DL. In this approximation the simplest interactions which produce electromagnetic waves, are *1. the scattering of the electric field of the DL off the fast electrons into an escaping high-frequency electromagnetic wave, and 2. the related process of double emission of DL field and electromagnetic field.* The process works if a particular resonance condition is fulfilled: the Doppler shifted (in the frame of the fast electron) frequency must equal plus (for scattering) or minus (for double emission) the Doppler shifted frequency of the electromagnetic wave

$$\omega^{DL}\left(\vec{k}^{DL}\right) - \vec{k}^{DL}\cdot\vec{v} = \pm\left(\omega^t(\vec{k}^t) - \vec{k}^t\cdot\vec{v}\right). \tag{18}$$

The electric field of the DL is static or slowly-changing on an electron transit time so that we can approximate $\omega^{DL} = 0$. Further, substantial amplification of radiation can only take place transverse to the applied electric field (cf. Eq.(15)). Eq.(18) then reduces for a nonrelativistic DL to

$$\omega^t \simeq k^{DL}v_f \simeq 2^{3/2}\pi u^{1/2}\omega_{po}/\ell, \tag{19}$$

where v_f is the electron beam velocity at the end of the DL and we have used $k^{DL} \simeq 2\pi/L$. From Eq.(19) it is clear that scattered radiation above the ambient plasmafrequency is possible. In the scattering process the small frequency of the DL electric field is upconverted by the energetic electrons into the relatively high frequency of the radiation. In a thermal plasma scattering of low-frequency waves off electrons is usually an inefficient process since the induced electron (or Thomson) scattering is practically cancelled by the simultaneous nonlinear scattering on the polarization clouds (opposite sign) of the electrons. Here however the electrons acquire a beamlike distribution in the strong DL, they loose the company of polarization clouds and induced Thomson scattering becomes important.

We have calculated the efficiency of the scattering (and double emission) process in a DL using the *weak turbulence approximation.* The calculations are presented elsewhere (Kuijpers, 1989b). Using an electron beam distribution function

$$f_o(\vec{p}) = \frac{n_f}{m}(2\pi)^{-3/2}v_{to}^{-3}\exp\left\{-[(v_z+v_f)^2+v_x^2+v_y^2]/2v_{to}^2\right\} \tag{20}$$

the calculated *growth rate* is

$$\gamma^t(\vec{k}^t) = \frac{u^{3/2}}{\ell^2}\frac{\omega_{po}^4}{\omega_t^3}y exp(-y^2/2), \tag{21}$$

where $y \equiv (v_f - \omega_t/\mid k^{DL}\mid)/v_{to}$ and we have used $\frac{1}{2}mv_f^2 = e\phi, W^{DL} \simeq \phi^2/8\pi L^2, n_f v_f = n_o v_{to}$. It is important to realize that this form of coherent radiation from a DL constitutes a *true maser* since the energy density of the electric field in the DL is fixed and maintained in principle by the much larger energy of the global electric circuit of which the DL forms part.

2.2.1 Frequency, Bandwidth and Polarization

The characteristic frequency of the radiation emitted in the maser is determined by the resonance condition (18). For a non- relativistic DL the frequency is given by Eq. (19) and the ratio of the frequency with respect to the plasmafrequency increases with the strength of the DL.

The bandwidth is determined by 1. the frequency dependence of the growth rate Eq. (21): $\gamma \alpha \omega_t^{-3} y exp(-y^2/2)$, which sets an upper limit to the bandwidth of order $u^{-1/2}$ and 2. the amplification length Δx in the source:

$$\frac{\Delta\omega}{\omega} \simeq \frac{c}{3\Delta x \gamma^t} \tag{22}$$

For an e-folding length of 50 (see below) the relative bandwidth of the emission is 4 10^{-3}, much narrower than the bunched emission of Section 2.1.

The strongest radiation is emitted in directions perpendicular to the DL axis and is *completely polarized in the ordinary mode.* This follows directly from the vanishing nonlinear cubic current for the case of a DL electric field parallel to the z-axis, escaping radiation in the x-direction and electric field of the radiation parallel to the y-direction (the extraordinary mode polarization).

2.2.2 Brightness Temperature, Flux and Angular Extent

The brightness temperature of the maser depends on the number of e-folding lengths over which the radiation can be amplified inside the DL and on the absorption outside the DL. The observed flux from a stationary DL of thickness $\ell\lambda_{Do}$ and radius R at distance D from the observer is related to the brightness temperature of the radiation by

$$S = \frac{f^2}{c^2} K T_b \frac{\ell \lambda_{Do} R}{D^2} \simeq 3.5\ 10^{-34} n_9^{1/2} T_6^{1/2} \frac{\ell}{100} \frac{R}{3km} \frac{T_b}{10^6} erg\ cm^{-2} s^{-1} Hz^{-1}, \tag{23}$$

where we have neglected intervening absorption. For a flux of $10^{-17} erg\ cm^{-2}s^{-1}Hz^{-1} = 10^6 Jy$ (or 100 solar flux units) and a DL with $\ell = 100, R = 3km$, an ambient density $n = 10^9 cm^{-3}$, and a temperature of $10^6 K$, a brightness temperature of 3 $10^{22} K$ is required.

Such a brightness temperature is reached when radio waves of an initial temperature of $10^6 K$ are amplified with the growth rate of Eq. (21) over a distance

$$\Delta x \simeq 38c/\gamma_t \leq 10^4 c/\omega_{po} \simeq 3km \tag{24}$$

at an ambient plasma frequency of 300 MHz (density of $10^9 cm^{-3}$) and for $u^{3/2}\ell^{-2} \geq 10^{-2}$.

A theoretical upper limit to the brightness temperature of the radiation is obtained from the condition that the *radiation energy density should not exceed the electron beam energy density*

$$T_b^t < \frac{1}{2} \frac{n_f m v_f^2}{K} \cdot \frac{(2\pi)^3 R}{k_t^3 \ell \lambda_{Do}} \simeq 1.6\ 10^{25} T_6^{1/2} \frac{R}{3km} \frac{u^{1/2}}{\ell/100} K, \tag{25}$$

where we have used the vacuum dispersion relation $k_t = \omega_{po}/c$. One can further compare the radiation brightness temperature with the brightness temperature of the DL electric field

$$T_b^{DL} = \frac{E^2}{8\pi} \frac{\lambda_{Do} R^2}{K} \simeq 10^{23} \frac{(u/10)^2}{(\ell/100)^2} \left(\frac{R}{3km}\right)^2 T_6^{3/2} n_9^{1/2} K, \tag{26}$$

where we have used a wave vector volume for the DL of $(\lambda_{Do} R^2)^{-1}$. Spontaneous decay of radiation into DL field becomes only important for a radiation brightness temperature in excess of $(\omega^t/\omega^{DL}) T_b^{DL}$. So Eq. (26) does not lead to a stronger limitation for the brightness temperature of the radiation than Eq. (25).

The radiation is only amplified inside the DL and with maximum growth rate transverse to the DL electric field direction. The angular extent of the radiation depends therefore primarily on the thickness of the DL in the direction of the electric field with respect to its transverse dimension. For typical numbers used above one finds a *source which is radiating into a narrow equatorial double cone* of angular extent

$$\frac{L}{R} \simeq \frac{\ell \lambda_{Do}}{3km} \simeq 1 arc\ min. \tag{27}$$

2.3 Relation with existing processes

The radiation treated above is a form of ***Linear Accelerator Radiation*** (Melrose, 1980; Krishan and Sivaram, 1983). We have shown that in the DL two coherent versions of this process exist, an antenna process and a maser. The maser mechanism is physically related to the processes of ***scattering and double emission***. Further it is also instructive to compare our process with the ***Free Electron Laser***, where low-frequency magnetic wiggles are upconverted by the (relativistic) electron beam into high- frequency radiation. In the case of the DL the low-frequency electric field takes over the role of the magnetic wiggles and the process operates already for nonrelativistic beams. In both cases ***the energy of the radiation derives from the electron beam.***

3 Applications and discussion

We have investigated the possible existence of coherent radiation from DLs primarily because such structures may be the site of magnetic energy release ***in solar and stellar flares*** (Kuijpers, 1989a) and magnetic interactions between magnetospheres and accretion disks (Aly and Kuijpers, 1989). If this is true, coherent radiation from DLs would permit us to observe directly the primary sites of particle acceleration.

Here we propose that some of the ***intense decimetric spikes*** which are observed during solar flares (Kuijpers et al., 1981; Allaart and van Nieuwkoop, 1989) are the result of coherent radiation of DLs in the maser version treated in Section 2.2.

Further we propose that the ***intense radio bursts of Type I*** observed during solar noise storms are coherent radiation from transient DLs occurring in a coronal electric circuit triggered by reconnection. Evidence that the sources of Type I bursts are indeed the locations of electron acceleration comes from simultaneously observed Type III bursts which are commonly interpreted as electron beams and which seem to originate on the low-frequency side (higher altitude) of the Type I bursts (Leblanc, 1989).

One may ask how DLs are generated under cosmic conditions. In the laboratory and in numerical "experiments" one way to produce them is by driving current densities over the critical level (electron drift speed surpassing the electron thermal speed). In my opinion this is highly unlikely under cosmic conditions as this would require a constriction of the current cross-section by a factor

$$\frac{\omega_{co}}{\omega_{po}} \frac{c/\omega_{po}}{r/2} \frac{c}{v_{to}},$$

where ω_{co} is the electron cyclotron frequency, ω_{po} the electron plasmafrequency, v_{to} the electron thermal speed and r, the radius of the original current channel, is put equal to the scale height of the magnetic structure. However a different way to produce DLs in laboratory and numerical conditions is to excite a huge voltage spike locally. Under cosmic conditions this may be easy to establish by reconnection: If reconnection starts locally in a strongly sheared magnetic field system the global current system will oppose the forced change of the current path by ***local inductive voltage spikes***. This may lead to the generation of a DL. The particle beams injected by this DL into the surrounding plasma may then cause anomalous resistivity which leads to enhanced dissipation, further voltage spikes and the generation of (transient) DLs. In this way local reconnection may trigger the global energy release in a cosmic electric circuit in the form of a multitude of transient DLs.

Coherent radiation from DLs in the maser version may also be relevant for some of the emission from ***planetary magnetospheres***, complementary to the cyclotron maser. Direct observations of the terrestrial magnetosphere indeed show the existence of weak DLs or in any case of strong low-frequency electrostatic fields (Bostrom et al., 1989; Hultqvist, 1989). Although in our calculation we have used a model DL structure intense coherent radiation of both kinds (Sections 2.1 and 2.2) is generally expected if electron beams traverse electrostatic fields of sufficient intensity. The advantage of a DL is that it creates the beam itself and that its energy is externally maintained by the much larger energy of a current system.

Finally the antenna version of coherent DL emission may be relevant to as yet unexplained intense *microwave emission from a DL plasma in the laboratory* (Lindberg, 1988).

4 Conclusion

We have shown that nonrelativistic DLs emit intense radiation both of the antenna and of the maser type. The antenna version is suggested to be relevant to observed radiation in laboratory experiments on DLs. The maser version leads to observable emission in solar flares. Our calculations use several simplifying assumptions, notably a homogeneous electric structure with electric field parallel to the magnetic field, a strong potential difference inside the DL, weak turbulence methods, and finally dispersion relations for transverse waves which are strictly valid only in homogeneous plasmas and which neglect the magnetic field. We therefore consider these calculations as a first and promising step on the way to coherent radiation from realistic 3-D, magnetized, relativistic and non-relativistic, transient DL cavities.

Acknowledgement: I wish to thank Lennart Stenflo, Lennart Lindberg, Michael Raadu and Staffan Torvén for stimulating discussions on coherent radiation processes and DLs during my visit to Umea and Stockholm. The calculations were done during a stay at Meudon. In particular I thank Gérard Trottet, Monique Pick, Nicole Vilmer, Ludwig Klein, Bengt Pedersen, Yolande Leblanc, Philippe Zarka and other members of DASOP for the stimulating discussions and their hospitality.

References

Alfvén, H.: 1958, Tellus **10**, 104.

Alfvén, H.: 1981, Cosmic Plasma, D. Reidel Publ. Cy., Dordrecht, Netherlands.

Allaart, M.A.F. and van Nieuwkoop, J.: 1989, Solar Phys. submitted.

Aly, J.J. and Kuijpers: 1989, Astron. Astrophys., in press.

Bostrom, R., Holback, B., Holmgren, G., Koskinen, H.: 1988 Physica Scripta **39**, 782.

Hultqvist, B.: 1989, Phil. Trans. Royal Soc. London A **328**, 209.

Jacobsen, C. and Carlqvist, P.: 1964, Icarus **3**, 270.

Krishan, V. and Sivaram, C.: 1983, Solar Phys. **84**, 125.

Kuijpers, J.: 1989a, Solar Phys. **121**, 163.

Kuijpers, J.: 1989b, in Plasma Phenomema in the Solar Atmosphere, eds. M.A. Dubois and D. Grésillon, Cargèse, France.

Leblanc, Y.: 1989, in Plasma Phenomema in the Solar Atmosphere, eds. M.A. Dubois and D. Grésillon, Cargèse, France, in press.

Lindberg, L.: 1988, Astrophys. Space Sci. **144**, 3.

Melrose, D.B.: 1980, Plasma Astrophysics II, Gordon and Breach Publ., New York, p. 349.

Melrose, D.B.: 1986, Instabilities in Space and Laboratory Plasmas, Cambridge Univ. Press, Cambridge, U.K., Ch. 6.

Raadu, M.A.: 1984, Proc. 2nd Symp. on Plasma Double Layers, eds. R. Schrittwieser and G. Eder, Innsbruck, p. 3.

Raadu, M.A.: 1989, Physics Reports, **178**, 25.

Raadu, M.A. and Rasmussen. J.J.: 1988, Astrophys. Space Sci. **144**, 43.

FAST SOLAR FLARE PROTON ACCELERATION BY MHD TURBULENCE

DEAN F. SMITH
Berkeley Research Associates
and Department of Astrophysical, Planetary, and Atmospheric Sciences
University of Colorado, Boulder
Boulder, Colorado 80309

ABSTRACT. Proton acceleration by short-wavelength Alfven (A) waves resonant at the first harmonic of the proton gyrofrequency is reconsidered, taking into account nonlinear wave-wave interactions, collisionless wave losses, and wave escape losses in the geometry of a model coronal loop. It is shown that for the A wave levels required for acceleration in the transrelativistic regime in the 1982 June 3 flare and for acceleration in the nonrelativistic regime in the 1980 June 7 flare, the nonlinear wave interaction of scattering on the polarization clouds of ions will be important. This interaction rapidly isotropizes the A waves which divide their energy with fast magnetosonic (M) waves with a negligible change in their frequency spectrum. Because of electron Landau damping and escape losses, the M waves are confined to two narrow cones about the magnetic field and the total (A+M) wave distribution is still highly anisotropic. The total (A+M) wave spectrum has the same acceleration efficiency as a pure A wave spectrum. There are two principal problems with models of this type. The first is that a large wave energy density is required in a fairly narrow range in k-space. The second is that the protons are effectively bottled up. This makes very impulsive behavior as in the 7 June 1980 flare difficult to explain because proton precipitation is relatively slow.

1. Introduction

Stochastic acceleration of ions by resonant Alfven waves is a viable mechanism with the restriction that there must be appreciable power in Alfven waves propagating both parallel and antiparallel to the magnetic field **B**. The basic paper on this process is Barbosa (1979). An extension of his results to the nonlinear regime appeared in Smith and Brecht (1989; hereafter SB). We review and extend these results here.

Miller and Ramaty (1987; hereafter MR) performed a Monte-Carlo simulation to extend the analytic results of Barbosa (1979) into the transrelativistic regime to explain the 1982 June 3 flare. Gamma-ray observations (Chupp *et al.* 1987) show that the 4.1–6.4 MeV emission due to ~30 MeV protons and the pion emission due to ~1 GeV protons peaked simultaneously to within 16 s at ~11:43.5 UT. This simultaneity implies an acceleration time τ_A from ~30 MeV to 1 GeV less than 16 s which leads to a total energy density in Alfven turbulence $W^A > 2.3$ erg cm^{-3} according to MR. Although this W^A is a small fraction of the energy density in a 100 or 300 G field, it is comparable to the thermal energy density for $n = 5 \times 10^{10}$ cm^{-3}, $T_e = T_i = 1.6 \times 10^6$ K. This relatively high value of W^A leads to the possibility of nonlinear interactions in which Alfven waves A are transformed into other Alfven waves A' or magnetosonic waves M and vice versa. Such interactions could lead to additional losses since M waves are more heavily damped than A waves. The most effective nonlinear interaction is scattering on the polarization clouds of ions which was considered by Livshits and Tsytovich (1970) and Kaplan and Tsytovich (1973). These processes can be represented schematically by

E. R. Priest and V. Krishan (eds.), Basic Plasma Processes on the Sun, 375–382.

$$A + i \leftrightarrows A' + i' \tag{1a}$$

$$A + i \leftrightarrows M + i', \tag{1b}$$

$$M + i \leftrightarrows M' + i' \tag{1c}$$

where i and i' represent the polarization cloud of the ion before and after the scattering respectively.

It turns out for cases of interest that the M waves are confined to fairly narrow cones along the direction of **B** (z-axis) so that the primary component of the wavevector **k** is k_z. Under these conditions the change in angular frequency ω for processes (1a, b, c) is negligible and the change in k_z is very small. This is fortunate because the assumed spectral form of Alfven waves $W^A(\omega)$ will not be significantly affected by these processes and the spectrum of magnetosonic waves $W^M(\omega)$ is known from the specification of $W^A(\omega)$. Along with MR we assume that $W^A(\omega)$ has a Kolmogorov spectrum $\sim\omega^{-5/3}$ which has some justification in the measurements of Matthaeus and Goldstein (1982). However, measurements from ω^{-1} to ω^{-2} have also been made (Sari and Valley 1976) and a variety of theoretical predictions have been given (Montgomery 1983).

Besides the possibility of enhanced losses due to nonlinear processes, there are wave escape losses in the geometry of a solar loop. The interaction region for waves and protons has a finite spatial extent which leads to losses for both A and M waves given in SB. The M waves are also subject to collisionless Landau damping by electrons at a rate given in SB. For simplicity we assume that the magnetic field lines follow a semicircular loop of length L and minor radius R, and that there are no currents flowing along **B**. We neglect convergence of magnetic field lines in the acceleration region which is assumed to be in the corona.

We summarize some results on the total energy in protons and their energy distribution to set limits on the volumes of the acceleration region. The proton distribution in a hard sphere scattering model is characterized by the parameter αT, a measure of the hardness of the distribution in a Bessel function fit (Ramaty and Murphy 1987). Here, $\alpha = V^2/\lambda c$, where V is the velocity of the scatterers and λ is the scattering mean free path. As explained in Ramaty (1986) and Hua and Lingenfelter (1987; hereafter HL) it is possible to determine αT and the number of protons N_p above a threshold energy, usually taken as 30 MeV, from the ratio of the prompt 4–7 MeV γ-ray fluence to the delayed 2.223 MeV γ-ray fluence. For example, for the 1980 June 7 flare HL derived $\alpha T = 0.021 \pm 0.003$ and N_p (> 30 MeV) = $(9.2 \pm 2.4) \times 10^{31}$. This leads to a total energy above 1 MeV of $\sim 2 \times 10^{29}$ erg.

The time over which there was significant γ-ray production was 50 s (Chupp 1983). Hence, 4×10^{27} erg s^{-1} was going into protons. Because the 1982 June 3 flare consisted of an impulsive phase followed by a more gradual phase due to shock acceleration, it is not possible to make deductions of comparable accuracy since the published number of protons N_p refers to the whole flare. However, approximately N_p (> 30 MeV) = 3×10^{32} and $\alpha T = 0.035 \pm 0.003$ for the impulsive phase (HL) which leads to a total energy above 1 MeV of $\sim 7 \times 10^{29}$ erg. The impulsive phase lasted 66 s for this flare. Hence, 1.1×10^{28} erg s^{-1} was going into protons. Although the region in which this energy is processed could be smaller as in reconnection models of flares, we will minimize wave escape losses by taking a volume of 10^{27} cm^3 and finding L and R. The volume of a loop is $\pi L^3 (R/L)^2$. Typical R/L values are 10% which leads to $R = 3.2 \times 10^8$ cm, $L = 3.2 \times 10^9$ cm, and a loop height $\sim 10^9$ cm. Smaller loop heights can be obtained in a multiple loop model, but SB give evidence that a single loop was involved in the 1980 June 7 flare. Thus, we limit the scope of this paper to developing a single loop model in as much detail as possible.

2. Model and Wave Results

The model consists of a semicircular loop with a uniform toroidal field in which Alfven waves initially propagating along the field are excited. Further details of the model are given in SB. All that is important for our purposes is that the waves have an energy density W^A.

To obtain a level W^A for waves in the 1980 June 7 flare necessary to accelerate protons to 30 MeV in 2 s (Chupp 1983), we use a formula for the acceleration time τ_A due to Melrose (1974) which was also obtained by Barbosa (1979)

$$\tau_A = \left\{ \frac{\pi}{4} \omega \left(\frac{\delta B}{B}\right)^2 \frac{v_A}{v} \right\}^{-1}, \tag{2}$$

where δB is the perturbation in B due to the presence of the waves. Thus, $(\delta B)^2/8\pi = W^A$, or more generally, $W^A + W^M$, where W^M is the energy density in M waves. With $\tau_A = 2$ s, $\omega = 1.3 \times 10^4$ s^{-1}, $v_A = 10^8$ cm s^{-1} and $v = 7.6 \times 10^9$ cm s^{-1} corresponding to a 30 MeV proton, $\delta B/B = 0.062$ which leads to $W^A = 1.5$ erg cm^{-3} ($W^M \approx 0$ initially).

An important problem for the amplification processes that will be considered next is the initial level of M waves, $W_o^M(\omega_{min})$. It is reasonable to suppose that no wave excitation mechanism will excite purely Alfven waves. Alternately, even in the case of pure A wave excitation, weak density gradients would lead to a low level of M waves (Wentzel 1989). Hence, we shall assume that 0.1% of the energy going into Alfven waves goes into M wave excitation and A wave excitation at other angles with the same spectral form. This implies $W_o^M = W_o^{A'} = 2.3 \times 10^{-3}$ (1.5×10^{-3}) erg cm^{-3} for the 1982 June 3 (1980 June 7) flare.

The equations describing the nonlinear interaction of A and M waves due to scattering on the polarization clouds of ions are given in SB. We concentrate on the results here. It follows from these equations that two A-waves with wave vectors in the same half-plane do not interact. Likewise, there is no interaction between two M waves propagating in the same direction. A waves amplify A and M waves propagating in the other half-plane of the same ω at a rate

$$\nu_{nl} = \frac{\pi}{3} \frac{\omega^2 W_\omega^A}{n m_i v_A{}^2 (1 + T_e/T_i)^2} . \tag{3}$$

The coefficient in equation (3) is valid for arbitrary distributions of A waves and for M waves with directions close to **B**. We shall see below that this restriction on M waves is well satisfied in our applications. If ν_{nl}^{-1} is small compared to time scales of interest so that several e-folds of amplification can occur, then initially one-dimensional A waves become almost isotropic. There is a very small wedge of $\lesssim \pm 0.5°$ around $\theta = 90°$ where A waves are Landau damped and do not grow. For the same condition, M waves grow up to a level where $W^M = W^A$ in two cones about z where $\nu_{nl} > \nu_{damping}$ given in equations (3) and (4) of SB.

Application of these ideas leads to the results in Table 1 from SB where $\nu_g = \nu_{nl} - \nu_{damping}$ is the net nonlinear growth rate and τ is the time for 6.2 e-folds of growth, the time for M waves to grow to half the initial level. This is the saturation level since energy is now divided equally between A and M waves.

TABLE 1. Total and net nonlinear growth rates
For both flares $n = 4.8 \times 10^{10}$ cm^{-3} and $T_e/T_i = 1$

Flare	$\omega_{min}(s^{-1})$	W^A_ω (erg cm^{-3} s)	v_A (cm s^{-1})	$\nu_{nl}(s^{-1})$	$\nu_g(s^{-1})$	$\tau(s)$
1982 June 3	2.8×10^4	5.4×10^{-5}	3×10^8	1.5	1.4	4.4
1980 June 7	1.3×10^4	7.9×10^{-5}	10^8	4.2	4.17	1.5

TABLE 2. Angular range and average damping of M waves
For both flares $v_e = 4.9 \times 10^8$ cm s^{-1}

Flare	$\omega_{min}(s^{-1})$	v_A (cm s^{-1})	θ_1	$\nu^M(s^{-1})$
1982 June 3	2.8×10^4	3×10^8	18.7°	0.82
1980 June 7	1.3×10^4	10^8	25.1°	2.1

It can be seen from the values of τ in Table 1 that nonlinear effects will be important in the first 16 s interval for the 1982 June 3 flare. Nonlinear effects will be marginally important in the first 2 s of the 1980 June 7 flare. The impulsive phase γ-ray emission of this flare lasts 50 s. Thus, nonlinear effects will certainly be important for most of the γ-ray emission to the extent that all the emission comes from one loop.

The angular range of the M waves and their average damping were calculated in SB and are given in Table 2, where θ_1 is the half-angle of the cone along **B** to which M waves are confined because of electron Landau damping.

The following general picture emerges for the parameters of these two flares. Nonlinear effects will be important for most of the impulsive phases of both flares to the extent that all the emission comes from a single loop. The initial A waves will be rapidly isotropized. The M waves will grow in two cones aligned along the field where their damping is sufficiently weak to a level with $W^M \approx W^A$. For both flares the principal loss is thermal Landau damping of M waves. We shall see in the next section that these losses are comparable to the energy leaving the acceleration region in accelerated protons.

3. Proton Acceleration

The overall acceleration picture follows closely the one of Barbosa (1979) with the caveat that one of the wave modes (M waves) are subject to other forms of damping. The process (1a) preserves ω^A which means that it preserves $k_z^A = k^A \cos\theta$. Thus, larger (smaller) values of k^A are created in the scattering process with momentum being obtained from (given to) the polarization cloud of the ion. An initial one-dimensional $k_z^{-5/3}$ spectrum with k_{min} does not lead to an isotropic $k^{-5/3}$ spectrum for which the minimum $k = 0$. By taking $|\cos\theta| = 2/\pi$, MR effectively took a one-dimensional (1–D) spectrum at $\theta = 50.4°$ which differs from the spectrum of Section 2 by the scale factor $\pi/2$. In process (1b), the k_z^A of the A wave becomes the k^M of the M wave since frequency is preserved. Since the M waves are concentrated in narrow cones about z with most of the power close to $\theta = 0$, the k_z spectrum of these waves is $k_z^{-5/3}$ to within an error of less than 5% as long as the M waves are in a cone of half-angle $\leq 26°$. Considering the degree of uncertainty of other parameters in the problem, it is reasonable to disregard this small error and write the resonance condition as (cf. MR)

$$\omega = \Omega_i v_A / \gamma v \, |\cos \phi| \quad (4)$$

for all the waves, A or M, where v is the velocity and ϕ the pitch angle of the proton, respectively, $\Omega_i = eB/m_i c$ is the nonrelativistic gyrofrequency and γ is the proton's Lorentz factor.

With the same resonance condition (4) and the same ω spectrum, the M waves have exactly the same acceleration properties as A waves, but the physical picture is slightly changed. The A wave is left-hand circularly polarized and the M wave is right-hand circularly polarized with the perturbation electric field 90° out of phase with that of the A wave. Because of the gyromotion of a proton, the same average force components are present as for an A wave and the proton can gain or lose energy depending on its cyclotron phase. An average over a random-phased, isotropic distribution of protons leads to cyclotron damping of the M wave. Just as for the A wave, scattering is a stronger process than acceleration and the isotropy of the proton distribution is automatically insured by the acceleration.

The injection problem of having some protons with $v > v_A$ remains. As discussed by SB, this problem is tied to the flare primary energy release and the source of the A or A and M waves. A loop with many tearing modes occurring in a small fraction of the total volume is a good candidate which leads to the acceleration of a small fraction (~ 1%) of the ions to $v > v_A$. The figures for energy above 1 MeV in Section 1 with a volume of 10^{27} cm^3 lead to the requirement of accelerating 0.1–1% of the protons to $v > v_A$ throughout the whole impulsive phase. This is consistent with the picture of tearing modes which instantaneously occur in a small fraction of the volume, but which sequentially cover or process the magnetic energy of a large fraction of the volume.

We consider the energetics of the process with A and M waves as compared to the pure A wave result. As given in Section 1, for the 1982 June 3 flare there is 11 erg cm^{-3} s^{-1} going into proton energy above 1 MeV. We found in Section 2 an average loss rate due to M waves $\nu^M = 0.82$ s^{-1}. With $W^M = 1.2$ erg cm^{-3} (half of 2.3 erg cm^{-3} which was the initial level), there is an additional loss of 1.0 erg cm^{-3} s^{-1}, about 9% of the proton energy loss. For the 1980 June 7 flare there is 4 erg cm^{-3} s^{-1} going into proton energy above 1 MeV (cf. §I). We found in Section 2 an average loss rate due to M waves $\nu^M = 2.1$ s^{-1}. With $W^M = 0.75$ erg cm^{-3} (half of 1.5 erg cm^{-3}, the initial level), there is an additional loss of 1.6 erg cm^{-3} s^{-1}, about 40% of the proton energy loss. Hence, these losses are of the same order of magnitude as the accelerated proton losses and should be included in any detailed energetic considerations. They are much larger than the pure A wave case.

4. Discussion

We have reexamined proton acceleration by short-wavelength Alfven waves resonant at the first harmonic of the proton gyrofrequency. Within the limitations of the assumptions made such as a uniform field in a semicircular loop, the results are the following. For levels of waves which have already been proposed for proton acceleration in the transrelativistic regime (MR) and necessary for a well known impulsive flare in the nonrelativistic regime, initially 1-D A waves will rapidly become almost isotropic and give half their energy to almost 1-D M waves. This result partially justifies MR's use of an isotropic distribution of A waves. However, the total distribution of A and M waves remains peaked along z due to the almost 1-D M wave angular distribution. Important results of SB are the demonstration of the importance of M waves and the identification of two substantial loss mechanisms for these waves, escape losses and thermal electron Landau damping losses. We have not considered variations in the initial A wave spectrum, but believe that spectra in the observed range from ω^{-1} to ω^{-2} (Sari and Valley 1976) would lead to similar results.

Although this type of model is promising, there are two problems which deserve mention. The first problem is that there is a large energy density $W^A + W^M$ in a fairly narrow range of ω- or k-space. There will be another potentially large loss due to cascading from small to large k. Unfortunately, our knowledge of MHD turbulence is sufficiently poor (Montgomery 1983) that there is no way to meaningfully evaluate this loss. We are simply assuming that acceleration occurs in part of the inertial range of MHD turbulence and that the acceleration does not significantly affect this range.

The second problem is that the protons are well confined by the levels of turbulence required for acceleration and the proton transport is diffusive. The mean free path for scattering is (Forman, Ramaty, and Zweibel 1986)

$$\lambda_S = \frac{m_i^2 \, v^2 \, c^2}{12\pi^2 \, e^2 \, W(k_R)} \tag{5}$$

where the resonant wavenumber k_R is given by the right hand side of equation (4) without V_A and

$$W^A + W^M = \int W(k) \, \frac{dk}{2\pi} \, . \tag{6}$$

As discussed by Barbosa (1979) the time for the protons to diffusively travel a loop half length L/2 is

$$\tau_d \equiv \frac{L^2}{4\kappa_{\|}} \tag{7}$$

where the parallel spatial diffusion coefficient

$$\kappa_{\|} = \frac{2}{3} v\lambda_S = \frac{m_i^2 \, v^3 \, c^2}{18\pi^2 \, e^2 \, W(k_R)} \, . \tag{8}$$

Results of using equations (5) and (8) for the parameters given above are given in Table 3 where the first two rows are for the flare of 1980 June 7 and the last two rows are for the flare of 1982 June 3.

TABLE 3. Proton Mean Free Paths and Propagation Times

E(MeV)	v(cm s^{-1})	λ_S(cm)	τ_d (s); L = 10^9 cm	τ_d (s); L = 10^8 cm
1	1.4×10^9	5.9×10^4	4.7×10^3	47
30	4.4×10^9	1.0×10^5	8.3×10^2	8.3
30	4.4×10^5	5.2×10^4	1.7×10^3	17
10^3	3.0×10^{10}	8.3×10^4	1.5×10^2	1.5

It can be seen that these times are too long to explain the most rapid (2s) variations in the 1980 June 7 flare and possible for the 1982 June 3 flare only for short ($L = 10^8$ cm) acceleration region lengths.

It is in principle possible to circumvent this problem by going to very short acceleration region lengths, but then many acceleration regions or a single very large area region is required. The latter possibility is inconsistent with current knowledge about loop R/L values. The problem with many acceleration regions is that an additional coordinating mechanism is required to produce the semi-periodic behavior that was observed in the 1980 June 7 flare.

Neither of the above problems is insurmountable and both problems should be considered as further restrictions on the use of MHD turbulence models.

This work was supported by NSF Grant ATM-8909845.

References

Barbosa, D.D.: 1979, *Ap. J.* **233**, 383.

Chupp, E.L.: 1983, *Solar Phys.* **86**, 383.

Chupp, E.L. *et al.*: 1987, *Ap. J.* **318**, 913.

Forman, M.A., Ramaty, R., and Zweibel, E.G. 1986, in *Physics of the Sun, Vol. II.*, ed. P.A. Sturrock (Dordrecht: Reidel), p. 245.

Hua, X.-M., and Lingenfelter, R.E.: 1987, *Solar Phys.* **107**, 351 (HL).

Kaplan, S.A., Tsytovich, V.N.: 1973, *Plasma Astrophysics*, trans. D. ter Haar (Oxford: Pergamon Press), p. 68.

Livshits, M.A., and Tsytovich, V.N.: 1970, *Nuclear Fus.* **10**, 241.

Matthaeus, W.H., and Goldstein, M.L.: 1982, *J. Geophys. Res.* **87**, 6011.

Melrose, D.B.: 1974, *Solar Phys.* **37**, 353.

Miller, J.A., and Ramaty, R.: 1987, *Solar Phys.* **113**, 195 (MR).

Montgomery, D.: 1983, in *Solar Wind Five*, ed. M. Neugebauer, NASA CP-2280, p. 107.

Ramaty, R.: 1986, in *Physics of the Sun, Vol. II.*, ed. P.A. Sturrock (Dordrecht: Reidel), p. 251.

Sari, J.W., and Valley, G.C.: 1976, J. *Geophys. Res.* **81**, 5489.

Smith, D.F., and Brecht, S.H.: 1989, *Ap. J.* **344**, 1004 (SB).

Wentzel, D.G.: 1989, *Ap. J.* **336**, 1073.

DISCUSSION

STEPANOV: What is the origin of Alfvén wave turbulence in a flare loop?

SMITH: I have not considered this problem. I have simply assumed that the Alfvén waves are there.

HOYNG: How sensitive are your results to the assumed spectral distribution of waves ($k^{-5/3}$)?

SMITH: I have only done calculations for the range k^{-1} to k^{-2} observed by Sari and Valley in the solar wind. Within this range the results are not sensitive to the assumed spectral distribution.

SIVARAM: (i) Would this process also accelerate other ions? One has a preferential acceleration mechanism for He–3 for instance?
(ii) Could the protons be accelerated to energies well over a GeV?
(iii) As neutrons are produced, would tritium form from D.

SMITH: (i) I have not investigated this question, but one could have resonant acceleration of other ions by Alfvén waves, but of different frequencies.
(ii) If you extend the spectrum to lower frequencies, protons could be accelerated above 1 GeV.
(iii) The cross-section for this process for solar densities is too small to be important.

SUPER–ALFVÉNIC BEAM–PLASMA INSTABILITIES IN SOLAR FLARES

Frank Verheest
Instituut voor theoretische mechanika, Rijksuniversiteit Gent
Krijgslaan 281
B–9000 Gent, Belgium

ABSTRACT. *Special type III radio bursts with significantly lower drifts have been attributed to proton beams at the flare site. These low energy beams (some MeV) are nevertheless super-Alfvénic. Recent reports deal with the stability of such beams against* resonant *scattering by waves below the proton gyrofrequency. Here a different mechanism is proposed, namely* nonresonant *Alfvén instabilities, which may have higher growth rates and can thus dominate. A selfconsistent multispecies plasma description of such instabilities, developed for solar wind and cometary plasmas, is applied to slowly drifting type III radio bursts. The instability growth rate is fairly insensitive to the beam velocity but decreases with the ratio of beam to ambient plasma densities. Even low beam densities can trigger these nonresonant and hence generic instabilities. Moreover, growth rates are sizeable fractions of the wave frequency, whereas resonant instabilities would give much lower values at low beam densities.*

1. Introduction

Conventional MHD describes magnetized plasmas as one single fluid. In many astrophysical applications, however, it is essential to distinguish between the different species making up a plasma, especially if there are different flow velocities in equilibrium. One of these cases is where a plasma or a beam flows through another (ambient) plasma, as in the case of type III radio bursts (caused by electron beams propagation in the solar corona) or of the recently reported class of special type III radio bursts with a significantly lower drift than normal and which are attributed to proton beams at the flare site (Simnett 1986, Benz & Simnett 1986). These are beams of fairly low energy protons (with energies of a couple of MeV) which are nevertheless super-Alfvénic. Similar situations are encountered in the solar wind flow around comets, and an extensive literature has attested the occurrence there of different types of low-frequency instabilities, from an observational, analytical and numerical point of view (see *e.g.* Sagdeev *et al.* 1987, Lee 1988). Recently Tamres *et al.* (1989) have reported on the stability of such super-Alfvénic proton beams in solar flare loops against *resonant* scattering by waves in the flare loop coronal plasma at frequencies below the proton gyrofrequency. We would propose here a different mechanism, namely that such beams may also be unstable to *nonresonant* Alfvén instabilities, as we know from cometary plasma physics that under certain conditions these may have the highest growth rate (Goldstein & Wong 1987) and thus are more likely to dominate. We will draw on the selfconsistent description of such instabilities developed for solar wind and cometary plasmas (Verheest 1977,1987, Lakhina & Verheest 1988, Verheest 1989) to illustrate their possible relevance for the said slowly drifting type III radio bursts. In the next section the

E. R. Priest and V. Krishan (eds.), Basic Plasma Processes on the Sun, 383–389.

theoretical framework is recalled, and then in section 3 applied to the special type III radio bursts under consideration.

2. Theoretical framework for nonresonant Alfvén beam modes

We start the selfconsistent multispecies plasma description from a set of transport equations for a number of cold plasma species with number densities n_s and fluid velocities $\mathbf{u}_s$ (Verheest 1977):

$$\begin{aligned} \frac{\partial n_s}{\partial t} + \nabla \cdot (n_s \mathbf{u}_s) &= 0, \\ \frac{\partial}{\partial t}\mathbf{u}_s + \mathbf{u}_s \cdot \nabla \mathbf{u}_s &= \frac{q_s}{m_s}(\mathbf{E} + \mathbf{u}_s \times \mathbf{B}). \end{aligned} \tag{1}$$

Pressure effects have been left out of the description, as these are not of importance for the kind of low-frequency waves we will consider. $\mathbf{E}$ and $\mathbf{B}$ are the electric and magnetic fields. Their changes and the collisionless coupling between the different plasma species obey Maxwell's equations, of which we only need the following:

$$\begin{aligned} \nabla \times \mathbf{E} + \frac{\partial}{\partial t}\mathbf{B} &= \mathbf{0}, \\ c^2 \nabla \times \mathbf{B} - \frac{\partial}{\partial t}\mathbf{E} &= \frac{1}{\varepsilon_0}\sum_s n_s q_s \mathbf{u}_s. \end{aligned} \tag{2}$$

The summation over s in the expression for the current is over all plasma constituents.

For the sake of simplicity, a homogeneous equilibrium is now assumed, with equilibrium densities N_s and bulk species velocities $\mathbf{u}_{s0} = U_s \mathbf{B}_0 / B_0$ (possibly zero for some species, depending upon the reference frame chosen). This implies that a local picture of the wave phenomena is taken. In a selfconsistent description bulk motion perpendicular to $\mathbf{B}_0$ is the same for all plasma species, namely $\mathbf{E}_0 \times \mathbf{B}_0 / B_0^2$, and can be eliminated if need be by going to a Hoffman–Teller frame, where the motional electric field $\mathbf{E}_0$ vanishes. Also, $\mathbf{E}_0 \times \mathbf{B}_0 / B_0^2$ does not influence the stability criteria to be derived furtheron and having the beams parallel to the flare magnetic field is quite consistent with the flux tube configuration of the solar flare plasma. In such a truly selfconsistent approach there is no net charge nor current in equilibrium:

$$\sum_s N_s q_s = 0 \quad , \quad \sum_s N_s q_s U_s = 0. \tag{3}$$

Due to their high mobility, the electrons are assumed to adjust themselves so as to maintain the required charge and current neutrality. In other words, they provide the necessary return current for any ion beam present in the system, much more easily than in the reverse situation where electron beams are thought of as being of primary importance.

We shall need several global quantities for the combined plasma as a whole. First of all, there is the global bulk velocity:

$$\overline{U} = \sum_s N_s m_s U_s / \sum_s N_s m_s. \tag{4}$$

As in a first approximation there are two distinct streaming velocities to be considered, say U_A for the speed of the ambient subplasma and U_B for the beam velocity of the drifting subplasma constituents, we can rewrite $\overline{U}$ as

$$\overline{U} = \frac{(\sum_A N_s m_s)\, U_A + (\sum_B N_s m_s)\, U_B}{(\sum_A N_s m_s) + (\sum_B N_s m_s)} = \frac{U_A + \sigma U_B}{1+\sigma}, \tag{5}$$

if σ stands for the ratio of the mass densities of drifting to ambient subplasmas. If both subplasmas are taken as hydrogen plasmas, then σ is but the ratio of number densities N_B/N_A.

Similarly, we define V_A as the Alfvén velocity of the whole plasma through

$$V_A^2 = \frac{B_0^2}{\mu_0 \sum_s N_s m_s} = \frac{V_{AA}^2}{1+\sigma}, \tag{6}$$

with V_{AA} the Alfvén velocity in the ambient subplasma. Also, the normalized kinetic energy in the relative parallel motions can be expressed as

$$W^2 = \frac{\sum_s N_s m_s (U_s - \overline{U})^2}{\sum_s N_s m_s} = \frac{\sigma (U_A - U_B)^2}{(1+\sigma)^2}. \tag{7}$$

Due to the requirements (3) of charge and current neutrality in equilibrium, the term in W^2 will only be nonzero in general if there are at least three species present (such as electrons and two kinds of ions), and if these species move at differing speeds, however moderate the differences. If all plasma constituents have the same drift velocities, then one can simply use results from stationary plasmas, taken over to a drifting frame. However, with ions or subplasmas moving at different speeds, there is no longer a natural frame of reference.

The low-frequency, long-wavelength waves which we will now consider are such that

$$\omega \ll |\Omega_s| \quad , \quad |k_\parallel U_s| \ll |\Omega_s|. \tag{8}$$

Here $\Omega_s = q_s B_0 / m_s$ are the gyrofrequencies, defined inclusive of the charge of the particles under consideration. The conditions (8) mean essentially that for each species the appropriate Doppler-shifted frequency $\tilde{\omega}_s = \omega - kU_s$ is small in magnitude compared to the gyrofrequency Ω_s. Then the dispersion laws for parallel and oblique propagation of such Alfvén waves are (Verheest 1977)

$$(\omega - k_\parallel \overline{U})^2 = k^2 V_A^2 - k_\parallel^2 W^2 \tag{9}$$

for the compressional or fast mode and

$$(\omega - k_\parallel \overline{U})^2 \simeq k_\parallel^2 (V_A^2 - W^2) \tag{10}$$

for the shear or slow mode. The last dispersion law is not valid close to perpendicular propagation, although the range to be excluded is very small, typically a couple of degrees around 90° (Verheest 1977). It is worth pointing out that the same dispersion laws can be obtained by starting from a full kinetic treatment instead of using the simpler cold fluid approach.

The Alfvén waves found sofar can go unstable provided

$$V_A^2 < W^2. \tag{11}$$

If this is the case, both Alfvén waves are unstable, the shear mode described by (10) at almost all directions of propagation, the fast mode described by (9) only for those directions of wave propagation obeying

$$\tan^2\vartheta = \frac{k_\perp^2}{k_\parallel^2} < \frac{W^2}{V_A^2} - 1, \tag{12}$$

where ϑ stands for the angle between the wavevector $\mathbf{k}$ and the external magnetic field $\mathbf{B}_0$. Looking at the criterion (11) or the dispersion laws (9–10), one sees that the W^2 term is destabilizing, while one can interpret V_A^2 as a confining effect of the equilibrium magnetic field, perpendicular to the field. Hence mass motions perpendicular to the magnetic field stabilize the waves, whereas similar effects parallel to the magnetic field tend to destabilize the Alfvén waves. The balance determines the final outcome of stability or instability, possibly enhanced by anisotropic pressure effects, had we included these, as in the familiar firehose instability.

We can rewrite the instability criterion (11) as

$$\frac{V_{AA}^2}{1+\sigma} < \frac{\sigma U_B^2}{(1+\sigma)^2}, \tag{13}$$

if we place ourselves for simplicity in a frame in which the ambient subplasma is at rest, with $U_A = 0$. With the help of the Alfvénic Mach number $M = U_B/V_{AA}$ we cast (13) as

$$\frac{1}{M^2-1} \leq \sigma. \tag{14}$$

There is thus a threshold $\sigma_{\text{thr}} = 1/(M^2-1)$ for the mass density ratio of both subplasmas, depending upon M. As in any case one needs $M > 1$, the relative streaming between both subplasmas has to be super-Alfvénic. In more complicated situations, one would have to define super-Alfvénic by (11).

We find so-called *purely growing modes*, but then seen as such in a frame drifting with the mean bulk of the combined plasma, first of all for the shear Alfvén mode with dispersion law (10):

$$\begin{aligned} \mathrm{Re}\,\omega &= k_\parallel \overline{U} = k_\parallel \frac{\sigma U_B}{1+\sigma}, \\ \mathrm{Im}\,\omega &= k_\parallel\sqrt{W^2 - V_A^2} = \frac{k_\parallel V_{AA}}{1+\sigma}\sqrt{\sigma M^2 - \sigma - 1} = \frac{\sqrt{\sigma M^2 - \sigma - 1}}{\sigma M}\mathrm{Re}\,\omega. \end{aligned} \tag{15}$$

This instability, and thus the growth rate, is nondispersive (Verheest 1987, Lakhina and Verheest 1988), in contrast to less selfconsistent treatments of the subject (Sagdeev *et al.* 1986 and Lakhina 1987). Furthermore, the parallel phase velocity of these modes is

$$\frac{\mathrm{Re}\,\omega}{k_\parallel} = \overline{U} = \frac{\sigma M}{1+\sigma}V_{AA}, \tag{16}$$

and hence can be very different from the usual V_A, depending in which reference frame one wants to specify the different drift velocities and the global bulk speed $\overline{U}$. Here a frame was chosen in which one of the subplasmas is at rest, as said already. At low beam densities the phase velocity could be quite smaller than V_{AA}. This has to be contrasted to the case of cometary plasma physics, where in a cometocentric frame the beam is represented by the solar wind, a high speed and high density plasma (compared with the plasma created from cometary material by different ionization processes), and where consequently the phase velocity can be about 5 times as large as V_A in the undisturbed solar wind.

If we now come to the fast Alfvén mode, with dispersion law (9), we find similarly that in the unstable regime its growth rate obeys

$$\mathrm{Im}\,\omega = \sqrt{k_\parallel^2 W^2 - k^2 V_A^2} = \frac{\sqrt{\sigma M^2 \cos^2\vartheta - \sigma - 1}}{\sigma M \cos\vartheta}\mathrm{Re}\,\omega, \tag{17}$$

while $\mathrm{Re}\,\omega$ is given by the expression in (15).

3. Application to solar flare plasmas

For simplicity we will take the beams *and* the waves to propagate parallel to the solar flare magnetic field, so that we need only consider one type of Alfvén mode with its associated conditions. With proton beam energies of the order of a couple of MeV and $m_p V_A^2/2$ typically in the range of 10 keV (with thus $V_A \simeq 1,400$ km/s), we find that $M \simeq 10$, and hence the threshold number density ratio is about 1 %. A proton beam of 1 MeV gives beam velocities of the order 14,000 km/s, which only requires 500 eV electrons for the necessary return current. Any beam density above this low threshold (compared to the ambient plasma density) will excite the nonresonant Alfvén instabilities. We give in the following Table some values for the growth rates in three different cases, that of a low-density beam ($\sigma \simeq 0.1$), the intermediate case with $\sigma \simeq 1$ and that of a high density beam ($\sigma \simeq 10$).

TABLE. Values for σ, V_A/V_{AA}, $\mathrm{Re}\,\omega/k_\parallel V_{AA}$ and $\mathrm{Im}\,\omega/\mathrm{Re}\,\omega$ for two different choices of M

σ	$\frac{V_A}{V_{AA}} = \frac{1}{\sqrt{1+\sigma}}$	$\frac{\mathrm{Re}\,\omega}{k_\parallel V_{AA}} = \frac{\sigma M}{1+\sigma}$	$\frac{\mathrm{Im}\,\omega}{\mathrm{Re}\,\omega} = \frac{\sqrt{\sigma M^2 - \sigma - 1}}{\sigma M}$
$M = 5$	$\sigma_{\mathrm{thr}} = 0.04$		
0.1	0.95	0.45	2.37
1.0	0.71	2.50	0.96
10.0	0.30	4.55	0.31
$M = 10$	$\sigma_{\mathrm{thr}} = 0.01$		
0.1	0.95	0.91	2.98
1.0	0.71	5.00	0.99
10.0	0.30	9.09	0.31

As one can see from this table, the relative phase velocity at a given Alfvénic Mach number M increases with the density ratio σ, whereas the relative growth rate decreases. The latter is fairly insensitive to changes in M and is determined almost exclusively by σ. So, if the beam densities are above the low threshold to trigger an Alfvén instability, then this instability has a fairly generic character. Moreover, its growth rate is a sizeable fraction of the real wave frequency, around 1 at equal ambient and beam subplasma densities. Resonant instabilities (not discussed here, but see Tamres *et al.* 1989) would seem to have much lower growth rates, certainly for low-density beam subplasmas.

The picture emerging is that the source of free energy, namely the relative streaming between the ambient and beam subplasmas parallel to the solar flare magnetic field, can certainly drive several low-frequency waves unstable. That is, of course, as long as nothing changes in the given parameters. However, as the unstable waves grow to large amplitudes, inducing large magnetic fluctuations, they take away some or most of the free energy. Consequently the beam distribution has a tendency to relax to a stable state via the excitation of Alfvén wave turbulence followed by the scattering of the proton beam by the excited waves.

To describe this process theoretically, a form of quasilinear theory must be used. It seems to us, however, that a proper and truly selfconsistent quasilinear theory has not yet been given for a multispecies plasma. In particular the heating of the ambient subplasma or the deceleration of the beam subplasma has not been considered in any detail, including the required feedback. This question is also of importance in similar contexts, wherever one plasma flows through another, as *e.g.* the solar wind in the neighbourhood of a cometary coma. So at the moment we have to be content with hand-waving arguments to arrive at a situation where the differences in bulk speeds between the ambient and the beam subplasmas are of the order of the Alfvén velocity or less. This effectively quenches the Alfvén instabilities discussed in the present paper. In this respect, the treatment by Isenberg and Hollweg (1982) would perhaps offer the best starting point to compute in a theoretical way the final density and velocity parameters for both ambient and beam subplasmas, when the Alfvén waves have reached finite amplitudes, but also there no feedback is included as yet. As amply demonstrated, the effects discussed in the present paper are not amenable to a single fluid description, and hence a fully multispecies approach in all its details is necessary, however more cumbersome.

Acknowledgements

It is a pleasure to thank the National Fund for Scientific Research (Belgium) for a research grant, which made the stay in India possible.

References

Benz A O & Simnett G M (1986) "Solar radio signatures suggestive of proton beams", *Nature* **320**, 508–509

Isenberg P A & Hollweg J V (1982) "Finite amplitude Alfvén waves in a multi-ion plasma: Propagation, acceleration, and heating", *J. Geophys. Res.* **87**, 5023–5029

Lakhina G S (1987) "Low-frequency plasma turbulence during solar wind–comet interaction", *Astrophys. Space Sci.* **133**, 203–218

Lakhina G S & Verheest F (1988) "Alfvén wave instabilities and ring current during solar wind–comet interaction", *Astrophys. Space Sci.* **143**, 329–338

Lee M A (1988) "Ultra-low frequency waves at comets", *Proc. Chapman Conf. Plasma Waves and Instabilities in Magnetospheres and at Comets* (in press)

Sagdeev R Z, Shapiro V D, Shevchenko V I & Szegő K (1987) "Plasma phenomena around comets: Interaction with the solar wind", *Invited Papers 18th Int. Conf. Phenomena Ionized Gases* (Swansea) **2**, 134–143

Simnett G M (1986) "A dominant role for protons at the onset of solar flares", *Solar Phys.* **106**, 165–183

Tamres D H, Melrose D B & Canfield R C (1989) "On the stability of proton beams against resonant scattering by Alfvén waves in solar flare loops", *Astrophys. J.* **34**, 1284

Verheest F (1977) "Alfvén instabilities in streaming plasmas with anisotropic pressures and their relevance for the solar wind", *Astrophys. Space Sci.* **46**, 165–173

Verheest F (1987) "Alfvén wave plasma turbulence during solar wind–comet interaction", *Astrophys. Space Sci.* **138**, 209–215

Verheest F (1989) "Linear and nonlinear Alfvén waves in cometary plasmas", *Invited Papers 19th Int. Conf. Phenomena Ionized Gases* (Beograd, in press)

DISCUSSION

KUIJPERS: Can you estimate how deep a realistic flare proton beam penetrates?

VERHEEST: From first estimates the e-folding length would seem to be a small fraction of the loop length, indicating that the beam would be destroyed before it could travel down a sizeable fraction of the loop.

ACCELERATION, TRANSPORT OF AND RADIATION BY ELECTRONS IN IMPULSIVE PHASE OF FLARES

Vahé Petrosian
Center for Space Science and Astrophysics and
Department of Applied Physics
Stanford University
Stanford, CA 94305
USA

ABSTRACT. I review here results of works that we have carried out over the past few years in connection with various manifestations of the impulsive phase of solar flares. The primary goal of this work has been the determination of the characteristics of the accelerated electrons. We take into account various possible interactions that electrons suffer as they traverse the flare plasma and evaluate the expected radiation via several mechanisms. The comparison of these with the observed x-ray, gamma-ray and radio emissions allows us to set constraints on the characteristics of the accelerated electrons and on some properties of the flare plasma. I will describe some of the results we have derived from comparisons with SMM and other observations of flares during the past solar cycle. In particular, I will show that we can set limits on the range and distribution of the accelerated electron spectrum and pitch angle distribution, and on the column depth and field convergence rate of the flaring loop. More complete data and high spatial resolution and polarization observations could be very useful for furthering this kind of work.

I. INTRODUCTION

It is generally agreed that the source of energy of flares is current carrying magnetic fields, but the exact mechanism of release of energy and acceleration of particles remains the central and unsolved problem of solar flares. There has been, however, considerable progress in determination of the characteristics of the flare radiation and physical parameters of the models. In particular, it is now generally accepted that the bulk of the radiation observed during the impulsive phase of flares is either due to direct emission by accelerated electrons and ions (primarily protons) or is the result of deposition of energy of the accelerated particles in the flare plasma.

The analysis and interpretation of these radiations, therefore, are the first important steps in learning about the characteristics of the accelerated electrons. X-ray and continuum gamma-ray radiation, which are believed to be due to bremsstrahlung provide the most direct information in this regard, especially in the thick target model which, as we show below, is the most viable model. The primary uncertainties in the analysis of these data arise from our lack of knowledge about the magnetic field topology (assumed generally to be in the form of loops of magnetic flux), the plasma density and the variation of these quantities along the loop. These uncertainties become more severe in the case of microwave

E. R. Priest and V. Krishan (eds.), Basic Plasma Processes on the Sun, 391–402.

and type III emissions, which are also intimately connected with the impulsive phase. The analysis of these radiations requires a knowledge of the strength of the magnetic field, a more precise idea about acceleration site with respect to the flaring loop and other field lines, and inclusion of effects due to electron beam-plasma wave dynamics as well as radiation transport effects (optically thick or thin emission).

The whole problem is further complicated because the radiative processes are not the main mechanisms of dispersion and energy loss for the electrons. Coulomb collisions dominate for high density plasmas and low energy particles. Synchrotron losses become important at higher particle energies and field strengths for lower plasma densities. And the degree of loss due to plasma wave-particle interactions is quite uncertain. This means that the radiation does not reflect the characteristics of the initially accelerated particles, but their characteristics after they are modified by the above interactions. It is, therefore, imperative that the particle transport is taken into account properly.

In a series of papers with various graduate students at Stanford, we have been carrying out a systematic study of the above kinetic and radiative processes. In the next section I will summarize some of these theoretical studies. I will outline the process of comparison of these models with observations in section III, and in section IV, I will present some results and conclusions that we reach from such comparisons.

II. KINETIC EQUATION AND BASIC MODEL

In most astrophysical plasmas, and in particular, for solar flare plasmas, the gyroradius of the particles about the magnetic field is many orders of magnitude less than other length scales in the problem. Thus, the charged particles are essentially tied to the magnetic field lines so that the distribution function in phase space can then be specified by the position along the field lines and two components of momentum (or energy and pitch angle with respect to the magnetic field). The particle density distribution function $f(E, \mu, s, t)$ is then a function of four variables, the kinetic energy $E = \gamma - 1$ (expressed in units of mc^2), the pitch angle α (or $\mu = \cos\alpha$), the position s along the field, and the time t. In the limit where the change in particle momentum due to individual scatterings is much less than the original momentum, we can make use of the Fokker-Planck expansion of the Boltzmann collision integral in the continuity equation for f (see e.g. Lifshitz and Pitaevskii 1981). The Fokker-Planck equation can then be written in the form

$$\frac{\partial f}{\partial t} = -\mu c\beta\frac{\partial f}{\partial s} - \frac{\partial}{\partial \mu}(\dot{\mu} f) - \frac{\partial}{\partial E}(\dot{E} f) + \frac{\partial}{\partial \mu}(D_{\mu\mu}\frac{\partial f}{\partial \mu}) + \frac{\partial}{\partial E}(D_{EE}\frac{\partial f}{\partial E}) + \frac{\partial}{\partial E}(D_{E\mu}\frac{\partial f}{\partial \mu}) + \frac{\partial}{\partial \mu}(D_{E\mu}\frac{\partial f}{\partial E}) + S(E, \mu, s, t), \tag{1}$$

where βc is the velocity and $S(E, \mu, s, t)$ is the source. The coefficients $\dot{E}$ and $\dot{\mu}$ are the systematic changes in energy and pitch angle cosine due to external forces, radiation, and scattering, while the diffusion coefficients D_{ij} are due only to scattering processes. Note that the diffusion coefficients are defined differently than in the usual Fokker-Planck formalism.

a) The Interactions. The forms of $\dot{E}$, $\dot{\mu}$, and D_{ij} due to interactions of electrons with the ambient plasma particles, waves and magnetic field, and due to external forces have been published in various publications and are summarized in Hamilton, Lu and Petrosian (1990). Here we briefly mention some important features of these interactions.

Coulomb Collisions: Leach and Petrosian (1981), and Leach (1984) give these coefficients for different compositions and for partially ionized plasmas. We consider energies $E < 10^4$ so that *bremsstrahlung* losses are always less than Coulomb losses.

Magnetic Mirroring: Systematic change in the pitch angle occurs due to variation of the magnetic field B and is calculated using the adiabatic invariance of flux through particle orbits (see McTiernan and Petrosian 1990a, and McTiernan 1989).

Synchrotron: Classical optically thin synchrotron emission is assumed so that there are no diffusion terms (see e.g. Petrosian 1985).

External Forces: We ignore drifts perpendicular to the magnetic field (McTiernan 1989) so that only the component of the force parallel to the magnetic field $F_{\|}$ needs to be considered.

Plasma Turbulence: The diffusion rates due to the scattering of electrons by fluctuations in the magnetic or electric fields (Alfvén or Langmuir waves) are taken mainly from Schlickeiser (1989) and are given in Hamilton et al (1990).

We do not consider*Inverse Compton* losses which are expected to be unimportant for solar flare conditions, or the effects of the *Reverse Currents* which when important lead to the destruction of the nonthermal beams of electrons, that are needed for production of most of the observed radiations.

b) The Source Term. The source term $S(E,\mu,s,t)$ describes the characteristics of the accelerated electrons. As a starting point, we make the simplest assumption that this is a separable function of its parameters: $S(E,\mu,s,t) = F(E)G(\mu)H(s)K(t)$. There is a strong indication that the energy spectrum can be approximated by a power law with a spectral index $\delta : F(E) = F_T(\delta - 1)(E/E_o)^{-\delta}/E_o$, where F_T is the total flux above energy E_o. For the pitch angle distribution we assume a gaussian form which peaks in a direction α_1 (or μ_1) and has a width α_o (or μ_o). The pitch angle distribution becomes important when we deal with anisotropic emission processes and observations with information about the directivity of the radiation. In general, we take $\alpha_1 = 0$ which means that the accelerated electrons are beamed along the magnetic lines of force, but in some cases we also use what we call pancake distributions with $\alpha_1 = \pi/2$. Values of $\alpha_o \gg \pi/2$ represent the isotropic distribution.

We have little direct information about the spatial distribution of the accelerated electrons except that the acceleration must occur above the transition region in the corona. We shall assume that the injection of the accelerated electrons occurs near the top of a symmetrical loop. Because most of the observations that we shall be concerned with are spatially unresolved, the exact spatial distribution is not important. Therefore, for simplicity, we assume a delta function injection $H(s) = \delta(s - s_o)$ with $s_o = 0$ corresponding to injection at the apex of the loop. For the time dependence we make two extreme assumptions depending on the observations. For observations with time resolution (or observed intensity variation time scale) much greater than the interaction time scale, or the time taken by the electrons to traverse the whole length of the loop ($\simeq 0.1$ s), we assume the steady state condition: $\partial f/\partial t = 0, K(t) =$ constant. In this case, the observed variations reflect directly the form of $K(t)$ up to the resolution of the observation. In the opposite limit, for treating high time resolution observations or rapid fluctuations, one must include the $\partial f/\partial t$ term. In general, for these cases, we initially assume impulsive injection at time t_o; $K(t) = \delta(t - t_o)$.

c) The Plasma. We assume a cold plasma so that the only plasma parameters we need to specify are the values of the plasma density and magnetic field along the whole length of the loop: $n(s)$ and $B(s)$. The two most important parameters are the column depth to transition region at $s = s_{tr}$; $N_{tr} = \int_o^{s_{tr}} nds$, and the rate of the convergence of the magnetic field, $d\ln B/ds$, in the corona for low energy particles and in the photosphere for higher ones. We shall sometimes use the mirror ratio which is the ratio of magnetic field at $s = 0$ to that at the "end" of the loop: $b = B_o/B_{max}$. For low energy particles $B_{max} \approx B_{tr}$ but for higher energies, it may be as large as the field at the photosphere. For higher energy particles, the synchrotron losses and pitch angle scattering by plasma waves become important. In this case we need the absolute value of the magnetic field and the plasma wave density.

Analytic solutions have been obtained for equation (1) for some restricted problems, in particular, for small pitch angles (Leach and Petrosian 1983, Leach 1984, and Lu and Petrosian 1988) and for extremely relativistic electrons suffering Coulomb collision and synchrotron losses (McTiernan and Petrosian 1990a). Numerical codes for solution of the steady state equation were first developed by Leach (1984) and have been extended by McTiernan (1989) to higher energies and field strengths, and for the general time dependent problem by Hamilton, Lu and Petrosian (1990). The results described in the next two sections are based on all of these solutions.

III. COMPARISON WITH OBSERVATIONS

Using the results from the Fokker-Planck equations, we calculate the characteristics of the expected radiations from the resultant electron distribution. Comparison with observations then allows us to determine the characteristics of the accelerated electrons (parameters $F_T, \delta, E_o, \alpha_1, \alpha_o$) and of the plasma ($N_{tr}, d\ln B/ds$ or the mirror ratio b),in addition to testing the validity of our assumptions related to the source term S and the loop model. In general, such a comparison can be done in two different ways. In one, which is generally useful in the initial stages of investigation of a phenomenon, a few strong and well observed events are studied in great details. This approach, the most commonly used one, especially with observation from SMM, has proven the reliability of the non-thermal thick target model as described above and has shown that most of the model parameters have very wide ranges (sometimes having pure power law distributions indicating indefinite range). This then demands the second approach, whereby from statistical analysis of the observed properties of a large number of events, one attempts to describe the distribution of (or occurrence rate of flares with) various parameters, which can be written as $\psi_e(F_T, \delta, \alpha_i)$, where α_i stands for $\alpha_1, \alpha_o, d\ln B/ds, N_{tr}$, etc.

This, of course, is an ambitious goal. However, there are sufficient channels and modes of observations so that, in principle, this is an achievable goal. The impulsive phase of flares can be studied through spectral, temporal, spatial and polarization observations of hard x-ray, gamma-ray, microwave at type III emissions. The first two aspects of these emissions have been observed extensively but there is very little data with high spatial resolution (some at low energy x-rays from HXiS on SMM and from Hinotori, and some from VLA; for a review see the proceedings of the SMM workshop, Kundu and Woodgate, 1985) and there is essentially no reliable data on polarization (except for few flares at microwaves and possibly one flare at low hard X-rays; see Leach, Petrosian and Emslie, 1985, and references cited there). We will present results from both kinds of comparisons with observations. This, by no means, will be a complete description because the space limitation will not allow us to describe all of our results and, more importantly, results from similar studies by many other workers in the field.

IV. RESULTS AND CONCLUSIONS

We now compare the models with various observations with the aim of setting constraints on the parameters and their distributions.

a) Hard X-rays. We start with analysis of flare emission from 20 to 300 keV which is due to electron bremsstrahlung. This is the energy range where there has been the most systematic observation with those from HXRBS providing the largest set of high quality data. For a recent review see Dennis (1988) and the references there. These data are consistent with the non-thermal thick target model and, in general, show a power law spectrum: $J_x(k) \propto k^{-\gamma}$. There is no strong evidence for correlation between the photon flux J_x and spectral index γ. The distribution of x-ray fluxes(or photon counts) is observed to be a power law: $\psi_x(J_x) \propto J_x^{-1.8}$, approximately in agreement with earlier observation (Datlowe, Elcan and Hudson 1974). On the other hand the distribution of spectral indices has a finite range and is peaked at $\gamma \simeq 4$ (see figure 1).

It is well known, and our calculations confirm, that in the thick target model the hard x-ray emission is nearly isotropic and that its flux integrated over angles and space, $J_x(k) \propto F_T$, the electron flux, and that the index $\gamma = \delta + 1 + \epsilon(\theta, \delta)$, where ϵ and its derivatives with respect to δ or direction of observation θ are $\ll 1$. We can immediately conclude that the distribution of the accelerated electrons, $\psi_e(F_T, \delta, ...) \propto F_T^{-1.8} h(\delta)$. Then using model calculations we derive a form of $h(\delta)$ (the dashed line in figure 1) which gives the solid line x-ray spectral distribution in agreement with the observed histogram. Moreover, extending these results to bremsstrahlung at higher energies (300 keV and 1 Mev), we evaluate the expected distribution of HXRBS fluxes and spectral indices of the subset of these flares which could also be detected by GRS on SMM and find good agreement with the observations as shown by the lower histogram in figure (1). The details of these results can be found in J. McTiernan's thesis (1989) or in McTiernan and Petrosian (1991).

b) Gamma-rays (> 300 keV). For most flares the gamma-ray spectrum in the range between 1 to 7 MeV is dominated or contaminated by nuclear line emission. Therefore, we have considered the range between 300 keV to 1 Mev and that > 10 Mev separately. The choice of the lower limit is arbitrary and is chosen to correspond to the threshold of the GRS on SMM. As pointed out earlier (Petrosian, 1985) as we go to higher energies, the calculations and, therefore, the interpretation of the data is simplified gradually in three respects. At higher energies (i) the pitch angle diffusion rate of electrons due to collision (or other processes) becomes less important relative to energy loss rate, (ii) the photon is emitted in the direction of momentum of the electron, and (iii) most of the radiation comes from deeper layers, certainly below the transition region, where the field lines will be most likely vertical (or at least have uniform direction). Because of these the gamma-rays will reflect the characteristics of the accelerated electrons more clearly than hard x-rays, for which most of these characteristics are obscured by collisional diffusion and spatial inhomogeneities. However, other consequences of the above mentioned differences is that the degree of convergence of the field lines begins to play an important role and two other complications come in at very high energies. These are the energy losses due to the synchrotron process, which can be accounted for easily (McTiernan and Petrosian 1990a), and the possibility of pitch angle diffusion by plasma turbulence (Miller and Ramaty 1989, and Hamilton, Lu and Petrosian 1990), which requires some knowledge about the type, spectrum and energy density of turbulence that may be present in the flare plasma.

We shall not consider the effects of turbulence which are quite uncertain but include all other processes in a steady state calculation of the spectrum and directivity of the spatially integrated gamma-rays. We assume the distribution of the electron fluxes and spectral indices derived above and vary the parameters α_i from model to model. Of these parameters the most important are the gaussian width of the pitch angle distribution and the rate of the convergence of the magnetic field as long as the column depth $N_{tr} \leq 10^{22}\,\mathrm{cm}^{-2}$. We have used beamed, isotropic and pancake distributions and two models for field convergence. In one, we assume $d\ln B/ds =$ constant, which emphasizes the field convergence in the corona. In the other, which emphasizes the convergence below the transition region, we set $B \propto n^{\epsilon}$ with $\epsilon = 0.2$. Note that equapartition or a constant β plasma would mean $\epsilon = 0.5$ because the temperature below the transition region is relativly constant.

The anisotropy of the expected radiation is compared with observations. The direct observation of the anisotropy via stereoscopic observations (Kane it et al. 1988) unfortunately suffer from large uncertainties and are not useful for constraining the models. Nevertheless they are consistent with the basic model (McTiernan and Petrosian, 1990b). More conclusive but indirect evidence has come from GRS on SMM which has discovered considerable center-to-limb variation (Vestrand *et al.* 1987) specially for flares with emission at > 10 Mev photon energies (Rieger *et al.* 1984). As emphasized before (Petrosian, 1985), interpretation of this kind of data requires a statistical analysis and therefore a knowledge of the distributions of the parameters. With the distributions derived above we now have the proper tools for analysis of this kind of data.

I would also like to emphasize that it is important to carry out the appropriate kind of analysis depending on the type of the distribution under consideration and the sensitivity or threshold of the observations. For example, consider a characteristic X with the distribution $\psi(X) \propto X^{-\alpha}$ that is known to be anisotropic, $X = X_o f(\theta)$. Two simple properties of a sample of events (for which X is known) are the variation with θ of the total number and average value of X:

$$N(\theta) = \int_{X_{th}}^{\infty} \psi(X(\theta))dX, \quad \overline{X}(\theta) = \int_{X_{th}}^{\infty} X\psi(X(\theta))dX/N(\theta) \tag{2}$$

where X_{th} is the threshold value for detection of X. It is easy to show that if $X_{th} > X_{min}$, where X_{min} is the minimum value (or the lower cutoff) of the real distribution $\psi(X)$, then $\overline{X}(\theta)$ = constant but $N(\theta) \propto [f(\theta)]^{1-\alpha}$. In the opposite case $N(\theta)$ =constant and $\overline{X}(\theta) \propto f(\theta)$, irrespective of whether $\psi(X)$ has a power law form or not. In the first case $N(\theta)$ and in the second case $\overline{X}(\theta)$ is the parameter that will reflect the anisotropy. Therefore, it is important to make the correct test.

As shown in McTiernan and Petrosian (1990b), as we go to higher energies, both flux and spectral index show stronger dependence on the angle of observation with respect to the field lines (which for vertical field lines is the same as the heliocentric longitude). Figure 2 shows these effects for some models. However, as we have seen in section II above, the distribution of fluxes has a power law form extending beyond the threshold of HXRBS and GRS so that in this case we must look at the center-to-limb variation of the number of flares and not that the average value of the flux. On the other hand because the distribution of spectral index appears to turn over at both high and low values in this case we deal with the center-to-limb variation of the average value of the spectral index. It should be noted that there may be some instrumental effects in thsee distribution, but we ignore them as they are considered to be secondary effects.

The first result we obtain is that the center-to-limb variation of N in most models agrees with observations, but the calculated average spectral index shows the same general trend as the observations (spectral flattening toward the limb) but has a larger absolute value than those observed by GRS on SMM (i.e. the model spectra are steeper). This forces us to alter our assumption that the electron spectrum is described by a single power law at all energies. Instead there must be a brake (or gradual flattening) around 300 keV. This statistical evidence has been seen in spectra of individual flares (Dennis, 1988) and in spectra of impulsive electron flares observed directly at one A.U. (Dröge *et al.*,1989).

We then include such a break and from comparison of the center-to-limb variation of the spectral index, the ratio of HXRBS to GRS fluxes and the number of flares we place limits on the range of two of our model parameters mentioned above. These limits are shown in figure (3) which clearly rules out pancake models but allows beamed models with various degrees of beaming (value of α_o). The allowed values are anti-correlated with the degree of field convergence (the mirror ratio b) such that the stronger the convergence, the smaller α_o or the stronger the beaming. A narrower range is found from the spectral index data than from the ratio of fluxes. The distribution of the flares with emission at > 10 Mev across the solar disk also can be reproduced by the models in the allowed region of the diagram. The details of these results can be found in McTiernan (1989) and McTiernan and Petrosian (1991).

c) Microwaves. The microwaves are believed to be due to the synchrotron emission by the same population of electrons that produces the hard x-rays because of the correlation between the fluxes and similarity of the time profiles of the emission at the two energy bands. Many authors have discussed the relative number of electrons needed for production of x-rays and microwaves (Peterson and Winkler 1959, Takakura and Kai 1966, Holt and Ramaty 1969, Gary 1985, Schmahl, Kundu and Dennis 1985, and Kai 1986) with different conclusions. Some of these differences are due to the use of different models (e.g., thin target vs thick target) or due to effects of absorption of the synchrotron radiation. It turns out that the predicted ratio of hard x-ray to microwave emission (at high frequencies where absorption effects can be neglected) is sensitive to the x-ray spectral index γ (or electron index δ) and has different dependence on γ for different models.

In a recent paper (Lu and Petrosian 1989), we have used this fact and compared with observations the prediction of the thick target, thin target and thermal (multi-temperature so that one obtains a power law x-ray spectrum) models. From this we conclude that the model most consistent with observation is the thick target model with loop length of $\simeq 2 \times 10^9$ cm and average magnetic field strength of 500 to 350 G. In the thick target model the same population of electrons produce both radiation but from different locations. The microwaves come from the coronal portion of the loop where the electron spent most of their lifetime (essentially like in a thin target model) while the x-rays originate primarily from the footpoints and below the transition region where the density is higher and where the lifetime of electrons is proportionately lower. This picture essentially agrees with the existing but limited high spatial resolution observations mentioned above (Canfield *et al.*, 1985). On the other hand this difference between the sites of emissions of the two radiations may imply existence of differences in their temporal evolutions.

We have investigated this problem using the time dependent code mentioned above in the framework of the thick target model just described (Lu and Petrosian 1990). Two types of differences have been observed. On time scales comparable to electron transport time, it has been observed that elementary impulsive spikes peak later by about a quarter of a second at microwave as compared to x-rays, while one would expect the microwave emission coming from the the top of loop to start earlier than the x-rays which come from the footpoints. However, if the magnetic field converges from the top to the footpoints of the loop, some of the particles will be trapped in the corona and continue to emit synchrotron radiation for a longer time. We show that delays of up to 200 millisecond are possible by this method. For longer delays we must change our basic assumption that all energy particles are accelerated with the same time profile. The higher energy, microwave emitting, electrons must be accelerated later than the lower energy, x-ray emitting, ones: For example $K(t) = \exp[-((t + 0.1 - 0.3\beta)/0.07)^2]$ results in delay of microwaves with respect to the x-rays by 350 milliseconds.

Another kind of observed difference between the two time profiles is that on longer time scales the microwaves, on average, peak about 2 seconds later than x-rays and that during the decay phase of the burst there is excessive microwave emission as compared to x-rays. We find that contribution to microwaves from other radiative processes (such as free free or thermal synchrotron), or even very strongly converging field geometries (for trapping of the electrons in the corona) fail to explain the observation with reasonable parameters. Thus we tentatively conclude that these observations are an indication of hardening of the accelerated electron spectra (e.g. change into a flatter power law at higher energies) during the latter stages of an impulsive burst.

It would be useful to make a similar comparison between the gamma-ray and microwave emissions which are produced by electrons with similar energies.

d) Other Processes. We have also studied other flare associated processes, specially those involving wave particle interactions. Because of space limitations I will not describe these in any detail here. We can analyze the effects on the transport of the electrons due to interaction with plasma particles, waves and converging fields simultaneously, an important improvement on past works in flares. For example, we have generalized (Hamilton and Petrosian, 1990) the work by White, Melrose and Dulk (1986) on the possibility of maser action by inclusion of the effects of collisions explicitly. We find that even at moderate column depths to the transition region, $N_{tr} \simeq 2 \times 10^{19}\mathrm{cm}^{-2}$, the electron velocity distribution becomes smoothe reducing the magnitude of the possible maser action.

In a preliminary study we have also evaluated the expected spectral distribution of stochastically accelerated electrons when the ambient plasma has finite density. We find that presence of collisions can produce the type of observed spectral brakes discussed above. This possibility was also demonstrated analytically by Steinacker, Dröge, and Schlickeiser (1988) in connection with the observed brakes in the spectrum of the interplanetary electrons mentioned above(Dröge *et al.*, 1989). These results could also explain the observed hardening of accelerated electron spectra at higher energies (item b) or at later times in the evolution of the impulsive phase (item c). These are preliminary results and more data and theoretical work is needed for the clarification of these important points.

5) Summary. In summary, we have demonstrated that important conclusions about the properties of the acceleration process and about the flare plasma can be reached by the type of analysis we have been engaged in. Some of these are: (i) distribution of fluxes and spectral indices of accelerated electrons, (ii) deviation from simple power law energy spectrum at higher energies, (iii) spectral hardening during the later stages of the impulsive spikes of long and short durations, (vi) limits on the possible range of th pitch angle distribution of the electrons, and (v) limits on the length, density and field convergence rate of flaring loops.

Acknowledgement: All of the results I have described above are outcomes of the thesis research of Russell Hamilton, Edward Lu and James McTiernan. I would like to thank J. McTiernan for help in the preparation of the figures and specially R. Hamilton for extensive help in the preparation of this manuscript. Our research on solar flares has been supported by NASA grant NAGW 1976 and NSF grant ATM 8705084.

REFERENCES

Canfield,R. et al, 1985, **Energetic Phenomena on the Sun**. Eds. M.R. Kundu and B. Woodgate, NASA Conf. Publ. 2439, Chapter 3.

Datlowe, D.W., Elcan, M.J. and Hudson, H.S. 1974, *Solar Phys.,* **39**, 155.

Dennis, B.R. 1988, *Solar Phys.*, **118**, 49.

Dröge, W., Meyer, P., Evanson, P., and Moses, D. 1989, *Solar Phys.*, **121**, 95.

Gary, D. 1985, *Ap.J.*, **297**, 799.

Hamilton, R.J., Lu, E.T., and Petrosian, V. 1990, (To be published in *Ap.J.* May 10, 1990).

Hamilton, R.J. and Petrosian, V. 1990, (Submitted to *Ap.J.*).

Holt, S. and Ramaty, R. 1969, *Solar Phys.* **8**, 119.

Kai, K. 1986, *Solar Phys.* **104**, 235.

Kane, S.R., Fenimore, E.E., Klebasadel, R.W. and Laros, J.G. 1988, *Ap.J.*, **326**, 1017.

Kundu, M. R., and Woodgate, B. 1985, **Energetic Phenomena on the Sun**. NASA Conf. Publ. 2439.

Leach, J. 1984, Ph.D. Thesis, Stanford University.

Leach, J., Emslie, A.G. and Petrosian, V. 1985, *Solar Phys.*, **96**, 331.

Leach, J. and Petrosian, V. 1981, *Ap.J.*. **251**, 781.

Leach, J. and Petrosian, V. 1983, *Ap.J.*, **269**, 715.

Lifshitz, E.M. and Pitaevskii, L.P. 1981, **Physical Kinetics**, (Oxford: Pergamon), Ch. 2.

Lu, E.T. and Petrosian, V. 1988, *Ap.J.*, **327**, 405.

Lu, E.T. and Petrosian, V. 1989, *Ap.J.* **338**, 1122.

Lu, E.T. and Petrosian, V. 1990, (To be published in *Ap.J.* May 10,1990).

McTiernan, J.M. 1989, Ph.D. Thesis, Stanford University.

McTiernan, J.M. and Petrosian, V., 1990a, (To be published in *Ap.J.* May 1990).

McTiernan, J.M. and Petrosian, V., 1990b, (To be published in *Ap.J.* May 1990).

McTiernan, J.M. and Petrosian, V., 1991, (To be submitted to *Ap.J.*).

Miller, J.A. and Ramaty, R. 1989, *Ap.J.*, **344**, 973.

Petrosian, V. 1985, *Ap.J.*, **299**, 987.

Peterson, L. and Winckler, J. 1959, *J. Geophys. Res.* **64**, 697.

Rieger, E., Reppin,C., Kanbach, G., Forrest, D.J., Chupp, E.L. and Shares, G.H. 1984, 18th International Cosmic Ray Conference, **4**), 79.

Schmahl, E., Kundu, M., and Dennis, B. 1985, *Ap.J.*, **299**, 1017.

Schlickeiser, R. 1989a, *Ap.J.*, **336**, 243.

Steinacker, J., Dröge, W., and Schlickeiser, R., 1988, *Solar Phys.*, **115**, 313.

Takakura, T. and Kai, K. 1966, *Pub. Astr. Soc. Japan*, **18**, 57.

Vestrand, W.T., Forrest, D.J., Chupp, E.L., Rieger, E., and Share, G.H. 1987, *Ap.J.*, **322**, 1010.

White, S.M., Melrose, D.G., and Dulk, G.A. 1986, *Ap.J.*, **308**, 424.

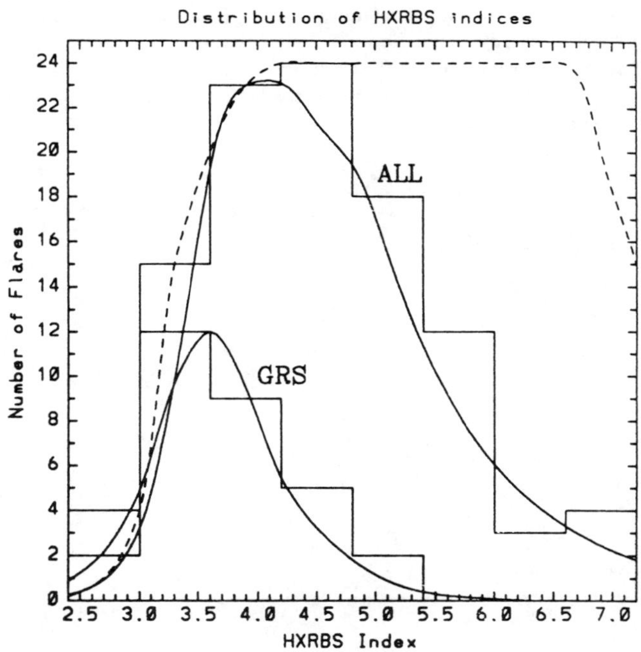

Figure 1. The distribution of hard x-ray spectral index, γ, of all flares with count rates > 1000 counts/s observed by HXRBS (histogram marked ALL) and the subset of this detected also by GRS. The two solid lines are calculated distributions assuming loops with uniform fields and accelerated electrons with power law spectral index δ. The dashed line depicts the distribution of $\delta + 1$ that gives the acceptable fits to obseervations shown.

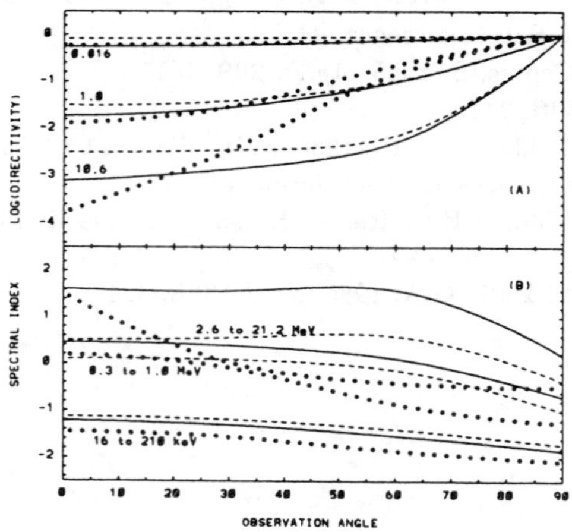

Figure 2. Directivity at the indicated photon energies in Mev (top panel) and photon spectral indices at the indicated energy range (lower panel) versus observation angle or heliocentric longitude (0 at the disk center and 90^0 at the limb). The high and low energy spectral indices are shifted up and down by one unit, respectively, for clarity. Solid lines are for uniform B field; dashed lines for a constant $d\ln B/ds$ and the dotted lines for $B \propto n^{0.2}$ with mirror ratios $b = 10$.

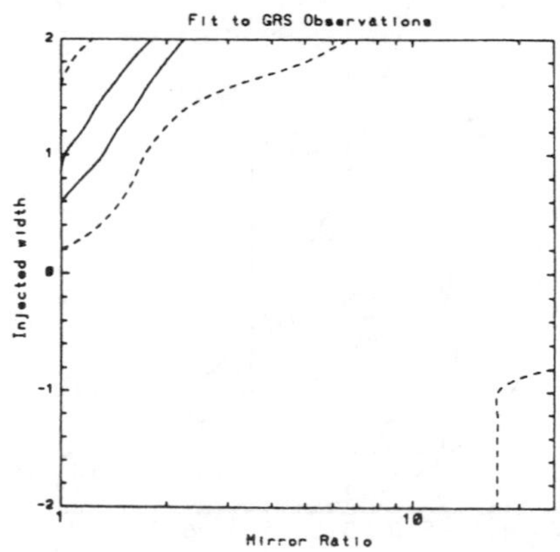

Figure 3. Constraints on the gaussian width [$\log(\alpha_o^2)$; positive for beam, 0 for isotropic and negative for pancake] of the pitch angle distribution and the degree of convergence of the magnetic field represented by the value of the mirror ratio b. The isotropic distribution is that of a model with $\alpha_o^2 = 4$ so that 1 and 2 represent $\alpha_o^2 = 0.4$ and $\alpha_o^2 = 0.04$, respectively. The area inside the solid line represents the region of the parameter space which agrees with the center-to-limb variation of the average spectral index as observed by GRS on SMM. The dashed line shows the same based on the variation of x-ray (HXRBS) to gamma-ray (GRS) flux ratio.

DIAGNOSING SOLAR PLASMAS FROM EUV AND X-RAY EMISSION LINES

BHOLA N. DWIVEDI
Department of Applied Physics, Institute of Technology
Banaras Hindu University, Varanasi-221005, India

ABSTRACT. The electron density and temperature measurement of solar plasmas from EUV and X-ray emission lines have been briefly discussed. Diagnostic EUV line ratios from MgVI ion and X-ray line ratios from MgVIII and SiX ions have been presented. A brief account of the general methodology for diagnostics of inhomogeneous structures has also been given.

Introduction

The determination of electron density N and temperature T of astrophysical and laboratory plasmas from diagnostic line ratios is a problem of universal importance. The physical significance of these diagnostics lies in the influence of N and T on plasma process being considered, whether thermonuclear fusion, transport coefficients, energy transport, magnetic reconnection, particle acceleration etc. Diagnostics of solar plasmas from EUV and X-ray emission lines have been extensively studied and several reviews are available (Gabriel and Jordan (1972), Dere and Mason (1981), Doschek (1985), Dwivedi (1988a) and references cited there). Density diagnostics are similar to that used by optical astronomers for deriving densities using the well known OII lines at 3726 and 3729 Å. Diagnostic techniques for an electron density determination usually rely on identifying transitions in which one or both levels are metastables. This causes the relative level population to be density dependent so that the intensity ratio of two collisionally excited lines becomes a function of density. The local electron temperature can be probed by selecting two lines both of which are collisionally excited by electrons from the same level. The excitation rates are dependent on the Boltzmann factor, exp $(-E_{ij}/kT)$, where E_{ij} is the energy of the transition. A ratio of line intensities yields the electron temperature for levels of sufficient energy separation. An elegant method for temperature diagnostics is to use the temperature dependence of the ratio of a line formed by dielectronic recombination to a line formed primarily by electron impact excitation.

E. R. Priest and V. Krishan (eds.), Basic Plasma Processes on the Sun, 403–407.

For a number of years we have been involved in an extensive series of N- and T- diagnostic calculations for solar plasmas. In this paper, some spectral line ratios which have hitherto received little attention are presented. A brief discussion on diagnostics of inhomogeneous structures is also presented.

Diagnostic Line Ratios

EUV emission lines from MgVI (4×10^5 K) ion have received little attention for diagnostic studies. Line ratios λ 387.9/λ 403.3 and λ387.9/λ 400.7 from MgVI ion have been used to infer electron density in sunspot plumes [Dwivedi (1989)]. Figure 1 shows some more EUV line ratios from MgVI ion which could be used as density monitors of the emitting region. Line ratios λ 349.1/λ 399.3 and λ 349.1/λ 403.3 are good density indicators in the range 10^7–10^{11} cm^{-3}. MgVIII (8×10^5 K) line ratio λ 430.5/λ 436.7 has been suggested to be an excellent candidate to infer density in the range 10^7–10^9 cm^{-3} and probe the quiet Sun and coronal holes [Dwivedi (1988b)]. Theoretical computations by us in the EUV region from ions of boron, nitrogen, oxygen and carbon sequences will be useful for density diagnostics, using observations that SOHO may provide [Dwivedi and Raju (1980, 1988), Raju and Dwivedi (1978, 1979)].

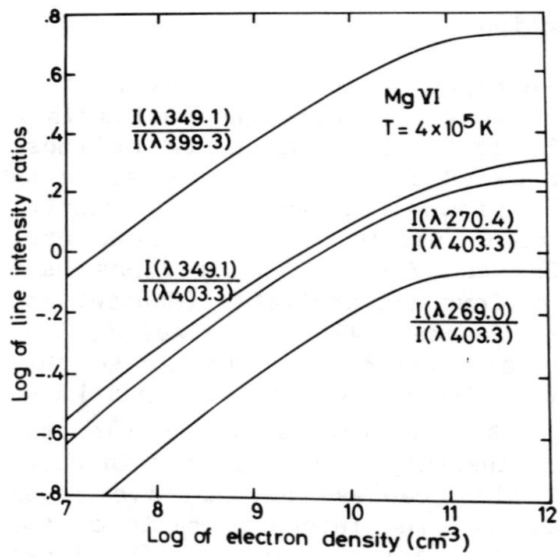

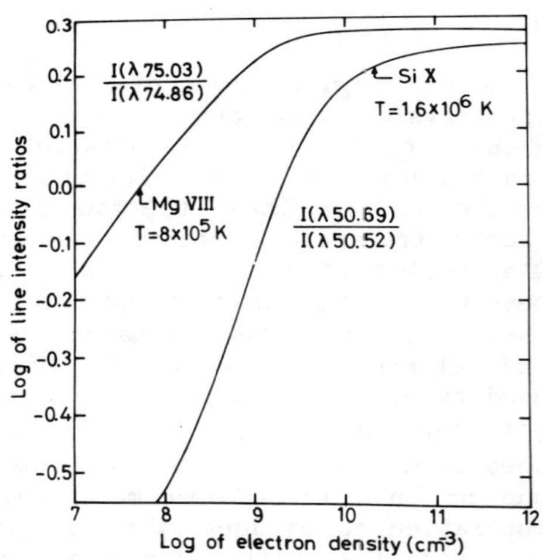

Figs. 1 and 2. Electron density dependence of EUV line ratios from MgVI ion and X-ray line ratios from MgVIII and SiX ions.

Figure 2 shows the variation of X-ray line ratios from MgVIII (8×10^5 K) and SiX (1.6×10^6 K) ions as a function of electron density. The line intensity ratio of the $(\frac{3}{2} - \frac{5}{2})$ and $(\frac{1}{2} - \frac{3}{2})$ lines of the multiplet ($2s^2\,2p\;^2P^o - 2s^2\,3d\;^2D$) is an electron density diagnostic. The $(\frac{3}{2}-\frac{5}{2})$ transition is usually blended with the $(\frac{3}{2}-\frac{3}{2})$ transition. X-ray line pair λ 75.03/λ 74.86 from MgVIII and λ 50.69/λ 50.52 from SiX are useful for diagnosing solar plasmas. However, observational data with improved spectral and spatial resolution are needed by future solar missions to probe solar structures.

Diagnostics of Inhomogeneous Structures

Almleaky et al. (1989) have discussed the problem and importance of interpreting density sensitive line ratios from plasmas which are isothermal but of non-uniform density. Brown et al. (1989) then develop the methodology of the diagnostics of inhomogeneous structures by considering arbitrary density and temperature distributions of which a brief discussion is presented here.

For an optically thin plasma of electron density $N(\vec{r})$, temperature $T(\vec{r})$ at position $\vec{r}$ in a volume V, the total power emitted in a spectral line α can be expressed as

$$P_\alpha = \iiint_V N^2(\vec{r})\, F_\alpha(N(\vec{r}), T(\vec{r}))\, dV \tag{1}$$

where F_α is determined by elemental abundance, ionisation and excitation balance calculations using atomic coefficients. The power emitted in resonance lines where F_α is independent of N and where proper choice of volume element dV is based on constant temperature surfaces S_T is given by Craig and Brown (1976):

$$P_\alpha^R = \int_{T=0}^{\infty} \xi(T)\, F_\alpha^R(T)\, dT \tag{2}$$

where

$$\xi(T) = \iint_{S_T} \frac{N^2(\vec{r})}{|\vec{\nabla} T(\vec{r})|}\, dS_T \tag{3}$$

is the emission measure differential in T which can, in principle, be derived from integral equation (2) using measurements of temperature-sensitive line strengths P_α^R. Likewise, the special case of density-sensitive line analysis for an isothermal plasma T_o has a natural formulation:

$$P_\alpha^D = \int_{N=0}^{\infty} \xi(N)\, F_\alpha^D(N, T_o)\, dN \tag{4}$$

where

$$\xi(N) = \iint_{S_N} \frac{N^2(\vec{r})}{|\vec{\nabla} N(\vec{r})|}\, dS_N \tag{5}$$

Consideration of these two cases shows that for the general case (1) we must deal with a volume element dV related to elements on which both N and T are constants. In first case, surfaces of constant N and of constant T are coincident. This is in fact the case which proves most tractable in terms of an inversion of the problem and which is probably of the greatest relevance. In second case, S_N and S_T are not coincident but intersect on a line $L_{N,T}$. Then the natural formulation of (1) into an integral expression over the variables N, T on which the line strengths depend, is in terms of the volume element as shown in Figure 3.

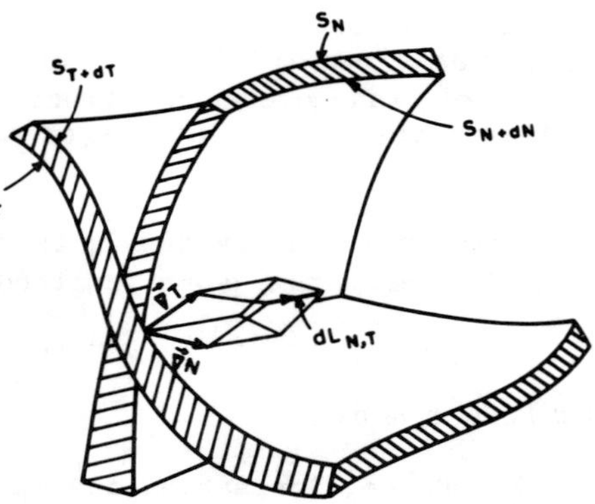

Fig.3 Representation of volume element in terms of surfaces of constant N and of constant T which are not coincident but intersect on a line $dL_{N,T}$ (Brown et al. 1989)

$$dV = dL_{N,T} \frac{dN\, dT}{|\vec{\nabla} N||\vec{\nabla} T|} (\sin \theta_{N,T})^{-1} \tag{6}$$

where $\theta_{N,T}$ is the local angle between the vectors $\vec{\nabla} N$ and $\vec{\nabla} T$ normal to surfaces S_N and S_T, respectively. If we then define an emission measure function differential in both density and temperature by

$$\psi(N,T) = \oint_{L_{N,T}} \frac{N^2(\vec{r})(\sin \theta_{N,T}(\vec{r}))^{-1}}{|\vec{\nabla} N(\vec{r})||\vec{\nabla} T(\vec{r})|} dL_{N,T} \tag{7}$$

We have for any line

$$P_\alpha = \int_{N=0}^{\infty} \int_{T=0}^{\infty} \psi(N,T)\, F_\alpha(N,T)\, dN\, dT \tag{8}$$

The general diagnostic problem is then seen as that determining $\psi(N,T)$ by inversion of double integral equation (8) for a set of measurements P_α of various α.

Concluding Remarks

EUV line ratios discussed here should be valuable to infer electron density using observations that SOHO may provide. X-ray line ratios will be useful for density determination when observational data with improved spectral resolution becomes available. The problem of diagnostics of inhomogeneous plasmas is of fascination and of great mathematical complexity.

Acknowledgements

I am grateful to Professor J.C. Brown and Dr. P.K. Raju for encouragement and to Professor V. Krishan for inviting me to present this work at IAU Symposium 142.

References

Almleaky Y.M., Brown J.C. and Sweet P.A.(1989) 'Density diagnostics and inhomogeneous plasmas I-Isothermal plasmas', Astron. Astrophys., In press.

Brown J.C., Dwivedi B.N., Almleaky Y.M. and Sweet P.A. (1989) 'Density diagnostics and inhomogeneous plasmas II - Non-isothermal plasmas', Paper in preparation.

Craig I.J.D. and Brown J.C. (1976) 'Fundamental limitations of X-ray spectra as diagnostics of plasma temperature structure', Astron. Astrophys. 49, 239-250.

Dere K.P. and Mason H.E. (1981) 'Spectroscopic diagnostics of the active region: Transition Zone and Corona', in F.Q. Orrall (ed.), Solar Active Region, Colo. Univ. Press, pp. 129-164.

Dwivedi B.N. (1989) 'On the electron density in sunspot plumes', Solar Phys. Lett., In press.

Dwivedi B.N. (1988a) 'Diagnostics of solar plasma using emission lines', Kodaikanal Obs. Bull. Ser. 9, 99-110.

Dwivedi B.N. (1988b) 'Electron density from MgVIII emission lines', Solar Phys. Lett. 116, 405-408.

Dwivedi B.N. and Raju P.K. (1988) 'Density diagnostics of solar emission lines from nitrogen-like ions', Adv.Space Res. 8, 11(179)-11(183).

Dwivedi B.N. and Raju P.K. (1980) 'Density dependence of solar emission lines of boron-like ions', Solar Phys. 68, 111-123.

Doschek G.A. (1985) 'Diagnostics of solar and astrophysical plasmas dependent on autoionization', in A. Temkin (ed.), Autoionization, N.Y. Plenum Publ. Corp., pp. 171-256.

Gabriel A.H. and Jordan C. (1972) 'Interpretation of spectral intensities from laboratory and astrophysical plasmas', in McDaniel E.W. and McDowel M.R.C.(eds.), Case Studies in Atomic Collision Physics II, North Holland, pp. 210-291.

Raju P.K. and Dwivedi B.N. (1978) 'Density dependence of solar emission lines of oxygen-like ions', Solar Phys. 60, 269-277.

Raju P.K. and Dwivedi B.N. (1979) 'Density dependence of solar emission lines of carbon-like ions', Pramana, 13, 319-328.

ELECTRONS AND X-RAY EMISSION OF SOLAR FLARES

V. G. KURT

Institute of Nuclear Physics Moscow State University
P. B. 119899 Leninskie Gory, Moscow, USSR

ABSTRACT
A statistical analysis of solar flare X-rays and interplanetary particle fluxes, measured onboard VENERA-13,14 Spacecraft, was performed. The correlation of fluences for different manifestations of solar flares is strong, especially for fast electrons and hard and soft X-ray emissions. Frequency dependence on fluence value ε_i for practically all kinds of solar flare emission can be described by power law $\nu\ (\varepsilon>\varepsilon_o) \sim \varepsilon^{-0.45\pm0.15}$ which does not change significantly with solar activity. For different H_α flare importances the values of ε_i were obtained. It is proposed that appearance of certain energy flare frequency is strongly dependent on some scale factor.

1. INTRODUCTION.

Complex experiments have been carried out during the last few years measuring various flare manifestations in a wide range of wave lengths. It was shown that the full energy impacted in the flare volume by fast electrons with kinetic energy above 25 keV can be comparable with the total emission energy of the low and high temperature flare regions. These results are very important and new. But one should remember that only for seven flares such experiments have been done, which is not enough to draw general conclusions [1].

But there is another way to solve the problem. For different kind of solar flares emissions the maximum burst amplitude (P), its duration (ΔT) and total photon numbers or fluence (ε) can be easily obtained from experiment. If we disregard the peculiarity of each event, we can obtain statictics for these features, which can be measured in a large number of experiments.

As a result of a statistical procedure we can obtain the frequency of the phenomena with known fluence and some size distribution functions. This statistical approach allows us to obtain if not precise then at least proper quantitative ratios, which relate the different emissions among themselves. These relations can be used to obtain the energy balance during the flare for determination of the role of fast electrons and construction of a phenomenological model of the flare process.

The present work summarizes our statistical investigations based on measurements of solar flare hard X-ray and energetic particles fluxes in the interplanetary space, which were carried out onboard the VENERA-13,14 Spacecraft in 1981-1983 [2-10]. The main results will be described below.

2. CORRELATION ANALYSIS. .

As a result of measuring N solar flares events we get the following values: Pxh = X-ray peak flux for Exh > 55 keV; Pxt = peak flux

E. R. Priest and V. Krishan (eds.), Basic Plasma Processes on the Sun, 409–413.

for 0.5 < Ex < 5 keV; PHα = Hα peak flux which is proportional to Hα importance, Pe = electrons peak flux; ε_{xh}, ε_{xt}, ε_H = the total quanta number or burst's fluence for Xh, Xt, Hα - emission and so on. For some couples (x, y) of these values we calculated the correlation coefficients. Some of these correlation diagrams are shown on Fig. 1.

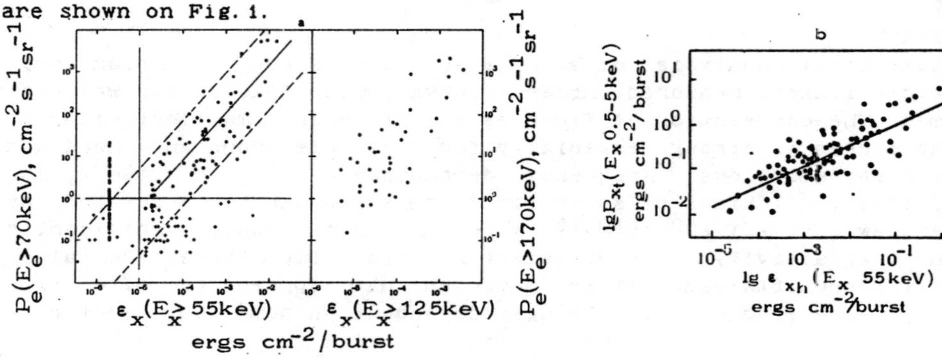

Fig. 1a. Correlation diagrams of the electron peak flux Pe with Ee>70 keV versus the X-ray with Ex > 50 keV fluence ε_{xh} (and for Ee>170 keV versus Ex > 125 keV). The vertical and horizontal lines show Pe and ε_{xh} values, within which selection effects exist.
Fig 1b. Scatter plots of the X_t peak flux Pxt versus the Xh (Ex>50 keV) fluence ε_{xh}.

By correlation analysis it was shown that if the number of events N exceeds 50, formally all of the correlation coefficients take the values: r_{oxy} > 0.5, and are "high-significant".

A list of results for strong correlations coefficients with r_{oxy} >0.75 is shown in Table 1. In this table we can see arrays x, y, value of r_{oxy} and the value of N for each of cross correlations.

Table 1. The results of cross-correlation analysis: y=a+bx.

No	X	Y	r_{oxy}	N	Reference
1.	lg Pxt	lg ε_{xh}	0.79	105	[9]
2.	lg Puv	lg ε_{xh}	0.84	101	[1]
3.	lg ε_e	lg ε_{xt}	0.80	-	[11]
4.	lg ε_{xh}	lg Pe	0.87	87	[8]
5.	lg ε_{xt}	lg Pp	0.83	55	[4]
6.	lg Pp	lg Pe	0.94	54	[10]
7.	lg Pe	lg ε_R	0.84	53	[10]
8.	lg Pp	lg ε_R	0.82	53	[10]

We include in Table 1 the results of Crannel [1, 11].

3. QUANTITATIVE RELATIONS BETWEEN SOLAR FLARES EMISSIONS.

Multiple regression analyses have allowed us to establish quantitative relations between Hα flare importance and the most likely value of the energy in optical, hard and soft X-ray emissions and the kinetic energy of the accelerated electrons [1, 9, 11] (see Table 2).

Table 2. Estimates of solar flare emissions energy.

The emission form	Hα importance SN	SB	1N	1B
$\varepsilon_{H\alpha}$	$(2-6).10^{27}$	$(0.4-1.5).10^{28}$	$(2-5).10^{28}$	$(0.4-1.5).10^{29}$
ε_{opt}	1.10^{28}	2.10^{28}	1.10^{29}	$(2-3).10^{29}$
ε_{xt}	$(2-3).10^{28}$	$(0.8-2).10^{29}$	$(2-4).10^{29}$	$(2-5).10^{29}$
ε_e (>25 keV)	$(3-9).10^{28}$	$(1-2).10^{29}$	$(2-3).10^{29}$	$(3-10).10^{29}$

The emission form	Hα importance 2N	2B	3B
$\varepsilon_{H\alpha}$	$(2-5).10^{29}$	$(0.5-1.5).10^{30}$	$(0.3-2).10^{31}$
ε_{opt}	$(1.5-2.5).10^{30}$	$(3-5).10^{30}$	$(3-5).10^{31}$
ε_{xt}	$(0.4-1.7).10^{30}$	$(0.9-3).10^{30}$	$(1-4).10^{31}$
ε_e (>25 keV)	$(1-3).10^{30}$	$(2-4).10^{30}$	$(1-3).10^{31}$

These quantitative relations were obtained in usual models of emission generation and the movement of charged particles.

4. DEPENDENCE OF THE NUMBER OF EVENTS ON THE FLUENCE were found for different kinds of flare emission (Fig. 2). All the dependence can be described by power law with an exponent $\beta = 1.45 \pm 0.15$ for wide range of fluences. Only for extremely high fluences the spectra become steeper with $\beta = 1.9-2.0$.

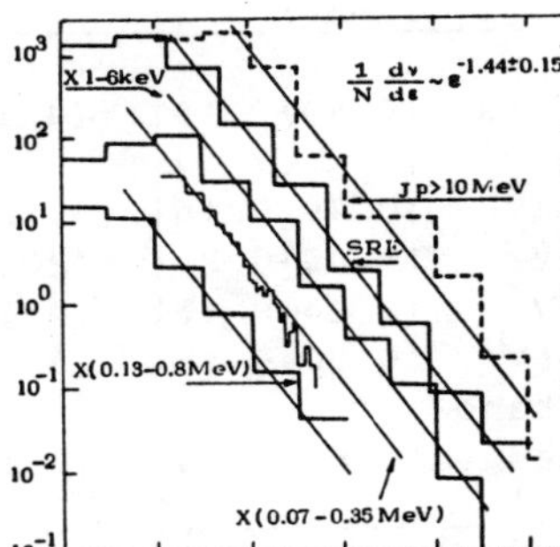

Fig. 2. Frequency dependence (differencial spectra) on fluence values: $(1/N)(d\nu/d\varepsilon)$ for different kinds of solar flare emissions.

5. DISCUSSION.

5.1. The total energy of the fast electrons in the impulsive phase of the flare and the energy of the hot plasma at the end of the impulsive phase, when the plasma temperature reaches its maximum, are practically functionally related: $r_o = 0.8$ [1, 9, 11].

5.2. The X-ray fluence (ε_{xh}) and electrons flux (Pe) are linearly connected (b=0.9 ± 0.10) in such events, where hard X-ray bursts and high energy interplanetary electrons are observed [8]. (Peak flux Pe ~ Ne - total electrons escaping number in the interplanetary space in all particles propagations models). One can explain such ratio if the total number of: hard quanta (Nx) and escaping electrons (Ne) are both proportional to the total number of the accelerated electrons (Ni). Electron escaping efficiency from the source into interplanetary space is described by the ratio Ke = Pe /ε_{xh}. This value K_e can be obtained directly from experiment without any additional suppositions (see Fig. 1a).

5.3. There is a strong correlation between the number of accelerated electrons (Ni) and protons (Mi) in the flare. The escape contidions into the interplanetary space are seemingly analogous [10].

The conclusion of the multifunctional correlation analysis is that the energy emmitted in the soft and hard X-ray range, the photons energy in the optical range of wave length and the energy of accelerated particles are a proportional measure of the total flare energy. From Table 2 can be seen that the energy inserted into the flare volume by energetic particles is comparable to the total emission energy of the low and high temperature plasma regions.

However for the continious spectrum of the accelerated electrons in their energy range Ee<10 keV [14], the true value for εe must be higher. Nearly the same energy can be inserted into the chromosphere by the accelerated protons with Ep>100 keV [15].

5.4. Table 3 shows the exponents of integral spectra obtained in different experiments during last 15 years. There is no reason to believe that the spectra vary significantly in time.

Table 3. Summary of β values for different solar flares emissions.

Emission, energy	Spectral index	Reference	Emission, energy	Spectral index	Reference
	electrons				
Ee>0.17keV	0.5±0.2	[16]	Rμ bursts	0.49±0.15	[4,19]
Ee>0.45keV	0.5±0.2	[17]	X-ray		
Ee>0.70keV	0.35±0.15	[8]	Ex 1-6keV	0.44±0.15	[4,12,13]
Ee>500keV	0.46±0.15	[17]	Ex>30keV	0.45±0.2	[20]
	protons		Ex>55keV	0.44±0.15	[4]
Ep>10MeV	0.15±0.1	[18]	Ex>300keV	0.50±0.15	[3,4,21]
Ep>10MeV	0.45±0.15	[17]	Optical band		
Ep>25MeV	0.35±0.15	[4]	of wavelength	0.80±0.10	[22]

Flare frequency with fluence higher than εo, has a simple dependence upon this energy $\nu\ (\varepsilon > \varepsilon_o) \sim \varepsilon^{-\beta}$, with rather narrow β range. For extreme values of flares energy β increases up to 0.9-1.5.

This universal relation shows the existence of some deep and common objective regularities. These regularities demonstrate that the appearance of certain energy flare frequency is strongly dependent on some scale factor.

Pustilnik suggested [23] a possible explanation of this relation. In existing theories of stars convection the turbulence velocity ν_{turb} is described by usual relation:

$\nu_{turb}(l) \sim l^{-s}$, l - the size of turbulence cell.

The spectrum index s varies within rather small values: from s=1/2 (for convection with magnetic field) and s = 1/3 (for Kolmogorov's turbulence) to s = 1/4 (spectrum of acoustic's and Alfven's waves turbulence, Kaplan). As a result Pustilnik obtained the dependence on flare frequency with energy higher than Eo:

$$\nu_{flare}(E>Eo) \sim \begin{cases} E^{-1.0+s/2} \sim E^{-(0.88-0.75)} & \text{(a)} \\ E^{-0.66+s/3} \sim E^{-(0.5-0.58)} & \text{(b)} \end{cases}$$

a) - fast magnetic field decrease with increasing altitude.
b) - slow magnetic field decrease with increasing altitude.

Thus we naturally come to distribution function of flare frequency $\nu_{flare}(E>Eo)\sim Eo^{-\beta}$, with index β which corresponds well to experimental data. This is not the only possible interpretation of solar (star) flare appearance frequency dependence. However, the possibility of obtaining experimental data which can give us information about the internal solar structure, make accumulation of such kind of data very important.

REFERENCES.

1. Crannel C.J., Hudson H.S. (1986)'Energetic phenomena on the Sun', Ed. Woodgate B.E. and Kundu M.R. (NASA CP-243a).
2. Barat C., Chambon G., Harley K., et al. (1981), Sp. Sci. Instr. 5, p 229.
3. Data from sol. hard X-ray monitor SIGNE-2M on VENERA-13,14 (1987) Sov. Geoph. Commit. A.N. of the USSR, Moscow, Ed. Kurt V.G.
4. Kurt V.G. (1989), In Sol. Plasma Physics, M. Nauka, p 138.
5. Melnikov V.F., Kurt V.G., et al. (1988), Sol. Max. Analysis (17-24 June, 1985, Irkutsk, USSR) Add. Issue, Novosibirsk, Nauka, p 366.
6. Daibog E.I., et al. (1987), 20th Int. Cosm. Ray Conf., V.3, p 45.
7. Daibog E.I., Kurt V.G., et. al. (1988), Izv. A.N. USSR, ser. fis. 52, p 2403.
8. Daibog E.I., et al. (1989), Sov. Cosm. Res. 27, p 113.
9. Kurt V.G., et al. (1989), Sov. Cosm. Res., 27, p 425.
10. Melnikov V.F., et al. (1990), presented at 21th ICR Conf., V5, Adelaida, Australia.
11. Crannel C.J., et. al. (1982), Ap. J. Lett., 329, p 969.
12. Teske R.G., Thomas R.J. (1972), Sol. Phys., 16, p 431.
13. Mayfield E.B., Lawrence J.K. (1985), Sol. Phys. 96, p 243.
14. Lin R.P., L. de Pan, Kane S.R. (1984), Sol. Phys., 91, p 345.
15. Daibog E.I., Kurt V.G. et al. (1983), Sov. Cosm. Res., 21, p 704.
16. Potter D.W., Lin R.P., Anderson R.A. (198), Ap. J. L., 226, L97.
17. Belovskiy M.N., Kurt V.G., et al. (1979), Sov. Cosm. Res., 17, p 906.
18. V. Hollebeck MAI, et. al. (1975), Sol. Phys. 41, p 189.
19. Neidig D.F. (1978), Sol. Phys., 57, p 385.
20. Pearson D.H. et al. (1989), Ap. J. 336, p. 150.
21. Vestard T. et. al., SMM observ. Gamma ray Astr. Group Phys. Dep. Univ. of Hampshire. Preprint N HO3224.
22. Kurochka L.N. (1987), Sov. Astronom. J. 64, p 443.
23. Pustilnik L.A. (1988), Sov. Astrom. J. Let., 14 p 940

Relationship between Solar Flares and Solar Cosmic Rays

M N Vahia
Tata Institute of Fundamental Research
Homi Bhabha Road, Bombay 400 005, INDIA

Abstract: We discuss the possibility that the plasma trapped inside the solar magnetic loop region may be the principal arbitrator in deciding whether a given active region will release particles into the interplanetary medium or will remain a solely electromagnetic event. We analyze the particle abundance patterns and show that the particle selection occurs in a cool region much below the corona. We discuss the particle acceleration in time varying magnetic field and show that the betatron process can be very effective under solar conditions. We then show that the electron density in the flare region may be the critical parameter in deciding the evolution of the active region. Finally we show that direct observations of flares in the interplanetary medium agrees well with this suggestion.

Introduction: It has been generally suggested that magnetic flux tubes form at the bottom of the convection zone, and at some stage become unstable and rise to the surface under magnetic buoyancy to give magnetic activity on the outer atmospheric layers of the Sun (cf Wilson, 1986). Once this upward motion has begun, it is generally assumed (Piddington, 1975) that the dynamo action ceases and the plasma within the tube remains isolated (Wilson, 1986). Above the photosphere, these loops build up in strength and become bright in Hα (cf Priest, 1981).

In the case of large flares, electromagnetic loops rise high into the upper corona of the Sun and then give particles from large open loop regions. In some cases, the flare loops may simply brighten in the electromagnetic band, giving rise to a large variety of radio and X-ray emission without any accompanying particle emission. Recent satellite measurements have also identified a class of events where small electromagnetic activity as well as particle emission occurs. Under such a condition, the particle abundances tend to be invariably peculiar. Since all this activity is, at its source magnetic, we investigate the possibility that the particles trapped in solar flare loops may mediate in the evolution of the all this activity. We show that for reasonable values of the parameters like the time scales and intensity of magnetic field variations in the flare region, the plasma trapped inside the magnetic loops can efficiently convert magnetic energy into electromagnetic energy then show that the physical conditions of the plasma trapped inside a magnetic loop may decide whether a given active region will remain purely electromagnetic event, or a small or large particle release event. We finally show that the above suggestion is well borne out by comparing the suggest made here with those made from direct observations of various of various types of flares.

Basic Selection and Acceleration process: Several studies of the large flares have established a common abundance pattern for solar cosmic rays (cf Biswas et al, 1983; Mason et al, 1986; Vahia et al, 1988) in large flares. In the literature the classification of elements is made according to their first

E. R. Priest and V. Krishan (eds.), Basic Plasma Processes on the Sun, 415–419.

ionisation potential (cf Biswas et al, 1983, Meyer, 1985). However, this classification is, at best, only a first approximation of the true processes that occur in the flare region since all elements have charge states greater than +1 at temperatures above 6 10^4 K. Hence, we take into consideration more realistic charge states of particles under solar conditions. Several studies have shown that carbon, nitrogen, neon, silicon and argon are only weakly enhanced at low energies (< 5 MeV/nucleon) in large flares. They also have relative abundances which are very similar to the photospheric abundances and show little variation from flare to flare. Based on the assumption that these characteristics are due to the fact that the source acceleration mechanism is unable to distinguish between these elements due to similarity in their charge to mass ratio in the ambient medium, a source region temperature of 6 ± 1 10^5 K is indicated (Vahia, 1987). This suggestion agrees well with the direct measurement of the solar cosmic ray charge state by Hovestadt et al (1984). The direct measurements indicate that the element charge states fall into two distinct components one with a mean temperature of 5 to 8 10^5 K and another of temperature around 2 to 5 10^6 K which corresponds to the temperature of particles picked up during passage through the corona. The fact that this is true for several large flares studied by a variety of techniques and at various times indicates that this is a common feature of large flares and therefore, by implication, that the large flares arise in regions of temperature about 6 10^5 K. This temperature is below the coronal temperature and therefore, the solar flare plasma separation must be occurring in lower chromospheric regions from which they rise into the coronal region.

The most common feature of the solar flares is the time varying magnetic field in the active regions. The most likely acceleration mechanisms in the active regions therefore are the stochastic and reconnection acceleration. We calculated the efficiency of the stochastic and the reconnection processes based on the assumption that the volume of the acceleration region is 10^8 cm^3, the magnetic field of the Solar flare acceleration region is 10^3 G and the electron density is 10^{14} p cm^{-3}. The basic formulation of Sorrell (1983) for the reconnection process and of Hillas (1984) for stochastic acceleration process are used. The calculations show that while the stochastic acceleration can give a maximum energy of 10^{13} eV, its time scales of 10^6 sec are far too large. On the other hand the time scales of reconnection, of the order of 50 seconds, agrees well with the time scales of the impulsive phase of solar flares and can give a maximum energy of 10^9 eV. This is more in agreement with the direct observations of particles up to 1 GeV in solar flares. We therefore discuss the reconnection process in some detail. In particular, in Vahia (1986) the details of the betatron process are discussed with the emphasis for accelerating solar flare particles. We use the boundary conditions that the magnetic field strength changes from 100 to 3000 G in 100 seconds. These values are within the observed variations of the magnetic field strengths in flare regions compiled by Priest (1984) which shows that flux changes of the order of 10^{19} Mx/hr can occur in active regions. From such calculations it can be seen that this results in a runaway acceleration of the particles and is most effective for temperature of 6 10^5 K (Vahia and Rao, 1987). However, in that work the energy loss processes were neglected. The comparison of the acceleration mechanism against the energy loss process are given in figure 1. The energy loss formulation is taken from Perez Peraza (1981) and the energy gain process is taken as the betatron process from Vahia (1986)(see also Vahia and Rao, 1987). Since the pp collisions have a crossection of about 40 mbarns at this energy, the energy loss by collision can be neglected. We discuss the implication of this figure in the three cases, when the electron density is less than 10^7 p cm^{-3}, when it is around this value and when it is greater than this value.

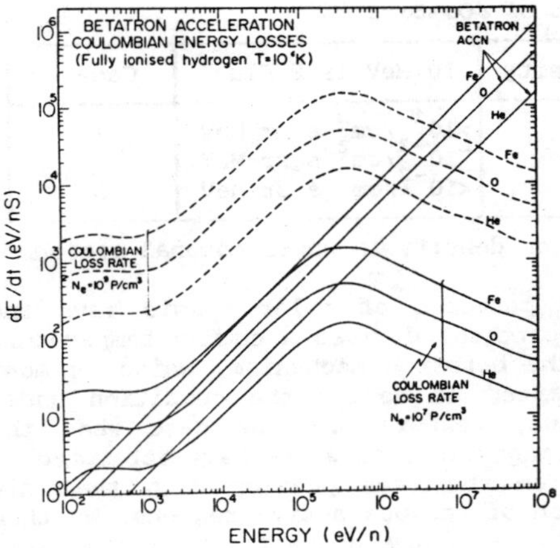

Fig. 1: Comparison of energy gain and loss processes in flare regions.

Case 1: Large Particle Flares: When the particle density is less than 10^7 p cm^{-3}, the particle acceleration rate exceeds the energy loss rate and, as a result, the active region will see a large release of particles as the plasma exceed the magnetic pressure. Since such low density is seen in upper coronal region only, the active regions must have large, stable and energetic loops that rise high into the corona before the final acceleration releases the particles. This feature is invariably seen in the active regions that release large flux of particles. The particle escape from such a loop can occur either from opening of the lines of force to the interplanetary medium or by rigidity dependant escape mechanisms (Vahia, 1986).

Case 2: Intermediate Flares: When the electron density is around the critical value of 10^7 p cm^{-3}, the region as a whole will not release particles but small heterogeneities in such a system can produce runaway acceleration and special selectivities that are characteristic of small flares.

Case 3: Thermal Flares: When the active region has electron density in excess of the critical value, practically no particle release would occur from such regions and all the magnetic energy would be converted into heat energy. However, in such cases also, the particles can continue to mediate in the active region in converting the magnetic energy into thermal energy.

Hence, if the above discussed processes have any relevance to the solar active regions, the particle population in such regions and the densities of such regions may be the critical deciding factors in forming a large particle flare, a small flare or a purely electromagnetic event. Also, if the above suggestion is true, the energy release in various types of events should show a clear correlation which can be derived from physical parameters that can be directly associated to the active regions.

Three kinds of classifications of the solar flares is made in the literature. The first, large flares are conspicuous by particle energy release of the order of 10^{31} ergs compared to electromagnetic radiation of the order of 10^{28} ergs (cf Hakura, 1976). The second class of events are associated with active regions that show a small amount of energy are known to show an enhancement in the He^3/He^4 ratio and are known as He^3 rich solar flares (cf Reames et al, 1988). The third class is that of active regions that show intense electromagnetic emission without any accompanying particle release. Detailed features of the three classes of events also agree well with the classifications made on the basis of direct observation by Bai (1986) for flares analogous to cases 1 and 3 and by Reames et al (1988) for case 2. The salient features of these observations are given in table 1.

Table 1 Observed features of Some classes of flares

Type of Event	Flare Size	Electron Density	10 MeV IP P Flux	Case
Gradual[1]	$>10^{9}$ cm	$<10^{7*}$ cm^{-3}	$>10^{1}/cm^{2}$ s sr MeV	1
He^{3} flare[2]	$\sim10^{9}$ cm	10^{8} cm^{-3}	$\sim10^{-2}/cm^{2}$ s sr MeV	2
Impulsive[1]	$<10^{9}$ cm	10^{11-12} cm^{-3}	$<10^{-2}/cm^{2}$ s sr MeV	3

[1]Bai(1986); [2] Reames et al (1988) *-Based on density in upper coronal region

<u>Conclusion:</u> We discussed the abundance patterns of solar cosmic rays and considering the similarity in abundance patterns derived a common temperature condition for them. We have then used the betatron mechanism, which is most effective at this temperature and attempted to derive the condition under which the acceleration becomes a runaway feature and the case when the acceleration may produce a purely electromagnetic flares. We have attempted to show that the entire feature arises from the same physics with the only difference being in the physical condition of various active regions. We then compared the observations and show that the flare classification suggested here is in good agreement with those made empirically from observational data.

<u>Acknowledgement:</u> The author wishes to express his gratitude to Dr A R Rao and Dr S Ramadurai for their valuable discussions during this work.

References

Bai T, 1986, *Astrophys J*, <u>308</u>, 912
Biswas S, Durgaprasad N and Vahia M N, 1983, *Solar Phys*, <u>89</u>, 163
Hakura Y, 1976, *Space Sci Rev*, <u>19</u>, 411
Hillas A M, 1984, *Ann Rev Astron Astrophys*, <u>22</u>, 425
Hovestadt D et al, 1984, *Astrophys J*, <u>261</u>, 463
Mason G M et al, 1986, *Astrophys J*, <u>303</u>, 849
Meyer J P, 1985, *Astrophys J*, <u>57</u>, 151
Perez Peraza J, 1981, in Nucleosynthesis, ed S Biswas, S Ramadurai and M N Vahia, Tata Institute of Fundamental Research, p 275
Piddington J H, 1975, *Astrophys Space Sci*, <u>34</u>, 347
Priest E R (ed), 1981, Solar Flare Magnetohydrodynamics, Gordon and Breach Science Publishers
Priest E R, 1984, *Adv Space Res*, <u>4</u>, no 7, 37
Rao A R and Vahia M N, 1986, *Adv Space Res*, <u>6</u>, no 8, 85
Reames D V et al, 1988, *Astrophys J*, 327, 998
Sorrell W H, 1986, in Radiation in Contemporary Astrophysics ed M M Shapiro, D Reidel Pub Comp, p 79.
Vahia M N, 1986, *Astron Astrophys*, <u>168</u>, 331
Vahia M N, 1987, *Astron Astrophys*, <u>173</u>, 361
Vahia M N, Biswas S and Durgaprasad N, 1988, *Astrophys Space Sci*, <u>149</u>, 241
Vahia M N and Rao A R, 1987, *Proc 20 Int Cosmic Ray Conf*, Moscow, <u>3</u>, 21
Wilson P R, 1986, *Solar Phys*, <u>106</u>, 1

DISCUSSION

SMITH: As far as I can see your generalized reconnection is nothing more than an E-Field along the B-Field, but you cannot cite a specific mechanism. Is this true?

VAHIA: The generalised calculations given here are an attempt to estimate - by order of magnitude - the efficiency of various processes. It is not an attempt to determine or identify a specific process.

KUIJPERS: You said that the particle density in the loop determines whether the particles are released or not into the interplanetary medium. What is the physical process which lets the particles escape from the loop and how does it depend on density?

VAHIA: The critical factor that decides whether particles are released or not is the energy gain to loss rates. If the energy gain rate exceeds the energy loss rate (when the density is less than 10^7 cm^{-3}) there will be a runaway acceleration and in this case the particles will escape; either when their gyroradius becomes larger than the loop diameter or when the lines of force open to the interplanetary medium. Conversely, if the energy loss rate exceeds the energy gain rate, then all the energy put into particles will be converted into electromagnetic radiation.

ENERGETIC PARTICLES IN A FLARE LOOP : SPECTRA AND RADIATION SIGNATURES

P. A. BESPALOV and V. V. ZAITSEV
Institute of Applied Physics
Gorky, USSR

A. V. STEPANOV
Crimean Astrophysical Observatory
Crimea, USSR

ABSTRACT. It has been shown that high energy particle spectra, particle dynamics, and radiation in a flare loop are determined by wave-particle interactions. The electron-whistler interaction occurs under conditions of strong pitch angle diffusion that makes the particle distribution function isotropic. The flare loop electrons retain information about the particle source spectrum. The interaction of energetic ions with Alfven waves is characterized by strong, moderate, and weak diffusion. The time delays in hard X-ray and gamma-ray emission during one-step acceleration processes might be understood in terms of a trap-plus-turbulent propagation model. The density of precipitating particles is less than or equal to the trapping one. Radiation signatures of flare loop electrons are discussed.

1. INTRODUCTION

Pitch angle diffusion plays a very important role in the dynamics and radiation of the energetic particles in a solar flare loop. The accelerated particles diffuse into a loss-cone due to pitch angle scattering and precipitate toward the loop footpoints. The interaction of electrons with E > 10 keV with chromospheric plasma near the loop footpoints leads to hard X-ray emission. Gamma-ray line emission is produced by precipitation of ions with E > 10 MeV. According to Bespalov and Trakhtengertz (1986) the three modes of pitch angle diffusion are possible :
(i) Weak diffusion is realized when $\sigma T_o < T_d$, where T_d is the diffusion time, σ is the mirror ratio, $T_o = L / 2v$ is the depletion time of the loss cone, L is the loop length, and v is the energetic particle velocity. The loss cone is always empty.
(ii) Moderate diffusion corresponds to the inequality $T_o < T_d < \sigma T_o$. Filling of the loss cone is essential and the particle distribution function is close to an isotropic one. These diffusion modes were defined by Kennel and Petschek (1966). Bespalov and Trakhtengertz (1986) have shown that there exists a still more efficient scattering process:

E. R. Priest and V. Krishan (eds.), Basic Plasma Processes on the Sun, 421–427.

(iii) Strong diffusion, $T_d < T_o$, occurs when the particle source is powerful enough. The particle changes direction several times during one transit time of the loop due to effective scattering on the waves and the particle propagation velocity may be much lower as compared with cases (i) and (ii). The pitch angle scattering may be due either to Coulomb collisions or to wave-particle interactions. In the trap-plus-precipitation model by Melrose and Brown (1976) collisions lead to weak pitch angle diffusion. In spite of wave-particle interaction in a flare loop only weak diffusion was considered in many papers (Aschwander et al., 1989; Hulot et al., 1989). Moderate diffusion in a flare loop due to Alfven wave-ion interaction was investigated by Zweibel and Haber (1983) and Bespalov and Zaitsev (1986). Bespalov et al. (1987) drew attention to the possibility of strong pitch angle scattering in a flare and suggested a trap-plus-turbulent propagation model.

In the present report we will show that the dynamics of the flare accelerated particles are determined mainly by wave-particle interactions. Moreover, for high energy electrons wave-particle interactions occur in the strong diffusion regime only. Particle diffusion on waves is responsible for particle spectra formation and affects flare radiation signatures.

2. CYCLOTRON INSTABILITY IN A FLARE LOOP

Energetic particles with a loss cone anisotropy may be unstable against small-scale electromagnetic wave excitation. In the context of pitch angle diffusion cyclotron resonant interaction with whistlers $\omega = \omega_e(kc/\omega_p)^2$ is the most important for the electrons. The waves grow at a rate (Kennel and Petschek, 1966)

$$\gamma \approx \frac{n_1}{n_0}\,\frac{\omega_e}{\sigma} \tag{1}$$

where n_1, n_0 refer to the densities of the energetic particles and the background plasma, and $\omega_e = eB/mc$. From the cyclotron resonance condition $\omega - k_{\parallel}v_{\parallel} = \omega_e$ and the anisotropy threshold $\omega < 3\,\omega_e/(2\sigma+1)$ we find that the energy of the electrons interacting with whistlers must exceed the value

$$E_{min} \approx 14\,\frac{(\sigma-1)^3}{(2\sigma+1)^2}\left(\frac{v_A\ \mathrm{cm\ s^{-1}}}{10^8}\right)^2\ \mathrm{keV}\ , \tag{2}$$

where v_A is the Alfven velocity. For $v_A \approx (1\text{-}3)\ 10^8$ cm s^{-1} and $\sigma \approx 10$, for example, from Eq. (2) we obtain $E_{min} \approx 20 - 200$ keV. The energetic particle density instability threshold under flare loop conditions is about $n_1/n_0 \approx 10^{-9} - 10^{-8}$ and determined by Joule dissipation or by

Landau damping.

Anisotropic ions generate Alfven waves $\omega = k_{\parallel} v_A \approx \omega_i v_A / v$ with a growth rate of the order of (Kennel and Petschek, 1966)

$$\gamma \approx \frac{n_1}{n_0} \frac{\omega_i^2}{\omega\sigma}, \quad \omega_i = \frac{eB}{Mc} \tag{3}$$

The instability threshold $n_1/n_0 \approx 10^{-8}$ is determined either by ion viscosity or by Landau damping. By assuming that in the magnetic loop with a volume 10^{26} cm^3 the number of accelerated particles is about 10^{30} - 10^{32} for a typical value of $n_0 \approx 10^{11} - 10^{12}$ cm^{-3}, we obtain $n_1/n_0 \approx 10^{-8} - 10^{-6}$. Thus, there is no problem to excite the whistlers and Alfven waves by high energy particles in a flare loop.

3. WAVE PARTICLE INTERACTION

Wave particle interaction is described by quasi-linear equations for the distribution function of the energetic particles f and for the spectral density of the wave energy W_k:

$$\frac{df}{dt} = \frac{\partial}{\partial v_i} D_{ij} \frac{\partial f}{\partial v_j} + j$$
$$\frac{dW_k}{dt} = (\gamma - \nu) W_k \tag{4}$$

where D_{ij} is the diffusion coefficient, j is the particle source, and ν is the damping rate. The condition $T_d < T_o$ is a necessary, but not sufficient condition for strong diffusion. The full criterion has been found by Bespalov and Trakhtengertz (1986) on the basis of Eqs. (4). In the strong diffusion limit the integral particle source power

$$I = \iint \frac{B_{max}}{B} j \, d^3vdz$$

exceeds the critical value

$$I_* \approx \frac{cB\sigma}{4\pi eL} \tag{5}$$

It should be noted that I_* does not depend on the particle species. Setting $L \approx 10^9$ cm, $B \approx 10^3$ G, $\sigma \approx 10$, into Eq. (5), we obtain $I_* \approx 5$ 10^{13} $cm^{-2}s^{-1}$. In the steady-state case the particle flux towards the loop footpoints is equal to the integral source power 2S = I. It follows from hard X-ray data that the typical value of electron flux is of the order of $S_e \approx 10^{15}$- 10^{17} $cm^{-2}s^{-1}$ >> S_*. The ion fluxes producing gamma-ray line emission are usualy two or three orders of magnitude less. This implies that for flare loop electrons the strong diffusion limit is formed easily and the electron distribution function is isotropic. For energetic ions strong, moderate, and weak diffusion limits are possible.

Under strong diffusion conditions instead of free particle propagation along the loop axis, turbulent propagation occurs with the wave group velocity, which is the whistler group velocity for electrons and Alfven velocity for ions. The particle turbulent propagation time from the top of the loop to the footpoint for electrons and ions can be written in the following form (Bespalov et al., 1987)

$$t^e_{turb} \approx \frac{m}{M}\frac{v}{v_A}\frac{\sigma L}{2v_A}, \qquad t^i_{turb} \approx \frac{\sigma L}{2v_A} \tag{6}$$

From Eqs. (5) and (6) we can estimate the critical density of energetic electrons n_* in the strong diffusion limit. Keeping in mind that $S_* \approx n_* v_{turb} \approx 2.5\ 10^{13}\ cm^{-2}s^{-1}$, we find $n_* \approx 6\ 10^3 cm^{-3}$. Hence, the value of n_* is very close to the whistler cyclotron instability threshold $n_1 \approx 10^3 - 10^4\ cm^{-3}$. This estimate indicates a very important point : under flare loop condition the whistler-electron interaction is realized in the form of strong diffusion.

The event of 4 August 1972 is a good example of strong diffusion. The hard X-ray time delay was smoothly increasing with the increase of photon energy up to 5 s until a photon energy of 150 keV where it increases suddenly to 15 s (Bai and Ramaty, 1979). This jump may be caused by the transition from collisional diffusion to strong difusion of electrons on whistlers when the electron energy exceeds the threshold value (2). As follows from Eqs. (6) the ion-to-electron turbulent propagation time ratio is $t^i_{turb}/t^e_{turb} \approx 10$.
Therefore, as both electrons and ions are accelerated simultaneously, the bulk of the ions will reach the footpoints several seconds or tens of seconds later than the electrons. As a result, gamma-ray line emission peaks will be delayed with respect to corresponding peaks of hard X-ray emission (Bespalov et al., 1987). This supports the idea that the electrons and ions are accelerated in a flare simultaneously (Chupp, 1983).

4. PARTICLE SPECTRA

The energetic particle spectra in a flare loop contain very important information on the flare acceleration process. Thus, the question arises : what is the difference between the spectra of particles interacting with waves and the particle source spectrum ? Bespalov and Zaitsev (1986) have shown that the spectra of proton fluxes for weak and moderate diffusion do not depend on the source spectrum but are determined by the index p in the wave damping rate $\nu = \nu_0(k/k_0)^{2p}$ only:

$$\begin{aligned} S_i &\sim E^{-p+1/2}, \text{ weak diffusion} \\ S_i &\sim E^{-p-3/4}, \text{ moderate diffusion ,} \end{aligned} \tag{7}$$

where p = 1 for ion viscosity damping and for Landau damping. In the strong diffusion limit the situation is different. The particle distribution function retains information on the source spectrum (Bespalov et al., 1990) :

$$f(E) \approx t_{turb} j(E) \sim \begin{cases} j(E) & \text{for ions} \\ E^{1/2} j(E) & \text{for electrons} \end{cases} \quad (8)$$

5. RELATION BETWEEN HARD X-RAY AND MCW RADIATION ELECTRONS

There exists a well known discrepancy between the number of electrons inferred from the hard X-ray emission N_X and the gyrosynchrotron radio emission $N_R \approx (10^{-3} - 10^{-5}) N_X$. This implies that the number of trapped electrons producing mcw radiation is much less than the number of precipitating ones. However, such a situation is impossible in a flare loop. Indeed, we have shown that for electrons with $E > E_{min}$ the typical pitch angle diffusion mode is strong diffusion which makes the electron distribution function isotropic. For an isotropic distribution the particle density is constant along the magnetic loop, i.e. $N_R \gtrsim N_X$. In case of weak or moderate diffusion $N_R >> N_X$. In order to resolve this discrepancy there exists only one way out : low emissivity of gyrosynchrotron emission due to a moderate value of $B < 300$ G (Gary, 1985).On the other hand Kai (1982) assumed that the electrons generating mcw emission have an anti-loss-cone distribution function. Such a distribution function cannot be formed in a flare loop because of strong diffusion.

6. CONCLUSIONS

We are apt to conclude this report by summarizing the results as follows:

1. The energetic particle spectra, particle dynamics and radiation in a flare loop are determined mainly by wave-particle interaction. 2. The strong pitch angle diffusion threshold implies that the flux of precipitating particles must be more than $10^{13} - 10^{14}$ cm^{-2}s^{-1} which is easily realized in a flare loop. The most probable diffusion mode for high energy electrons interacting with whistlers is strong diffusion. For energetic ions the strong, moderate, and weak diffusion on Alfven waves are possible.
3. Under strong diffusion energetic particles retain information on the particle source spectra.
4. Time delays in hard X-ray and gamma-ray emission can be explained in the framework of a one-step acceleration process due to turbulent propagation of particles under strong diffusion conditions.
5. The density of precipitating particles in a flare loop is less than or equal to the density of trapped ones. The discrepancy between mcw and hard X-ray radiating electrons can be resolved by low emissivity of gyrosynchrotron emission.

It should be noted also that the mcw radiating electrons in a flare loop are almost isotropic due to strong diffusion. Nevertheless, isotropic electrons with a gap may be responsible for electron cyclotron maser emission (Blanken and Kuckes, 1969).

REFERENCES

Aschwander, M.J., Benz, A.O. and Kane, S.R. (1989)"Correlation of solar radio pulsations with hard X-ray emission", Astron. Astrophys. in press.

Bai, T. and Ramaty, R. (1979)"Hard X-ray time profiles and acceleration processes in large solar flares", Ap. J., 227, 1072-1081.

Bespalov, P.A. and Trakhtengertz, V.Yu. (1986)"Cyclotron instability in the Earth radiation belts", in Reviews of Plasma Physics, vol. 10, Plenum Publ. Corp., New York.

Bespalov, P.A. and Zaitsev, V.V. (1986) "Formation of energetic proton spectra in solar flares" in Solar Maximum Analysis, VNU Sci. Press, p.247-253.

Bespalov, P.A., Zaitsev, V.V. and Stepanov, A.V. (1987) "On the origin of time delays in hard X-ray and gamma-ray emission of solar flares", Solar Phys. 114, 127-140.

Bespalov, P.A., Zaitsev, V.V. and Stepanov,A.V. (1990) "Consequences of strong pitch angle diffusion of particles in solar flares", Ap.J. Suppl. in press.

Blanken, R.A. and Kuckes, A.F. (1969) "Synchrotron deceleration of electrons in plasma and the generation of electromagnetic radiation" Plasma Phys. 11, 321-331.

Chupp, E.L. (1983) "High energy particle acceleration in solar flares - observational evidence", Solar Phys., 86, 383-393.

Gary, D.E. (1985) "The numbers of fast electrons in solar flares as deduced from hard X-ray and microwave spectral data", Ap.J., 297, 799-804.

Hulot, E., Vilmer, N. and Trottet, G. (1989) "Relative timing of solar prompt gamma-ray line and X-ray emission expected from a trap plus precipitation model for protons and electrons", Astron. Astrophys., 213, 383-396.

Kai, K. (1982) "An explanation of discrepancy of required electron numbers between concurrent mcw and hard X-ray bursts", in Hinotori Symposium on Solar Flares, Tokyo, pp. 162-167.

Kennel, C.F. and Petschek, H.E. (1966) "Limit of stability trapped particle fluxes", J. Geophys. Res. 71, 1-22.

Melrose, D.B. and Brown, J.C. (1976) "Precipitation in trap models for solar hard X-ray bursts", Mon. Not. R.A.S. 176, 15-30.

Zweibel, E.G. and Haber, D.A. (1983) "The propagation of energetic ions in magnetic loops and gamma-ray emission from solar flares", Ap. J., 264, 648-659.

DISCUSSION

SMITH: If one tries to accelerate the protons by resonant Alfvén waves in the 7 June 1980 flare, one finds that the protons propagate in strong diffusion. However, this bottles up the protons so effectively that you cannot explain the rapid time-scales in this flare.

STEPANOV: If the Alfvén waves are generated not by energetic ions but by an ambient source the situation will be different. In order to determine the diffusion regime we need the level of Alfvén waves instead of the particle source power.

SIVARAM: (i) What are the energies of the gamma-ray lines and what are they caused by?
(ii) What is the energy of the hard X-rays? What is the time delay?

STEPANOV: (i) The energy of the solar gamma-ray line emission is 0.2-6.4 MeV. This emission is caused by interaction of > 10 MeV ions with photospheric plasma.
(ii) The energy of hard X-ray emission is 10-500 keV. The time delay is several seconds typically.

THE INTERACTION OF COMETARY PLASMA WITH INTERPLANETARY MEDIUM - A POST-HALLEY VIEW

Debi Prasad
Physical Research Laboratory
Navrangpura
Ahmedabad-380 009, India

ABSTRACT

The strong interaction of cometary plasma with the interplanetary medium results the disconnection events and other features observed in plasma tail and the coma of comets. The understanding of these features provide direct probe of the heliospheric magnetic field. However the interpretation of these events is not without ambiguity, as revealed in the recent work with comet Halley. In this paper a review of the "Post-Halley era" developments of the cometary plasma physics is presented.

INTRODUCTION

The study of cometary plasma have contributed significantly in understanding the nature of interplanetary medium (IPM). Based on the observation of plasma tail of comets in 1951 Biermann /1/ concluded the existance of continuous solar wind in the IPM at all solar latitudes. By far, the largest volume in the interplanetary space and largest span of time are covered by comets with plasma tails. Today, about more than 150 comets with plasma tails are known, with the frequency of about one or even two new discoveries per year. In comparison with several methods, comets can be used to study a large volume of interplanetary medium /2,3/. Therefore the understanding of the interaction of cometary plasma with the IPM is an essential part of the study of the basic plasma processes on the sun.

The comets are frazile coglomorates of various types of ices, made during the early stages of the solar system formation. On close approach to sun these bodies sublimate and strongly interact with solar wind producing the large coma and long plasma tails. The plasma tails consists of the ionic molecules such as CO^+, CO_2 and H_2O^+, appearing blue in color due to the characteristic emission

E. R. Priest and V. Krishan (eds.), Basic Plasma Processes on the Sun, 429–434.

(mostly of CO^+ and CO_2). As a result of the complex interaction of solar wind with the cometary atmosphere, a number of features in the form of kinks, knots and condensations are formed in the tail and coma of the comets. The observations of various comets during last decade from the ground and from the space have enriched our knowledge in understanding many of these processes, with a single major understanding due to comet Halley in 1985-86.

THE PLASMA BOUNDARIES

The cometary plasma boundaries are excellently reviewed by Craveus /4/. A brief outline is given here. Figure 1 outlines various plasma boundaries. The solar extreme ultraviolet radiation ionizes the cometary neutrals, which slows down the solar wind due to the mass-loading. As a consequence the bow shock is generated with Mach number 2 (weak shock) at a distance of about 10^6 km for Comet Halley (during the encounters) /5/. The solar wind continues to be mass-loaded and to slow down, after being shocked and ultimately stagnates for cometocentric distances less than about 10^5 km (for Halley's comet during encounters). Consequently the magnetic field strength increases, leading to the development of a magnetic barrier (or pile up region) in the vicinity of the stagnation region. Eventhough this process is described quite well by global, three-dimensional MHD models, the measurements made by the instruments on various spacecrafts have shown that the structure of the stagnation region is rather complicated /6/. Hence presently this region is called cometopause. The imaging Fabry-Perot studies have resulted the line-of-sight velocity maps /7/ in H_α emission at 6563 A°, during an occassion when the H_α intensity of this region was above the cometary continuum. These observations are valuable in understanding of the complex dynamics of cometopause. The theoretical studies involving the velocity map of this region is in progress /8/. As the cometocentric distance decreases the plasma becomes increasingly slow (the flow speeds less than 20 km/s), dense (electron density about 10^2-10^4 cm^{-3} between cometopause to 1000 km cometocentric distance) and cold (with electron temperature of 100 ev at cometopause to 0.1 ev at 1000 km cometocentric distance). This region is chemically complex and have magnetic field of about 60 at 10^4 km due to declaration of the flow. In this region therefore, the collisional processes become important. The collisionpause is defined as the boundary within which the transport or dyhamical processes become less important than the collisional processes. The location of collisionopause for a given process is the region where the transport time τ_T equal the characteristic collision time τ_c. Therefore there exist a number of collisionpauses for different ionic species in the region between the cometopause and diamagnetic cavity, generally in the vicinity of 2×10^4 km. In the innermost portion of the magnetic barrier, as a consequence of the balance between an outward ion-neutral frictional force on the plasma and an inward-directed magnetic pressure gradient force the diamagnetic cavity exists at a distance of about 4000 km from the nucleus (as observed by Giotto 4468 km in bound, (4150 km out bound) /9/. The comet tails are a consequence of mass loading processes which take place in the head of the comet and slows down the plasma flow in the sunward side, thus "hanging up" the field lines. The ICE spacecraft measured at about 8000 km downstream of comet Giacobini-Zinner, detecting the narrow layer of cold dense plasma in the middle of the tail and magnetic tail lobes, which were obviously due to drapped magnetic field lines /10/.

Apart from these boundaries which are "Permanent" in nature, the boundaries intrinsic to solar wind carried into the cometary environment and some small scale features which are due to transient disturbances associated with plasma instabilities are also observed in comets. The later is poorly understood.

MASS MOTION VERSUS WAVE MOTION

Displacement of features in cometary plasma tails, obtained from a sequence of photographs, have been used to infer plasma motions with tailward velocities in the range of about 20 to 250 km/s /11, 12/. However, from these observations it is not clear whether these apparent motions represent propagating disturbances such as magneto hydrodynamic waves moving down the tail or actual mass motions of the cometary ions or combination of both. This question can be resolved with Doppler measurements of ionic emissions. Huppler et al. /13/ have reported the doppler velocities of about 20-40 km/s by measuring the line shift in H_2O^+ emission in comet Kohoutek - thereby indicating the existance of mass motion. Debi Prasad et al. /7/ have observed with an imaging Fabry-Perot spectrometer, the motion of a plasma region with doppler velocities of about 30±10 km/s. The Fabry-Perot observations of Comet Halley in H_2O^+ emission by Scherb et al. /14/ have revealed the ionic accelerations about 37 to 300 cm sec^{-2}. From these measurements, it can be concluded that the tail particles are accelerated while moving in the plasma tail. However, it is still not clear whether the motion of the features represent the mass motion, since they show systematically lower acceleration in comparison with the Doppler measurements. In a future occasion it will be valuable to observe the Doppler velocities in the tail features. The ground based observations, specially with imaging Fabry-Perot spectrometers in CO^+ emission lines will prove useful in these experiments.

THE SHORT LIVED PLASMA FEATURES

Historically the visual observations of the great comet of 1882 by Bernard /15/ can be regarded as the earliest record of the cometary transient features. During last two decades several such features have been identified /7,16,18/ especially during the Comet Halley observations. Lynch and Russel /16/ have reported the observation of a blob like stucture at about 32000 km from the nucleus in the plane of sky, which did not show up in A'Hearn's /17/ measurements, thereby requiring a dissipation time scale of an hour. The image was obtained at λ_{eff} 5800 A° with a bandwidth of about 2000 A°. There is an evident of comet passing through the interplanetary sector boundary, prior to these observations. From three IUE low resolution spectra taken during 18-19 March 1986 Fedman et al. /18/ reported the sudden enhancement of CO^+ intensity in the tailward side, at a distance of about 150 000 km from the nucleus. The enhancement lasted only for 37 minutes. The calculations show that the photoionization and electron impact ionization processes cannot satisfactorily explain the event. It can be noted here that the comet crossed the interplanetary sector boundary on March 16.5 UT. With the imaging Fabry-Perot spectroscopic observations Debi Prasad et al. /7/ have recorded a short lived enhanced brightness region at position angle 160° at a nucleocentric distance of about $2x10^5$ km. The size of the region was 30,000x 20,000 km in the plane of sky. The brightness was prominent in H_2O^+ emissions at 7000 A°. A Fabry-Perot interferogram is recorded in $H\alpha$ (6563 A°) emission lines, following the H_2O^+ imagines at the enhanced brightness region. Analyzing the interferogram, these authors have obtained the spatio-kinematic informations of this region. It was found that the radial recession velocity of the enhanced brightness region is 30±10 km/s (cometocentric). The Doppler widths of the lines are 5 to 22 km/s. The relative differential velocity map is structured as shown in figure 2, with high & low velocity regions clustered. Line splittings are seen in the sunward side. From the interferogram data, it is inferred that the dispersal time of the blob is about 30s, which is further supported by the contemporary observations. From the Sakigake and Giotto measurements and considering the motion of the comet, it is inferred that the southern part of the Haley's comet dipped first into the away sector in the changing polarity region. Therefore it is more likely that this event is a "tail disconnection type event" resulted due to the reconnection of field lines. Lutz and Wagner /19/ have reported that comet IRAS-Araki-Alcock 1983d showed an intense brightening event on 10 May 1983. The flare-like event manifested itself in increase of flux in continuum and emission band in 20 minutes during 0618 & 0638 UT, preceded & followed by periods of normal stability. The hypothesis of the ejection of rock-like particles and the encounter of interplanetary shock as the precursor of this event are unsatisfactory. Since the sun was quite the solar cause can be rulled out. It is proposed that venus like interaction of interplanetary magnetic field with the coma can cause such events /22/. In absence of any observational evidence for supporting this mechanism, the reconnection mechanism, while the comet passes through the interplanetary magnetic field can also be considered.

Reviewing the well documented transient plasma events therefore indicate that the magnetic field reconnection process must be responsible for most if not all of the events. However, detailed theoretical studies of these events are yet to be done.

ACKNOWLEDGEMENTS

The author would like to thank Drs.J.N.Desai and T.Chandrasekhar for their critical comments.

REFERENCES

1. Biermann, L., 1951, Z.Astrophys., 29, 274.
2. Debi Prasad C., 1988, Ph.D thesis, Guj.University, India.
3. Hoeksema, J.C., 1989, Adv.Space Res., 9, (4) 141.
4. Cravens, T.E., 1989, Adv.Space Res., 9, (3) 293.
5. Anderson, K.A., Carlson, C.W., Curtis, D.W., Lin, R.P., Re'me, H., Sanvaud, J.A., d'Uston, C., Korth A., Richter, A.K., and Mendis, D.A., 1987, Astron. Astrophys., 187, 290.
6. Re'me, H., Sanvaud, J.A., d'Uston, C., Cros.A., Anderson, K.A., Carlson, C.W., Curtis, D.W., Lin, R.P., Korth A., Richter, A.K., and Mendis, D.A., 1987, Astron. Astrophys., 187, 33.
7. Debi Prasad C., Chandrasekhar T., Desai, J.N., Ashok N.M., 1989, J. Astron. Astrophys., 10, 1.
8. Khara A., Personal communication.
9. Ip.W.H., and Axford, W.I., 1987, Nature, 325, 418.
10. Meyer-Vermet, N., Conturier, P., Hoang, S., Perche, C., Steinberg, J.L., Fainberg, J. and Meetre, C., 1986, Science, 232, 370.
11. Jockers, K., 1985, Astron.Astrophys., Suppl. Ser., 62, 791.
12. Celnik, W.E. and Th.Schmidt-Kaler, 1987, Astron. Astrophys., 187, 223.
13. Huppler, D., Reynolds, R.J., Roesler, F.L., Scherb, F., Trauger, J., 1975, Ap.J., 202, 276.
14. Scherb, F., Magee-Sauer, K., Roesler, F.L. and Harlander J., 1989, Icarus, submitted.
15. Bernard, E.E., 1893, Astron. Assoc., J., 9, 59-63.
16. Lynch and Russell, 1988, Pub. Astro. Soc. Pac., 100, 1122.
17. A'Hearn, M.F., Personal Communication.
18. Feldmann et al., 1986, Nature, 324, 646.
19. Lutz, B.L., Wagner, R.M., 1986, Ap. J., 308, 993.
20. Russel, C.T., Luhmann, J.G., Baker, D.N., 1987, Geophys. Res., Lett., 14, 991.

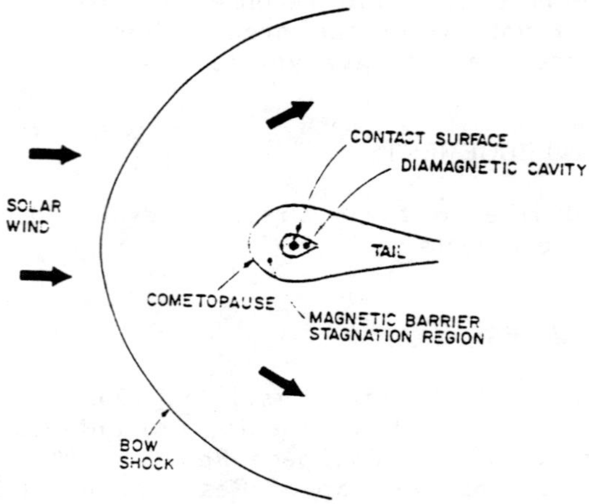

Fig. 1 : Schematic of plasma boundaries of an active comet.

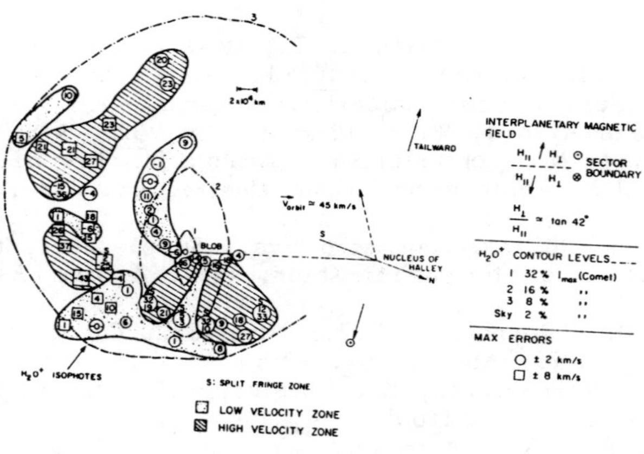

Fig. 2 : Doppler shift dispersion velocity distribution in the H_2O^+ blob region. H_2O^+ isophotes contour levels are drawn from the cometary image taken earlier to the interferogram.

MAGNETIC FIELD-CHROMOSPHERIC PLASMA INTERACTION AND THE PROBLEM OF THE BRAKING FORCE IN SURGE DYNAMICS

V. Dermendjiev

Department of Astronomy and NAO
72 Lenin Blvd.
1784 Sofia
Bulgaria

1. INTRODUCTION

An important stage in the evolution of solar flares is the generation of a number of dynamic phenomena, greatly varying in size and velocity, but actually representing plasma ejections. They are responsible for the transport of mass and energy and are usually considered to be result of the development of local magnetic features accompanied by a magnetic energy increase in the solar atmosphere. One of these phenomena is the surge prominence. In this contribution, we propose a possible explanation for the nature of the braking force acting in the decelerated ascending phase of surge dynamics. The possibility of structuralization of the chromospheric plasma during surge formation and the plasma action on the external, colimating magnetic field is taken into account.

2. OBSERVATIONAL AND THEORETICAL CONSIDERATIONS

This type of prominences is described in [1] as a colimate stream of chromospheric plasma. In [2], ballistical maintenance by dynamic effects was proposed. Some observations [3] suggest that the surge is a cluster of a large number of fine jets with a diameter 1", apparently connected with the bright knots, or Ellerman bombs in the lower chromosphere. Likewise, rotational motions in kernel structures of the surges are observed. On the other hand the behaviour of Ellerman bombs in H-alpha [4] suggests their association with bipolar emerging magnetic flux. In [5] it was indicated that the nonlinear effects during small-scale magnetic field penetration in the lower chromosphere lead to temperature and density increase, sufficient to create a gas dynamic driving force for a surge-like jet.

One of the peculiarities in surge dynamics is the so-called braking or retarding effect. It means (see for example [3]) that the deceleration of the outward-moving plasma which takes place after the maximum velocity has been reached is stronger than when gravity alone is acting.

Using the above mentioned observational and theoretical results, we make an attempt at constructing a model of the surge phenomena

E. R. Priest and V. Krishan (eds.), Basic Plasma Processes on the Sun, 435–437.

explaining the braking effect in surge dynamics.

3. THE NATURE OF THE BRAKING FORCE

We adopt a surge model which represents the phenomenon as a cluster of surge-like jets, each generated by a temperature gradient on the top of a chromospheric bright point. Some hydrodynamical considerations (see for example [6]) lead to the idea that each surge-like jet could be treated as a buoyant vortex structure starting as a forced plume. The velocity profile

$$\dot{z}(\tau) = \Gamma(4\pi R)^{-1}\left[\ln\left(8R/2(\nu\tau)^{1/2}\right) - 1/4\right]$$

of such an isolate buoyant vortex ring structure is proportional to $\ln(\tau^{-1/2})$,
where

$$\tau = \int_0^t R\,(t')dt'/R\,(t),$$

(R is the radius of the ring) and differs significantly from the mean surge velocity curve obtained from observations. This suggest that the hydrodynamical effects alone cannot explain surge formation and dynamics, so that it is necessary to take into account some hydromagnetic effects, as well.

Some lab plasma experiments (see for example [7]) show that relatively stable plasma structures, called dynamically stable current filaments, can be formed. In the most general case these structures originate in toroidal vorticies and exibit proper longitudinal and azimuthal magnetic fields of approximately equal value. It is quite possible that in the magnetic flux tubes of the collimating external magnetic field, surge-like jets consist of similar MHD-structures which behave as plasmoids expelled by the action of the diamagnetic forces. In the electric field of the helix shaped cord of such MHD structures, electron acceleration is possible.

As for the braking force in the decelerated ascending phase, a simple explanation can be made if we take into account the fact that the effective size of the vortex core

$$\delta = 2(\nu\tau)^{1/2},$$

where ν is the kinetic viscosity, grows rapidly with the height z. Besides, from the most general hydrodynamic consideration it can be shown [8] that in a moving fluid with a mean velocity V of the velocity profile, pulsations will be generated when the pressure gradient dP/dy transverse to **V** is proportional to V^2. In [3] it was shown that the acting opposing force increases proportionally to the first or perhaps to the second power of the velocity of the surge.

If we assume that the surge-like jet plasma is moving in a magnetic flux tube (see the Figure), then the action of dP/dy will lead to the broadening and dipping of the tube and to the appearance of a small component **b** of the magnetic field, normal to **V**. This disturbance of the magnetic tube will correspond to a local electric curent source. The generated Lorentz force **F** = (1/c)**j** x **b** on the moving plasma will be directed in an opposite direction of **V** and will act as a braking force.

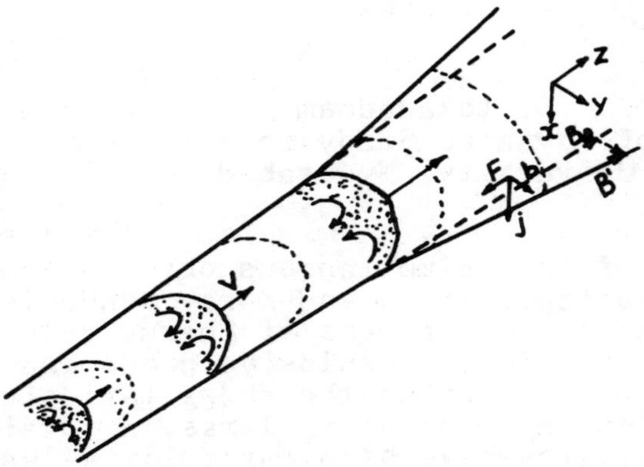

Figure. A sketch illustrating the action of the braking force on the surge plasma.

REFERENCES

1. Tandberg-Hanssen, E. (1974) Solar Prominences, D. Reidel Pbl. Co., Dordrecht.
2. Zirin, H. (1978) 'The ecology of prominences' in E. Jensen, P. Maltby and F. Q. Orrals (eds.), IAU Colloquium 44 'Physics of Solar Prominences', pp. 193-208.
3. Roy, J.-R. (1973) 'The dynamics of solar surges'. Solar Phys. 32, 139-151.
4. Zachariadis, Th., Alissandrakis, C. E., Banos, G. (1985) Lecture Notes in Physics 233, Springer-Verlag, pp 304-306.
5. Dermendjiev, V. N., Buyukliev, G. T. (1987) 'Magnetic turbulent diffusion and the problem of solar surge formation' in L. Hejna and N. Sobotka (eds.) The Sun, Proc. of 10th European reg. astron. meeting of the IAU, Praha, pp. 113-116.
6. Turner, J. S. (1973) Buoyancy Effects in Fluids, Cambridge University Press.
7. Komelkov, V. S., Skvortsov, Y., Tereshenko, V. N., Zerevitinov, S. S. (1962) 'A dynamical current filament' in H. Maecker (ed.) Proc. of the Fifth Int. Conf. of Ionization Phenomena in Gases, vol. II, North-Holland Publ. Co., Amsterdam, pp. 2191-2200.
8. Lamb, H. (1945) Hydrodynamics. Dover Publications, New York.

Observations of Energetic Electrons in Solar Flares

B. Lokanadham
Centre of Advanced Study in Astronomy
Osmania University, Hyderabad 500 007

A study of the simultaneous observations of solar flares in optical, radio and X-ray bands is important in understanding the process of energy release and particle acceleration in the explosive phenomena of solar flares. In order to determine the characteristics of such energetic electrons in solar flares, a total number of 50 two-ribbon flares have been carefully selected during the period 1979-89 having simultaneous observational data in the optical, X-ray and radio bands.

About 60 % of these solar flares have been found to be associated with high energy proton events of $\geqslant$10MeV. From the analyses of their optical Hα, X-ray and microwave radio data it has been found that:

a) two-ribbon flares are the most powerful sources of particle acceleration.

b) maximum particle acceleration occurs in the early impulsive phase of solar flares.

c) high-resolution Hα, radio and X-ray recordings of a few flares showed good correspondence in their time profiles within one second.

d) radio frequency spectra of the microwave bursts are characterised by a U shape and increases in intensity in the transition from 3 GHz to higher frequencies.

e) the hard X-ray spectra are power law spectra with cut off energies $<$10 KeV with γ= 4-5.

From these observations it is concluded that the two-ribbon flare process is more favourable characteristic for particle acceleration in most high energy events. The most rapid acceleration occurs during the initial impulsive phase of the flare with continued acceleration and injection by flaring-loops originating due to magnetic reconnections occurring throughout the life of a solar flare.

E. R. Priest and V. Krishan (eds.), Basic Plasma Processes on the Sun, 438.

ROLE OF BEAM FOIL SPECTROSCOPY IN UNDERSTANDING BASIC PLASMA PROCESSES ON THE SUN

G.KRISHNAMURTY, P.MEENAKSHI RAJA RAO, P.SARSWATHY AND B.N.RAJA SEKHAR
Spectroscopy division
Bhabha Atomic Research Centre
Trombay, Bombay- 400085
INDIA.

Beam-Foil spectroscopy(BFS) has proved to be a valuable technique for the determination of radiative lifetimes of excited atomic levels leading to the evaluation of the transition probabilities. The time- resolved nature of the decay process in a collisionless environment is a unique characterstic of the beam-foil light source. The relevance of BFS to astrophysics comes from the importance of radiative transition probabilities in the quantitative analysis of optical spectra. Stellar abundances are obtained from the intensity of a spectral line which essentially is a product of the abundance of the element in the source and the probability of the transition.Thus the evaluation of accurate values of transition probabilities contribute significantly to stellar abundance analysis.

The basic principle of BFS is that if a monoenergetic beam of fast ions is directed through a thin solid foil (often made of carbon having a typical thickness of 5-20$\mu g/cm^2$), the ions undergo excitation as well as further ionization or electron capture as a result of inelastic collisions between the atoms of the foil and fast ions. The emerging particles from the excited states decay spontaneously downstream of the foil. The ions are practically monoenergetic after the foil and their velocity v is known from the beam energy. In view of the excellent time resolution,each point x from the foil corresponds to a known time t after the excitataion. Thus the lifetimes τ of excited levels can be measured simply by recording the spectral line intensities I(x) as a function of x. If the excited level i decays only to one final level k, then the lifetime τ_i is simply the inverse of transition probability A_{ik}. If several final states are possible then:

$$\tau_i^{-1} = \sum_k A_{ik} \qquad \text{...........(1).}$$

In such a case the individual transition probabilities can be obtained from the branching ratios (1). One of the most serious problems that affect lifetime measurement is cascading which can be corrected by multiexponential program. The absorption oscillator strength (f-value) is related to transition probability as follows:

E. R. Priest and V. Krishan (eds.), Basic Plasma Processes on the Sun, 439–440.

$$f = 1.499\ \lambda^2\ (g_i/g_k)\ A_{ik}\ \ ....(2).$$

where λ is the wavelength (Å) of the transition, g_i and g_k are the statistical weights of levels i and k.

The spectroscopy division of BARC has developed the basic instrumentation for BFS experiments making use of the availble accelarators in BARC and TIFR. It consists of a half metre monochromator in a Czerny-Turner mount and one metre vacuum monochromator in a Seya-Namioka mount which was designed and fabricated indigenously(2) for carrying out the experiments in the Visible,UV and VUV region. The emitted light from the beam is viewed by the monochromators in the perpendicular direction to the beam. The signals from the monochromator exit slit are detected by the photomultiplier EMI 9789 QB using a peltier cooled housing. The output of the photomultiplier is analysed by a photon counting system consisting of a spectroscopy amplifier,discriminator and dual counter timer. In the VUV region a sodium salicilate coated photomultplier is used. The capability of this experimental set up was already demonstrated by investigating the beam-foil spectrum of the ions of oxygen produced in the 2 MV Tandem accelarator indigenously built in BARC. The details of the results can be seen in reference(3).

There is a strong local maximum of abundances of elements in the immediate neighbourhood of iron with iron itself at the peak in solar plasmas. Thus, these elements need lot of investigation,so that the data will be very valuable for understanding of basic plasma processes in the Sun. We wish to pursue the BFS studies further keeping the intersts of astrophysics in mind.

REFERENCES

1. Bashkin,S.(1976)"Application to Astrophysics" Beam-Foil Spectroscopy in **Bashkin(Ed.) ,Topics in current physics, Springer Verlag pp.179-207**
2. Krishnamurty,G.,Sarma,Y.A.,Meenakshi Raja Rao,P.,and Bhattacharya,S.S.(1983) "Design and Fabrication of a Vacuum ultraviolet monochromator using Seya-Namioka mount",BARC-1218.
3. Mythili,R.,Singh,P.,Bhattacharya,P.K.,Parui,A.,Betigeri,M.G., Meenakshi Raja Rao,p.,Krishnamurty,G.,and Raju,V.S.,(1983) "Measurement of excitation functions and radiative lifetimes of oxygen ions" Z-phys D.7,311-318.

ROLE OF PLASMA SPECTROSCOPY IN UNDERSTANDING PLASMA PROCESSES ON THE SUN

P.Meenakshi Raja Rao, P.Sarswathy, B.N.Raja Sekhar and G.Krishnamurty
Spectroscopy Division
Bhabha Atomic Research Centre
Trombay, Bombay- 400085
INDIA

Emission spectroscopic methods are very useful in determining the plasma parameters such as electron density, electron temperature, chemical abundances and energy levels of atoms and ions. A knowledge of the above mentioned parameters and collision cross sections provides an insight into various plasma processes on the Sun. As one passes from photosphere to chromosphere and corona the temperature as well as the electron density changes drastically (Te $\sim$4500 - $2X10^6$ ^{0}K; $n_e \sim 10^8 - 10^{13}$ cm^{-3}) (1). Hence the solar spectrum, excited by different mechanisms and different equilibrium conditions, extends from vacuum ultraviolet to visible and infrared regions. For example the spectrum in the region between 3000-1300Å is produced by the upper photosphere and lower chromosphere. In this region the temperature is in the range of 6000-10000 ^{0}K. This region is characterised by several emission and absorption lines superimposed over continuum. Below 1600Å consists of emissions from highly ionised atomic species originating from chromosphere and corona (2). A correct interpretation of the spectral features is possible only after understanding the influence of various factors on spectral line shapes and intensities. They are 1) damping by collisions with neutral atoms; 2) collisions by charged particles leading to linear and quadratic Stark effects on atomic lines of hydrogen and helium; 3) thermal Doppler broadening 4) Doppler broadening or shift due to microturbulent velocity field.

In this context a detailed analyses of spectra (including line shapes and intensities) generated in laboratory plasmas under controlled conditions and evaluation of plasma parameters help a great deal in understanding astrophysical plasmas.

BARC has an ongoing programme in this field (3). The plasma sources used were Z-pinch devices such as vacuum spark and plasma focus developed in our research centre. The spectra from different zones of plasma were photographed in the visible region (4000-6000Å) on a indigenously built 1M Czerny-Turner spectrograph. In the case of vacuum spark the plasma was generated in helium atmosphere and plasma temperature was evaluated which was 1eV. From extensive work done on spectral emissions from 2KJ plasma focus device, where the plasma was generated in hydrogen atmosphere, the following important observations were made.

E. R. Priest and V. Krishan (eds.), Basic Plasma Processes on the Sun, 441–442.

Observations

1. The instrumental width evaluated from hollow cathode spectra was 0.5Å
2. The spectral lines due to the hydrogen atom were very broad ($\Delta\Lambda_{1/2}$ for H_β and H_γ lines = 20-40Å)

3. The spectral lines due to heavier elements like copper and zinc originating from electrode material also were broadened but to a lesser extent. ($\Delta\Lambda_{12}$=1.0 - 5.5 Å)

4. No anomalies were observed in the intensity patterns.

These observations were made use of to evaluate the following plasma parameters.

1. Using copper as a thermometric species the plasma temperature was evaluated by the line intensity ratio method. ($T = 9800^0$ K in the focus region). At this temperature the Doppler width is 0.3Å for hydrogen lines0.05Å for Cu and Zn lines. Thus the contribution from instrumental and Doppler broadening to the observed line widths was not significant.

2. The atomic lines of hydrogen were subject to the linear Stark effect due to the high electron density environment. Hence from the observed half widths the electron density was evaluated. (n_e= $9.9X10^{16}$ cm^{-3} in the focus region.)

3. CuI and ZnI lines were influenced by quadratic Stark effect. From the observed half widths the Lorentzian component was evaluated and Stark coeffiecents (C_4) for the quadratic Stark effect were evaluated.

REFERENCES:

1. Engvold.O and Hange.O (1970) "Solar abundance determination"., Nuclear instruments and methods, 90, 351-362.

2. Heroux.L (1976) "Applications of beam-foil spectroscopy to solar ultraviolet emiision spectrum", Bashkin, S(Ed). "Beam-foil Spectroscopy", springer Verlag, 193-208.

3. Meenakshi Raja Rao P Saraswathy.P, Krishnamurty, G., Rout, R.K., Auluck, S.K.H., Shyam, A., Kulkarni, I.V., Oza, H.D., (1989) "Line broadening studies in low energy plasma Focus", Pramana-J.Phys.32, 627-639 and references therein.

DENSITY DIAGNOSTICS OF SOLAR EMISSION LINES FROM THE NITROGEN-LIKE Mg VI ION

P.K.Raju and R.Vasundhara
Indian Institute of Astrophysics
Bangalore-560034, India

Abstract The variation of theoretical line intensity ratios with electron density for Mg VI ion are presented. This study indicates that the line intensity ratios for Mg VI can be used to infer electron density for solar plasma.

Introduction

The solar Extreme Ultraviolet spectrum is rich with emission lines in the range 200 Å to 2000 Å. These lines are representative of the solar, chromosphere-corona transition region and the corona. The analysis of such a spectrum would enable us to know the physical parameters, like electron density and temperature, within the emission regions. Knowledge of these parameters are essential to understand the physical processes within the solar plasma. In recent years, an important result from the study of emission line solar spectrum is the development of techniques which allow direct determination of the physical parameters independently of the detailed model of the solar atmosphere. One such technique involves the effect of electron density on line intensities. In this respect, lines belonging to nitrogen-like ions have received little attention as compared to lines of ions belonging to other isoelectronic sequences. In this brief study we have considered the feasibility of using emission lines from Mg VI ion for electron density determination. The detailed investigation on the nitrogen-like ions Ne IV, Mg VI and Al VII would be published elsewhere. The results for the ions Si VIII and S X and the references to the earlier literature are given in the paper by Dwivedi and Raju (1988).

Discussion

The electron density determination, by using emission lines of nitrogen-like ions, becomes possible because the lines are excited by electron collisions with the metastable levels of the ground term. The collisional and radiative de-excitation rates among these levels would become comparable for appropriate electron densities. In such cases the population for these levels and in turn the line intensity ratios become sensitive to density. Fig.1 for Mg VI ion illustrates this point. The schematic energy level diagram

E. R. Priest and V. Krishan (eds.), Basic Plasma Processes on the Sun, 443–444.

for nitrogen-like ions is shown in Fig.2.

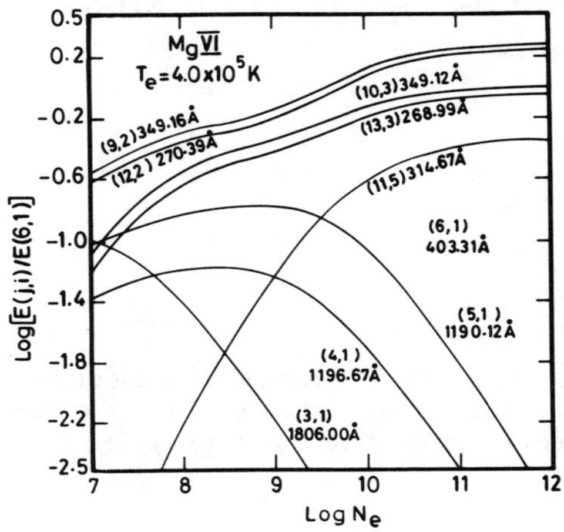

Fig.1. Line intensity ratios E(j,i)/ E(6,1) as a function of electron density N_e. T_e is the temperature for maximum relative ion abundance for Mg VI.

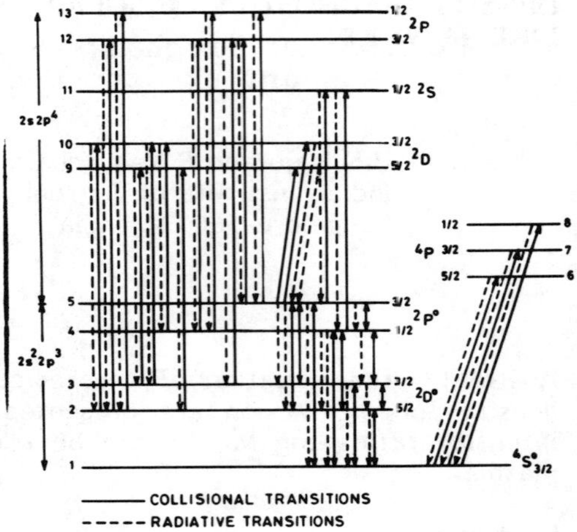

Fig.2. Energy level scheme for nitrogen like ions.

Conclusion

The line intensity ratios for Mg VI, as discussed here, should be useful to infer electron densities in the emission regions of the solar atmosphere.

Acknowledgements

The authors are grateful to the referee for valuable comments and suggestions.

References

Dwivedi,B.N. and Raju,P.K. (1988) 'Density diagnostics of solar emission lines from introgen-like ions', Adv. Space Res. 8, 11 (179)-11 (183).

Analysis of Prognoz - 9 Solar Flare Hard X-ray Data-support for the Non-thermal thick target Model

R.R.Rausaria, Ranjana Bakaya & P.N.Khosa
Department of Physics, Regional Engineering College, Hazratbal, SRINAGAR - 190 006

Solar flare hard X-ray data obtained by Prognoz-9 spacecraft (Abrosimov et al 1988) in the energy range 10-200 keV are analysed. In examples of events which we consider here, high energy X-ray pulses appear earlier than low energy ones, which is contrary to many other events where the low energy X-ray peak emission takes place earlier. The variation of the spectral index was dynamical.

The evolution of energy and angular distributions of electrons at different depths in the solar atmosphere have been studied by combining analytically treated small-angle scattering of electrons with a large angle monte carlo treatment (Haug et al 1985) for initially non-thermal and thermal distributions. Using these distributions and a Sauter bremsstrahlung cross-section, we have computed the energy spectra of X-rays for different column densities and photon energies. The change in the spectral indices of the generated X-rays are graphed in Figure (1) at different column densities (10^{21} - 10^{22} cm^{-2}) for a non-thermal electron distribution. For the case of a thermal model the change in spectral indices is plotted in Figure (2) for different temperatures Nocera, et al 1985. In this case the change in the spectral indices are not as dynamical as in the case of non-thermal model. The spectra are softer for all the temperatures considered. Various curves of Figure (2) can be taken as representative of the variation of the photon spectral index at different depths in the solar atmosphere. For an electron source temperature less than 50 keV, the variation in the photon spectral index will be somewhat greater. However, for high energy photon emission (100-200 keV) such temperatures indicated on the curves will not be possible, which is contrary to observations. Therefore, it may be concluded that non-thermal processes play an important role in hard X-ray characteristics.

E. R. Priest and V. Krishan (eds.), Basic Plasma Processes on the Sun, 445–447.

References:

Abrosimov, A.T. et al., : 1988, "Catelogue of solar hard X-rays bursts as detected on Prognoz-9 satellite", Soviet Geophysical Committee, Academy of Sciences of the USSR, World Data Centre-B.

Haug, E., Elwert, G. and Rausaria, R.R. : 1985, Astron Astrophys. 146, 159.

Nocera, L., Skrynnikov, Yu I and Somov, B.V. : 1985 Solar Phys 43, 415.

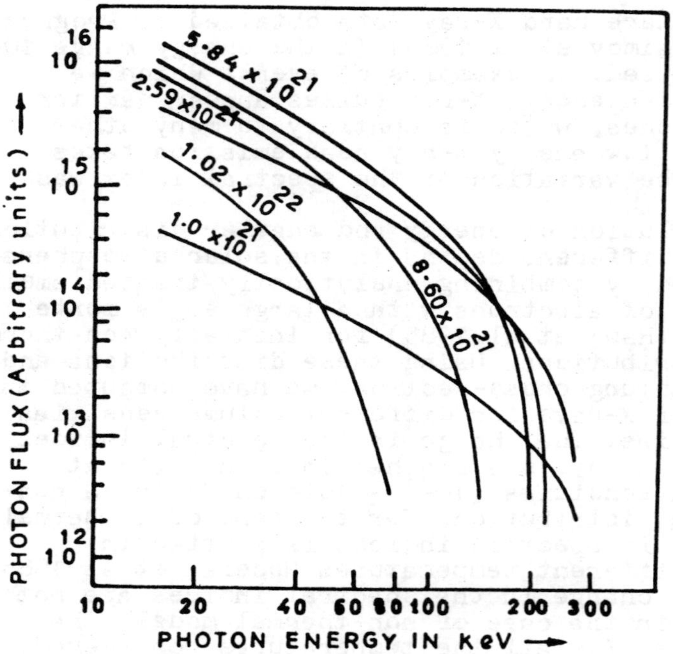

Fig.1. Variation of X-ray flux at different column densities with photon energy for incident electron energy 300 keV and consequent changes in spectral indices.

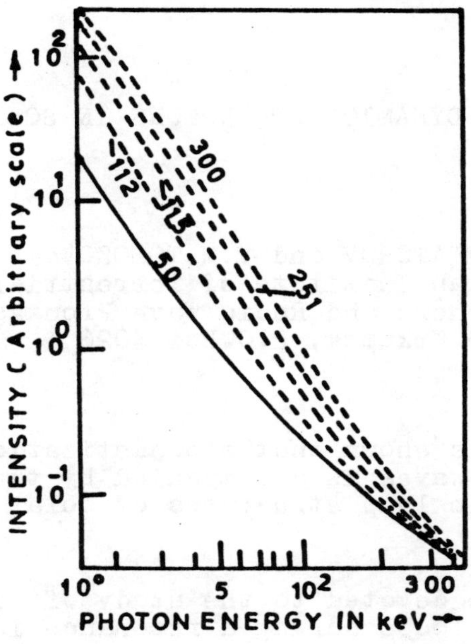

Fig.2. Variation of X-ray intensity and spectral indices for initially thermal electron distribution.

ON STOCHASTIC DYNAMICS OF PROTONS IN SOLAR MAGNETIC LOOPS

V.N.SENATOROV and V.M.TOMOZOV
Siberian Institute of Terrestrial Magnetism,
Ionosphere and Radio Wave Propagation (SibIZMIR)
664033 Irkutsk, P.O.Box 4026
USSR

ABSTRACT. It is shown that stochastization of particles due to Alfven waves is accompanied by their effective acceleration in loop structures of solar magnetic fields.

This report is devoted to the study of ion acceleration in solar coronal loops during a resonance interaction of energetic particles (protons) with Alfven waves in terms of a theory presented by Zaslavsky and Sagdeev (1988). The particle stochastization condition in a plane-wave field (Chirikov's criterion) (Zaslavsky and Sagdeev, 1988) in our problem can be written as: $K = 1/\tau^2\Omega^2 \gg 1$, where $\tau = (m/eEk_z)^{1/2}$ is the period of phase oscillations of a particle with mass m and charge e in the wave's potential well, E is the wave's electric field, and $\Omega \sim |\omega - \omega_{Hi}A/v_z|$. Here ω and A are the frequency and velocity of Alfven waves, respectively, and ω_{Hi} and v_z are the gyrofrequency and the proton velocity projected onto an external magnetic field, respectively. The boundary of the stochasticity region can be determined from the relationship $K \sim 1$. The largest width of the stochastic layer is formed near the separatrix separating trapped and transient particles in the MHD wave field (Chirikov, 1978).

The physics behind the appearance of the stochasticity effect is associated with the presence of special points on the phase plane such as points of self-intersection of the separatrix, near which the integrals of motion are violated. A particle, trapped into a potential well, is subjected to a stochastic instability under the action of any other wave that plays the role of a disturbance. The motion of the particle in a stochastic layer of the wave becomes similar to Brownian motion, which leads to the fact that the particle's energy, on the average, grows with the time as:

E. R. Priest and V. Krishan (eds.), Basic Plasma Processes on the Sun, 448–449.

$$\mathcal{E} \sim \delta\mathcal{E}_0 (T/\tau)^{2/3} \quad (1)$$

where $\delta\mathcal{E}_0$ is the variation of the particle's energy during a period τ, and T is acceleration time (Zaslavsky and Sagdeev, 1988).

Qualitatively, the physics of the acceleration mechanism is thus: the variation in energy of the particle associated with the wave's electric field, is $\delta\mathcal{E}_0 \sim eE\ell_{eff}$, where ℓ_{eff} is the effective length of resonance interaction between the particle and the wave. It is clear that ℓ_{eff}, in the case of the usual Landau resonance and in the presence of stochastization, are different, and ℓ_{eff} in the latter case is significantly larger, which leads, accordingly, to a greater growth of the particle energy.

Based on formula (1) we shall make an estimate of the proton energy gain for the solar flare of 7 June 1980 described by Smith (1989), who, as in our case, treats the flare coronal magnetic loop as a semicircular magnetic trap filled with MeV-ions and MHD waves with the wave energy density $W_A \sim 1.5$ erg$\cdot$cm^{-3} comparable with the thermal plasma energy density for $n \sim 5\cdot10^{10}$ cm^{-3} and $T_i \sim 1.6\cdot10^6$K. The Alfven wavelength $\lambda \sim 10^7$ cm, $\omega_{Hi} \gg 10^4$ s^{-1}, $A \sim 10^8$ cm$\cdot$s^{-1}, and $v_z \sim 10^9$ cm$\cdot$s^{-1}. By taking for estimations the Alfven wave amplitude of $\sim 10^{-1} - 10^{-2}$G and the ratio $T/\tau \sim$ $\sim 5 - 30$, we find that the growth of the MeV-ion energy makes up from 3 to 10 MeV. (The value of the T/τ-ratio was obtained by estimating τ and the lifetime of protons in the coronal loop). The share of the particles heated during stochastization in the coronal loop usually is $\lesssim 10\%$.

Unlike Smith (1989), the more effective proton acceleration discussed here is connected mainly with stochastic dynamics of protons (the mean energy and the share of accelerated protons is greater, owing to the contribution of transient particles). Therefore, this acceleration mechanism is of interest for interpretation of the gamma-radiation of solar flares.

REFERENCES

Chirikov, B.V.: 1978, Fizika plazmy 4, 3, 521-541.

Smith, D.F. and Brecht, S.H.: 1989, Astrophys.J. Sept. 13 (in press).

Zaslavsky, G.M. and Sagdeev, R.Z.: 1988, Introduction to nonlinear physics. Moscow: Nauka, 368 p. (in Russian).

RELATIONSHIP OF CORONAL MASS EJECTION EVENTS WITH SOLAR FLARES AND CORONAL HOLES

V K Verma

U.P. STATE OBSERVATORY, NAINI TAL - 263 129, INDIA

ABSTRACT : We present an analysis of the relationship between coronal mass ejection (CME) events with solar flares and coronal holes. Out of 79 CME events whose location and span are known, 10 CME event circles contain solar flares. On the other hand out of 71 CME events (selected for the study with coronal holes) 40 CME event circles contain coronal holes. We have examined the above associations by calculating the Yule's coefficients of association and found that the CME events and solar flares are statistically not associated while CME events and coronal holes are statistically associated. This does not necessarily indicate that CMEs come from coronal holes as claimed by Hewish and Bravo (1986).

OBSERVATIONAL DATA, ANALYSIS, RESULTS AND DISCUSSIONS

To study the relationship of the CME's with the solar flares and coronal holes, the CME data has been taken from a paper by Sheeley *et al* (1984). The solar flares related to CME's observed during 1979-1982 were identified by Sheeley *et al* (1984) and by Robinson *et al* (1986). The coronal hole locations (10830 A) for the above period are taken from a paper by Stewart *et al* (1985).

For elucidating the study of the temporal and spatial relation between CMEs and solar flares, and CMEs and coronal holes, we have drawn a circle of radius equal to the span of CMEs, say ±10, (i.e. a circle of radius 10 degrees) location in latitude say S25, (i.e. 25 degree south of the solar equator). If the CME circle drawn above contains a solar flare or coronal hole we infer that the CME event and coronal hole or solar flare are related to each other. In Figure 1 we have shown a synoptic map of the Sun for the period May 23 - June

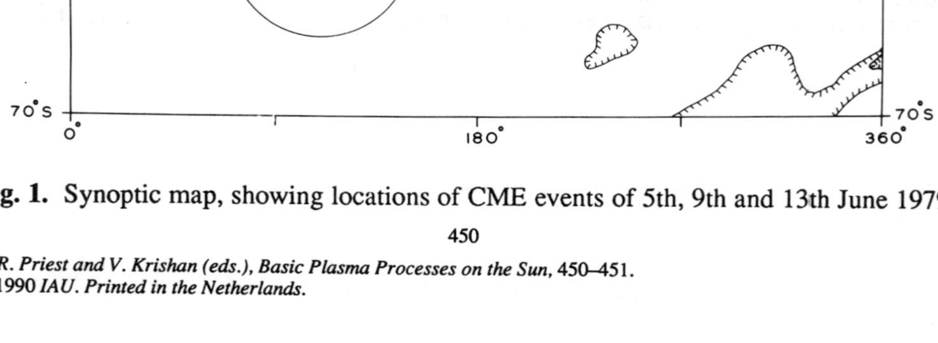

Fig. 1. Synoptic map, showing locations of CME events of 5th, 9th and 13th June 1979.

E. R. Priest and V. Krishan (eds.), Basic Plasma Processes on the Sun, 450–451.

20, 1979 (CR-1682). In Figure 1 we have shown CME circles for 3 events viz. 5 June 1979, 9 June 1979 and 13 June 1979. As it is clear from Figure 1 that 5th June 1979 CME circle contains a coronal hole while the flare temporarily associated with this CME as identified by Sheeley *et al* (1984) and by Robinson *et al* (1986) is located outside the CME circle. The location of the associated flares is shown in Figure 1 by a cross mark inside a small circle. From Figure 1 it is clear that the 5th June 1979 CME event is associated with the coronal hole and associated with the solar flare. The 9th and 13th June 1979 CMEs are associated neither with coronal holes nor solar flares. Several figures similar to Figure 1 are drawn to carry out the analysis.

Out of 79 CMEs whose location and span are known 10 CME event circles contain solar flares. On the other hand out of 71 CMEs (selected for the study with coronal holes), 40 CMEs are found to be related with coronal holes. The detailed method for selecting events are described in a paper by Verma (1989). We have examined the above association by calculating the Yules coefficients of association and found that CMEs and solar flares are not statistically associated while CMEs and coronal holes are associated.

The overall aim of the present study is to understand better the origin of CMEs. Earlier in a detailed study of three interplanetary events Hewish *et al* (1985) found that mid-latitude coronal holes lay near the sources of the disturbances on the Sun and it was suggested that the eruption of a short-lived stream, or the acceleration of an existing stream from the hole, provided an explanation of the disturbances. CMEs originate from coronal holes. However, recent coronagraph observations from the Solar Maximum Mission have shown clearly that CMEs occur when pre-existing closed structures such as coronal arcades or helmet streamers lose equilibrium and erupt (Hundhausen *et al*, 1984; Low, 1986; Priest, 1988). The association with coronal holes can therefore be explained if the opening of the field during the CME creates a coronal hole or if it is the closed region to one side of a coronal hole that erupts.

REFERENCES

Hewish, A, Bravo, S : 1986, Solar Phys. **106**, 185.
Hewish, A, Tappin, S J, Gapper, G R : 1985, Nature, **314**, 137.
Hundhausen, A J *et al* in Solar Terrestrial Physics : Present and Future (ed D Butler and K Papadopoulos) NASA CP-2439, Chapter 6.
Low, B C : 1986, Highlights in Astron. **7**, 243.
Priest, E R : 1988, Astrophys. J. **328**, 848.
Robinson, R D *et al* : 1986, Solar Phys. **105**, 239.
Sheeley, Jr N R *et al* : 1984, Astrophys. J. **279**, 839.
Stewart, R T *et al* : 1985, Proc. Astron. Soc. Australia, **6**, 231.
Verma, V K : 1989, Astron. Astrophys, to be submitted.

ION CYCLOTRON INSTABILITY AND ELECTRON ACCELERATION IN CORONAL MAGNETIC FLUX TUBES *

Xu Min-jian, Li Ding-yi**, Wang De-yu***, Tsai Shih-tung
Association for Plasma Studied of China, Chinese Astronomical Society, Institute of Physics Academia Sinica, Beijing 100080, China

Part of the energy of low solar corona may be released and converted to plasma energy that may be the dominant source of solar flares [1]. Recently, many physical mechanisms have been proposed to accelerate ions and electrons [1]–[3]. In the magnetic mirror device of fusion research , ion cyclotron instability and the energetic ion losses have been discovered and the plasma becomes electron rich.

In this paper, we discuss the ion cyclotron instability and electron acceleration in coronal magnetic flux tubes, which are very similar to what happened in magnetic mirror devices .

These accelerated electrons and ions are nonthermal particles. They are injected into magnetc flux tubes which form a magnetic mirror configuration during the time of solar flare activity in the lower corona.

The plasmas in the Magnetic flux tubes are composed by a low-temperature thermal background and a loss cone distribution plasma with larger perpendicular kinetic energy.

The free energy of nonthermal plasma can excite quasi-electrostatic cyclotron instabilities. Electrostatic cyclotron waves have a rather high frequencies ($w \approx l w_{cs}$) and a perpendicula wavelength $\lambda_{\perp} < a_s$, where l represents the harmonic number, w_{cs} and a_s are the cyclotron frequency and gyroradius of the particles of species s.

The threshold of the cyclotron instability can be estimated as:

$$w_{pe}^2 > l^2 w_{cs}^2 \tag{1}$$

where w_{pe} is the electron plasma frequency. The range of density is given by equation (1). Besider equiation (1) there exist other two necessary conditions for cyclotron instability, which can be obtained:

---

* Support by the National Science Foundation of China Grant.

** Zhenjiang Shipbuilding Institute, Zhenjian, Jiangsu

*** Purple Mountain Observatory, Academia Sinica, Nanjing, Jiangsu

E. R. Priest and V. Krishan (eds.), Basic Plasma Processes on the Sun, 452–453.

$$a_s > 2\lambda_\perp^2 \tag{2}$$

and

$$1 - q_s > (1 - q_s)_{min} \tag{3}$$

where $q_s = n_{0s}/(n_{0s}+n_{fs})$, n_{0s} and n_{fs} are the densities of the thermal background plasma and the nonthermal plasma respectively, n_{fs}/n_{0s} is much less than 1. Equation(2) is consistent with the condition described in paper[6] and [7].

Our numerical results show:

(1) The higher harmonics, the instability occurs easier than the lower harmonics.

(2) The ion cyclotron mode is easier to be excited than the electron cyclotron mode because the density of energetic ions is much larger than that of energetic electron.

After the ion cyclotron instability takes place, an evolution of the velocity distribution of nonthermal particles occurs [8].Ions diffuse quickly into the loss cone and lost, therefore the electric potential of the plasma becomes negative. Electrons are continuously accelerated by electric field and lost from ends of the magnetic flux tube.

In the process that the growth rates approach saturation, except the very short time at the begining, the electric potential increases slowly. The energy of the accelerated electron can reach several hundred keV. These electrons can produce a microwave bursts, with time scale in the order of second, and radiate hard X rays. The mechanism of electron acceleration can also be applied to the physical processes of magnetic flux tubes of the lower corona.

References

[1] Vlahos, L. etal., Solar Maximum Mission Workshop Report of Team B on Particle Acceleration in Flare,1984,p.47.

[2] Wang Deyu., Science Bulletin 27(16), 995(1982)

[3] Toptyghin, I. N., Space Sci. Rev., 26(1) 157(1980)

[4] Cordey, J. G. etal., Proc. 3rd Internation Conf. of Plasma Phys. and Controlled Nuclear Fusion Research, Noyosibirsk, Vol. 2. 1968, p.267.

[5] Damm, C.C. etal., ibid. p. 253.

[6] Hortton. C. W. etal., Phys. Fluids, 14(10), 2019(1971).

[7] Post, R. F. etal., Phys. Fluids, 9(4), 730(1966).

[8] Wu. C. S. etal., Astrophys. J., 248(1),384(1981).

MODELLING A SOLAR FLARE FROM X-UV AND RADIO OBSERVATIONS

F. CHIUDERI DRAGO
Department of Astronomy and Space Science, University of Florence
Largo Enrico Fermi 5 - 50125 Florence, Italy

B.C. MONSIGNORI FOSSI
Arcetri Astrophysical Observatory
Largo Enrico Fermi 5 - 50125 Florence, Italy

A slow evolving flaring loop has been observed by the UVSP, XRP and HXIS instruments on board SMM on June 10, 1980. Simultaneous radio observations from Toyokawa (Japan) are also available. The former instruments have an angular resolution ranging from 3" to 30" which allows the determination of the loop structure. The XRP:FCS rasters permit to analyze the flare only during three time intervals of about 10 minutes each.

It appears that the observations cannot be accounted for by a single loop although with a variable temperature and pressure. The presence of a hot and less dense isothermal plasma is necessary to explain the harder emission (HXIS).

X-UV data are very well reproduced using a 2 component model consisting of a loop of constant section in which the differential emission measure varies as power law of the temperature and a hot isothermal plasma ($T \cong 2\ 10^7$ K) component. A full model of the loop is deduced from continuity, momentum and perfect gas law equations together with the best fit of the differential emission measure. Considering the very slow flux variation, a stationary state during each of the three time intervals is assumed. The temperature, density and velocity profiles are deduced (Chiuderi Drago and Monsignori Fossi, 1990).

The model is then checked with radio data. The comparison between computed and observed fluxes shows that they agree within a factor 2 at all frequencies.

An estimate of the total energy input in the hot component and of the energy deposition along the loop is also given. In this latter the heating function can be approximated with a T^{-2} decay above $T \cong 2.5\ 10^5$ K, while, at lower temperature, the heating presents some fluctuations due to the trend of the radiative losses. The total energy input in the loop increases from $2.8\ 10^{27}$ erg s^{-1} to $7.5\ 10^{27}$ erg s^{-1} from the beginning to the maximum of the flare.

A comparison with the results relative to the hot component shows that the amount of energy that must be supplied to this latter is negligible with respect to that necessary to the loop.

REFERENCES

Chiuderi Drago, F. and Monsignori Fossi, B.C.: 1990, in preparation.

E. R. Priest and V. Krishan (eds.), Basic Plasma Processes on the Sun, 454.

IX. SOLAR RADIO EMISSION

Millimeter and Microwave Activity of the Sun

M. R. Kundu and S. M. White
Astronomy Program
University of Maryland, College Park, MD

Abstract

The emission of solar flares at millimeter wavelengths is of great interest both in its own right and because it is generated by the energetic electrons which also emit gamma rays. Since high-resolution imaging at gamma-ray energies is not presently possible, millimeter observations can act as a substitute. Except for that class of flares known as gamma-ray flares the millimetric emission is optically thin. It can be used as a powerful diagnostic of the energy distribution of electrons in solar flares and its evolution, and of the magnetic field. We have carried out high-spatial-resolution millimeter observations of solar flares this year using the Berkeley-Illinois-Maryland Array (BIMA), and report on the preliminary results in this paper (Kundu et al 1990; White et al 1990). We also report some recent results obtained from multifrequency observations using the VLA (White et al 1990).

Millimeter Bursts

Most millimeter-wave observations of solar flares have been hampered both by poor spatial resolution (because of the lack of synthesis interferometers) and by poor sensitivity (since the flux from the Sun's thermal emission is so high at millimeter wavelengths). The number of reported observations of bursts at millimeter wavelengths is relatively small, and there are none for which true imaging data have been reported.

The BIMA array, presently consisting of three antennas but being expanded to six to produce an aperture synthesis imaging instrument with 15 baselines, has a spatial resolution of $\gtrsim$ 1" arc, and time resolution as good as 0.3 sec. In its present configuration there are too few baselines to image rapidly time-variable phenomena on the Sun, but the range of spacings available at BIMA (2" - 60") is ideal for the study of flare sources; with data from three baselines we can determine the spatial scales involved and study the time evolution at high sensitivity and resolution. If good phase information is available, we can in principle obtain positions of burst sources.

E. R. Priest and V. Krishan (eds.), Basic Plasma Processes on the Sun, 457–465.

In March 1989 an extremely productive solar active region crossed the disk of the Sun and produced numerous large flares. At short notice the array was used to track this active region during daylight hours. The array was in a wide configuration with interferometer fringe spacings from 2" - 5", and due to maintenance work only one baseline was available much of the time. Using sidereal observing programs the best available time resolution was 10 seconds. We observed over 15 flares at ~ 86 GHz during the observing period, ranging from sub-flares to the largest class of flares (GOES soft-X-ray class "X"). Examples of these are shown in Figure 1.

Our sensitivity is about 0.01 - 0.02 sfu. Note that the patrol telescopes in Bern and Nobeyama presently operating at these wavelengths have sensitivities of 10 sfu at best. Thus the BIMA observations are roughly 1000 times more sensitive, and provide spatial scale and phase information as well. Subsequently for solar observations we were able to use a much better time resolution available (0.32 seconds) and in addition obtain total power information to supplement the correlated amplitudes. The high-time-resolution mode was used during an international campaign in June of 1989, and again a number of flares were detected. Examples of these are given in Figure 2.

We present the time profiles of several bursts during the March period in Fig. 1. The vertical axis in this figure is in solar flux units, while the horizontal axis is in UT hours. The time axes are all on the same scale. This figure clearly illustrates that while most of the bursts were impulsive and short-lived, we also saw a number of longer duration bursts.

All the bursts are fairly weak, with none exceeding 1 sfu of correlated amplitude. A number have rise times less than the available resolution of 10 seconds. Due to the short fringe spacing (generally about 4", i.e. sensitive only to structures which are smaller than 2"), we believe that in a number of these bursts the correlated flux is well below the total power radiated. This is particularly true for the longer duration bursts, which generally have larger source sizes, and we believe that the spiky profiles of the longer bursts are evidence that they are over-resolved.

On occasions when we were able to use three baselines we found that sometimes the shortest baseline was seeing much greater flux than the longer baselines, and therefore the spatial scale of the burst is believed to be longer than the largest resolution available (about 2" here). However, there are bursts for which there is evidence for a complicated asymmetry, such that the shortest baseline does not see the largest flux at all times.

The flares shown in Fig. 2(a) and Fig. 2(b) (White et al 1990) were observed on June 23 and June 24, 1989 using .32 second time resolution. The vertical axis is in arbitrary units (the peak flux is about 1 sfu), and the horizontal axis is in UT seconds. The time profile is that of a simple impulsive microwave burst. The rise time is resolved: in the 5.7 second duration of the rise (start-to-peak) are some 19 data points. The decay appears to be exponential. The solid line on the figure is the correlated flux (good phase measurements were obtained but are not shown here). The smooth profile suggests that the source structure is very

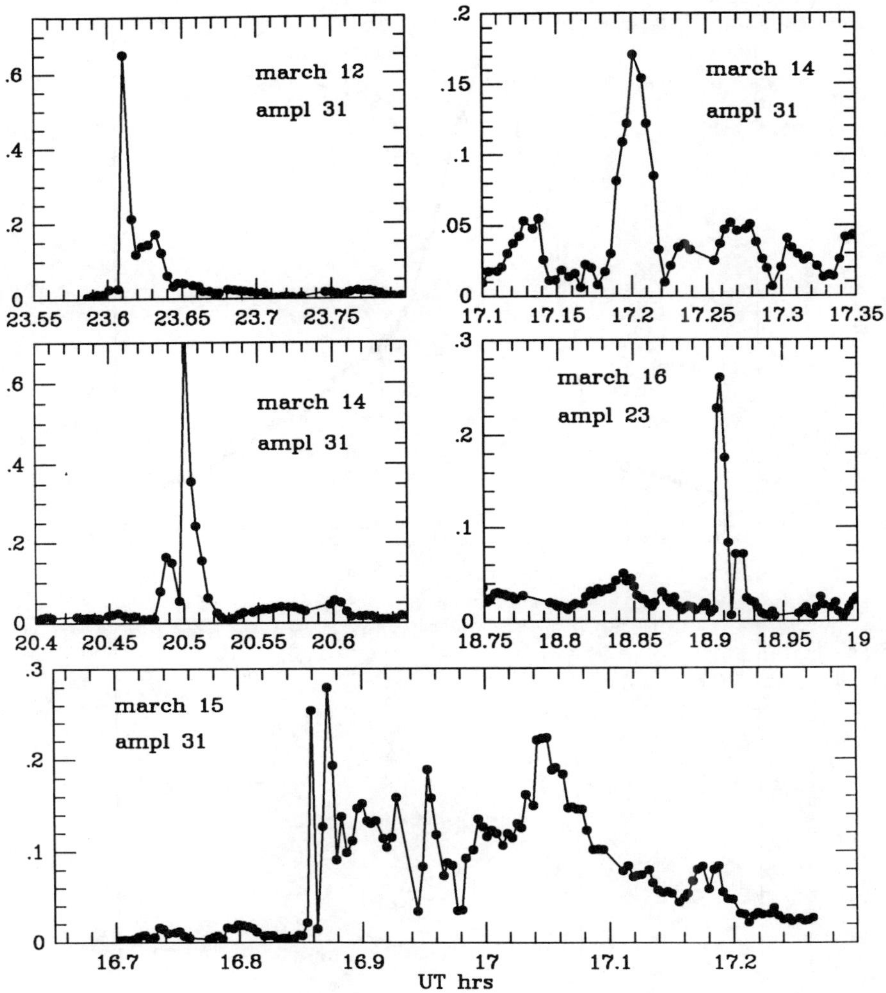

Figure 1 Time profiles of some 89 GHz bursts observed in March 1989 (from Kundu et al 1990). See text for details.

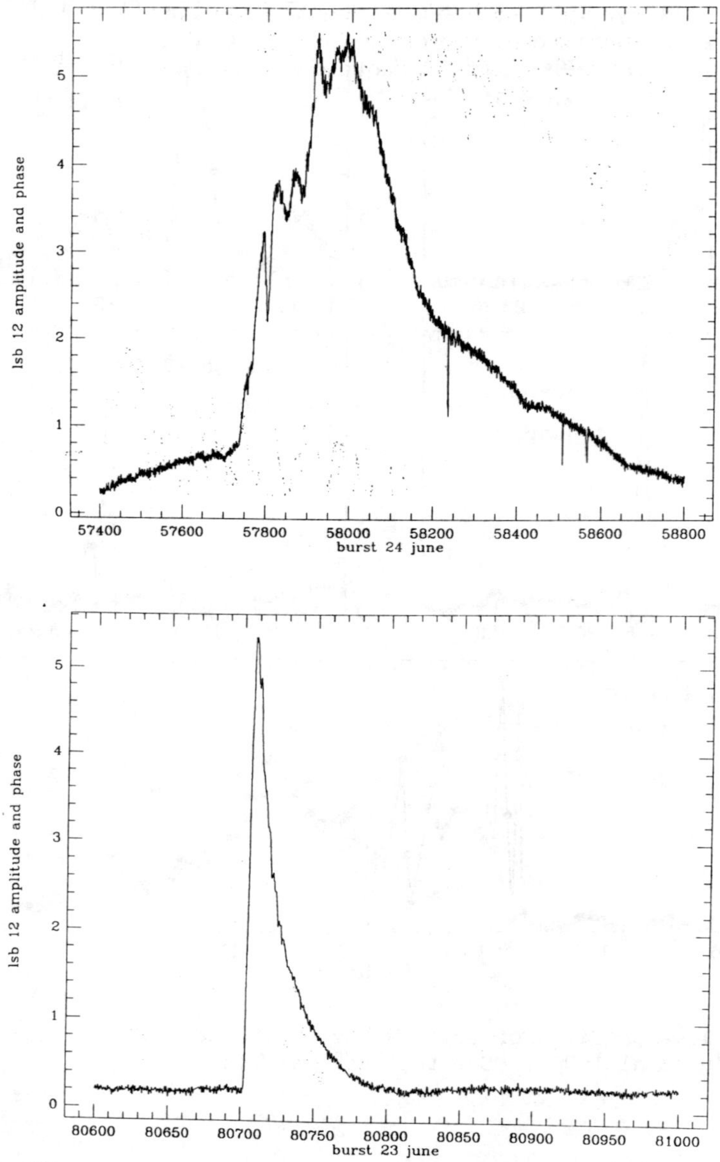

Figure 2 Time profiles of an impulsive flare (top) and long duration flare (bottom) observed in June 1989 with 0.32 sec time resolution (from White et al 1990a). See text for details.

simple, and that all the flux is resolved. This makes the source size about 5 arc seconds or smaller.

By contrast with the impulsive flare shown in Fig. 2(a) the June 24 event (Fig. 2(b) shows a gradual rise and emission continuing for 20 minutes. On the basis of the smooth profile we again argue that most of the flux is being seen by the interferometer, and therefore the burst source is smaller than ~ 4 arc seconds. The relative stability of the phase measurements (not shown here) are further evidence for this.

One important conclusion that emerges from our March 1989 observations is that the millimeter bursts cannot be thermal emission. Any thermal gyrosynchrotron spectrum would be well below our detection threshold at these high frequencies, because thermal emission falls off roughly as f^{-8}. The spectral indices implied by the March detections are in the range -3 to -5.

Comparing the March observations, with long baselines, and the June observations with a shorter baseline, we conclude that typical source sizes at 89 GHz are in the range 2 - 5 arcseconds.

Flares such as that in Figure 2(b), if all the emission is due to energetic particles, clearly imply particle acceleration continuing throughout the flare.

The rather long bursts with many spikes such as that shown in Fig. 1 (event of March 14) are over-resolved. This implies that the spikes are not due to genuine time variation, but rather due to spatial scales; sequential brightening of many flaring kernels will manifest itself as time structure. Similar behavior is sometimes observed in Hα.

Microwave Activity

During the International solar month (September 1988) we observed the Sun with the Very Large Array (VLA) on four days in the period September 11-17. The VLA was in its most compact configuratiaon (D). which is ideal for studying large scale coronal structures, although with relatively poor resolution (40 arc sec at 1.5 GHz, 3 arc min at 0.33 GHz, 15 arc sec at 5 GHz, 10 arc sec at 8.4 GHz and 5 arc sec at 15 GHz). The data are being used for several studies. The White et al paper in this volume discusses some large scale features using the full Sun disk pictures at 1.5 GHz and 0.333 GHz. At higher frequencies we do not have full disk pictures. However, several active regions were mapped quasi-simultaneously at both higher (5-15 GHz) and lower (1.5 - 0.333 GHz) frequencies.

The most striking result that has come out of these observations is that there is exceptionally good correspondence between the 1.5 GHz radio sources and the regions which are dark in He 10830 images and bright in Ca images. Dark regions in the He 10830 images indicate the presence of hot material in the corona. Even very small weak features in the radio maps appear to be associated with dark regions on He images. A set of four full-disk pictures at 1.5 GHz is shown in the paper by White et al in this issue. Using the observations at 5 different frequencies it is possible to construct a 3-dimensional picture of the solar atmosphere above an active region. Figure 3 shows an example of such 3-dimensional picture. It should be noted while the regions at 15 to 1.5 GHz are quiet

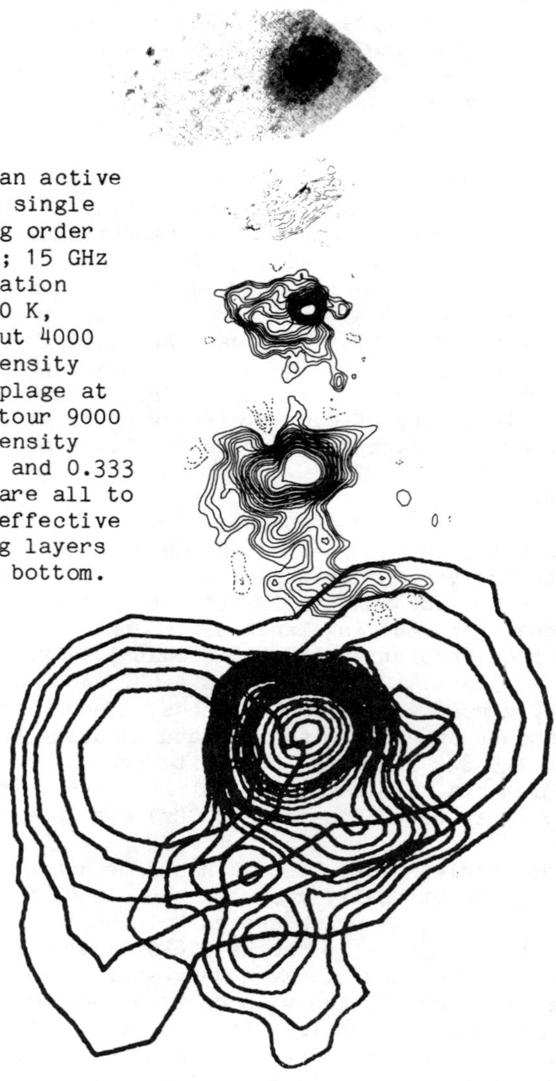

Figure 3. Images of an active region dominated by a single spot, in the following order from the top: optical; 15 GHz left circular polarization (plage at about 30,000 K, bottom contour at about 4000 K); 8.4 GHz total intensity (peak 2.8 million K, plage at 800,000 K, bottom contour 9000 K); 1.5 GHz total intensity (peak 1.8 million K); and 0.333 GHz. The radio data are all to the same scale. The effective height of the emitting layers increases from top to bottom.

active regions, the region at 0.333 GHz is a weak storm center. The 15 GHz map shows only left circular polarization: the right circular polarization map shows a point source at the location of the left-hand peak, but twenty times as strong.

Studies of three-dimensional structures of active region have been done in the past by Shevgaonkar and Kundu (1984) using the VLA at 15, 5 and 1.5 GHz. The main results of these studies indicated that the radiation at 2 cm is dominated by thermal bremsstrahlung even in the case where the observed degree of polarization is very high, ~ 80%. The radiation at 6 cm is always due to optically thick gyroresonance absorption, whereas 20 cm radiation could be either due to gyroresonance absorption or due to optically thin thermal bremsstrahlung, depending upon the observed brightness temperature. If T_B is $\leq 10^6$ K, the thermal bremsstrahlung mechanism dominates, but for $T_B \gtrsim 2 \times 10^6$ K, the radiation is due to gyroresonance absorption.

At 6 and 3.3 cm the peak brightness temperature probably corresponds to the x mode being optically thick at the third gyroresonance harmonic. The peak brightness temperature at these frequencies represent the temperature in the layers where the magnetic field is 600 and 1000 gauss, respectively. The temperature drops as we go out (up in wavelength) from 3.3 cm to 20 cm, implying that at different frequencies we are looking at different loop structures. In this case the lower-lying loops (those at 3.3 cm) have the greatest temperature. At the higher frequencies we can clearly distinguish between a spot source and the "plage" source associated with the active region trailing the spot, but at 6 and 20 cm it is difficult to do so, because the plage emission is probably free-free and its brightness temperature increases rapidly with wavelength, while the gyroresonance emission from the spot decreases. Thus the relative contributions of "spot" and "plage" sources to the slowly varying continuum is a strong function of frequency.

Our studies are based upon 5 frequency observations and as such we have the possibility of observing much more details as a function of height. This structure is obviously related to the structural details of loops and therefore of magnetic fields between the heights of ~ 5,000 to 100,000 km above the photosphere.

This research was supported by NSF grant ATM 87-17157 and NASA grant NAG-W-1541.

References

Kundu, M.R., White, S.M., Gopalswamy, N., Bieging, J.H. and Hurford, G., 1990, Ap. J. Letters (submitted).

Shevgaonkar, R.K. and Kundu, M.R., 1984, Astrophys. J., **283**, 413.

White, S.M., Kundu, M.R., and Bieging, J.H., 1990a, in preparation.

White, S.M., Kundu, M.R., Gopalswamy, N., and Schmahl, E.J., 1990b, in this issue.

DISCUSSION

SMITH: As you probably know the 1-10 MeV range is not very good for mm waves as a proxy for γ-rays, because in this range there is a major component due to excitation of trace elements by protons and it is impossible to separate this component from the bremsstrahling component due to electrons. Is there any hope of extending the observations to > 10 MeV where we know that the γ-rays are due solely to electrons?

KUNDU: We can probably go to ~ 1.2 mm - even there with reasonable magnetic fields in flaring regions, the electron energy required for mm emission will be of the order of 1 MeV. However, we can test other effects of γ-ray flares, for example anisotropy (limb excess of γ-ray flares in contrast to disk distribution). If we find a similar effect, then we shall be on solid ground regarding mm bursts as a proxy for γ-ray bursts. If we do not find such an anisotropy, then we shall have to deal with different mechanisms to explain γ-ray anisotropy.

ALURKAR: (i) For the over-resolved burst event, what was the spatial separation of the burst structure?
(ii) What is the possibility that the observed structure is a temporal variation of the burst radiation?

KUNDU: (i) Spatial separation of individual burst source structures cannot be determined with the present system. At present all we can say is that there are several structures each of order 1 arcsec within the overall burst source.
(ii) If there is no phase jump associated with spikes, I shall consider the time structures as genuine time structures.

STEPANOV: What is the origin of superthermal electrons in moving type IV source?

KUNDU: There may be several possible origins of nonthermal particles for type IV: reconnections taking place driving the onset of a CME or an accompanying flare; shock waves piston-driven by a CME; or a blast wave generated by the flare explosion accompanying the CME. If you are referring to slow-mode shock acceleration of particles responsible for slow type IV's associated with slow CME's, I really do not know the answer. However, in the paper on the February 17, 1985 CME event (*Ap.J*, in press 1989), we suggested current-driven lower-hybrid waves, which can operate at both fast and slow shocks, as the source of the energetic electrons for a moving type IV burst.

VERMA: It was found that some flares show that low-frequency microwave (mw) bursts start earlier than the high-frequency mw bursts and some flares show the reverse. Please explain the reason for the same.

KUNDU: It is possible to explain the relative delay between low-frequency, high-frequency microwave bursts and hard X-ray bursts by a combination of Joule heating and an acceleration mechanism (for details, see G. Holman,*Ap.J*,1985).

KALKOFEN: Radiation at mm wave lengths is also emitted from the middle chromosphere. The radiation is due to free-free transitions of H^-, with the source function given by the Rayleigh-Jeans function in LTE. Would you comment on a periodic signal and on the magnitude of the temperature fluctuations?

KUNDU: The non-flare mm emission is certainly due to free-free transition. The emission from quiet regions appears to have some quasi-periodicity, with a period of about 180 sec. Searches for other periods using single dish observations have so far not been successful. The problem has to be attacked again using interferometric observations.

ELECTRON BEAMS AND LANGMUIR TURBULENCE IN SOLAR TYPE III RADIO BURSTS OBSERVED IN THE INTERPLANETARY MEDIUM

R. P. Lin
Space Sciences Laboratory and Department of Astronomy
University of California
Berkeley, CA 94720 USA

ABSTRACT. The ISEE-3 spacecraft has provided *in situ* observations of electron beams, plasma waves, and associated solar type III radio emission in the interplanetary medium near 1 AU. These observations show that electron beams are formed by the faster electrons arriving before the slower ones, following an impulsive injection at the Sun. The resulting bump-on-tail in the reduced one-dimensional distribution function, $f(v_{\parallel})$, is unstable to the growth of electrostatic electron plasma (Langmuir) waves. The Langmuir waves are observed to be highly impulsive in nature. The onset and temporal variations of the observed plasma waves are in good qualitative agreement with the wave growth expected from the evolution of measured $f(v_{\parallel})$. However, far higher Langmuir wave intensities are predicted than are detected. In addition, the lack of obvious plateauing of the bump-on-tail suggests that the waves have been removed from resonance with the beam electrons by some wave-wave interaction. Bursts of low frequency, 30–300 Hz (in the spacecraft frame) waves are often found coincident in time with the most intense spikes of the Langmuir waves. These low-frequency waves appear to be long-wavelength ion acoustic waves, with wave number approximately equal to the beam-resonant Langmuir wave number. The observations suggest several possible interpretations: modulational instability, electrostatic decay instability, and electromagnetic decay instability; but none of these are fully consistent with the observations. Microstructures, too short in duration to be resolved by present experiments, have been invoked as an explanation of the phenomenon. Experiments are currently being developed to study these processes using fast wave-particle correlation techniques.

INTRODUCTION

Type III solar radio bursts (Figure 1) are produced by fast particles ejected from the Sun (*Wild,* 1950*a,b*). The primary radiation is believed to occur at either the fundamental, f_{p-}, or the harmonic, $2f_{p-}$, of the local electron plasma frequency. The bursts are characterized by a rapid drift from high to low frequency which is attributed to the decreasing electron plasma frequency f_{p-} encountered by the fast particles as they

E. R. Priest and V. Krishan (eds.), Basic Plasma Processes on the Sun, 467–481.

move outward through the solar corona. It has been confirmed that electrons with energies from a few keV to a few tens of keV, often from solar flares, are responsible for the type III radio emission (*Lin,* 1970, 1974; *Alvarez et al.,* 1972; *Frank and Gurnett,* 1972; *Lin et al.,* 1973).

Gurnett and Anderson (1976, 1977) have also observed intense electron plasma oscillations in association with type III radio bursts, apparently confirming a basic mechanism first proposed by *Ginzburg and Zheleznyakov* (1958). In this mechanism, type III radio bursts are produced by a two-step process in which intense electrostatic electron plasma oscillations are first excited by a two-stream instability caused by the fast electrons (Figure 2*a*), and then these plasma oscillations decay into electromagnetic radiation because of nonlinear interactions. Electrons slightly faster (slower) than the wave phase velocity $v_\phi = \omega/k$ will lose (gain) energy to (from) the plasma wave. Thus if a bump is present on the tail of the distribution (Figure 2*b*) so that there are more electrons supplying energy to the wave than taking energy away (i.e., a positive slope in the distribution) the wave will grow. If the plasma waves grow rapidly enough, the bump will be flattened out, a process known as "quasi-linear relaxation" or "plateauing." *Sturrock* (1964) pointed out that in a homogeneous beam-plasma model the fast particle stream, i.e., a bump-on-tail distribution, is so rapidly diffused in velocity space by the intense plasma waves that a coherent stream would only be able to travel a few thousand kilometers from its source before it is plateaued. A major question then for type III burst theory is: What allows the stream to propagate from the Sun to well past 1 AU, as observed?

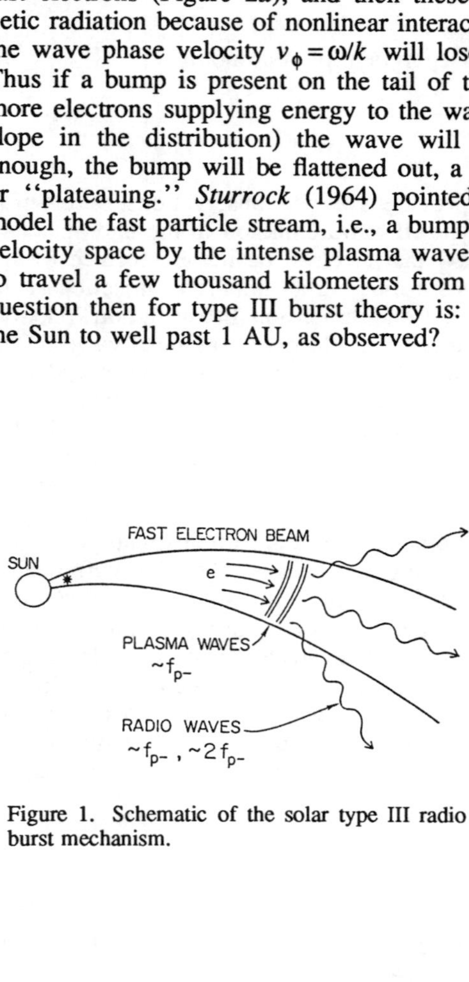

Figure 1. Schematic of the solar type III radio burst mechanism.

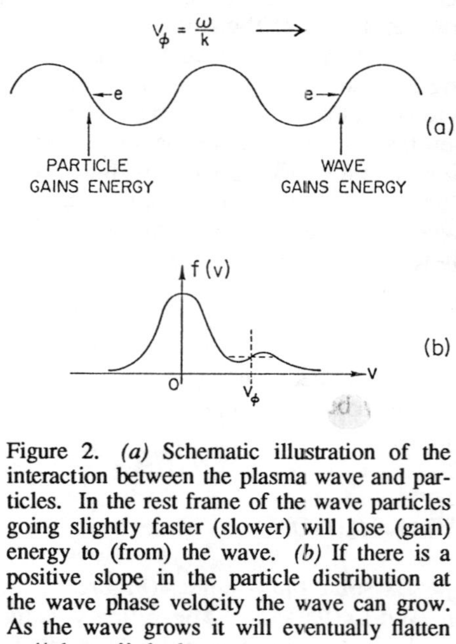

Figure 2. *(a)* Schematic illustration of the interaction between the plasma wave and particles. In the rest frame of the wave particles going slightly faster (slower) will lose (gain) energy to (from) the wave. *(b)* If there is a positive slope in the particle distribution at the wave phase velocity the wave can grow. As the wave grows it will eventually flatten or "plateau" the bump.

Observations show that the stream is highly inhomogeneous with the bump-on-tail distribution produced by the faster electrons running ahead of the slower ones. At a given distance from the fast particle source, the bump in the distribution moves from high to low velocity with time. Thus, one possibility is that plasma waves produced by the positive slope portion of the distribution may not build up to sufficient amplitudes to interact with the bump before the bump has moved to lower velocities. Also, the plasma waves produced with a given phase velocity at time t_1 by a positive slope may be reabsorbed at a later time t_2 when a negative slope is present. Several numerical model calculations suggest that the quasi-linear relaxation (plateauing) will occur, but reabsorption of the plasma waves is important in enabling the stream to propagate large distances without catastrophic energy loss (*Magelssen and Smith,* 1977; *Takakura and Shibahashi,* 1976; *Grognard,* 1980).

Another possibility is that various nonlinear wave-wave interactions may be important in shifting the plasma waves to different wave numbers where they are no longer resonant with the beam electrons; this would allow the electrons to travel large distances without catastrophic energy loss. These mechanisms include induced scattering of the Langmuir waves off ion clouds (*Kaplan and Tsytovich,* 1968), parametric instabilities such as oscillating two-stream instability (*Papadopoulous et al.,* 1974; *Papadopoulous,* 1975; *Bardwell and Goldman,* 1976), and strong turbulence processes such as soliton collapse (*Zakharov,* 1972; *Nicholson et al.,* 1978; see also reviews by *Goldman,* 1983, 1984).

In addition, density irregularities and ion acoustic waves already present in the solar wind can significantly affect the growth and saturation of beam-driven Langmuir waves, to the point of stabilizing the beam (*Escande and de Genouillac,* 1978; *Goldman and DuBois,* 1982; *Russell and Goldman,* 1983; *Muschietti et al.,* 1985).

Because type III bursts propagate to 1 AU and beyond, direct *in situ* observations of the electron beam, Langmuir waves and low frequency waves, and associated type III radio emission in the interplanetary medium near 1 AU are available from spacecraft. In this paper I review these observations and their implications for the plasma processes underlying type III bursts. Although type III bursts are among the best studied of any solar plasma phenomena, we shall see that it is still not well understood, and that a variety of plasma phenomena may be involved. Finally, we discuss briefly observations planned for the future.

OBSERVATIONS OF ELECTRON BEAMS AND LANGMUIR WAVES

The primary observations were obtained from the International Sun Earth Explorer (ISEE)-3 spacecraft, which was located in the solar wind approximately a million miles upstream from the Earth. The radio and plasma wave measurements are from the joint TRW-JPL-Iowa plasma wave instrument on ISEE-3, and the electron measurements are from the University of California, Berkeley, energetic particle instrument on ISEE-3.

Figure 3 shows an example of an impulsive electron event with associated type III radio emission and *in situ* electron plasma waves (*Lin et al.,* 1981). The top panel shows the electric field intensities in four frequency channels from 17.8 to 100 Hz. The smooth intensity variations characteristic of a type III radio burst, consisting of a rapid rise followed by a slow monotonic decay, are clearly evident in the 56.2 and 100 kHz channels. The very intense irregular electric field intensity variations in the 31.1 kHz channel from about 20:00 to 21:30 UT are narrow band electron plasma (Langmuir) waves. Because of the filter overlap, somewhat weaker Langmuir wave intensities are also evident in the 17.8 and 56.2 kHz channels. The electron plasma frequency, f_{p-}, computed from the measured solar wind density of 7 cm^{-3}, is 24 kHz.

The large difference between the peak and average electric field intensities indicates that the Langmuir waves are very impulsive. Figure 4 shows the plasma wave

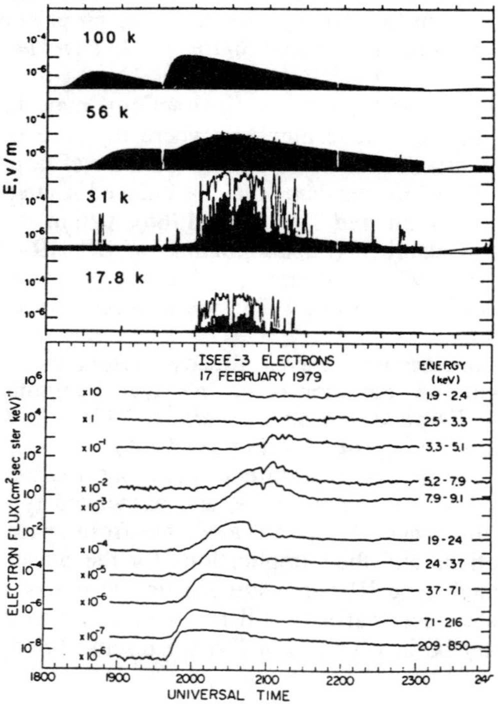

Figure 3. The top panel shows the electric field intensity measured from ISEE-3 (from *Lin et al.*, 1981) on 17 February, 1979 in four broad frequency bands for the event of interest. The black areas show the intensity averaged over 64 s. The solid lines give the peak intensity measured every 0.5 s. The smoothly varying profiles in the 100 and 56.2 kHz channels show two solar type III radio bursts. The second one is of interest here. The intense highly impulsive emissions observed in the 31.1 and 17.8 kHz channels are the electron plasma (Langmuir) waves. The lower panel shows the omnidirectional electron fluxes from 2 keV to >200 keV. Velocity dispersion is clearly evident.

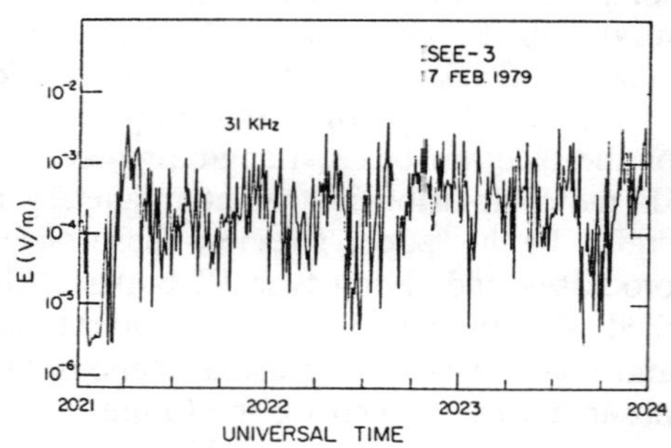

Figure 4. An expanded plot of the 31 kHz channel at 0.5 s resolution for 3 minutes near the maximum plasma wave intensity in the event. Note the extremely impulsive nature of the waves. The peak electric field appears to have an upper limit of 3–4 mV m^{-1}.

amplitude at 0.5 sec time resolution for three minutes near maximum. Intensity changes of over two orders of magnitude can occur within 0.5 s and single bursts rarely last more than one or two 0.5 s readouts. The peak amplitudes appear to be limited to a level of a few mV m^{-1}. These very impulsive fluctuations are characteristic of previous observations of Langmuir waves associated with type III radio bursts. Although more intense bursts have been observed closer to the Sun, this is one of the most intense Langmuir wave events observed at 1 AU.

The bottom panel of Figure 3 shows the spin-averaged low-energy electron intensities simultaneously detected by ISEE-3. The solar flare electrons are first detected above 200 keV energy at about 19:40 UT. Electrons are subsequently detected in progressively lower and lower energy channels at later and later times with very clear evidence of velocity dispersion. Simple comparison of the plasma wave and electron intensities suggests that the Langmuir waves are associated with electrons in the energy range of about 3 to 30 keV.

Figure 5 shows the measured pitch angle distributions for the electrons. Typically at energies above ~15 keV the FWHM of the pitch angle distribution is >60−80°. In contrast, the electrons below ~10 keV are much more beamlike with angular FWHM of <25°. This contrasting behavior of the >15 keV and <10 keV electrons is observed in most solar impulsive electron events. These data are used to construct the distribution of the electrons as a function of $v_{\parallel}$, their velocity parallel to the magnetic field. The electron distribution $f(\overline{v})$ is assumed to by gyrotropic. $f(\overline{v})$ is integrated over $v_{\perp}$ for all particles with a given $v_{\parallel}$ to obtain $f(v_{\parallel})$. These are shown in Figure 6 at 5-minute intervals through the event. Each succeeding distribution is shifted to the right by 2×10^9 cm s^{-1} in velocity.

The solid dots in Figure 6*a* show $f(v_{\parallel})$ for the background population just prior to the onset of the electron event. The level of this long non-thermal tail background population, which is observed at all times although at varying flux levels, is important in determining when a bump will form on the tail of the distribution. The electron distributions develop a well-defined bump-on-tail by ~2000 with the bump progressing to lower velocities with time. In contrast to the expectations of the inhomogeneous quasi-linear numerical models, there is no obvious plateauing of the bump.

There is good qualitative agreement between the Langmuir wave growth and the evolution of $\partial f/\partial v_{\parallel}$. The positive slopes are typically $\partial f/\partial v_{\parallel} \approx 10^{-28}-10^{-24}$ cm^{-5} sec^2 with width $\Delta v/v \approx 0.2-0.3$. $\partial f/\partial v_{\parallel}$ is greater than 10^{-25} cm^{-5} sec^2 for ~10 minutes at a given velocity. Since $\partial f/\partial v_{\parallel} > 10^{-25}$ cm^{-5} sec^2 implies a growth rate $\gamma \gtrsim 1$ sec^{-1} the plasma wave levels should have grown to far higher levels than those measured. At the observed wave levels the bump-on-tail in the electron distribution should have been plateaued in ~0.1 sec to ~1 min., much shorter than the ~10 minutes persistence time for the bump. These measurements indicate that the Langmuir waves have been shifted out of resonance with the electron beam.

EVIDENCE FOR NONLINEAR WAVE-WAVE INTERACTIONS

Various nonlinear wave-wave interactions have been suggested for shifting the plasma waves out of resonance with the electron beam (see *Goldman,* 1983, for review). In general, the dominant Langmuir wave transfer processes are those which involve a low-frequency excitation in addition to another Langmuir wave (*Tsytovich,* 1970). *Lin*

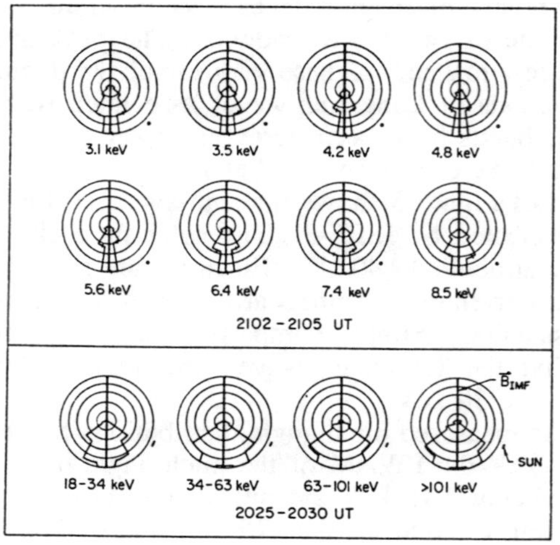

Figure 5. Polar plots of the pitch angle distributions for electrons in different energy intervals measured from ISEE-3 on 1979 February 17. Note the broad, >90° FWHM, flat-topped and sharp-sided distributions at energies above ~20 keV, as compared with the one-dimensional <25° FWHM beamlike character at low energies.

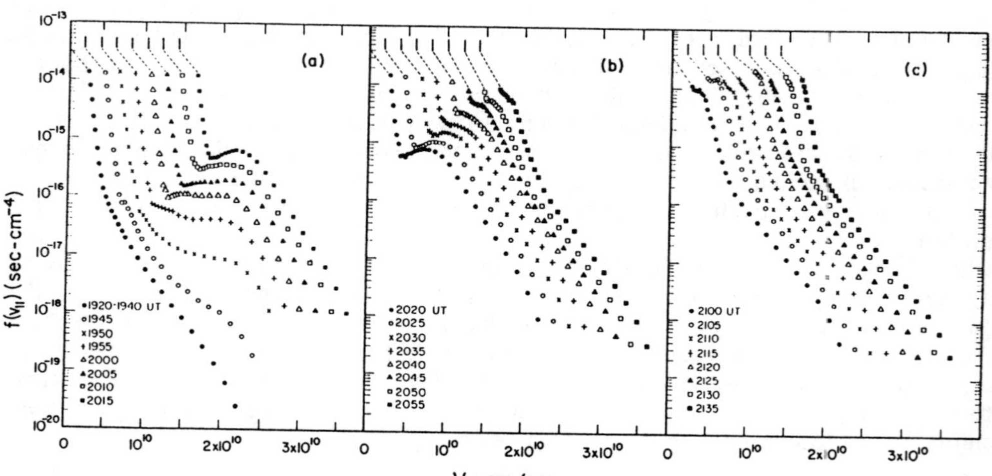

Figure 6. The comprehensive measurements over energy and angle taken from ISEE 3 on 1979 February 17 have been used to construct the one-dimensional velocity distribution function of the electrons every 64 s. The distribution averaged over 20 minutes prior to the event onset is indicated by the solid dots in panel *(a)*. The 64 s measurements of the distribution during the event are shown every 5 minutes. Note the positive slope which develops and moves to lower velocities with time.

et al. (1986) searched for low frequency fluctuations in association with the Langmuir waves observed in type III radio bursts by the ISEE-3 spacecraft. Figure 7 shows a type III solar radio burst with clearly associated Langmuir waves and energetic electrons. The smooth intensity variations characteristic of a type III radio burst, consisting of a rapid rise followed by a slow monotonic decay, are clearly evident in the 31.6, 56.2, and 100 kHz channels. The electron plasma oscillations are observed in the 17.8 kHz channel (with some overlap into the 10 kHz channel) from about 1110 to 1150 UT, in good temporal agreement with $f(v_{\parallel})$ which exhibits a bump-on-tail from ~1109 to ~1142. The electron plasma frequency computed from the solar wind plasma density of ~2 cm^{-3} (Table I) is 12.7 kHz. From ~316 Hz down to 31.6 Hz there are highly impulsive bursts at the same time as the most intense bursts of Langmuir waves. In the 56.2 and 31.6 Hz channels, these are the only bursts detectable above the background.

Table I. Plasma, Beam, and Wave Parameters for 11 March 1989 Event

Solar wind plasma:[a]	
Solar wind density, n	2 cm^{-3}
Solar wind velocity, V_{SW}	480 km s^{-1}
Angle of magnetic field to solar wind, θ_B	139°
Electron temperature, T_e	2×10^5 K
Ion temperature, T_i	4×10^4 K
Debye length, λ_D	2.2×10^3 cm
Electron plasma frequency, f_{p-}	13 kHz
Ion plasma frequency, f_{p+}	3×10^2 Hz
Fast electrons:	
Beam velocity, v_b	~3.5×10^9 cm s^{-1}
Beam density, n_b	~7×10^{-5} cm^{-3}
Positive slope, $\partial f/\partial v_{\parallel}$	~10^{-25} cm^{-3} s^2
Beam width, $\Delta v_b/v_b$	~0.1–0.2
Langmuir pump waves:	
Beam resonant wave number, k_0	2.3×10^{-5} cm^{-1}
Maximum wave amplitude, $E_{0_{max}}$	~1 mV m^{-1}
Maximum normalized energy density, $W_{max}=E_0^2/8\pi n\kappa T_e$	8×10^{-7}
Long wavelength ion acoustic waves:	
Wavenumber, k_I (typical)	1.8×10^{-5} cm^{-1}
Ion acoustic speed, c_s	5.2×10^6 cm s^{-1}
Ion acoustic frequency, f_I	15 Hz
Maximum electric field, $E_{I_{max}}$	~40 μV m^{-1}

[a] Solar wind parameters provided by J. Gosling and W. Feldman, and magnetic field by B. Tsurutani, private communication (1984).

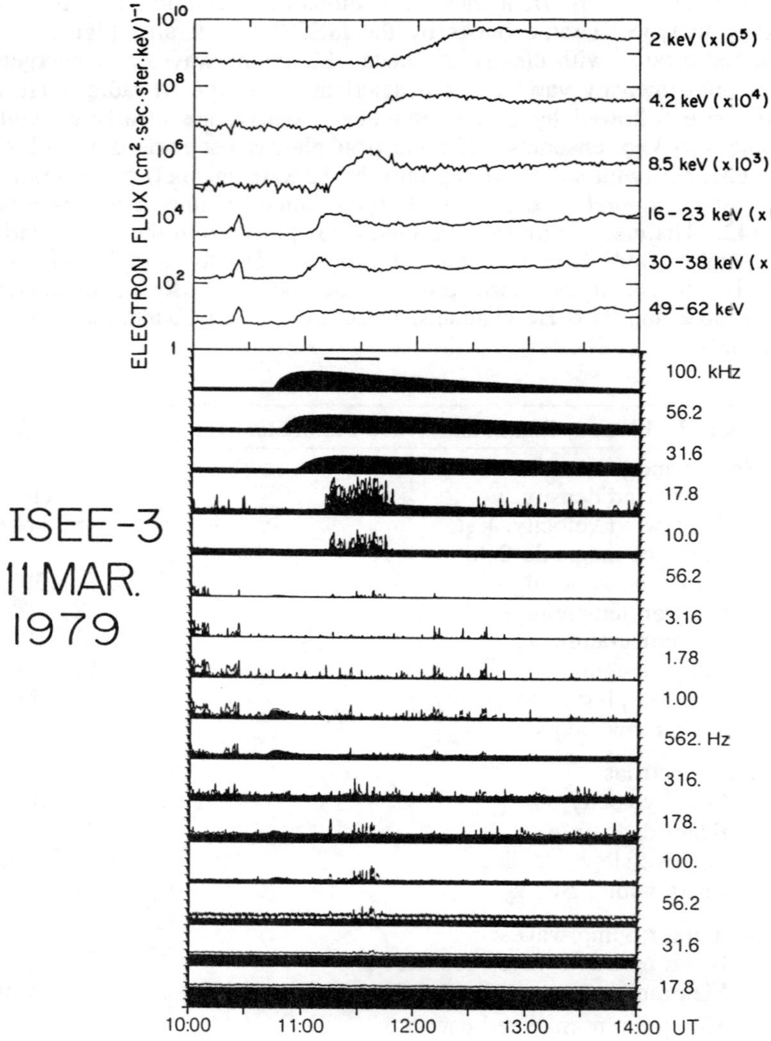

Figure 7. The top panel shows the spin-averaged flux of electrons from 2 to 62 keV for the type III solar radio burst of 11 March, 1979. The spike at 10:20 UT and smaller spikes which occur simultaneously in several energy channels are due to energetic particles flowing upstream from the Earth's bow shock. The lower panel shows the electric field intensity measured in 16 broadband channels from 100 kHz down to 17.8 Hz. The horizontal bar at the top indicates times of positive slope in the electron reduced velocity distribution [$f(v_{\parallel})$]. The smoothly varying profiles (100, 56.2, and 31.6 kHz) show the type III radio burst. The intense, highly impulsive emissions at 17.8 kHz and 10 kHz are electron plasma waves. The sporadic bursty emissions between 3.16 kHz and ~316 Hz have been previously identified as short wavelength ion acoustic waves. The impulsive emissions at frequencies from 316 to 31.6 Hz which occur from ~1115 to 1145 UT, simultaneous with the electron plasma waves, are believed to be long wavelength ion acoustic waves (from *Lin et al.*, 1986).

Figure 8 plots the highest time resolution (0.5 s) data for the 17.8 kHz Langmuir wave channel and the 100 Hz electric field channel for the period 1136–1140 UT. Most, but not all, of the Langmuir bursts which exceed ~0.1 mV m^{-1} are accompanied by a corresponding burst at 100 Hz. The maximum electric field at ~100 Hz is ~0.04 mV m^{-1}, almost two orders of magnitude above the background level. The fact that this low frequency noise is not observed in the 100 Hz magnetic channel suggests that it is a low-frequency electrostatic mode, most likely a long wavelength ion-acoustic wave.

Representative spectra of these low-frequency bursts are shown in Figure 9. the bursts show broad peaks at $f_I \sim 100$ Hz and very rapid cutoff at high frequencies; the 1/10 maximum point occurs at ~250 Hz. The frequency spectrum measured by the spacecraft is almost certainly determined by Doppler shifts, since the solar wind velocity is much larger than the phase velocity of all the known low-frequency electrostatic modes. If these are long-wavelength ion acoustic waves, Doppler-shifted by the solar wind velocity and aligned with the magnetic field, then their wave numbers k_I are

$$k_I \approx \frac{2\pi f_I}{V_{SW} \cos\theta_B + c_s},$$

where V_{SW} is the solar wind speed, c_s is the ion sound speed, and the angle between V_{SW} and the magnetic field is θ_B. For $f_I \approx 100$ Hz and the solar wind parameters in Table I, $k_I \approx 1.8\times10^{-5}$ cm^{-1}. The wave number of the ion acoustic waves is thus comparable to the wave number of the beam resonant Langmuir waves, $k_0 \approx 2.3\times10^{-5}$ cm^{-1}. Table I summarizes the parameters of the solar wind plasma, electron beam, and plasma waves.

Figure 10 plots the low-frequency wave amplitude versus the Langmuir wave amplitude for the bursts shown in Figure 8. As mentioned earlier, there are a few intense bursts of Langmuir waves which appear to have little or no effect above background at low frequencies (points at bottom right). For most Langmuir bursts the electric field of the correlated low-frequency burst appears to increase approximately linearly with Langmuir wave amplitude.

DISCUSSION

The close temporal correlation of the 30–300 Hz low frequency noise bursts with the most intense Langmuir wave bursts suggests that they are the result of a nonlinear wave-wave interaction. Theoretical work and numerical simulations suggest that strong turbulence phenomena such as modulational instability and self focusing of Langmuir waves (soliton collapse) should become important at a threshold of $W = E^2/8\pi nkT_e \approx 10^{-5}$, where E is the Langmuir wave electric field (see *Goldman*, 1983, for review). The 11 March 1979 event (Table I), as well as almost all the type III events observed at 1 AU, however, have maximum values of W substantially below 10^{-5}. In addition, the low frequency wave spectrum associated with modulational instability appears inconsistent with the observed spectrum (*Lin et al.*, 1986). Although very intense plasma waves confined to very small spatial regions might be missed by the plasma wave instrument, it appears unlikely that strong turbulence phenomena play

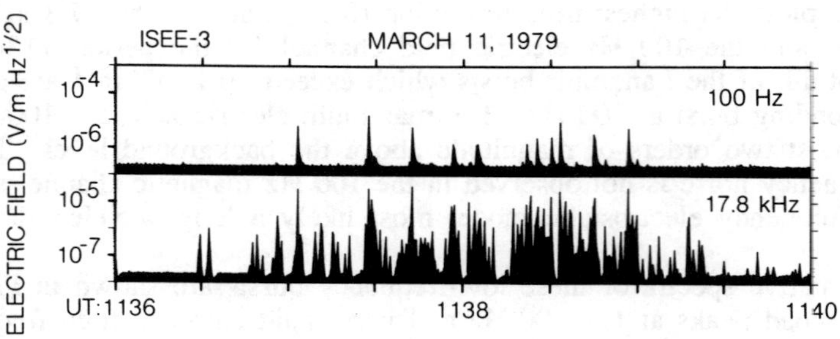

Figure 8. High time resolution (0.5 s) plots of the Langmuir wave channel (17.8 kHz) and the 100 Hz long-wavelength ion acoustic wave channel for the 11 March, 1979 event (from *Lin et al.*, 1986). Note the close correspondence between the most intense Langmuir wave spikes and the 100 Hz spikes.

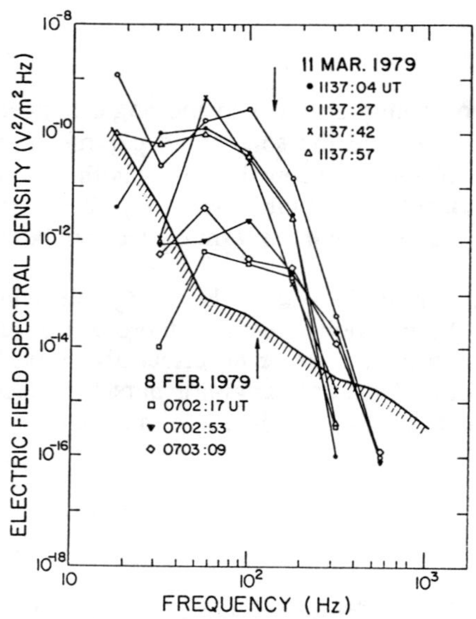

Figure 9. Spectra of the low-frequency bursts which are coincident with the intense Langmuir wave spikes. Background levels are shown by the hatched line. Each spectrum is taken at a single 0.5 s point. The arrows indicate the frequency at which ion acoustic waves would be observed in the spacecraft frame if they had wave number equal to that of the beam resonant Langmuir waves.

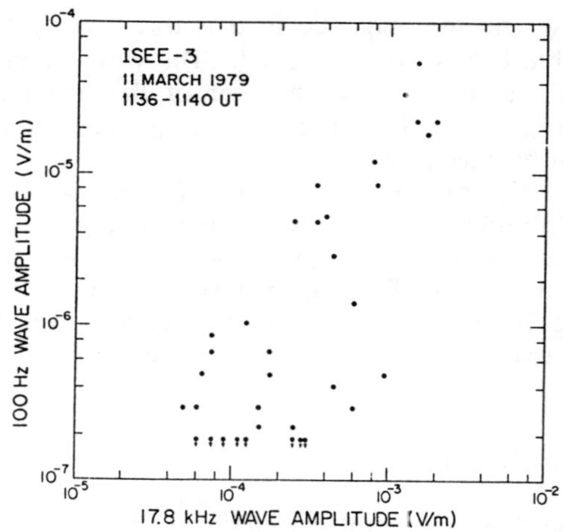

Figure 10. Peak electric field amplitude measured in the low-frequency (100 Hz) channel vs. the Langmuir wave (17.8 kHz) amplitude for the bursts of Fig. 8.

a significant role in type III bursts with the possible exception of a few very intense events.

If the low frequency bursts observed simultaneously with type III Langmuir wave bursts are due to long wavelength ion acoustic waves, these waves will not be very strongly damped because $T_e/T_i \approx 5$ (Table I). Thus a parametric decay instability, rather than induced scattering, might be the appropriate description of the situation for type III radio bursts observed at 1 AU.

Two three-wave decay processes appear relevant: the decay of a pump Langmuir wave into a daughter Langmuir wave and an ion acoustic wave, $L \rightarrow L' + I$; and into a daughter transverse electromagnetic wave and an ion acoustic wave, $L \rightarrow T + I$. The low frequency wave spectrum would be consistent with either of these decay processes. The transverse waves from the electromagnetic decay process would provide a natural source for type III radio burst emission at the fundamental, i.e., near f_{p-}. *Kellogg* (1980) and *Dulk et al.* (1984) find that the radio emission at the onset of interplanetary type III is most likely at the fundamental.

Detailed calculations show that the threshold for the electrostatic decay instability, for parameters of Table I, is about an order of magnitude higher than the observed Langmuir wave levels. The threshold for the electromagnetic decay instability, if one ignores convective effects, is very low and easily exceeded by typical type III events. If the instability is convective, then the large group velocity of the electromagnetic waves may severely limit the growth of both the ion acoustic and transverse waves. It would be difficult in this case to interpret the observed low-frequency bursts in terms of an electromagnetic decay instability. Alternatively, the electromagnetic wave energy might be localized by solar wind density fluctuations; either through wave trapping by refraction in large-scale density fluctuations (*McDonald,* 1983), or wave localization by cumulative phase reinforcement due to multiple scattering by random, small-scale density fluctuations (*Escande and de Genouillac,* 1978; *Escande and Souillard,* 1984; *Levedahl,* 1987).

The density inhomogeneities observed in the solar wind (*Celnikier et al.,* 1983; 1987) may also strongly affect the growth rate of the Langmuir waves. *Muschietti et al.* (1985) showed that the effective damping rate due to angular diffusion of the Langmuir waves by the observed density fluctuations exceeds the growth rate inferred from the particle data. The clumpy character of the Langmuir waves may reflect the fact that the waves can be amplified only along certain favored paths (*Smith and Sime,* 1979).

In an attempt to reconcile these observations with the present theoretical models of the plasma processes for type III bursts, *Melrose and Goldman* (1987) proposed that there are localized microstructures where $\partial f/\partial v_\parallel$ is larger than those measured by a factor of $\gtrsim 10^2$. These microstructures would pass by the spacecraft on time scales of $\lesssim 1$ sec, so the present electron distribution measurements, which have a temporal resolution of ~1 minute, will miss them. The proportionately larger growth rates implied by such $\partial f/\partial v_\parallel$ would overcome the damping by density fluctuations and allow growth to an initial level of Langmuir waves above the threshold for electrostatic decay instability and perhaps also above the threshold for modulational instability and soliton collapse.

The microstructures develop because of the threshold for wave growth imposed by the density fluctuations. Velocity dispersion will tend to steepen positive $\partial f/\partial v_\parallel$

with time until it exceeds that threshold; then waves will grow very rapidly to high intensities. These waves then plateau the electron distribution, slowing down the fast electrons to provide a local enhancement of the fast electron density. The cycle then can start all over.

New, much faster measurements will be required to test these ideas. Experiments currently being developed for the International Solar Terrestrial Physics (ISTP) Global Geospace Science (GGS) mission Wind spacecraft, to be launched in late 1992, will have the capability to measure the electron distribution function in the region of the bump-on-tail with temporal resolution of $\lesssim 10^2$ ms. Furthermore, a wave-particle correlator will search for bunching of the beam electrons in the wave electric field. If the waves are resonance with the beam (see Figure 2*a*) then the electrons should bunch in phase with the wave. If the electrons are losing energy to the wave the peak of the electron bunching should be on the ascending portion of the wave. These measurements should provide a new window on the wave-particle and wave-wave processes in type III bursts.

ACKNOWLEDGMENTS

This research was supported in part by NASA grant NAG5-376 and NASA contract NAS5-30366. The travel to this symposium was supported in part by an IAU travel grant.

REFERENCES

Alvarez, H., F. T. Haddock, and R. P. Lin, *Solar Phys*. **26**, 468, 1972.
Bardwell, S., and M. V. Goldman, *Astrophys. J.* **209**, 912, 1976.
Celnikier, L. M., C. C. Harvey, R. Jegou, M. Kemp, and P. Moricet, *Astron. Astrophys*. **126**, 293, 1983.
Celnikier, L. M., L. Muschietti, and M. V. Goldman, *Astron. Astrophys*. **181**, 138, 1987.
Dulk, G. A., J. L. Steinberg, and S. Hoang, *Astron. Astrophys*. **141**, 30, 1984.
Escande, D. F., and G. V. de Genouillac, *Astron. Astrophys*. **68**, 405, 1978.
Escande, D. F., and B. Souillard, *Phys. Rev. Lett*. **52**, 1297, 1984.
Frank, L. A., and D. A. Gurnett, *Solar Phys*. **27**, 446, 1972.
Ginzburg, V. L., and V. V. Zheleznyakov, *Sov. Astron.–AJ* **2**, 653, 1958.
Goldman, M. V., *Solar Phys*. **89**, 403, 1983.
Goldman, M. V., *Rev. Mod. Phys*. **55**, 709, 1984.
Goldman, M. V., and D. F. DuBois, *Phys. Fluids* **25**, 1062, 1982.
Grognard, R. J.-M., *Solar Phys*. **81**, 173, 1980.
Gurnett, D. A., and R. R. Anderson, *Science* **194**, 1159, 1976.
Gurnett, D. A., and R. R. Anderson, *J. Geophys. Res*. **82**, 632, 1977.
Kaplan, S. A., and V. N. Tsytovich, *Sov. Astron.–AJ* **11**, 956, 1968.
Kellogg, P. J., *Astrophys. J.* **236**, 696, 1980.
Levedahl, K., Ph.D. thesis, University of California, Berkeley, 1987.
Lin, R. P., *Solar Phys*. **12**, 266, 1970.

Lin, R. P., *Space Sci. Rev.* **16**, 189, 1974.
Lin, R. P., L. G. Evans, and J. Fainberg, *Astrophys. Letters* **14**, 191, 1973.
Lin, R. P., D. W. Potter, D. A. Gurnett, and F. L. Scarf, *Astrophys. J.* **251**, 364, 1981.
Lin, R. P., W. K. Levedahl, W. Lotko, D. A. Gurnett, and F. L. Scarf, *Astrophys. J.* **308**, 954, 1986.
Magelssen, G. R., and D. F. Smith, *Solar Phys.* **55**, 211, 1977.
McDonald, S. W., Ph.D. thesis, University of California, Berkeley, 1983.
Melrose, D. B., and M. V. Goldman, *Solar Phys.* **107**, 329, 1987.
Muschietti, L., M. V. Goldman, and D. Newman, *Solar Phys.* **96**, 181, 1985.
Nicholson, D. R., M. V. Goldman, P. Hoyng, and J. C. Weatherall, *Astrophys. J.* **225**, 605, 1978.
Papadopoulous, K., *Phys. Fluids* **18**, 1979, 1975.
Papadopoulous, K., M. L. Goldstein, and R. A. Smith, *Astrophys. J.* **190**, 175, 1974.
Russell, D. A., and M. V. Goldman, *Phys. Fluids* **26**, 2717, 1983.
Smith, D. F., and D. Sime, *Astrophys. J.* **233**, 998, 1979.
Sturrock, P. A., *AAS-NASA Symposium on the Physics of Solar Flares,* ed. W. N. Hess (NASA SP-50), p. 357, 1964.
Takakura, T., and H. Shibahashi, *Solar Phys.* **46**, 323, 1976.
Tsytovich, V. N., *Nonlinear Effects in Plasmas*, New York, Plenum, 1970.
Wild, J. P., *Australian J. Sci. Res.* **A3**, 541, 1950*a*.
Wild, J. P., *Australian J. Sci. Res.* **A3**, 399, 1950*b*.
Zakharov, V. E., *Soviet Phys.–JETP* **35**, 908, 1972.

DISCUSSION

MONTGOMERY: The velocity-space distribution you are talking about is the one-dimensional velocity distribution function, $f_e(v_{||})$. This is an integral of the fully three-dimensional distribution $f_e(\mathbf{v})$. I do not have a good picture of the three-dimensional presumed behaviour of $f_e(\mathbf{v})$. Alterations of $f_e(\mathbf{v})$ which "flatten" (stabilize) it for one direction of $\mathbf{k}$ often destabilize it for some other direction of $\mathbf{k}$. Can you be sure that an $f_e(\mathbf{v})$ exists that has the properties you require of it, and if so, could you speculate upon what it looks like?

LIN: We have constructed the 3-D distribution from our angular measurements assuming gyrotropy. This has then been integrated to obtain $f_e(v_{||})$. We do not see a truly plateaued (flat) $f_e(v_{||})$ and we have not analyzed the 3-D distribution in detail yet.

HOLLWEG: Are there proxy indicators for the presence of the intense Langmuir clumps? For example, how much would they scatter the background (thermal) solar wind electrons, and could the consequent isotropization of f be observed?

LIN: The effect of these Langmuir clumps on the thermal solar wind electrons would be minimal and probably unobservable, unless the wave levels are much more intense than presently measured.

KUNDU: (i) How valuable will it be to conduct similar experiments near the Sun, for example with a solar probe?

(ii) What would you expect to happen, when we make similar experiments at about 4 solar radii, especially from the point of view of conversion of electrostatic waves into electromagnetic waves?

LIN: (i) It would be extremely interesting to conduct similar experiments close to the Sun. It may be that conditions change rather drastically with distance. Certainly the type III radio emissivity varies substantially with frequency.
(ii) I can only speculate that with the very different conditions at $4R_0$ the conversion mechanisms could be quite different.

VAN BALLEGOIJEN: The background electron distribution shows a non-thermal tail. Do these tail electrons contribute significantly to the heat flux?

LIN: The non-thermal tail above 5 keV generally does not contribute a large fraction of the total electron heat flux.

LANG: How often do type III bursts on the Sun generate electron beams that reach the vicinity of the Earth? A related question is how long was the ISEE-3 spacecraft in outer space detecting these events? You showed two events detected ten years ago in 1979, so one wonders if these are typical of many such events actually detected.

LIN: Our ISEE-3 experiment operated continuously in the interplanetary medium for about 15 months. Over 300 impulsive electron events were obtained, each of which is associated with interplanetary type III bursts.

KRISHAN: While studying parametric instabilities, have you considered the broad-band nature of the pump, for it would raise the thresholds and lower the growth-rates?

LIN: We have not considered the broad-band nature of the pump. It would be interesting to see how the thresholds and growth rates would change.

HOYNG: What is the physical origin of the microstructures postulated by Melrose and Goldman (1987).

LIN: Melrose and Goldman assumed an initially spatially localized group of electrons. As they free-stream along the field line a positive slope grows larger, but since there are ambient density fluctuations, the growth of Langmuir waves is suppressed until the positive slope exceeds a threshold. At that point Langmuir waves grow rapidly to large amplitudes and plateau the electrons. In the process of plateauing the electrons are bunched spatially. The entire sequence then repeats.

SEMIKOZ: You have just shown some elementary processes for *strong* turbulence, particularly you have shown a decay instability ($t \rightarrow \ell + s$, *etc.*). I have not seen the induced Compton process (type: $\ell + t \rightarrow \ell + t$) with a four-waves interaction. Why?

LIN: We have only treated the simple 3-wave decay processes. Perhaps 4-wave processes should also be considered. However, the low-frequency waves are only observed coincident with the most intense Langmuir spikes, and are not present otherwise.

GANGADHAVI: Do you expect a parametric decay of radio waves into electron and ion Langmuir waves in the interplanetary medium and corona?

GANGADHAVI: Do you expect a parametric decay of radio waves into electron and ion Langmuir waves in the interplanetary medium and corona?

LIN: I suspect that the threshold for this decay is probably too high in the interplanetary medium, but this has to be calculated.

DIAGNOSTICS OF THE SOLAR PLASMA USING RADIO OBSERVATIONS WITH THE RATAN-600

G.B.Gelfreikh, Leningrad, USSR
Central (Pulkovo) Astronomical Observatory
Academy of Sciences of the USSR
196140 Leningrad, Pulkovo
U.S.S.R.

ABSTRACT. This paper summarises the methods and main results of studying physical parameters and processes in different structures of the solar atmosphere from the chromosphere to the outer corona as based on the observations with a 600-meter radio telescope the RATAN-600 (1975-1989). Diagnostics of the solar plasma have been made using the spectral-polarization measurements in the wavelength range 0.8-32 cm with high spatial resolution. The most important results concern the structure and strength of the magnetic fields at different levels of the solar atmosphere and detection of localized regions of long-lasting (several days) energy release in the lower corona.

1. INTRODUCTION

Solar observations using the world's largest radio telescopes (VLA, WSRT, RATAN-600 et al.) have shown during the Solar Maximum Year program that the radio observations yield valuable information about the magnetic fields and particle acceleration in the solar corona. Such diagnostics depend on detailed spectral analysis and accurate polarization measurements with high spatial and temporal resolution.

2. SCIENTIFIC PROBLEMS

An understanding of structuring in the solar atmosphere in connection with magnetic fields and their evolution as well as MHD-oscillations, constitutes the main objective of studying the Sun with the RATAN-600. We hope to understand the nature of the plasma structures as well as their important role in the storing, transfer and release of energy in the solar atmosphere. The RATAN can uniquely measure magnetic fields in a range of heights in the solar atmosphere from the higher chromosphere to the low corona. Regions of nonthermal release of energy are also localized in different regions of the corona. Structures in the plasma (and their generation) are also of special interest for recent

E. R. Priest and V. Krishan (eds.), Basic Plasma Processes on the Sun, 483–488.

developments non-linear physics and mathematics; the Sun is capable of giving many important examples of such structures.

3. MEASUREMENTS OF SOLAR MAGNETIC FIELDS

We hope developed three methods of measuring magnetic fields in the upper chromosphere and lower corona of active regions.

3.1. Thermal Bremstrahlung Mechanism

In the presence of magnetic field we have an increase of opacity for the extraordinary mode and a decrease for the ordinary one This results in a difference of the effective levels of generation of the two modes in an optically thick plasma (Bogod & Gelfreikh, 1980). In the presence of a temperature gradient, the radiation becomes circularly polarized with degree of polarization

$$P = (nf/f_H)\cos(\alpha), \tag{1}$$

where the spectral index

$$n = -d(\ln T_B)/d(\ln f), \tag{2}$$

T_B - brightness temperature, f - frequency of observation, f_H- gyrofrequency, α - angle of the magnetic field. From (1) we get a formula for measuring longitudinal component of the magnetic field H.

$$H = 107\ P\%\ /(n\ \lambda), \tag{3}$$

where λ [cm]- wavelength. In the case of an optically thin plasma we have n = 2, and (3) becomes

$$H = 54\ P\%\ /\lambda\ . \tag{4}$$

The method has been successfully used for measurements of flocula magnetic fields, for prominences (at the limb) and coronal condensations. Flocula magnetic fields (of the order of 20 G) proved to be in reasonable agreement with the photospheric magnetograms of the same region. We have also found the magnetic field strength of about 20 G at the height of 70 000 km in a coronal condensation.

3.2. Thermal Cyclotron Emission

The most elaborate study using this method was made for the radio cores above the umbra of sunspots. Both the emission at the second and third harmonics are essential. The spectra of these types of sources, obtained with the RATAN-600, have shown the practically linear increase of polarized intensity (Stokes parameter V) with wavelength (Akhmedov et al., 1986). One can find from observations the shortest wavelength

λ, where the emission is observed, thereby estimating the maximum value of the magnetic field strength, H, penetrating the corona as

$$H = 3570 / \lambda \tag{5}$$

which corresponds to the emission at the third harmonic of gyrofrequency. For the second harmonic we have

$$H = 5400 / \lambda \tag{6}$$

The limiting wavelength λ for emission of the second harmonic can also be evaluated from a spectra, though with somewhat lower accuracy.

A number of interesting results have been obtained using this method:

- strong magnetic fields amounting to 2000 G or more may penetrate the corona above a large sunspot;
- the corona above a large sunspot is very low above the photosphere (often not more than 1-3 thousand km);
- the solar atmosphere above a sunspot is strongly inhomogeneous, so the emission in hydrogen lines and CIV lines originate in the comparatively cold plasma amounting to the level of 1000-1500 G, which is probably higher than the level of generation of radio emission;
- the ratio of the radio and optical photospheric magnetic fields is normally about 0.7-0.8 (a good correlation exists!); in some cases, however, a strethening of the magnetic field in the corona above a sunspot preceeding development of the active region has been observed, so radio data may be an important parameter for forecasting development of an active region.

A more detailed analysis of the magnetic field by this method, especially with the coverage of shorter wavelengths (λ<2 cm), is needed for construction of a more comprehensive model of magnetic fields and structure of the chromosphere-corona above different sunspots. Investigation of temporal variation may give important information on the three dimensional structure of magnetic field of sunspots when they emerge or dissipate. Of special interest are observations of quasiperiodic variations of these sources reflecting the flux of MHD waves in a fluxtube of a sunspot. Ring and horseshoe structures of the sunspot-associated sources are of great interest for a study of the spatial distribution of thermal energy above a sunspot in the corona. The same method of measuring magnetic fields is applicable to some other sources (e.g. magnetic loops).

3.3. Inversion of Polarization in QT-region

The RATAN observations also demonstrated the change of the sign of polarization both with wavelength and time (Gelfreikh et al., 1987). This is due to propagation through a region of quasi-transverse magnetic field and may be used succesfully for measurements of magnetic fields in the corona at the heights up to 120,000 km. We may estimate the magnetic field strength using the formula (in QT-region)

$$H = 1.57\times10^{8} / (\lambda \sqrt[3]{\lambda NL}) \quad \text{or roughly} \quad H = 43/\lambda, \tag{7}$$

where λ is wavelength of inversion (P = 0), N - electron density and L - the scale of the magnetic field. On the basis of modelling computations we can try to restore the full picture of the magnetic field in the corona of an active region. Using this method it was found that the magnetic fields are in reasonable agreement with potential extrapolation of the photospheric fields. Some important departures, however, may occur, especially at lower heights.

4. PECULIAR SOURCES

Peculiar sources have been detected with the RATAN in the most developed active regions with high flaring activity (Akhmedov et al., 1986). In many respects their observational properties are similar to those of the sunspot associated cores but yet could be definetly distinguished using the following characteristics:
(i) The position above a neutral magnetic line on the photospheric level (often in connection with magnetic delta-configuration of the sunspot group);
(ii) the size (and height) of about (10-20)x1000 km;
(iii) high brightness temperature amounting 10^{7} K at some wavelength in the range of 2-4 cm;
(iv) steep spectrum with the index amounting n =10;
(v) moderate degree of circular polarization (P < 30%);
(vi) life-time of 2 to 5 days and rather good stability of emission;
(vii) strong photospheric currents were detected near the position of one of peculiar sources (in July 1982).

The mechanism of generation of peculiar sources is still uncertain. We may consider the following possibilities:
(i) thermal gyroresonance emission at 2-4th harmonics of gyrofrequency. (One needs the strong magnetic field H>1000 G in this case);
(ii) thermal cyclotron emission at higher harmonics (one here needs very high temperature $T=10^{8}$K and high electron density $N=10^{12}$);
(iii) nonthermal cyclotron emission (then a stable and effective mechanism of acceleration of electrons is necessery);
(iv) thermal bremsstrahluhg of a strongly inhomogeneous plasma (electron density should amount $>10^{12}$ cm^{-3}).

Whatever their possible nature is, the peculiar sources localize the region of long-lasting and stable nonthermal source of energy release in the corona above an active region.

5. DECIMETRIC HALO

An analysis of spectra of local sources taking into account their structure shows the growth of their flux (intensity) approximately proportional to λ or faster, the maximum being achieved near λ=10cm.

The brightness temperatures for λ> 10 cm are near coronal value (10^6K) but they can be as high as 10^7K. High resolution (VLA) observations at 20 cm have show some bright (sometimes loop-like) structures but their impact to the general flux of the active region generally is not significant (a few per cent or less). The above spectrum needs a more complicated interpretation taking into account possibly such nonthermal components as accelerated particles and plasma turbulence.

6. CONCLUSIONS

The above results demonstrate the importance of microwave observations. High resolution observations combined with spectral and polarization measurements yield new information on structure of coronal magnetic fields and locolize sources of the release of nonthermal energy and probably of plasma turbulence as well. Future important conclusions will result from joint radio, optical, EUV and X-ray observations. Having in mind new opportunities provided by the international programs of the 22-nd cycle of solar activity (such as STEP, MAX^91, FLARE 22 etc) we have begun a modernization of the radio telescope RATAN-600 to improve its spectral, spatial and time resolution (Bogod et al., 1988). This work is now under way and may (we hope) result in much more detailed diagnostics of the plasma structure on the Sun.

REFERENCES

Akhmedov, Sh. B. et al.: 1982, Solar Phys., 79, 41.
Akhmedov, Sh. B. et al.: 1986, Astrophys. J., 301, Part 1, 460.
Bogod, V. M. et al.: 1980, Solar Phys., 67, 29.
Bogod, V. M.et al.: 1988, SAO AS USSR Preprint, No. 22.
Bogod, V. M. et al.: 1982, in Sun and Planetary System. Reidel Publishing Company, 109.
Borovik, V. N.: 1981, Astrofiz. issled., 13, 17.
Borovik, V. N. et al.: 1988, Soln. dannye, No. 11, 110.
Gelfreikh, G. B.: 1986, Contributions of the astronomical observatory Skainate Pleso, 15, 33.
Gelfreikh, G. B., and Lubishev, B. I.: 1983, SIBIZMIR preprint, 21-83.
Gelfreikh, G. B. et al.: 1987, Solar Phys., 108, 89.

DISCUSSION

HOLLWEG: Can one use radio observations to determine how much the coronal magnetic field is tangled or braided as in the models of Parker, Heyvaerts and Priest, van Ballegooijen, and others?

GELFREIKH: We can measure magnetic field strength in the corona in a number of special situations such as: gyroresonance at the second, third and probably higher harmonics; well determined bremsstrahling (both optically transparent or not), where the ray is crossed by a transverse magnetic field. Our program is to combine radio data on magnetic fields with extrapolation of the photospheric magnetic field to check and correct the latter. Some special complicated magnetic field structures with annihilation of magnetic fields and current sheets could be found from high-resolution spectral-polarization radio observations (and most probably are found), since some peculiarities in the radio maps of the Sun are not easy to interpret in generally accepted thermal models of the S-component. Such radio sources are connected with long-lasting acceleration of particles, plasma turbulence and other non-thermal total releases of energy.

KUNDU: Maybe I can answer Joe Hollweg's question from my own point of view. It is not possible to see the braided field lines directly in the radio. However, if there are braided field lines, the angle between the line of sight and magnetic field will be close to 90° since the gyro-resonance absorption coefficient increases by a couple of orders of magnitude as the angle increases from 0° to 90°, we shall see bright clumps in regions of braided fields, because of the increase in optical thickness.

KNEER: You gave us a value of 2000 Gauss measured above sunspots. Could you tell us the height above photospheric levels to which this measurement refers and give us also the magnetic flux density in the photospheric layers of the sunspot?

GELFREIKH: Certainly, yes. We have found statistically for a large number of sunspots that the strength of magnetic fields from radio data (at the base of the corona) is some 20% lower than at the photospheric level for the *same* sunspot. The level of radio emission used for these measurements has been found from the projection effect (the displacement of a polarized radio source relative to the sunspot umbra); it is normally not more than 3,000 km above the photosphere. We assume that the corona above a sunspot may begin at very low heights (but the transition region must be very inhomogeneous to exclude contradictions with the measurements of magnetic fields from CIV EUV from SMM).

SHEVGAONKAR: Using multifrequency observations from the VLA the evolution of active regions as well as flares has been studied in the past. The change in the polarization characteristics during a flare indirectly supports the presence of new magnetic flux probably emerging flux loops or deformation of the magnetic field, as in the model of Heyvaerts, Priest and Rust.

GELFREIKH: The observations made with the RATAN-600 have lower resolution than the data obtained with the VLA and more limited opportunities to study the development of a radio source on the Sun. At the same time, the RATAN has better wavelength coverage (10 wavelengths in the range from 0.8 cm to 31 cm at the moment), better polarization sensitivity and much longer periods of observations of the Sun. That suggests that the data from the two instruments concern different aspects of the structure of magnetic fields and other physical parameters of active regions. Cooperative programs of observations with both instruments would result in a better understanding of the nature of the plasma structures observed in the solar atmosphere.

DUAL FREQUENCY VARIABILITY STUDY OF AN ACTIVE REGION

R. K. SHEVGAONKAR[1] and M. R. KUNDU[2]

1 Electrical Engineering Department, Indian Institute of Technology, Powai, Bombay 4oo o76, India
2 Astronomy Program, University of Maryland, College Park, Maryland 20742, USA

ABSTRACT. A short-time variability study of a solar active region simultaneously at 6 and 2 cm wavelengths was carried out using the VLA. The observations show interesting uncorrelated brightness temperature variation at the two wavelengths. The observed low brightness temperatures indicate that the emission is mainly originating from the chromosphere - corona transition region.

A transition region model with constant pressure and power-law temperature variation as a function of height has been assumed to analyse the data. The uncorrelated variation of the observed brightness temperature at the two wavelengths suggest different dominant emission mechanisms (bremsstahlung at 2cm and gyro-resonance at 6cm) operative at the two wavelengths. It is shown that an independent variation of a few percent in the magnetic field (900 ±45 G) and a factor of two variation in the density (2 to 4 x 10^{10} cm^{-3}) over a time scale of few hours is required to explain the uncorrelated brightness temperature variations at the two wavelengths.

1. INTRODUCTION

Optical observations indicate that solar active regions undergo continual structural changes. These changes should manifest in the emission at radio wavelengths. A time variability study of meter wavelength emission from active regions has recently been carried out (Shevgaonkar et al, 1988). However, this study was confined only to variability at the coronal level. In this paper we present dual frequency centimetric observations of an active region and study its variability. The centimetric emission mainly originates from the transition region and the lower corona where temperature and density gradients are sharp. The brightness temperature is therefore very sensitive to the changes in these parameters.

E. R. Priest and V. Krishan (eds.), Basic Plasma Processes on the Sun, 489–493.

A study of the brightness temperature variability at centimeter wavelengths will help us in better understanding the dynamics of active regions at lower heights in the solar atmosphere.

2. OBSERVATIONS AND RESULTS

An active region at S26E54 heliographic coordinates was observed with the VLA on June 8, 1987. The full-day 6 and 2 cm maps of the active region with angular resolution of 18" x 18" and 6" x 6" respectively are shown in Fig.1(a,b). The degree of circular polarization at both wavelengths was less than ~10% and it is therefore assumed that for all practical purposes the emission is unpolarized.

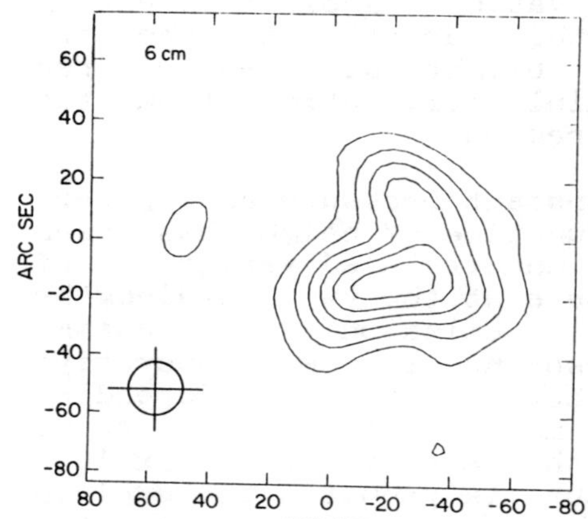

FIG. 1a

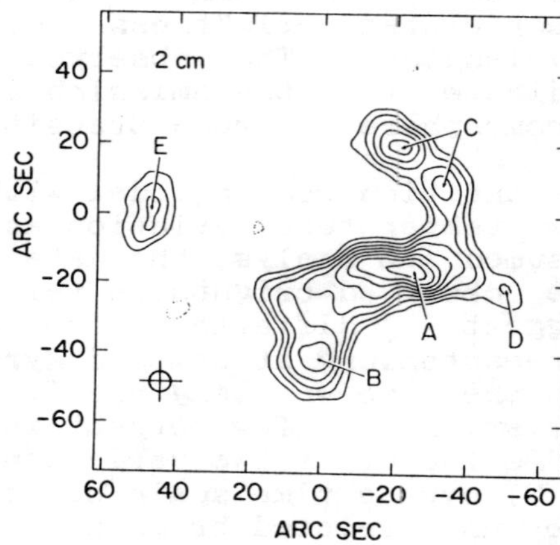

FIG. 1b

Figure 1a-b Full-day synthesis maps of the active region at 6 and 2 cm wavelengths. Contour intervals are 3.4×10^3 K and 333 K respectively for 6 and 2 cm maps.

After applying the appropriate corrections (For details see Shevgaonkar and Kundu, 1989) the peak brightness temperatures at 2 and 6 cm wavelengths are found to be 12 - 18 x 10^3 K and 4 - 11 x 10^4 K respectively. To study the variability of the active region, 1-hr duration snap-shot maps were produced. All the sources marked in the full-day 2 cm map show short time variability. However, the brightness temperature variation only for the strongest source A (Fig.1.b) is shown here in Fig.2. It is evident that the brightness temperature shows uncorrelated variations at the two wavelengths. At 2 cm the brightness temperature varies between 15 and 18 x 10^3 K and at 6 cm it varies between 4 and 7 x 10^4 K.

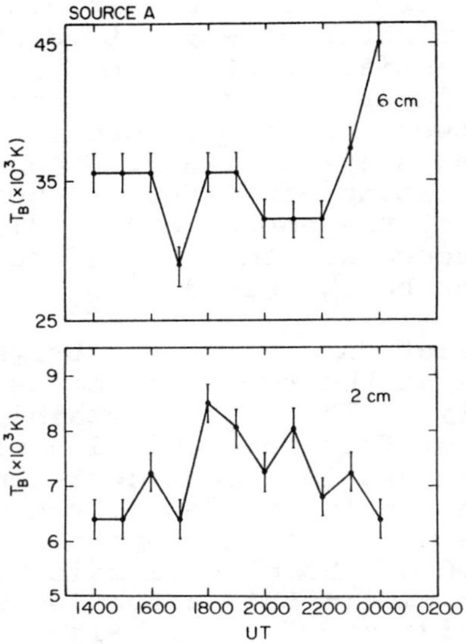

Figure 2 Variability of brightness temperature of source A as a function of time at 2 and 6 cm wavelengths.

3. DISCUSSION

From the computations carried out by Rao and Kundu (1977) and Kundu, Melozzi and Shevgaonkar (1986) it can be seen that the electron temperature T_e in the transition region has almost a power-law variation with height h above the chromosphere (we assume the height of chromosphere to be about 2000 Km). Therefore let us assume that the temperature in the transition region can be written as

$$T_e(h) = 10^4 \; (h / h_{min})^{\alpha} \tag{1}$$

where α is a constant, and at $h = h_{min}$, that is at the bottom of the transition region, the electron temperature is 10^4 K. For low heights it is also reasonable to assume the pressure to be constant as a function of height. If h_{max} is the height at which the temperature reaches coronal temperature, the optical depth upto a layer at height h_1 due to free-free emission is given as

$$\tau = \frac{14\times10^{-6} \; N_o^2 \; h_{min}^{3.5\alpha}}{2 - 7\alpha} \{ h_1^{-3.5\alpha+1} - h_{max}^{-3.5\alpha+1} \} \tag{2}$$

where N_o is the density at a height where temperature is 10^4 K. Now since $\alpha > 0$ and $h_{max} >> h_{min}$ and/or h_1, we can see that most of the opacity comes from layers close to h_1 or h_{min}. Taking a typical value of $N_o = 3 \times 10^{10}$ cm^{-3} (for details see Shevgaonkar and Kundu, 1989), the heights of τ = 1 layers for 2 and 6 cm emissions come out to be 2250 km and 4250 km above the photosphere respectively. Therefore the height difference between the emitting layers at the two wavelengths is only about 2000 km. To explain the uncorrelated variation of the brightness temperature at the two wavelengths, it is essential that the physical conditions in the two emitting layers change independently. Since the height difference between the two layers is small, these type of independent changes are unlikely. Even if we assume some kind of localized heating at different heights, thermal conduction will transport the energy over a distance of 2000 km within a few seconds, and therefore slow uncorrelated variability can not be explained.

To obtain uncorrelated variation in the brightness temperature the layers emitting at the two wavelengths must move up or down independently. This is certainly not possible within the frame work of free-free emission without magnetic field. However, if we take into consideration the presence of the magnetic field the independent movement of the emitting layers can be obtained. If the magnetic field density is adequate (e.g. $\sim$1000 G) the 6 cm emission could be due to gyro-resonance mechanism. In this situation we see that the 6 cm emission will originate from a height where the observing frequency equals the 2nd or 3rd harmonic of the gyro-frequency, whereas the 2 cm emission will originate from a layer which is optically thick due to free-free emission. Any fluctuation in the electron density will change the height of the 2 cm layer (free-free emission) without affecting the height of the 6 cm layer. On the other hand, any change in magnetic field will change the height of the 6 cm (gyro-emission) layer without significantly changing the height of the 2 cm emission layer. Taking exponential variation of the magnetic field as a function of height, the observed variation in the 6 cm brightness temperature corresponds to a few percent change (900 $\pm$ 45 G) in the magnetic field. On the contrary the variation in 2 cm brightness temperature requires a change in density by a factor of 2, that is, from 2×10^{10} cm^{-3} to 4×10^{10} cm^{-3} at the bottom of the transition region.

4. REFERENCES

Kundu, M.R., Melozzi, M., and Shevgaonkar, R.K. (1986) 'A study of solar filaments from high resolution microwave observations', Astron. Astrophys., 167, p. 166.

Rao, A.P. and Kundu, M.R. (1977), Solar Phys., 55, p. 161.

Shevgaonkar, R.K., Kundu, M.R., and Jackson, P.D. (1988) 'Variability of metric emission from the sun', Astrophys. J., 329, p. 982.

Shevgaonkar, R.K. and KUndu, M.R. (1989) 'Time variability of solar active regions at centimeter wavelengths', Astrophys. J., 342, p. 586.

DISCUSSION

GELFREIKH: (i) Why do you refer in your models to the chromospheric instead of the photospheric level?
(ii) Do you not suppose that the chromospheric level in an active region may be quite different?

SHEVGAONKAR: (i) The model essentially is for the transition region where there is a high temperature gradient. The heights therefore are measured from the level above which the temperature increases steeply. Since in the chromosphere the temperature is more or less constant, the power-law model does not apply to the chormosphere. The heights therefore are estimated from the chromosphere rather than from the photosphere.
(ii) The height of the chromosphere could very well be different over an active region. Depending upon the chromospheric level the height of the emitting layer above the photosphere will change accordingly.

FORBES: I would like to make sure I understand your conclusion. Am I correct in stating that you have ruled out the possibility that the lack of correlation between the 2 cm and 6 cm level is due to a hot magnetic loop which extends to the 2 cm height but not to the 6 cm height?

SHEVGAONKAR: Yes, we have ruled out the possibility of low-level magnetic loops to some extent. The magnetogram shows essentially unipolar magnetic field (as we have considered). I would say that in this particular case, the presence of low-level hot magnetic loops is probably not plausible. However, I agree with you that a small magnetic loop which extends to the 2 cm level but not to the 6 cm height can produce uncorrelated variations in the brightness temperatures at the two wavelengths.

CLARK LAKE RADIO OBSERVATIONS OF CORONAL MASS EJECTIONS

N. Gopalswamy
Astronomy Program
University of Maryland
College Park, Maryland 20742

Abstract

We review some recent studies of mass ejections from the Sun using 2-D imaging observations of the Clark Lake multifrequency radioheliograph. Radio signatures of both fast and slow coronal mass ejections (CMEs) have been observed using the Clark Lake radioheliograph. Using temporal and positional analysis of moving type IV and type II bursts, and white light CMEs we find that the type II's and CMEs need not have a direct cause and effect relationship. Instead, the type II seems to be generated by a "decoupled shock", probably due to an associated flare. The moving type IV burst requires nonthermal particles trapped in magnetic structures associated with the CME. Since nonthermal particles can be generated independent of the speed of CMEs, moving type IV bursts need not be associated only with fast CMEs. Specific examples are presented to support these views.

1. Introduction

Coronal mass ejection (CME) has become a topic of primary interest over the past decade because of its interaction with coronal structures such as streamers and its association with shock waves (manifested as type II bursts) and other distrubances such as plasmoids or magnetic arches (manifested as type IV bursts). Such magnetic structures can be inferred from multifrequency radioheliographic observations (Gopalswamy and Kundu 1989a,b,c). Only a handful of simultaneous observations of CMEs and radio bursts are available and hence their exact physical relationship (e.g., the location of the radio sources with respect to the overall white light transient event) is poorly understood. Extensive reviews of CME radio burst associations are available in literature (e.g. Dulk, 1980; Stewart, 1985; Bougeret, 1985; Hildner et al, 1986). Here we concentrate only two aspects of this association: (i) Is the type II burst due to a shock piston driven by CME or by a blast wave from flare? (ii) Is there any CME speed limit to be associated with radio burst generation? We use specific events observed by the Clark Lake radioheliograph (Kundu et al 1983) and simultaneously by the Coronagraph

E. R. Priest and V. Krishan (eds.), Basic Plasma Processes on the Sun, 495–500.

Polarimeter aboard SMM satellite (SMM-C/P) to address these questions.

2. CME - Type II Association

Based on Skylab observations, there was no clear picture of the relative location of type II bursts and CMEs, which is crucial to understanding their association. MacQueen (1980) showed an example of a type II burst located in front of the leading edge of the CME, suggesting that it was being driven by the CME. Using observations from the P78-1 Solwind Coronagraph and the Culgoora radio heliograph, Sheeley et al (1984) and Robinson and Stewart (1985) found that some type II bursts lay at the leading edge of the CME, and others behind it, but rarely were they seen to precede the leading edge of the CME.

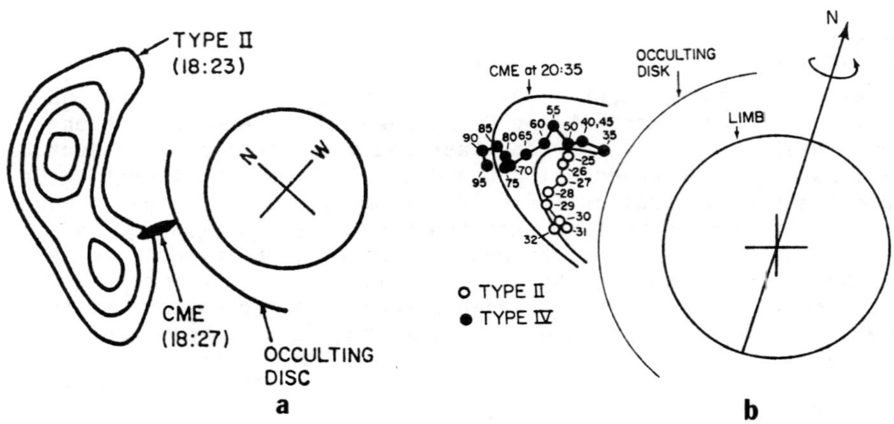

Figure 1. a) Superposition of type II centroids on the SMM-C/P difference image for the June 27, 1984 event. b) Superposition of type II and moving type IV centroids on the SMM-C/P CME loop for the February 17, 1985 event.

In order to understand the physical relationship between CME and type II, one has to consider the following: (i) CME - type II source timing (ii) relative location, (iii) relative velocities and (iv) directions of motion. A type II burst observed (Gary et al 1984) well below the associated SMM-CME leading edge was interpreted as due to blast waves produced by flare that followed the initiation of mass ejection (Wagner and MacQueen, 1983). However, Steinolfson (1984) presented an alternative model to the same event arguing that the shock was driven by the CME and attributing the type II-CME location to the specific geometry needed for the shock-drift acceleration proposed by Holman and Pesses (1983). Steinolfson's (1984) model explains a lower projected height for the type II relative to CME leading edge, but it was outside the CME loop all the time, attaining larger relative height at later times.

In Fig. 1(a) a type II burst at 30 MHz is superposed on the SMM-C/P image of June 27, 1984 CME (Gopalswamy and Kundu, 1987). The CME is a single clump of material with a small speed (350 km s^{-1}). The type II has a large extended structure with two prominent centroids, one ahead of and the other far north of it. The centroids are separated by a distance of ~ 2.2 $R_\odot$. The nearest type II centroid is ~ 0.4 $R_\odot$ ahead of the CME and the farthest one is ~ 1.5 $R_\odot$ away. The overall size of the type II burst is ~ 20 times bigger than that of the visible ejection. Clearly a bow shock ahead of such a small CME is inconsistent with a huge shock implied by the type II. If the shock were generated during the associated flare (which took place ~ 30 min after the initiation of CME) then it takes ~ 9 min to reach the observed height implying a speed of ~ 1500 km s^{-1}. This is ~ 5 times larger than the CME speed and hence the type II might have overtaken the CME. Type II bursts with speeds and trajectories different from those of CMEs have been reported earlier (Gergely, Kundu and Hildner, 1983).

Another example (Kundu et al, 1989) of a type II burst with speed and trajectory different from those of a CME which took place on February 17, 1985 is shown in Fig. 1(b). The filled circles are the centroids of a moving type IV burst. The open circles represent the centroids of 50 MHz type II burst. The speed of the CME and moving type IV are ~ 200 km s^{-1} in the radial direction whereas the type II centroid moved nearly parallel to the limb with a speed of ~ 1100 km s^{-1}. The type II burst started near the northern leg, but moved southward. The location of the type II burst is ~ 0.57 $R_\odot$ below the leading edge of the CME loop. Therefore, the relative locations, directions of motion and speeds of the CME and type II burst strongly suggest that the shock was not generated by the CME. Apparent metric burst source heights tend to be greater than the actual source heights, due to ducting (e.g. Duncan, 1979). Since in this case the apparent height of the type II lies below the CME, we can be confident that the actual height is also below the CME. Therefore, the type II shock might not have been piston-driven by the CME. The best alternative is that the associated flare, (evidenced by metric type III bursts and sharp rise in GOES soft X-rays - Kundu et al 1989) generated the shock. In addition to these examples none of the SMM-CME observations associated with metric type II bursts indicate a cause and effect relation between CMEs and type IIs (Wagner, 1984; Gopalswamy and Kundu, 1989b).

3. Radio Signatures of Coronal Mass Ejections

Based on Skylab CMEs and their temporal association with metric type II/type IV bursts, Gosling et al (1976) concluded the following:

a) If the speed of the CME, V_{CME} < 400 km s^{-1}, no type II or type IV bursts can be associated with the CME.

b) If 400 km s^{-1} < V_{CME} < 700 km s^{-1}, either a type II or a type IV burst can be associated with the CME.

c) If V_{CME} > 700 km s^{-1}, a type II - type IV burst pair can be associated with the CME.

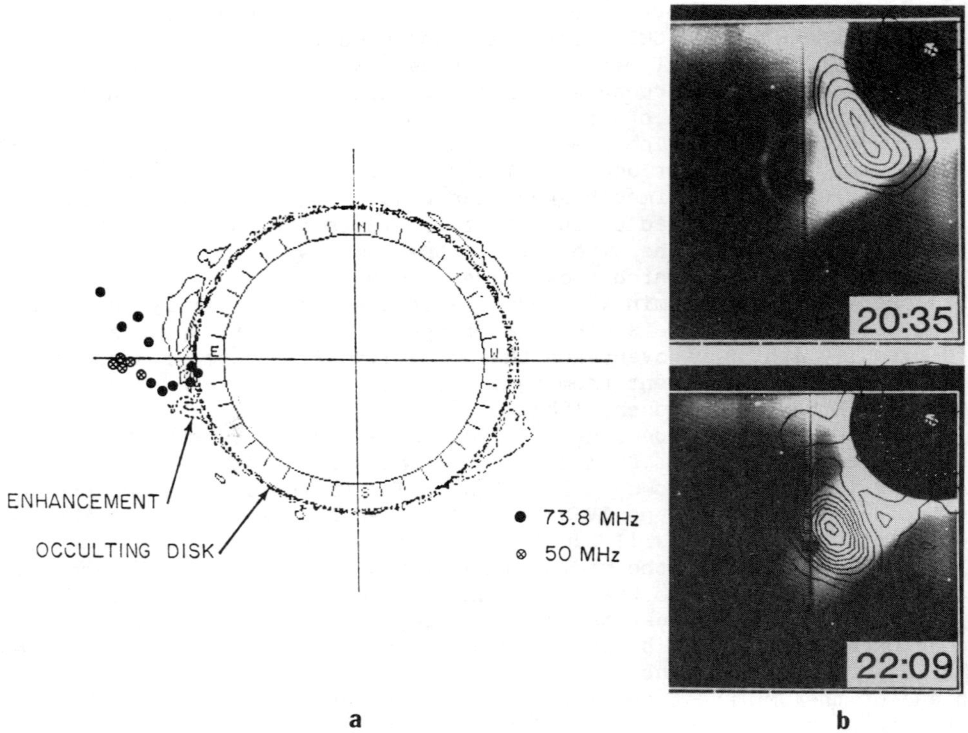

Figure 2. a) Superposition of moving type IV centroids on the Mauna Loa K-coronameter coronal enhancement picture for the February 2, 1986 event. b) Superposition of type II and moving type IV contours on the SMM-C/P CME for the February 17, 1985 event.

This result was based on temporal association. All positional comprisons were done by extrapolating radio and white light observations. During the period of operation of SMM, there were several simultaneous observations which compared the white light features of CMEs with imaging observations of radio bursts (Wagner et al 1981; Gary et al, 1984; Hildner et al 1986; Gopalswamy and Kundu, 1987; 1989b). Most of these events were high speed CMEs, except for the June 27, 1984 event (Gopalswamy and Kundu, 1987) which had a CME speed of ~ 350 km s^{-1} (still close to the 400 km s^{-1} limit), and the results obtained were in agreement with Gosling et al.'s (1976) conclusions. Observations with the SMM-C/P instrument during the sunspot minimum years 1985-1986 have indicated that slow CMEs with speeds $\lesssim$ 100 km s^{-1} occur far more commonly than previously realized (Hundhausen, 1988). Since the Clark Lake multifrequency radioheliograph had a sensitivity far superior to what was available with other instruments, we find that radio bursts of both type

II and type IV could be associated with slow CMEs.

Moving type IV bursts need not have any speed restriction because they primarily depend on non-thermal particles trapped in moving magnetic structures such as plasmoids and loops that accompany the CMEs. Therefore, even in principle, the speed requirement of Gosling et al (1976) seems inconsistent. If there are no non-thermal particles generated during a CME event, then the only radio signature possible is the enhanced thermal emission due to the denser coronal material contained in the CMEs as the thermal bremsstrahlung depends on the thermal electron density. Such an enhanced thermal emission was first reported by Sheridan et al (1978) for a Skylab era CME.

On February 2, 1986 a moving type IV and a continuum were observed associated with a coronal enhancement observed by the Mauna Loa K-coronameter and a filament eruption. The speed of the moving type IV burst was ~ 140 km s^{-1} with almost identical speed for the associated erupting filament (Gopalswamy and Kundu, 1989a). There was no SMM-C/P observations for this day. However, it is known that the Hα filament material found in the cores of several CMEs move only slightly slower than the associated CMEs (Hundhausen, 1988). Hence we believe that the CME speed may not greatly exceed 140 km s^{-1}. Fig. 2(a) shows the centroids of moving type IV bursts associated with the coronal enhancement.

During the February 17, 1985 event (Kundu, 1987; Kundu et al, 1989), both moving IV and tupe II bursts were associated with a slow CME (speed ~ 200 km s^{-1}). In Fig. 2(b) we have superposed the moving type IV burst on the SMM-C/P image for two instants showing that the type IV was always confined to the densest part of the CME (see Fig. 1(b) for the motion of type IV centroids). While the CME and type IV bursts had nearly the same speed and direction, the type II burst location was behind the CME and the motion was transverse. Therefore, only the type IV source can be directly associated with the slow CME. These events show that moving type IV and type II bursts can be associated with slow CMEs with speeds as low as 200 km s^{-1}, contrary to the earlier belief.

4. Conclusions

Temporal association between type IIs and CMEs is quite likely in the corona. However, present imaging studies of type II-CME association do not seem to indicate any causal relationship between CMEs and type II shocks. Type II shocks seem to be generated by flare explosions associated with CMEs. Our observations lead to the conclusion that the association of moving type IV bursts with CMEs depends on the availability of non-thermal particles and the ability of the instrument which detects these emissions. The CME speed requirement to be associated with radio emission does not seem to hold.

Acknowledgements

Research reported in this review, was performed in collaboration with Prof. M.R. Kundu and Drs. E.J. Schmahl, S.M. White, P. Cargill and E. Hildner. This work was supported by NSF grant ATM 88-16008 and by funds from the University of Maryland Astronomy Program, College of Computer, Mathematical and Physical Sciences and the office of Graduate Studies.

References

Bougeret, J.-L.: 1985, in "Collisionless Shocks in the Heliosphere: Review of Current Research", B.T. Tsurutani and R.G. Stone (eds.), American Geophys. Union, Washington, D.C., p. 13.

Dulk, G.A.: 1980, in "Radio Physics of the Sun' (eds: M.R. Kundu and T.E. Gergely), p. 419.

Duncan, R.A.: 1979, Solar Phys. 63, 389.

Gary, D.E., Dulk, G.A., House, L., Illing, R., Sawyer, C., Wagner, W.J., McLean, D.J., and Hildner, E.: 1984, Astron. Astrophys. 134, 222.

Gergely, T.E., Kundu, M.R., and Hildner, E.: 1983, Astrophpys. J. 268, 403.

Gopalswamy, N. and Kundu, M.R.: 1987, Solar Phys. 111, 347.

Gopalswamy, N. and Kundu, M.R.: 1989a, Solar Phys. 122, 145.

Gopalswamy, N. and Kundu, M.R.: 1989b, Solar Phys. 122, 91.

Gopalswamy, N. and Kundu, M.R.: 1989c, Solar Phys. (submitted).

Gosling, J.T., Hildner, E., MacQueen, R.M., Munro, R.H., Poland, A.I., and Ross, C.L.: 1976, Solar Phys. 48, 389.

Hildner, E. and 20 co-authors: 1986, in "Energetic Phenomena on the Sun", M.R. Kundu and B. Woodgate (eds.), NASA Conference Publication No. 2439, p. 6-1.

Holman, G.D., and Pesses, M.E.: 1983, Astrophys. J. 267, 837.

Hundhausen, A.J.: 1988, in the Proc. Sixth International Solar Wind Conference (ed. V.J. Pizzo, T.E. Holzer and D.G. Sime), NCAR Technical Note 306, Vol. I, p. 181.

Kundu, M.R., Erickson, W.C., Gergely, T.E., Mahoney, M.J., and Turner, P.J.: 1983, Solar Phys. 83, 385.

Kundu, M.R.: 1987, Solar Phys. 111, 53.

Kundu, M.R., Gopalswamy, N., White, S.M., Cargill, P., Schmahl, E.J. and Hildner, E.: 1989, Astrophys. J. 347, 505.

MacQueen, R.M.: 1980, Phil. Trans. Roy. Soc. Lon. A297, 605.

Robinson, R.D., and Stewart, R.T: 1985, Solar Phys. 97, 145.

Sheely, N.R., Stewart, R.T., Robinson, R.D., Howard, R.A., Koomen, M.J., and Michels, D.J.: 1984, Astrophys. J. 279, 839.

Sheridan, K.V., Jackson, B.V., McLean, D.J., and Dulk, G.A.: 1978, Proc. Astron. Soc. Australia 3, 249.

Steinolfson, R.S.: 1984, Solar Phys. 94, 193.

Stewart, R.T.: 1985, in "Solar Radio Physics", Cambridge Univ. Press, Cambridge.

Wagner, W.J.: 1984, Ann. Rev. Astron. Astrophys. 22, 267.

Wagner, W.J., and MacQueen, M: 1983, Astron. Astrophys. 120, 136.

Wagner, W.J., Hildner, E., House, L.L., Sawyer, C., Sheridan, K.V. and Dulk. G.A.: 1981, Astrophys. J. 244, L123.

DISCUSSION

FORBES: In my numerical simulation of a magnetically driven CME, the fast shock first appears just above the prominence, and below what one might think of as the top of the CME. So I believe your interpretation of the location of the shock relative to the CME is consistent with my simulation.

VLA OBSERVATIONS OF THE CORONAL PLASMA

KENNETH R. LANG
Department of Physics and Astronomy
Tufts University
Medford,MA 02155
U.S.A.

ABSTRACT. VLA observations at 20-cm wavelength specify the brightness temperature and magnetic structure of plasma constrained within coronal loops in solar active regions. Comparisons with simultaneous SMM observations at soft X-ray wavelengths lead to measurements of physical parameters like electron density, electron temperature and magnetic field strength. Such comparisons also indicate coronal loops can be detected at either radio or X-ray wavelengths while remaining invisible in the other spectral domain, and that the dominant radiation mechanisms can be thermal bremsstrahlung or thermal gyroresonance radiation. VLA observations at the longer 90-cm wavelength reveal the thermal emission of a hot transition sheath enveloping a cooler, underlying Hα filament seen in absorption. The 20-cm VLA observations indicate that the precursor, impulsive and post-flare components of solar flares originate in spatially separated and resolved sources.

I. INTRODUCTION.

This paper provides a brief overview of recent discoveries by the Tufts University group using the Very Large Array (VLA) to study the coronal plasma under the support of the AFOSR and NASA (Section 5). More detailed accounts can be found in the references given in this introduction and in Section 6. Ground-based VLA observations at 20-cm wavelength can detect the hot coronal plasma previously detected by space-bourne X-ray telescopes; detailed comparisons of simultaneous data (SMM and VLA) indicate that physical parameters can be obtained (Section 2), but that some coronal loops are invisible in either spectral domain (Lang, Willson, Smith and Strong 1987a,b). At the longer 91.6-cm wavelength, the VLA detects more extensive emission interpreted as a hot 10^5 K interface between cool, dense Hα filaments and the hotter enveloping rarefied corona (Section 3, Lang and Willson, 1989). The unparalleled spatial resolution of the VLA at 20-cm wavelength has shown that the precursor, impulsive and post-flare components of solar bursts originate in nearby, but separate, coronal loops or systems of loops (Section 4, Willson, Lang and Liggett, 1990).

E. R. Priest and V. Krishan (eds.), Basic Plasma Processes on the Sun, 501–507.

2. QUIESCENT 20 CM AND X-RAY EMISSION FROM CORONAL LOOPS

The development of aperture synthesis telescopes like the Very Large Array (VLA) has permitted ground-based observations of coronal loops at 20-cm wavelength; the quiescent, or non-flaring, loop emission has brightness temperatures, T_B, comparable to the million-degree coronal electron temperature. A comparison with simultaneous soft X-ray images of comparable angular resolution and field of view (Fig. 1) indicates that X-ray coronal loops can be imaged at 20 cm.

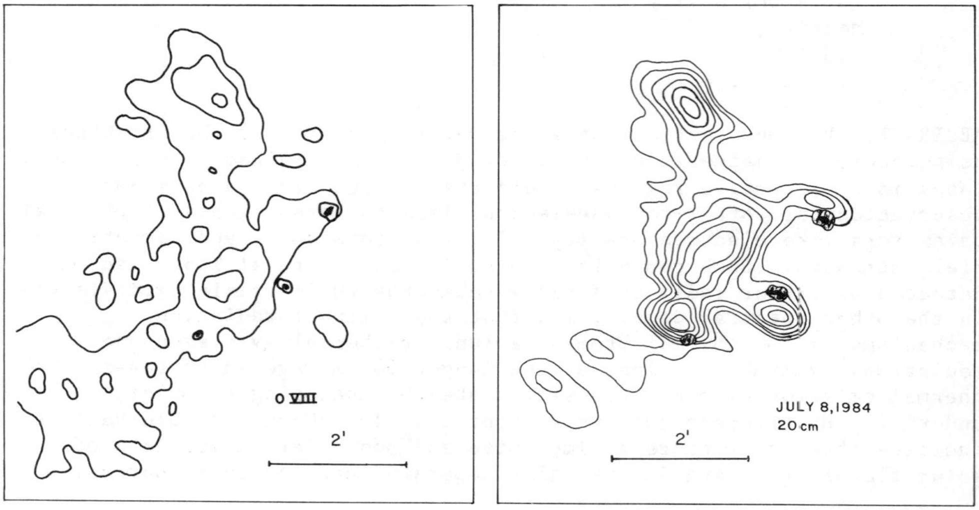

Figure 1. A comparison of soft X-ray (SMM FCS - left) and 20-cm (VLA-right) images of an active region. Here the sunspots are denoted by small black dots with a circle around them (cf. Lang et al. (1987a)) Soft X-ray spectral lines have been used to determine the electron temperature, T_e, and electron density, N_e, of the X-ray emitting plasma that spatially coincides with the 20-cm radiation; for the data shown in Fig. 1, average values of $T_e = 3.0 \pm 0.1 \times 10^6$ K and $N_e = 2.4 \pm 0.4 \times 10^9$ cm^{-3} were obtained. The quiescent X-ray radiation is attributed to thermal bremsstrahlung. The 20-cm emission is optically thin with optical depths $\tau = T_e/T_B = 0.3$. The thermal bremsstrahlung of the X-ray emitting plasma ought to be optically thin at 20 cm, but its thermal gyroresonance radiation should be optically thick at this wavelength. Thermal gyroresonance radiation must account for the intense 20-cm radiation near and above sunspots where no X-ray radiation is detected (also see Fig. 1).

3. QUIESCENT 90 CM EMISSION FROM FILAMENTS

Although 20-cm VLA observations of the quiet Sun reveal the ubiguitous coronal loops anchored within solar active regions, the VLA results at the longer 91.6-cm wavelength reveal quiescent emission from more extensive structures (angular sizes $\theta = 3'$) that are not associated with active regions (Fig. 2 right and left). The 91.6-cm features are similar in shape, position, elongation and orientation to dark Hα filaments; but the radio structures are wider and longer, and they are detected in emission rather than absorption. This 91.6-cm emission has been interpreted as the thermal bremsstrahlung of a hot (10^5K), transition sheath that envelopes the cooler Hα filaments and acts as an interface with the hotter surrounding corona. This transition sheath has been previously detected in absorption with lower brightness temperatures at shorter wavelengths of 6 and 20 cm (Rao and Kundu, 1980; Kundu, Melozzi, and Shevgaonkar, 1986).

Variable physical parameters of the transition sheath can explain controversial reports of the detection of, or the failure to detect, the meter-wavelength counterpart of Hα filaments. If the sheath were substantially thinner, then the optical depth would not be large enough for detection of its 91.6-cm bremsstrahlung; and if the electron density of the sheath was much higher than $N_e = 10^9$ cm^{-3}, then its plasma frequency would exceed our observing frequency and the sheath radiation could not propagate out to be observed.

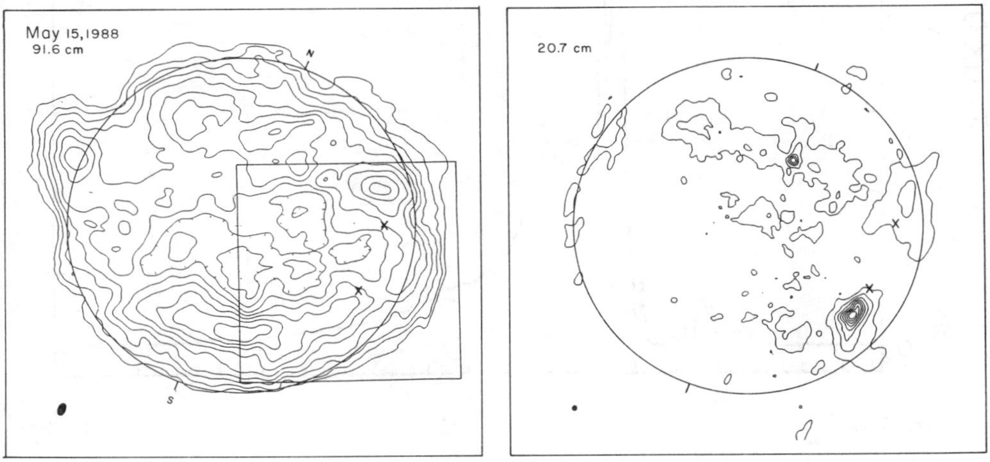

Figure 2. The quiescent 91.6 cm (left) and 20.7 cm (right) emission from the Sun. The 20.7-cm contours delineate active-region coronal loops with a peak brightness temperature of $T_B = 2.5 \times 10^6$ K, while the 91.6-cm contours show more elongated structures that are not associated with active regions and have a peak $T_B = 7.8 \times 10^5$ K. (Adapted from Lang and Willson (1989)).

4. RESOLVING FLARE COMPONENTS

The classical time profile for the microwave or radio emission of solar flares, or eruptions, consists of a low-level precursor, and a rapid powerful impulsive component, followed by a more gradual post-burst, or decay, phase (see Fig. 3). The VLA has now been used to map these flaring components at 3-second time intervals, showing that they originate in spatially separated sources (Fig. 4).

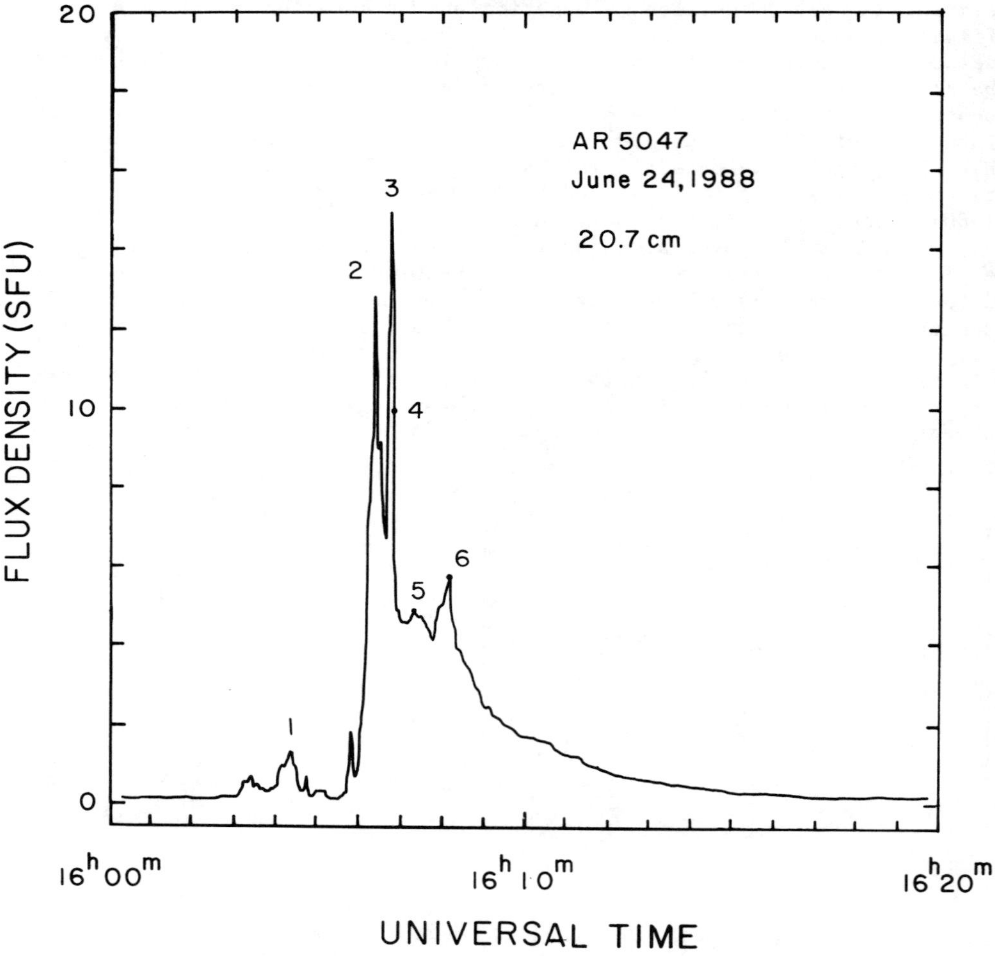

Figure 3. The time profile of a 20.7-cm burst showing the preburst (1), impulsive (2,3), and post-burst (4, 5, 6) phases. The VLA snapshot maps given in Fig. 4 indicate that these three phases originate from spatially separated, resolved components. (Adapted from Willson, Lang and Liggett (1990)).

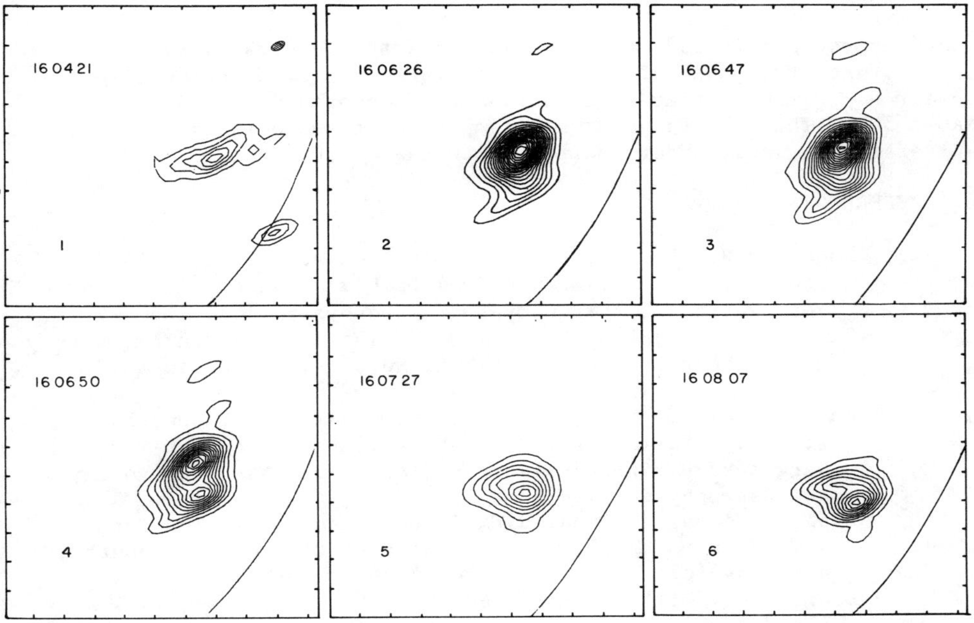

Figure 4. VLA snapshot maps of the total intensity, I, at 20.7-cm wavelength during 3.3-second intervals at the times denoted by 1 to 6 in Fig. 3. The precursor (1) is spatially separated from the subsequent impulsive bursts (2 and 3). The post-burst, or decay, phase (5, 6) originates in another spatially separate source that first becomes detectable during the end (4) of the second impulsive burst (3). The angular scale can be inferred from the 60" spacing between the fiducial marks on the axes. The contour intervals are in units of equal brightness temperature, T_B, with an outermost contour and contour interval of $T_B = 4.4 \times 10^5$ K for 1, and $T_B = 5.5 \times 10^6$ K for the others five images. The maximum value is $T_{Bmax} = 4.4 \times 10^6$, 1.0×10^8, 1.3×10^8, 7.7×10^7, 4.4×10^7 and 5.5×10^7 K for 1, 2, 3, 4, 5 and 6 respectively. (Adapted from Willson, Lang and Liggett (1990)).

These results suggest that solar flares are triggered by interacting coronal loops; emerging coronal loops may interact with adjacent ones, leading to the explosive release of magnetic energy stored within them. Evidence for the magnetic triggering of solar flares by interactng and reconnecting coronal loops has been provided by previous VLA observations at 20 cm and 6 cm wavelength (Kundu et al., 1982; Kundu and Lang, 1985). The system of coronal loops then relaxes to a spatially-different configuration during the decay phase.

5. ACKNOWLEDGMENTS.

Radio astronomical studies of the Sun at Tufts University are supported under grant AFOSR-89-0147 with the Air Force Office of Scientific Research. Related solar observations are supported by NASA grant NAG 5-501. The Very Large Array is operated by Associated Universities, Inc., under contract with the N.S.F.

6. REFERENCES.

Kundu, M.R., Melozzi, M., and Shevgaonkar, R.K. (1986) 'A study of solar filaments from high resolution microwave observations', Astronomy and Astrophysics 167, 166-172.

Kundu, M.R. and Lang, K.R. (1985) 'The Sun and nearby stars - microwave observations at high resolution', Science 228, 9-15.

Kundu, M.R., Schmahl, E.J., Velusamy, T. and Vlahos, L. (1982). 'Radio imaging of solar flares using the Very Large Array - new insight in flare processes', Astronomy and Astrophysics 108, 188-194.

Lang, K.R. and Willson, R.F. (1989) 'Radio emission from quiescent solar filaments at 91.6 centimeter wavelength', Astrophysical Journal (Letters) 344, L73-L75.

Lang, K.R., Willson, R.F., Smith, K.L, and Strong, K.T. (1987a) 'Simultaneous SMM flat crystal spectrometer and Very Large Array observations of solar active regions', Astrophysical Journal 322, 1035-1043.

Lang, K.R., Willson, R.F., Smith, K.L., and Strong, K.T. (1987b) 'Solar active region physical parameters inferred from a thermal cyclotron line and soft X-ray spectral lines', Astrophysical Journal 322, 1044-1051.

Rao, A.P. and Kundu, M.R. (1980) 'Synthesized map of a solar filament at 6 cm with ≈ 15" resolution', Astronomy and Astrophysics 86, 373-376.

Willson, R.F., Lang, K.R., and Liggett, M. (1990) 'Impulsive microwave burst and solar noise storm emission resolved with the VLA', Astrophysical Journal, February 20, 1990.

DISCUSSION

CHIUDERI DRAGO: The filament brightness temperatures given by Kundu *et al.* are not the observed ones, but they have been corrected for the coronal contribution. The coronal parameters used by Kundu *et al.* correspond to a coronal pressure $p \simeq 0.2$ dyne/cm^2, one order of magnitude larger than the pressure at the top of the transition region, in their "variable pressure model" - this inconsistency does not affect very much the data at short ($\lambda < 21$ cm) wavelengths, but it would affect very much the T_b at 90 cm. If the parameters you used for the coronal contribution subtraction are the same as those used by Kundu *et al.*, the agreement with the "variable pressure model" gives an inconsistency. If you take, for the computation of the coronal contribution, parameters in agreement with the variable pressure model, then you cannot compare your data with the Kundu *et al.* ones.

VAN BALLEGOOIJEN: Have you ever observed polarization in filament radio emission?

LANG: Our observations of filament emission at 90 cm wavelength were unpolarized, probably because the thermal radiation of the 10^5 K plasma was optically thick. However, these were the first such observations. So, when the Sun becomes less active, we hope to observe some emission that is optically thin and circularly polarized. This will let us determine the shape and strength of the filament magnetic field for the first time. We also hope to carry out strict tests of the theory by observing the same filament at several wavelengths, say 2, 6, 20 and 90 cm, and for varying angles between the line of sight and the magnetic field.

GELFREIKH: How do the size and shape of the radio (90 cm) filaments compare with optical (Hα) filaments?

LANG: The 90 cm filaments have the same size, shape, position and elongation as the underlying Hα filaments, but the radio emission is wider and longer. The transition sheath is thought to be about 100 km thick. We should be able to measure this thickness in future.

Type III bursts traced from the solar surface to 1 AU

Yolande LEBLANC
Observatoire de Paris, 92195 Meudon CEDEX, France
Bernard V. JACKSON
Center for Astrophysics and Space Sciences
San Diego, California 92093, USA

ABSTRACT We trace type III radio bursts from the sun until they can be observed as *in situ* electrons at the ISEE-3 spacecraft. Our study extends over the period of operation of the electron experiment on ISEE-3 from August 1978 to November 1979. Our observations include data from solar flares, kilometric type III burst spectra from ISEE-3, and *in situ* measurements of low energy electrons from ISEE-3. By carefully restricting the data sets involved, we find a peak of 20° width in the number of flares associated with *in situ* electrons near 60° west solar longitude. This peak shows that the electron beams of type III bursts are not spread over a great range of longitudes as earlier studies indicate, but are fairly limited in spatial extent. However, even in that longitude range only 40% of type III-flare events are associated with *in situ* electrons.

I. INTRODUCTION

Type III radio bursts are one of the most basic probes of solar particle acceleration mechanisms and interplanetary plasma radiation process. At metric wavelengths these radio bursts are often temporally and spatially associated with solar flares (Wild et al., 1959, Jackson and Sheridan, 1979, Kundu et al., 1982, Poquerusse, 1988). The actual radio-wave production is caused by the formation of radio emission at the local plasma frequency from a passing stream of electrons accelerated to energies between 10 to 100keV. To prove the point, type III bursts can be traced outward from the sun in metric and kilometric radiation until at 1AU where the electrons which cause the bursts are detected along with oscillations of the local plasma *in situ* (Lin et al., 1974; Fainberg and Stone, 1974; Gurnett and Anderson, 1977).

The rigidity of electrons of these energies will usually cause them to travel along pre-existing coronal and interplanetary magnetic field lines (Anderson, 1981), and thus should confine them to the field lines emanating from near the flare site. In theory, it should therefore be a simple matter to use the positions of solar flares and observations of electrons *in situ* to determine the width of the electron beams which cause type III radio bursts. In several studies, *in situ* electrons have been related to earlier-observed flare longitudes (eg., Lin, 1971; Wang, 1972; Lin et al., 1974). All of these studies find a general association of flares at west longitudes associated with the *in situ* electrons although the random coincidence level is very high, as demonstrated in next section. None of these studies use type III kilometric (km) bursts to restrict the timing of the flare relative to the observed *in situ* electrons.

E. R. Priest and V. Krishan (eds.), Basic Plasma Processes on the Sun, 509–512.

Lin (1974) reported that ~80% of his electron flares were associated with radio type III emission. Alvarez et al. (1972) first attempted to trace flare-associated type III bursts by using kilometric (Km) bursts observed by OGO-5 (1968 - 1971) and electrons observed by the Imp 5 and the Explorer 35 spacecrafts. Their results showed that the associated flare longitudes of type III *in situ* electrons are spread over a wide range of west longitudes. However, their data set is small; only 26 *in situ* events were associated with the km type III bursts and flares they observed. No peak in the number of flares at 50° to 70° west longitude was observed. Thus in this scenario, the electron beam must spread over a large distance from the flare site.

We believe that previous studies have often misidentified a given flare as the origin of type III electrons observed *in situ*. Only when strict limits are placed on the association of these flares with type III bursts and the corresponding *in situ* electrons that the number of random coincidences decreases to the point where a true average beam shape can be better discerned. In section II, using observations from the UCal particle experiment and the radio experiments from the ISEE-3 Spacecraft we place strict limits on flares associated with *in situ* electrons and km bursts. Using these associations we determine the longitudinal distribution of the associated flares, and discuss the implications of this result.

II. SOLAR RADIO BURSTS AND ASSOCIATED *IN SITU* ELECTRONS

For the purpose of this study, we are interested in keeping random coincidences to a minimum. Therefore we consider only those events where a flare had an isolated kilometric type III burst onset in progress between the flare onset and its maximum (flare window). Our observations of flare times are obtained from *Solar Geophysical Data (Prompt and Comprehensive Reports)*. We found that approximately one-quarter of the events in our sample were multiply defined: two or more flares differing greatly in longitude fulfilled our association criteria for the same *in situ* electron event. These events were deleted from the histograms in the following sub-sections. We found a high random coincidence level of flares and subflares with electron events. Under ideal observing conditions there was approximately one flare or subflare every 48 minutes during the time interval of the electron events in 1978-1979. In addition, we found that each flare averages approximately 6 minutes in duration from onset to maximum.

II.1 Association of flares to *in situ* electron events

In our study we consider only the *in situ* electrons which occur within 15 to 60 minutes after the flare window. We thus estimate that a purely random coincidence occurs (45+6)/48 or of about 100% of the time. Therefore the histograms giving the flare distribution of electron events must be considered with caution (Lin, 1971 ; Wang, 1972 ; Lin, 1974). Figure 1 shows the distribution of solar flares associated with electron events. Although we estimate a high random coincidence level (dashed line), a slight East-West assymetry is observed.

II.2 Association of flares to kilometric radio bursts.

To further limit the random concidence level, we consider the timing of associated km type III burst. We use the onset of the burst and a 3 min leeway at each end of each flare observation. Thus, the random coincidence level that a type III burst is associated with a flare is (6+6)/48, i.e. 25%. Figure 2 shows the distribution in longitude of flares associated with km type III bursts. This distribution is symetrical with respect to the central meridian. Poquerusse (1988), who undertook a similar study with *metric* type III bursts, found a broad maximum approximately 30° east of the central meridian.

II.3 Association of flares to km bursts and to electron events.

If we assume that only flares with observed km bursts give rise to *in situ electrons*, and we limit the *in situ* electron events to those flares where we observe a km burst, then we cut the random coincidence level to at least 25%. For purely random associations of flares, km bursts and *in situ* events the coincidence level reduces to 100%×25% or 25%. The hatched area in Figure 2 represents the distribution in longitude of the flares associated with km type III bursts and electron events. There are 106 events in this area and 702 events in the overall curve representing the flares associated with km bursts, i.e. a ratio of one over seven. This result is in good agreement with our estimation of the occurrence of km burst, every 2.9 hours, and the occurrence of an electron event every 20.5 hours. The electron-km-flare-histogram shows clearly a peak from 50° to 70° west longitude. However, we observe electrons from east limb flares, and the number of these are well above the random coincidence level. Figure 3 compares the percentage of electron-km-flare events to km-flare events for each 20° longitude, and shows how the propagation along magnetic field lines near the sun and interplanetary space broadens the electron beam in longitude. We note also that even in the preferred longitude range, only 40% of km-flares are associated with *in siti* electrons.

III. CONCLUSION

In this study we have shown that out of 702 flares with associated km radiation, only 106 *in situ* electron events could be anambigiously associated with the preceeding flares – about one third of them being observed between 50° and 70° west longitude. The implication of this result is that the electrons do not spread over a great range of longitudes as earlier studies indicate, but are fairly limited in angular extent. The observed peak in flare longitude distribution is directly related to the most direct path that electrons can take from the sun to earth along the magnetic field Archimedean spiral.

Acknowledgements: B. Jackson would like to thank the Observatory of Paris for supporting his stay during the month of February 1989. Special thanks are also due to R. Lin and J.L. Steinberg for making available to us the electron and radio data, and to S. Dubus for her help in reducing the data.

References

Alvarez, H., Haddock, F. and Lin, R.P.: 1972, *Solar Phys.* **26**, 468.
Anderson K.A., McFadden, J.P. and Lin, R.P.: 1981, *Geophys. Res. Lett.* **8**, 831.
Fainberg, J. and Stone, R.G.: 1974, *Space Sci. Revs.* **16**, 145.
Gurnett, D.A. and Anderson, R.R.: 1977, *J. Geophys. Res.* **82**, 632.
Jackson, B.V. and Sheridan, K.V.: 1979, *Proc. Astron. Soc. of Australia* **3**, 383.
Kundu, M.R., Gergely, T.E. and Kane, S.R.: 1982, *Solar Phys.* **79**, 107.
Lin, R.P.: 1971, *12th Int. Conf. on Cosmic Rays* **5**, 1805.
Lin, R.P.: 1974, *Space Sci. Revs.* **16**, 189.
Poquerusse, M. : 1988, IAU Colloquium NO 104 on "Solar and Stellar flares"
Wang, J.R.: 1972, *Goddard SFC Document X-661-72-76.*
Wild, J.P., Sheridan, K.V. and Neyland, A.A.: 1959, *Australian J. Phys.* **12**, 369.

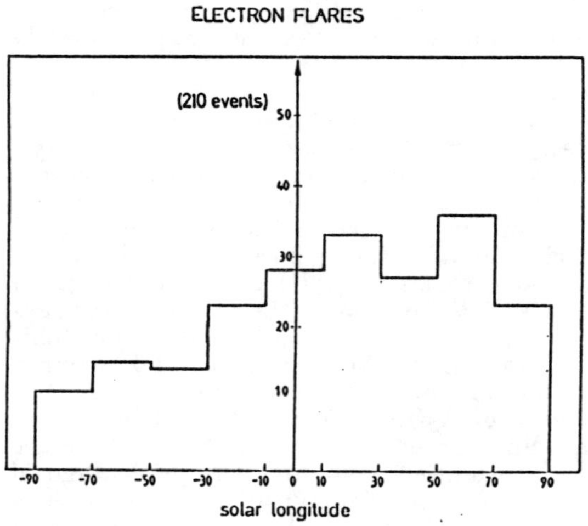

Fig. 1. Longitudinal distribution electron events temporally associated with solar flares.

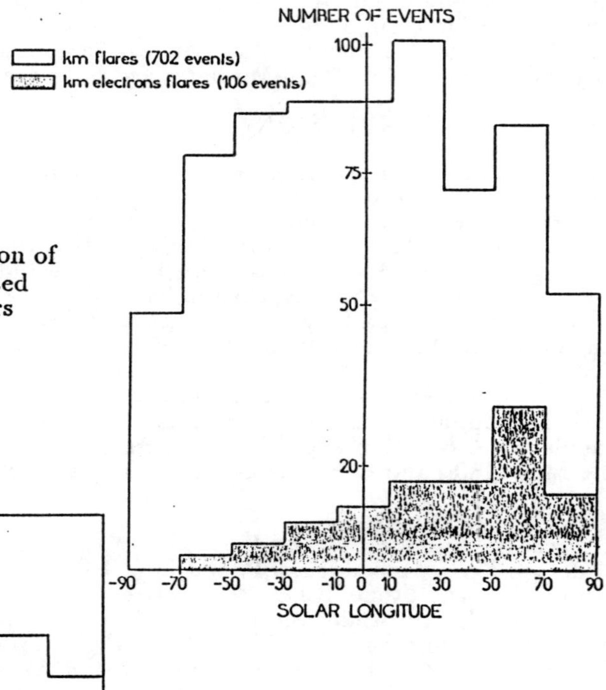

Fig. 2 Longitudinal distribution of km bursts temporally associated with flares. The grey area refers to km bursts associated with flares and electrons events.

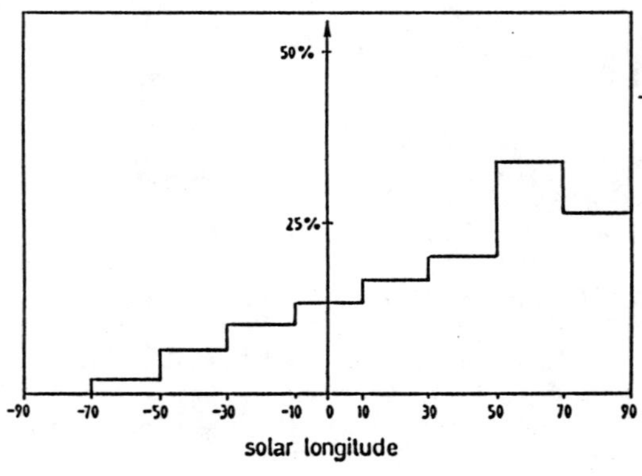

Fig. 3. Longitudinal distribution of the percentage of *in situ* electrons from each 20° longitude.

Observations of solar continuum emission at decameter Wavelengths

Ch.V.Sastry
Indian Institute of Astrophysics
and
Raman Research Institute
Bangalore, INDIA

Abstract

We observed the continuum emission from the radio sun when there is no burst activity at $\lambda = 8.7$ m with the large decameter wave radio telescope at Gauribidanur (Latitude 13° 36' 12" N and 77° 27' 07" E) with a resolution of 26'/40'. A compound grating interferometer with one dimensional resolution of 3' is also used. These observations are made during August 1983 and June 1986. The brightness temperature at the center of the sun varied from $0.2\ 10^6$ K to $0.8\ 10^6$ K during these periods on time scales of several hours to a day. Since the sun is absolutely quiet during these periods we believe that the radiation is purely thermal in nature. In this case the observed brightness temperature variations imply large scale density variations by more than a factor of three if the corona is optically thin at these wavelengths. Alternatively if the corona is optically thick the observations imply real electron temperature variations with or without accompanying density variations.

(A full account of this paper will be published elsewhere).

DISCUSSION

GELFREIKH: Dr Borovik at Pulkovo Observatory has detected a decrease of solar radio radius at dm wavelengths when a coronal hole is at the limb. Did you try to detect a similar effect with your instrument?

SASTRY: Yes. We have observed similar effects at 34.5 MHz. A comparison of our scans with those of Nancay (169 MHz) reveals that the effect is more pronounced at decametric wavelengths.

LANG: Is the displacement observed between the 34.5 MHz emission peak and the centre of the Sun concentrated toward the equator or above active regions?

E. R. Priest and V. Krishan (eds.), Basic Plasma Processes on the Sun, 513–514.

KUNDU: What Sastry has shown is that at the time of solar minimum, there is a small displacement of the image (at 34.5 MHz) centred from that of the Sun centre. If one assumes that this small displacement is due to ionospheric refraction, then there must not be any contribution from condensations to the decameter emission at solar minimum. This implies that only the quiet Sun thermal emission is being considered.

SASTRY: Yes, I agree.

BEHAVIOUR OF WHISTLERS IN CORONAL MAGNETIC TRAPS AND ITS RELEVANCE TO A NEW FINE STRUCTURE IN SOLAR TYPE IV RADIO BURSTS

G.P.CHERNOV
IZMIRAN, 142092 Moscow region, Troitsk, USSR

A possible contribution of whistlers to fine structure production due to the coalescence process with plasma waves l + w --> t at frequencies $\omega^l \pm \omega^w = \omega^t$ is estimated. A whistler ray tracing in the solar corona with the twice Newkirk density model and with a dipole magnetic field using a standard numerical integration of the Haselgrove two-dimensional equations is performed.

In the corona whistlers can canalize along density and magnetic field inhomogenieties (ducts) with a width $\lesssim 2 \cdot 10^8$ cm. Sufficiently narrowband whistlers (obliquely propagating) can undergo numerical reflections (Maltseva and Chernov, 1989a). Local kinetic increments, decrements by thermal plasma (cyclotron and Landau) and integral growth rates along trajectories of whistlers propagating under arbitrary angles to the magnetic field are computed for the distribution function of a beam of hot electrons with a loss-cone and temperature anisotropies. Predominant Landau damping at frequencies $f^w/f_{He} \lesssim 0.1$ with a maximum at moderate angles to the magnetic field 40-50 is discovered (Maltseva and Chernov, 1989b).

A consideration of whistler scattering by thermal ions and electrons and ion-sound wave coalesence for whistlers (s + s\`--> w) permits us to distinguish the following whistler manifestations.

A whistler spectrum formed by means of coupling processes with s-waves in a pulsating regime can yield millisecond radioemission pulsations and spike-bursts in dm- and metric ranges. It is proposed to distinguish pulsations by whistlers from an MHD-model with FMS-waves and a plasma model with the aid of a modulation depth dependence versus a pulsation period (Chernov, 1989).

A whistler cyclotron instability at the normal Doppler frequency shift and whistler upward duct propagation along a magnetic trap causes fiber bursts with intermediate frequency drift. Periodic whistler wave packets generated with the anomalous Doppler cyclotron resonance (due to bounce-movements of hot electrons at magnetic loop tops) yields zebra-pattern stripes.

A periodic loss-cone instability during the annigilation of magnetic islands after a flare in reconnection regions inside a small trap between fast shock fronts (the model of Forbes and Priest, 1983 and Hick and Priest, 1989) explains the formation of a new fine structure, namely narrow band ropes of fiber-bursts (see Fig.1a and Aurass et al., 1987).

E. R. Priest and V. Krishan (eds.), Basic Plasma Processes on the Sun, 515–516.

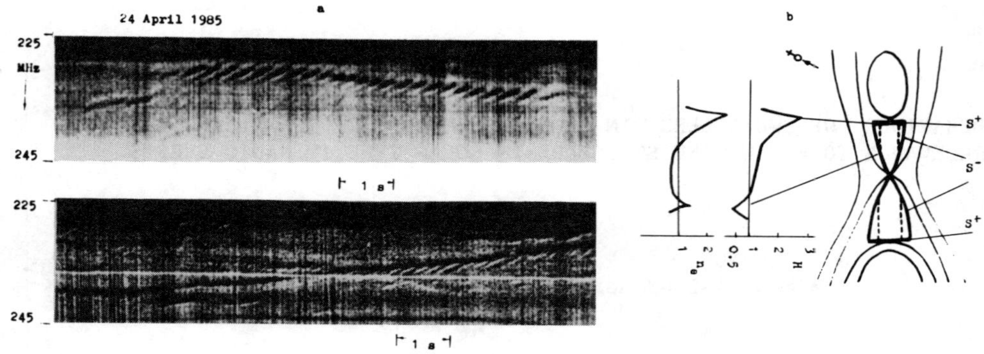

Fig.1. Rope-like chains of fiber bursts in the range 225-245 MHz (a); b- qualitative scheme of a quasistationary reconnection during the magnetic loop restoration after a flare (Forbes and Priest, 1983) and approximate variation of density (n_e) and magnetic field (H) in a reconnection region (see Sidneva and Semenov, 1985).

During a quasistationary reconnection with a neutral point of type X two pair of slow shocks and two fast shocks are formed. Between fast shock fronts (S^+) a new magnetic trap produces (Fig.1b). Near a fast shock front the magnetic field grows more than twice, maximal whistler increments may achieve enough high values $\gamma/\omega_{He} \sim 10^{-2}$ and the periodic whistler instability develops due to the bounce motions of fast electrons between shock fronts (S^+) distant at about 10^9 cm. Hence it follows the more high periodicity of fibers in a single rope ~ 5-$6\ s^{-1}$, than usual one ~ 1-$2\ s^{-1}$. Whistler wave packets can propagate from the fast fronts up to the slow shock fronts (S^-) where the magnetic field strength decrease in some times (see Sidneva and Semenov, 1985) and whistlers decay by the cyclotron damping. This distance (~ 2-$3 \cdot 10^8$ cm) defines the bandwidth of rope-like chains ($\sim$2-3 MHz).

The proposed scheme explains not only rope-like fibers but also other accompanying fine structures, namely slow drifting fibers in absorption only due to the screening of the emission from rarefaction regions by the slow shock fronts and millisecond pulsations according to the model of a pulsating regime of whistlers and ion-sound waves in a reconnection region (Chernov, 1989).

References

Aurass H., Chernov G.P., Karlicky M., Kurths J. and Mann G. (1987) Sol. Phys., 112, 347-357.

Chernov G.P. (1989) Astronomichesky J., 66, N.6, 1258-1270 (in Russian).

Forbes T.G. and Priest E.R. (1983) Sol. Phys., 84, 169-188.

Hick P. and Priest E.R. (1989) Sol. Phys., 122, 111-129.

Maltseva O.A. and Chernov G.P.(1989) a- Kinematic. Physic. nebesnyh tel, 5, N.5, 32-43; b- ibid, 44-54.

Sidneva M.V. and Semenov V.S. (1985) Geomagnetism i aeronomija, 25, N.5, 806-811 (in Russian).

HIGHER HARMONIC PLASMA RADIATION IN TYPE II SOLAR RADIO BURSTS

V.V.FOMICHEV, I.M.CHERTOK, R.V.GORGUTSA, A.K.MARKEEV
IZMIRAN, 142092 Moscow region, Troitsk, USSR
B.KLIEM, H.AURASS, A.KRUGER, J.KURTS
Zentralinstitute fur Astrophysik, AdW der DDR,
DDR - 1591, Potsdam
H.URBARZ
Astronomiches Institute der Universitat Tubingen
Aussenstelle Weissenaw 7980 Rasthalde/Ravensburg, BRD

The large behind -limb (W 140, 0900 UT) flare on February,16, 1984 was followed by a complex of radio bursts - type II, III, IV and V bursts.

TWO MAIN PECULIARITIES:
1. U - SHAPED DYNAMIC SPECTRUM - the propagation of shock wave along the coronal loop, connecting AR 4408 and AR 4410(W 95). Radio source - quasi parallel shock wave.
2. SIMULTANEOUS EMISSION OF THREE (OR FOUR) NEARLY HARMONICALLY RELATED BANDS (A - D).

	t_1	t_2		t_1	t_2		t_1	t_2
f_A/f_D	5.1	5.1	f_A/f_C	3.2	2.9	f_A/f_B	1.8	1.6
f_B/f_D	2.9	3.3	f_B/f_C	1.8	1.9	f_B/f_C	1.8	1.9
f_C/f_D	1.6	1.7				f_C/f_D	1.6	1.7

Frequency ratios for t_1 =09:00:50 UT and t_2 =09:03:50 UT.

Several possibilities of a connection between the bands:

2.1. Splitted F - H burst. Bands C, D are the fundamental component and bands A, B are the second harmonic component of type II burst with band splitting. This interpretation is inconsistent with the frequency ratios.

2.2. Two F - H bursts. Two coaligned emitting loops should be exist, and their densities differ by factor $(f_B/f_D)^2 \sim 10$.

2.3. Simple harmonik system. If the basic band is band D, then the fourth harmonic is missed. If the basic band is band C, the correspondence to the observed frequensy ratios is good, band D must be another type II burst. The herringborn structure in band B is not a duplicate of the structure in band C.

E. R. Priest and V. Krishan (eds.), Basic Plasma Processes on the Sun, 517–518.

2.4. Double resonanse. Generation at the levels where the condition $f_{uh}/f_{ce}=n$ (n=1,2...) is fulfilled. Bands at lower frequency should be exited with a time delay during the outward propagation of the shock wave, which is not observed.

2.5. Successive harmonics. Each of bands A,B,C is the near-harmonic of its low-frequency neighbour (B,C,D respectively). The generation of band B would then require the following interaction

$$t(\leq 2f_{pe}) + t(\leq 2f_{pe}) \pm \mu \longrightarrow t(\leq 4f_{pe} \pm f_{\mu})$$

μ is unspecified, probably e.m. mode.
However, the process with one transverse wave near f_{pe} is also possible. Two bands near f_B would result. Moreover, excitation of band A by coalescence of waves in band B would be negligible in comparison to coalescence of waves in band B with waves in the much more intence bands C and D. We have thus to abandon this hypothesis.

2.6. Wave decay and coalescence. This mechanism includes generation e.m. waves at 3/2 f_{pe} according to scheme

$$t(f_{pe}) \longrightarrow l(f_{pe}/2) + l(f_{pe}/2)$$
$$t(f_{pe}) + l(f_{pe}/2) \longrightarrow t(3f_{pe}/2)$$

Such a sheme works also for e.m. waves at $2f_{pe}$ to produce emission at $3f_{pe}$. Four emisssion bands can result with frequency ratios 1:1,5:2:3, which differ clearly from the the registrated ratios.

2.7. Modified harmonic system. The bands A-D are a harmonic system produced in part by four-wave interaction with appropriate shift off the integer harmonics. Suppose that, besides the Langmuir waves (or possibly uh-waves) an e.s. wave mode σ at frequencies $f_{pi} << f < f_{pe}$ is excited to a nonthermal level in the source volume. Then the four-wave interaction

$$l + t(\approx n\, f_{pe}) + \sigma \longrightarrow t(\approx (n+1)f_{pe}),\ n \geq 2$$

may be most effective in producing higher harmonics. More preciesely, transverse waves near $f = nf_{pe} + (n-2)f_{\sigma}$ are upconverted to transverse waves near $f = (n+1)f_{pe} + (n-1)f_{\sigma}$ for $n \geq 2$.
Test of hypothesis (2.7). The most probable candidate for mode σ are the e.s. electron cyclotron harmonic waves (CHW). Since instabilities of the CHW show the tendency to possess the lowest thresholds and highest growth rates in the lowest passbands $n < f/f_{ce} < n+1$; n=0,1, we put $f_{\sigma} \sim (0.5-1.5)f_{ce}$. In order to obtain a marked frequency shift, the ratio f_{pe}/f_{ce} must not be too large; we choose $f_{pe}/f_{ce} \sim 3$. The combination of one of the processes $l \pm s \longrightarrow t(\approx f_{pe})$ or $l \longrightarrow t(\approx f_{pe})$ for fundamental emission with the process $l + l' \longrightarrow t(2f_{pe})$ for the second harmonic generation and four-wave interaction for higher harmonic generation is capable to yield rough agreement with all observed ratios. However, this conformity is obtained only for $f \geq 0.5\ f_{pe}$.

CONCLUSION. The possibility of frequency-shifted higher harmonic plasma radiation, discussed here, where the occurence and frequency shift of the higher harmonics are both related to the nonthermal excitation of an appropriate mode at "intermediate" frequencies, should revive the search for higher harmonics in solar radio bursts.

ABSORPTION OF ELECTROMAGNETIC WAVES IN ASTROPHYSICAL PLASMAS

GANGADHARA, R.T
DEPARTMENT OF PHYSICS.
INDIAN INSTITUTE OF SCIENCE, BANGALORE-560012, INDIA
and
KRISHAN, V
INDIAN INSTITUTE OF ASTROPHYSICS,
BANGALORE-560034, INDIA

ABSTRACT: We study Parametric Decay Instabilities(PDI) using the kinetic description, in the homogeneous and unmagnetized plasmas. These instabilities cause anomalous absorption of the incident electromagnetic (e.m)radiation. The maximum plasma temperatures reached are functions of luminosity of the non-thermal radio radiation and the plasma parameters.

1. INTRODUCTION

Quasars, seyfert galaxies and pulsars are among the most luminous objects and emit strongly at radio frequencies. Plasma clouds of electron density $10^4 \leq n_e \leq 10^{10}\ cm^{-3}$ exist in their atmosphere. The conditions for PDI are known to be satisfied in the emission line regions (Krishan 1987). We extend the study of PDI in the entire range of plasma and pump parameters.

2. PARAMETRIC DECAY INSTABILITIES.

The incident e.m pump wave (ω_0, K_0) excites a high frequency electron plasma wave (ω_e, K_e) and a low frequency ion-acoustic wave (ω_i, K_i) such that $\omega_0 = \omega_e + \omega_i$ and $K_0 = K_e + K_i$. The PDI can be described using the dispersion relation (Liu & Kaw 1976)

$$\frac{1}{\chi_e} + \frac{1}{\chi_i + 1} = -(\mathbf{K}.\mathbf{V}_0)^2 \left[\frac{1}{(\omega-\omega_0)^2\, \varepsilon(\mathbf{K}, \omega-\omega_0)} + \frac{1}{(\omega+\omega_0)^2\, \varepsilon(\mathbf{K}, \omega+\omega_0)} \right] \qquad \text{...........(1)}$$

Where $\chi_e(\mathbf{K},\omega)$ & $\chi_i(\mathbf{K},\omega)$ are electron & ion susceptibilities respectively. ε is the dielectric function. $v_0 = eE_0/m_e\omega_0$ is the oscillation velocity of electrons in the pump field E_0. $E_0 = (L/Rc^2)^{1/2}$: $L = L41 \times 10^{41}$ erg/sec and $R = rpc \times 3.10^{18}$ cm are the luminosity and distance between the source and plasma respectively. We assume $k_e \simeq k_i = k, k_0 \simeq 0$ and $\omega_0 \simeq \omega_e + \omega_i$. Consider $k\lambda_{De} \ll 1$, $\omega_{pi} \gg \omega_i \simeq kc_s$, and neglecting $\varepsilon(K, \omega+\omega_0)$ as being off-resonant. ω_{pe} & ω_{pi} are electron & ion plasma frequencies respectively. λ_{De} is the Debye length. c_s is the ion acoustic velocity.

E. R. Priest and V. Krishan (eds.), Basic Plasma Processes on the Sun, 519–520.

The equation (1) is solved for $kv_e \gg \omega \geq kv_i$ & $\omega = \omega_r + i\gamma$ using $k=3\times10^{-3}cm^{-1}$ & $n_e=10^6 cm^{-3}$.

When $\omega_0 > \omega_{pe}$, pump e.m wave excites a electron plasma wave and a (I) ion-acoustic wave. (curve 1 in figs 1,2 & 3)
We found for $L41 \leq 10^2$ & $Ti=Te/20$, the growth rate $\gamma \ll \omega_r \simeq kc_s$. At lower $Te \leq 6\times10^6$ K collisional damping dominates which decreases with increasing Te and therefore threshold luminousity L41t decreases. For $Te \geq 10^7$ K Landau damping takes over and therefore L41t increases. But for $6\times10^6 \leq Te \leq 3\times10^7$ K nither collisional nor Landau damping is significant & hence minimum L41t is needed.

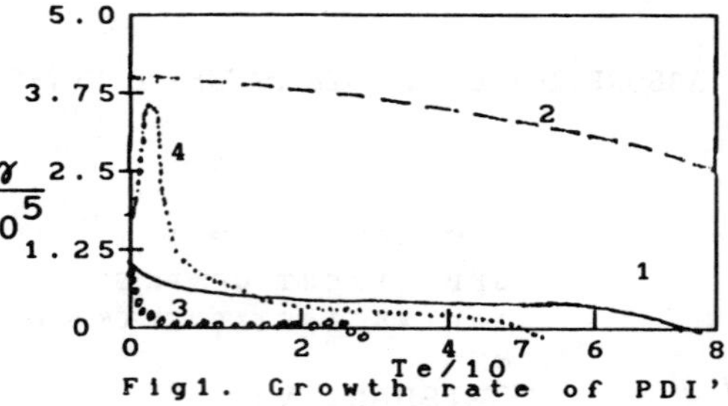

Fig1. Growth rate of PDI's

(II) reactive quasi-ion mode. (curve 2 in figs 1,2 & 3)
For $L41 \geq 10^2$ ion mode grows largely: $\omega = \omega_r + i\gamma$, $\gamma \gg kc_s$ & the ratio $\gamma/\omega_r \longrightarrow 3^{1/2}$.

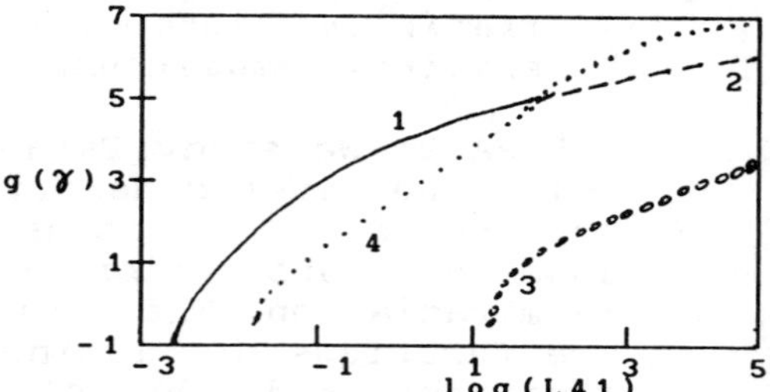

Fig2. Growth rate of PDI's

(III) resistive quasi-ion mode. (curve 3 in figs 1,2 & 3)
This mode excites when $T_e \leq T_i$: $Te=Ti/10$. Since $c_s \simeq v_i$, ion mode under goes large nonlinear Landau damping. Therefore this mode excites at large L41.
When $\omega_0 < \omega_{pe}$, pump e.m wave excites electron plasma wave & a non-oscillating ion mode. This case of PDI is known as oscillating two-stream instability. (curve4 in figs 1,2 & 4).

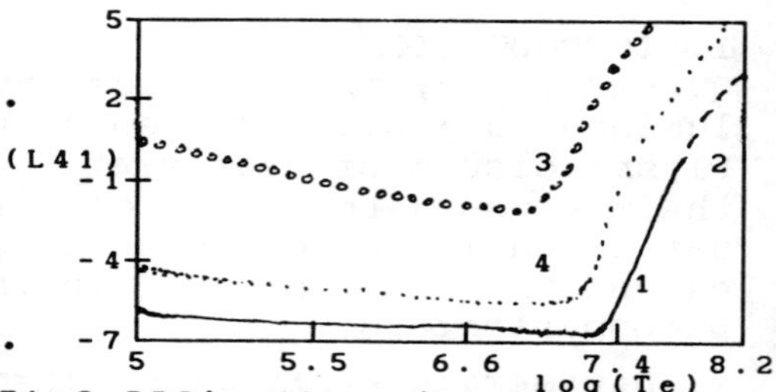

Fig3 PDI's thresold luminosity vs electron temperature Te.

Results and Conclusions.

1) Growth rates are maximal when $K \| E_0$ and negative for $K \perp E_0$.

2) The absorption rate of e.m wave (rate of PDI) is proportional to the ratio $L41/rpc^2$ and it is much larger than the free-free absorption rates. 3) T_e increases by a factor of several tens for moderate radio luminosities.

Decay instability may be the mechanism for the formation of hot lower density corona adjoining each photoionized dense region. In the case of quasars, the radio luminosity is found to be less thqn that expected from the extrapolation of the spectrum from high frequencies to low frequencies. Therefore absorption is certainly there and anomalous absorption might modify the spectrum in a significant way.

Referrences.

1 Krishan,V., 1987 Mon. Not. R. A. S. , **226**,629

2 Liu,C.S. & Kaw,P.K., 1976 Advances in plasma physics, 6,83

Microbursts at Meter-Decameter Wavelengths

G. Thejappa*, N. Gopalswamy and M.R. Kundu
Astronomy Program, University of Maryland,
College Park, MD 20742

I. Summary of Data Analysis

A statistical analysis of Clark Lake microbursts was performed using a data base consisting of ~ 100 bursts. The results are: (1) Microburst sources are well correlated with coronal streamers and active regions. (2) The peak brightness temperature (T_b) of the microbursts is in the range 6×10^5 K to 6×10^7 K (see Fig. 1; see also Kundu et al, 1986; White et al, 1986; Gopalswamy et al, 1987) whereas it is quite high for normal type III bursts ($10^7 < T_b < 10^{12}$ K - Dulk and Suzuki, 1980). (3) The source size varies over a wide range of 5.5 - 10.5 arc min at 73.8 MHz, 8.5 - 13.5 arc min at 57.5 MHz, 10.5 - 16.5 arc mins at 50 MHz and 12.5 - 22.5 arc min at 38.5 MHz. (4) The drift rate lies mainly in the range 4.5 to 6.5 MHz s^{-1}, with an average value of ~ 5.5 MHz s^{-1}.

II. Interpretation and Discussion

Most of the existing type III models consider the coherent emission of plasma waves by electron beams due to two-stream instability. Here we study both coherent and non-coherent emission of plasma waves by the electron beam in relation to microbursts. The intensity (I) of plasma waves emitted from a uniform plasma layer of thickness L can be approximated as: $I = a/\mu$ where a is the emissivity of the plasma waves and μ is the absorption coefficient which is the sum of collisional (μ_{coll}), beam (μ_b which is negative since it is growth) and Landau damping by ambient electrons μ_L. The μ_L can be obtained by calculating the distance that a plasma wave can travel in smoothly varying corona before it is effectively Landau damped. Using the spontaneous emission due to the electron beam distribution, we obtain the total energy density of plasma waves from I as:

$$W/n_eT_e \approx 2 \times 10^{-2}\, n_b/n_e\, 1/\gamma \qquad (1)$$

where γ is the effective damping due to all the above effects; n_e, n_b are

*On leave from Indian Institute of Astrophysics, Bangalore, India.

E. R. Priest and V. Krishan (eds.), Basic Plasma Processes on the Sun, 521–522.

background and beam electron densities; T_e is electron temperature. If the radiation is at fundamental plasma frequency due to Rayleigh scattering, the brightness temperature T_B is obtained as:

$$T_B \approx 6.0 \times 10^{13}\; n_b/n_e \,\frac{1}{\gamma} \qquad (2)$$

T_B reaches maximum when growth balances collisional damping. At this threshold of $n_b/n_e \approx 9.2 \times 10^{-9}$, $T_B \approx 1.2 \times 10^7$ K. T_B is plotted as a function of density in Fig. 2. As n_b/n_e approaches the threshold, T_B rises sharply. It is also clear that the entire brightness spectrum of microbursts can be explained by the spontaneous emission mechanism. Thus the microbursts are due to spontaneously emitted electron beams in the range of densities $0.5 \times 10^{-8} \leq n_b/n_e \leq 1 \times 10^{-8}$ with $V_b \approx \Delta V_b \approx 0.2$ c; the normal type III bursts are entirely in the coherent regime of Langmuir wave emission (see Thejappa et al 1989 for details).

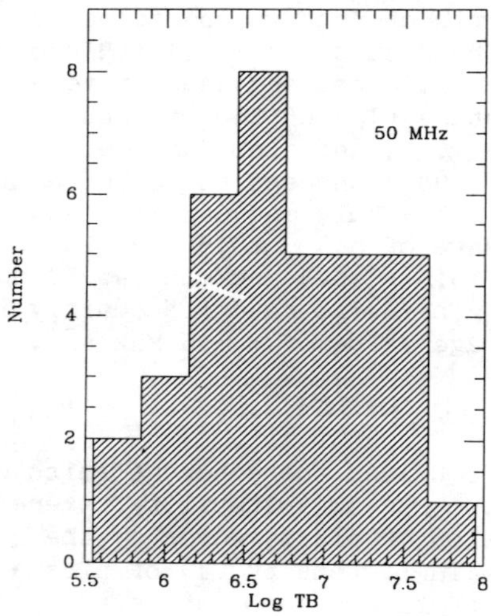

Fig. 1. Distribution of brightness temperature.

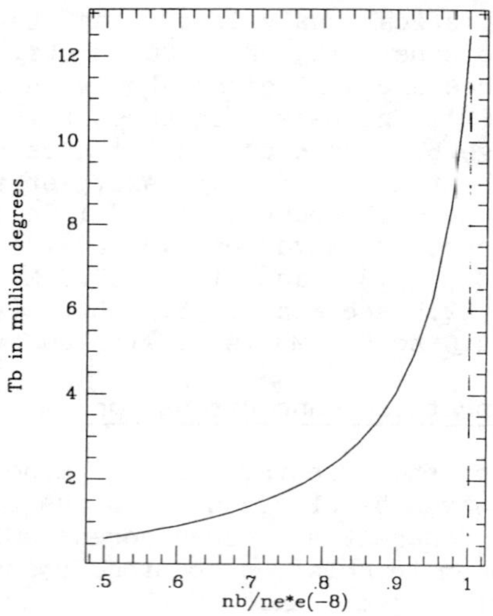

Fig. 2. Brightness temperature as a function of beam density. The region to the right of the dotted lines corresponds to coherent emission.

References

Dulk, G.A. and Suzuki, S.: 1980, Astron. Astrophys. 88, 203.

Gopalswamy, N., Kundu, M.R. and Szabo, A: 1987, Solar Phys. 108, 333.

Kundu, M.R., Gergely, T.E., Szabo, A., Loiacono, R., and White, S.M.: 1986, Astrophpys. J. 308, 436.

Thejappa, G., Gopalswamy, N. and Kundu, M.R.: 1989, Solar Phys. (in press).

White, S.M., Kundu, M.R. and Szabo, A.: 1986, Solar Phys. 107, 135.

THE SUN AT THE VLA's METRIC AND DECIMETRIC WAVELENGTHS

S. M. White, M. R. Kundu, N. Gopalswamy and E. J. Schmahl
Astronomy Program
University of Maryland, College Park, MD 20742

During September 1988 (International Solar Month) we observed the Sun with the Very Large Array on 4 days in the period Sep. 11-17. The VLA was in its most compact configuration, which is ideal for studying large-scale coronal structures. Here we summarize some preliminary results of the observations at 0.333 and 1.5 GHz. Despite the presence of numerous active regions the Sun was actually very quiet, with no flares during our observing, and this allowed us to make high-dynamic-range maps.

There is exceptionally good correspondence between the 1.5 GHz radio sources and regions which are dark in the He10830 images and bright in the Ca images. This extends to very small weak features in the 1.5 GHz radio maps, which nonetheless are found to have associated dark regions on the He images. Several filaments show up as depressions in active regions or as empty lanes. The 1.5 GHz sources are generally very extended, appearing as a canopy straddling several nearby regions in one hemisphere. None of the active regions was particularly bright at 1.5 GHz. The peak brightness temperatures of the active region sources on the four days are 1.7, 1.6, 1.5 and 1.8 x 10^6 on Sep. 11, 12, 13 and 17 respectively. However, on all days the range of peak brightness temperatures in different active regions was remarkably small. Typically there were about eight active region peaks present, each in the range 1.0 - 1.6 x 10^6 K, and with a low degree of ciruclar polarization.

Generally there are remarkably few structural changes in the active region sources from one day to the next, suggesting that structural evolution is relatively slow. We present contour maps at 1.5 GHz for each of the four days in the Figure. However one sharp change from the 11th to the 12th can be seen in the region at about S20W10 on the 11th. This region shows a highly polarized feature trailing the primary peak over the active region on the 11th. By the 12th this location has developed into a strong eastward-pointing extension, in conjunction with the rapid emergence of new positive magnetic flux in a region of predominantly negative flux. Another interesting change occurs on the 13th, when an ephemeral region appears as a strong compact source close to disk centre.

E. R. Priest and V. Krishan (eds.), Basic Plasma Processes on the Sun, 523–524.

Two noise storms were present at 0.33 GHz all week and were all highly polarized: the R-polarized storm in the north on all days is a single source, which however shows individual bursts shifting in position within the envelope of the storm. The noise storms in the south are more interesting. The maps from the 17th show them to be of opposite polarizations but well separated spatially, corresponding rather well to the simple underlying bipolar magnetc structure evident in the magnetogram. Clear evidence for the degree of divergence of magnetic field lines in the corona can be seen in the structure of this region. There is definite limb brightening present in the .33 GHz maps on all days.

This research was performed with support from NSF grant ATM-87-17157, NASA grant NAGW-1541 and NASA contract NAG 5-969. The very Large Array is a facility of the National Radio Astronomy Observatory, which is operated by Associated Universities, Inc., under contract with the National Science Foundation.

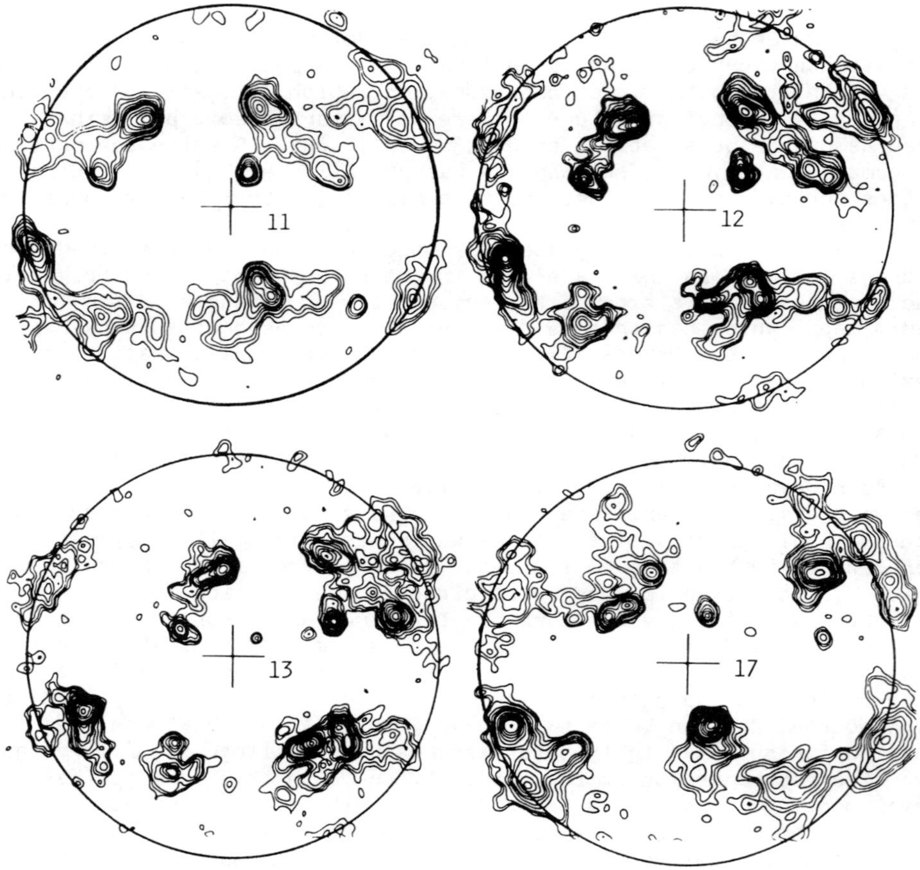

VLA-PHOENIX OBSERVATIONS OF A NARROW-BAND DECIMETRIC BURST

Robert F. Willson
Dept. Of Physics and Astronomy
Tufts University
Medford, MA 02155
USA

Arnold. O. Benz
Institute of Astronomy
ETH, CH-8092
Zurich
Switzerland

ABSTRACT We discuss observations of a highly-circularly polarized multiply-impulsive microwave burst detected by the Very Large Array and the Phoenix Digital Radio Spectrometer. The VLA was used to resolve the burst in two dimensions, while PHOENIX provided high time resolution information about its spectral properties. During part of the burst, positive frequency drifts were detected, suggesting inwardly propagating beams of electrons emitting Type III-like radiation.

1. INTRODUCTION

It is generally believed that gyrosynchrotron radiation from nonthermal electrons can explain the properties of most solar microwave bursts, including their brightness temperature, polarization and bandwidth. Recent observation using wide-band dynamic spectrographs have, however, discovered multiply-impulsive ($\tau < 100$ msec) bursts including spikes, pulsations and decimetric type III bursts whose narrow bandwidths and frequency drifts require a different explanation such as plasma radiation or electron-cyclotron maser emission (eg. Stähli and Benz 1987). Practically nothing is known about the spatial properties of these events since they have only been detected with single dish instruments with poor angular resolution. Here we discuss some spatially resolved observations of a narrow band decimetric burst using the VLA and the Phoenix Digital Radiospectrometer.

2. COORDINATED OBSERVATIONS OF MICROWAVE BURSTS

The Very Large Array was used to observe the Sun between 1300-2300 UT on June 25 and 27, 1989 as part of the Max 91 Campaign. Observations were made simultaneously at 1446 MHz (20.7 cm) and 327 MHz (91.6 cm) with a time resolution of 1.67 s. They were coordinated with observations at the Nancay Radioheliograph (164 and 327 MHz) and with the PHOENIX Digital Radio Spectrometer (ten frequencies each between 301 and 356 MHz and 1365 and 1525 MHz- 20 msec time resolution) in order to study the spatial, temporal and spectral properties of decimetric bursts.

Figure 1a shows an example of a burst detected by the VLA and PHOENIX at 1446 MHz. It consists of several impulsive spikes, each of 20-30 seconds in duration, covering a period of a few minutes. There was no

E. R. Priest and V. Krishan (eds.), Basic Plasma Processes on the Sun, 525–526.

evidence for burst emission at 327 MHz, nor was there a soft or hard X-ray burst detected by the SMM or GOES satellites. Snapshot maps (Fig. 1b) show that the bursts are compact (θ= 25-30") and highly circularly polarized (ρ_c = 90-100 %). Co-registration of these maps with a SOON Hα image taken at 1617 UT showed that the burst was located above a faint optical flare lying between two active regions (~N25, E21)

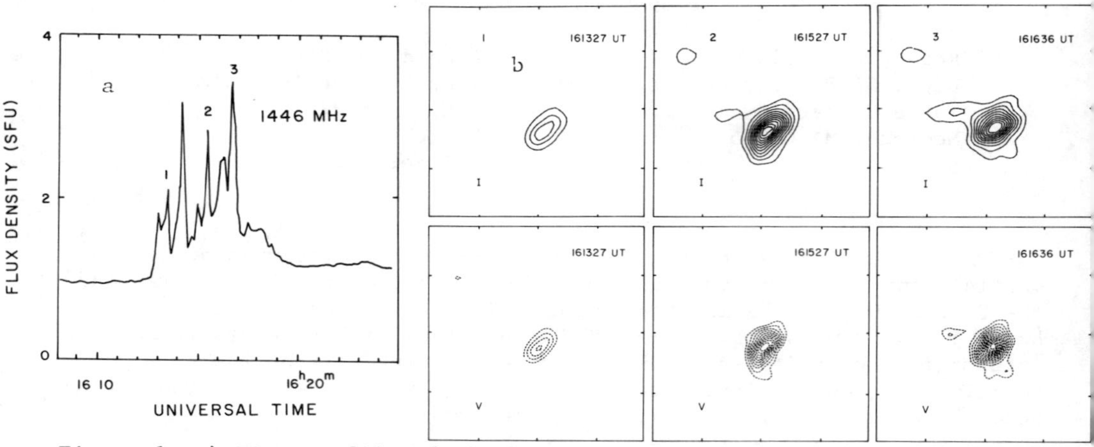

Figure 1. a) Time profile of the burst detected with one of the short VLA baselines on 1989 June 25. b) VLA 3.3 s snapshot maps at times 1, 2 and 3 shown in Fig. 1a. The brightness temperature contours for the I and V maps are drawn at intervals of $T_b = 8 \times 10^5$ and 4×10^5 K, respectively. Fiducial marks on the axes are drawn every 60".

The PHOENIX spectral data showed that the emission drifted from high to low frequencies near the time of the most intense peak (3) (~+750 MHz/sec) suggesting an inwardly-propagating beam of electrons. Broadband emission was detected at most other times. Although the positions of the bursts shifted from peak to peak (Δθ~10"), these changes were not systematic, as would be expected for a source moving through the corona. Nevertheless, this event may be related to the decimetric type III events (Stähli and Benz 1987) whose narrow bandwidths and frequency drifts suggest a coherent emission mechanism, such as plasma radiation or electron-cyclotron maser emission.

3. ACKNOWLEDGEMENTS

Radio astronomical studies of the Sun at Tufts University are supported under grant AFOSR-89-0147 with the Air Force Office of Scientific Research, and by NASA grant NAG-5-501.

4. REFERENCE

Stähli, M., and Benz, A.O. (1987) 'Microwave Emission of Solar Electron Beams', Astronomy and Astrophysics 175, 271-276.

SUMMARY LECTURE

N.O. WEISS
Department of Applied Mathematics and Theoretical Physics
University of Cambridge, Cambridge CB3 9EW, UK

This has been a memorable meeting. First of all, we have had an opportunity of coming to India and getting to know some of its people. We have discovered how attractive Bangalore is and we have glimpsed a little more of Karnataka: none of us will forget the colossal statue of Gomateshvara or the temples at Belur and Halebid. Then we have had our scientific proceedings and contacts with colleagues from across the world. Here I appreciated the unusual distribution of people attending the Symposium. I was especially pleased to see such a high proportion of women present at the meeting and delivering excellent contributions. It was also a welcome change – reflecting the spirit of glasnost – to have more delegates from the Soviet Union than from the United States, including many young physicists whom I congratulate on their skill in speaking here in English. Above all, we had a chance of meeting so many bright young Indian astrophysicists, while they have been able to see us and to observe our shortcomings. I am sure that we are all impressed by the range of research institutes, in Bangalore and elsewhere in India, as well as by the quality of research being done in them. In particular, we learnt more about the Indian Institute of Astrophysics with its commitment to research in solar physics. As Professor Bhattacharya reminded us, it has a two-hundred year history, with a long and distinguished tradition of observations at Kodaikanal coupled to new standards being developed here in Bangalore.

As you all know, this meeting would never have taken place without the persistent efforts of Professor Vinod Krishan. Her original concept involved a much wider coverage of plasma astrophysics: this was cut down to solar physics by the IAU but something of the original scope remains. This range has been immensely stimulating and truly interdisciplinary but no-one, least of all me, could be familiar with all the topics covered. So it is impossible (and probably undesirable) to produce a systematic summary of what we have achieved. That can be gleaned from the Proceedings. What have impressed me most are the quality of research, the wide range of activity and the number of young scientists coming into the subject. In what follows I shall only offer a very selective personal response, with apologies to any participants who feel that their own work has been unjustly neglected.

E. R. Priest and V. Krishan (eds.), Basic Plasma Processes on the Sun, 527–532.

Recent Developments

First of all I want to draw attention to various recent advances and their future prospects, affecting observations from the ground and from space and their theoretical interpretation. I begin with solar observations, go on to the solar system and to other stars and then come round to theory.

Solar oscillations. Precise measurements of the frequencies of p-mode oscillations with periods around five minutes have been inverted to yield amazingly precise measurements of the variation of density, angular velocity (and perhaps magnetic fields) within the Sun. In the future, continuous worldwide coverage will be achieved by the GONG Network, which includes a station at Udaipur. So we expect to investigate fine details of the internal structure and properties of the Sun.

Neutrino flux. The discrepancy between solar models and the neutrino fluxes measured at the Homestake mine and Kamiokande still remain. Is this a problem of stellar structure or high energy physics? Will the apparent correlation with magnetic activity persist – or disappear, like correlations between the solar cycle and the weather? Here we must be patient and wait for results from the gallium experiments that are being prepared.

High resolution photospheric observations. Striking images are being obtained from new ground-based telescopes at La Palma and Tenerife and through improvements at the Pic du Midi and Sacramento Peak. The most important advance came through realising that images were distorted by effects on the Sun and not in the earth's atmosphere. Once the five-minute oscillations were filtered out subgranular scales could be observed with unprecedented clarity. The Lockheed group is now making simultaneous measurements of intensity, velocity and magnetic fields on scales much less than an arc-second. In the future, with adaptive optics and LEST, we can expect to attain even higher resolution.

Radio observations. Radio emission provides increasingly detailed information on flares, coronal mass ejections, type III and type IV bursts and also, through scintillation measurements, of interplanetary magnetic features. The resolution obtained with the VLA and the new millimetre array will be enhanced as other instruments, including the new GMRT near Poona, come into operation.

Rocket flights. The remarkable X-ray images obtained this year by Golub and his colleagues show magnetic features with amazing resolution. There is fine structure on all scales down to 0.1 arc-sec. This shows what can be achieved with new technology and augurs well for future X-ray satellites.

Space missions. The next decade will see an impressive sequence. SOHO is a cornerstone of the ESA programme and will make a range of observations including measurements of solar oscillations. The USSR will launch the CORONAS mission in 1992 to make a suite of observations covering the solar atmosphere in a wide range of frequencies. The Japanese SOLAR-A mission will make uv and X-ray measurements and we expect NASA to launch OSL as a free-flyer, following up on Spacelab-2 with uv and X-ray observations as well as imaging with the Lockheed

SOUP device.
Solar wind and magnetosphere. Going beyond the Sun itself, we may learn from observations within the solar system. Continuing studies of interplanetary magnetic fields, of MHD waves and of reconnection near the magnetopause are improving our understanding of the solar wind and the earth's magnetosphere. Meanwhile, Voyager, Vega and Galileo have explored magnetic fields of planets and the plasma sheaths surrounding comets.
The solar–stellar connection. Although individual magnetic features can only be resolved on the Sun we can discover the effects of varying parameters, such as the rotation rate, by studying other stars. Their magnetic activity can be monitored by recording chromospheric Ca II emission, coronal X-ray emission or radio signals emitted by flares. In this way we have gained a much better understanding of cyclical activity and of the evolution of the Sun's magnetic field as it spins down owing to magnetic braking. Moreover, we can study a continuum of magnetic activity that extends from the Sun through active stars to white dwarfs and neutron stars, or from the corona through accretion discs to active galactic nuclei.
Numerical experiments and simulations. Moving to theory, we note the increasing role of computation. The availability of supercomputers has given a great boost to numerical modelling of plasma processes on the Sun. Until now much of the effort has gone into highly idealized models, many of them two-dimensional, but sophisticated three-dimensional simulations are now feasible. Current research is directed either at detailed atmosphere modelling, aimed at reproducing observations, or at probing turbulent convection in the solar interior.
Nonlinear dynamics. Recent mathematical developments have significant implications for processes in the solar atmosphere and interior. Nonlinear waves may propogate as solitons in flux tubes. The formalism of KAM theory, applied to chaos in Hamiltonian systems, can be used to describe the structure of magnetic fields or to discuss dynamo action by flows with chaotic streamlines. Bifurcation patterns associated with chaos in dissipative systems are also attracting considerable attention as part of a new approach to nonlinear convection and stellar dynamos. In addition, quantitative analysis of aperiodic behaviour is made possible by measuring Lyapunov exponents or computing the dimension of the attractor.
Modelling. These powerful new techniques are exciting in themselves and offer great new opportunities, since plasma processes are typically nonlinear. Nevertheless, they are only relevant in that they enable us to model and interpret the underlying physics of magnetic fields in the Sun. That understanding must be what we aim for.

Impressions.

It may be appropriate to begin by recalling the poster for this meeting. It shows the Sun with its atmosphere extending outwards and full of detail. We can observe this region over 10 decades in frequency, from X-rays to radio, and we know almost too much about it. So it is not surprising that it has received the most attention –

but it is as well to remember that the atmosphere contains only 10^{-10} of the solar mass. The structure of this atmosphere is dominated by magnetic fields which originate in the convection zone, containing $0.03M_\odot$. This can be probed directly, using magnetic flux tubes or solar oscillations. The radiative interior, where energy is generated, contains 97% of the mass and can only be explored with neutrinos or through helioseismology.

Solar interior. The internal structure of the sun can now be determined with extraordinary precision by inverting helioseismological data. Rotational splitting yields measurements of the internal angular velocity, as a function of radius and latitude, which have upset our preconceived ideas: the surface differential rotation apparently persists to the base of the convection zone, with a gradual transition to uniform rotation deeper in the radiative zone. The resulting shear must have a profound effect on the magnetic field.

Dynamo theory. This subject has grown more confusing in the last few years. We still need to establish the location of the solar dynamos and to construct a convincing phenomenological model. Mean field dynamo theory provides a convenient parametrized description of the generation process and some qualitative features can be studied with low-order nonlinear models. Some processes, such as magnetic braking or losses through magnetic buoyancy, can be isolated and studied in some detail but large-scale nonlinear calculations are still badly needed.

Convection and magnetoconvection. These phenomena are thoroughly nonlinear. Several recent numerical simulations of three-dimensional compressible convection agree in showing a cellular structure at the surface, giving way to rapidly sinking isolated plumes below. These results are impressive but the deep structure remains mysterious. Most studies of compressible magnetoconvection have so far been confined to two-dimensional models. Careful treatments of MHD turbulence in incompressible fluids show that oversimplified models (and simplistic postulates) have limited validity.

Surface motions. Differential rotation and meridional flows provide large-scale effects. There is evidence for meridional circulations from sunspot proper motions, as well as for fluctuations in rotation rate over the solar cycle. On a smaller scale, behaviour of photospheric flux tubes is controlled by granular, mesogranular and supergranular velocities. High resolution measurements of granular proper motions have confirmed the existence of mesogranules but the underlying dynamics is uncertain: are they thermally driven or produced by self-organization from a nonlinear cascade?

Flux tubes. The formation of isolated flux tubes with intense magnetic fields is understood but has not yet been modelled in detail. Once formed, these tubes support waves whose properties have been thoroughly described. Moreover, these waves can maintain the supply of energy to isolated bright points in the chromosphere. Pores and sunspots appear deceptively simple and it has proved surprisingly difficult to model the overall structure of a sunspot.

Nonlinear waves. Nonlinear waves crop up in several contexts, in the photosphere and corona as well as in the solar wind. Solar physics provides examples of fundamental nonlinear behaviour: solitons travel along flux tubes and coronal wave interactions can lead to chaos. In addition there are physical processes involving resonance absorption and particle acceleration which need to be described.
Heating the solar atmosphere. Acoustic heating provides an omnipresent background but other processes have to be invoked. Is the corona heated by hydromagnetic waves or by magnetic dissipation? This remains a controversial issue though the argument has grown more tolerant. Perhaps a clue can be obtained by studying fine structure in the latest X-ray images of the corona and assessing whether filaments are heated by local electric currents or by waves travelling along them.
Reconnection. This subject began over 40 years ago with the pioneering work of Dungey and Sweet. Since then modelling of prominences and flares has become increasingly detailed and sophisticated but we still need mechanisms for *rapid* reconnection. (It seems frustrating when speculations are damped by excessive observational constraints: life might be easier for theoreticians if we only knew about stellar flares.) A crucial issue concerns the evolution of force-free fields in the atmosphere as photospheric convection moves the footpoints of the flux tubes. Does lack of equilibrium lead to current sheets or are there only intermittent but continuous structures, as suggested by numerical experiments? Finally, when catastrophic reconnection occurs, which of several mechanisms – strong electric fields, nonlinear waves or turbulence – leads to particle acceleration?

This selective catalogue could continue to include spectroscopy and space plasmas. I end, however, with two quotes. Vinod Krishan warned us: "Do not underestimate the unknown devil." At the time I wondered what she meant but then Franz Kneer reminded us of the distinction between theoreticians and observers: an observer trusts everyone's results except his own while a theoretician does not believe any results except his own. Perhaps that is what she had in mind. If so, you will have to decide for yourselves what to accept in my summary of the meeting.

Conclusion

This Symposium has certainly demonstrated the characteristic flavour of solar physics. Our subject differs from, say, extragalactic astronomy in that theoretical ideas are heavily constrained by detailed observations. Yet those observations are incomplete and a proper understanding can only be gained by detailed modelling. Although the plethora of observational constraints sometimes appears to hinder progress we must acknowledge that without such observations we could never have invented the rich behaviour caused by magnetic activity on the Sun: magnetic cycles, isolated flux tubes, sunspots, rapid reconnection, flares, coronal heating. What makes our subject challenging and fascinating is (to use Parker's words) that "Nature is cleverer than we are."

Finally, on behalf of everyone I thank all those who have made this meeting so successful. This includes our sponsors, especially the Indian Institute of Astrophysics and Professor Bhattacharya, and the Raman Institute and Professor Radhakrishnan, the Scientific Organizing Committee, the local Organizing Committee and the staff who have so efficiently looked after our arrangements. Above all, Professor Vinod Krishan deserves our thanks: she corresponded with us, organized the meeting, arranged our sessions, provided advice, settled problems and found time to present her own work. We are very grateful.

LIST OF PARTICIPANTS

BELGIUM	
Verheest, F	Institute Theoretische Mechanica, Rijksuniversiteit
BULGARIA	
Dermendjiev, V N	Bulgarian Academy of Science
CHINA	
Zhou, A	Purple Mountain Observatory, Nanjing
Chou, C-K	National Central University, Taiwan
Xu, M J	Chinese Academy of Sciences, Beijing
CZECHOSLOVAKIA	
Rusin, V	Slovak Academy of Sciences
FRANCE	
Heyvaerts, J	Observatoire de Paris
Leblanc, Y	Observatoire de Paris
Paillere, P	Centre d'Etudes de Limeil, St Georges
GERMANY	
Kneer, F	Universitats-Sternwarts, Gottingen
Ulmschneider, P	Institut fur Theoretische Astrophysik, Heidelberg
INDIA	
Radhakrishnan, V	Raman Research Institute, Bangalore
Dwivedi, B N	Banaras Hindu University, Varanasi
Vats, H O	Physical Research Laboratory, Ahmedabad
Girish,T E	Government Victoria College, Palghat
Verma, V K	Uttar Pradesh State Observatory, Naini Tal
Bondal, K R	Uttar Pradesh State Observatory, Naini Tal
Sinha, K	Uttar Pradesh State Observatory, Naini Tal
Gaur, V P	Uttar Pradesh State Observatory, Naini Tal
Joshi, G C	Uttar Pradesh State Observatory, Naini Tal
Joshi, A	Uttar Pradesh State Observatory, Naini Tal
Lokanadham, B	Osmania University, Hyderabad
Uberoi, C	Indian Institute of Science, Bangalore
Narain, U	Meerut College, Meerut
Alladin, S M	Osmania University, Hyderabad
Janardhan, P	Physical Research Laboratory, Ahmedabad
Bobra, A D	Physical Research Laboratory, Ahmedabad
Narasimhan, K S V S	The New College, Madras
Somasundaram, K	Gandhigram Rural Institute, Gandhigram
Venkataraman, S	Gandhigram Rural Institute, Gandhigram
Rudraiah, N	Central College, Bangalore
Leony, M N	Osmania University, Hyderabad
Bhonsle, R V	Physical Research Laboratory, Ahmedabad
Ishwar, B	Bihar University, Muzaffarpur
Srivatsava, P K	Govertment New Science Collefe, Rewa
Prabhakaran, S R	Kerala University, Trivandrum
Khairozzaman, A E Md	B N College, Assam
Ambastha, A	Udaipur Solar Laboratory, Udaipur
Bhatnagar, A	Udaipur Solar Laboratory, Udaipur
Srivastava, N	Udaipur Solar Laboratory, Udaipur
Bose, S K	University of Gorakhpur
Alurkar, S K	Physical Research Laboratory, Ahmedabad
Tiwari, P B	Indian Institute of Science, Bangalore

D'silva, S	Indian Institute of Science, Bangalore
Krishan, S	Indian Institute of Science, Bangalore
Manoharan, P K	Radio Astronomy Centre, Ootacamund
Choudhuri, A R	Indian Institute of Science, Bangalore
Kumar, N	Indian Institute of Science, Bangalore
Kotcherlakota, L N	Shivaji University, Kolhapur
Shevgaonkar, R K	Electrical Engineering Department, Powai, Bombay
Sharma, A K	Physical Research Laboratory, Ahmedabad
Buti, B	Physical Research Laboratory, Ahmedabad
Swarup, G	Tata Institute, Poona University Campur, Pune
Aleem, P S M	Indian Institute of Astrophysics, Kodaikanal
Narayanan, R S	Indian Institute of Astrophysics, Kodaikanal
Gupta, S S	Indian Institute of Astrophysics, Kodaikanal
Vanajakshi, C T	Indian Institute of Astrophysics, Kodaikanal
Raman, K S	Indian Institute of Astrophysics, Kodaikanal
Nagabhushana, B S	Indian Institute of Astrophysics, Kodaikanal
Venkatakrishnan, P	Indian Institute of Astrophysics, Bangalore
Sasidharan, K	Indian Institute of Astrophysics, Bangalore
Javaraiah, J	Indian Institute of Astrophysics, Bangalore
Hiremath, K M	Indian Institute of Astrophysics, Bangalore
Gangadharan, R T	Indian Institute of Science, Bangalore
Vasundhara, R	Indian Institute of Astrophysics, Bangalore
Raju, P K	Indian Institute of Astrophysics, Bangalore
Kapoor, R C	Indian Institute of Astrophysics, Bangalore
Gokhale, M H	Indian Institute of Astrophysics, Bangalore
Sastri, J H	Indian Institute of Astrophysics, Bangalore
Bhattacharyya, J C	Indian Institute of Astrophysics, Bangalore
Hasan, S	Indian Institute of Astrophysics, Bangalore
Rajamohan, R	Indian Institute of Astrophysics, Bangalore
Ramesh, K B	Indian Institute of Astrophysics, Bangalore
Kariyappa, R	Indian Institute of Astrophysics, Bangalore
Mohan Rao, D	Indian Institute of Astrophysics, Bangalore
Rangarajan, K E	Indian Institute of Astrophysics, Bangalore
Peraiah, A	Indian Institute of Astrophysics, Bangalore
Bagare, S P	Indian Institute of Astrophysics, Bangalore
Prasad, N D N	Indian Institute of Astrophysics, Bangalore
Sivaraman, K R	Indian Institute of Astrophysics, Bangalore
Sivaram, C	Indian Institute of Astrophysics, Bangalore
Krishan, V	Indian Institute of Astrophysics, Bangalore
Sastry, Ch V	Indian Institute of Astrophysics, Bangalore
Saha, S K	Indian Institute of Astrophysics, Bangalore
Subramanian, K R	Indian Institute of Astrophysics, Bangalore
Ebenezer, E	Indian Institute of Astrophysics, Bangalore
Pati, A K	Indian Institute of Astrophysics, Bangalore
Madhava Rao, B R	Indian Institute of Astrophysics, Bangalore
Paranjype, A	Indian Institute of Astrophysics, Bangalore
Anantharamaiah, K R	Raman Research Institute, Bangalore
Sharma, J K	Department of Science and Technology, Bangalore
Srinivasan, G	Raman Research Institute, Bangalore
Trehan, S K	Punjab University, Chandigarh
Narasimham, N A	18 Saras Baug, Deonar, Bombay
Varma, R K	Physical Research Laboratory, Ahmedabad
Pande, M C	Uttar Pradesh State Observatory, Naini Tal
Narlikar, J V	Inter University Centre for Astronomy and Astrophysics, Pune
Das, I M L	University of Allahabad, Allahabad
Kumar, M	Meerut College, Meerut
Asthana, M	Ravishankar University, Raipur

Guha, S	Ravishankar University, Raipur
Singh, R N	Banaras Hindu University, Varanasi
Rausaria, R R	Regional Engineering College, Srinagar
ISRAEL	
Cuperman, S	Tel Aviv University
Steinitz, R	Ben-Gurion University of Negev, Beer-Sheeva
ITALY	
Chiuderi, C	University of Florence, Firenze
Drago, F	University of Florence, Firenze
Fofi, M	Osservatorio Astronomico di Roma, Rome
Fossi, M	Osservatorio Astrofisico Arcetri, Firenze
Pallavicini, R	Osservatorio Astrofisico di Arcetri, Firenze
JAPAN	
Unno, W	Kinki University, Higashi-Osaka
KOREA	
Hong Sik, Y	Seoul National University
THE NETHERLANDS	
Hoyng, P	Sterrekundig Instituut, Utrecht
Kuijpers, J	Sterrekundig Instituut, Utrecht
SAUDI ARABIA	
Al-Khashlan, A	KACST, Riyadh
SWEDEN	
Falthammar, C G	The Royal Institute of Technology, Stockholm
UNITED KINGDOM	
Jianke, L	Astronomy Centre,University of Sussex, Brighton
Mestel, L	Astronomy Centre,University of Sussex, Brighton
Priest, E R	University of St Andrews, Fife
Roberts, B	University of St Andrews, Fife
Weiss, N	DAMTP, Silver Street, Cambridge
USA	
Lang, K R	Tufts University, Medford
Chaturvedi, P K	Naval Research Laboratory, Washington
Ballegooijen, A V	Center for Astrophysics, Cambridge
Montgomery, D	Dartmouth College, New Hampshire
Forbes, T G	University of New Hampshire, Durham
Smart, R N	National Solar Observatory, New Mexico
Hollweg, J V	University of New Hampshire, Durham
Karpen, J T	NRL, Washington
Kane, S R	University of California, Berkeley, Washington
Kalkofen, W	Center for Astrophysics, Cambridge
Howard, R F	National Solar Observatory, Tucson
Dryer, M	Space Environment Laboratory, Colorado
Smith, Z	Space Environment Laboratory, Colorado
Detman, T R	Space Environment Laboratory, Colorado
Grubb, R N	Space Environment Laboratory, Colorado
Kundu, M	University of Maryland
Smith, D F	University of Colorado
Chitre, S M	NASA/Goddard Space Flight Centre
Lin, R P	University of California

Mendis, D A	University of California at San Diego
Petrosian, V	Standford University, California
Martens, P C	Center for Astrophysics, Cambridge
Mahajan, S M	University of Texas at Austin
Willson, R F	Tufts University, Medford

USSR

Chernov, G P	IZMIRAN, Troitsk
Kuznetsov, V D V	IZMIRAN, Troitsk
Ruderman, M	Academy of Sciences, Moscow
Krivodubskij, V N	Kiev Shevchenko University, Ukraine
Stepanov, A	Crimean Astrophysical Observatory
Kosovichev, A G	Crimean Astrophysical Observatory
Zaitsev, V V	Academy of Sciences, Gorky
Zlotnik, E Y	Academy of Sciences, Gorky
Zheleznyakov, V V	Academy of Sciences, Gorky
Ryutova, M P	Academy of Sciences, Novosibirsk
Taktakishvili, A	Abastumani Observatory, Pavlov
Obridko, V N	Academy of Sciences, IZMIRAN
Fomichev, V V	Academy of Sciences, IZMIRAN
Zhurin, V V	Academy of Sciences, IZMIRAN
Semikoz, V B	Academy of Sciences, IZMIRAN
Somov, B	Academy of Sciences, Moscow
Pavlov, A K	Physical Technical Institute, Leningrad
Kocharov, L	Physical Technical Institute, Leningrad
Pataraya, A	Abastumani Observatory, Pavlov
Mogilevsky, E I	Academy of Sciences, IZMIRAN
Gelfreikh, G B	Pulkovo Observatory, Leningrad
Bardakov, M	Academy of Sciences, Irkutsk
Makarov, V I	Pulkovo Observatory, Leningrad
Tomozov, V	SibIzmir, Irkutsk
Kurt, V V G	Academy of Sciences, Moscow
Kononovich, V	Moscow University
Oraevskij, O V	Academy of Sciences, IZMIRAN
Kotov, Y D	Moscow Engineering Physical Institute, Moscow

INDEX